UNSTEADY AERODYNAMICS AND AEROELASTICITY OF TURBOMACHINES AND PROPELLERS

Proceedings of the Fifth International Symposium
September 18–21, 1989
Beijing, China

Organized by

Beijing University of Aeronautics and Astronautics

International Academic Publishers
A Pergamon-CNPIEC Joint Venture

Pergamon Press
Oxford · New York · Beijing · Frankfurt
São Paulo · Sydney · Tokyo · Toronto

DISTRIBUTORS:

U.K.	Pergamon Press plc, Headington Hill Hall, Oxford OX3 0BW, England
U.S.A.	Pergamon Press, Inc., Maxwell House, Fairview Park, Elmsford, New York 10523, U.S.A.
PEOPLE'S REPUBLIC OF CHINA	International Academic Publishers, Xizhimenwai Dajie, Beijing Exhibition Centre, Beijing 100044, People's Republic of China
FEDERAL REPUBLIC OF GERMANY	Pergamon Press GmbH, Hammerweg 6, D-6242 Kronberg, Federal Republic of Germany
BRAZIL	Pergamon Editora Ltda, Rua Eça de Queiros, 346, CEP 04011, Paraiso, São Paulo, Brazil
AUSTRALIA	Pergamon Press Australia Pty Ltd., P.O. Box 544, Potts Point, N.S.W. 2011, Australia
JAPAN	Pergamon Press, 5th Floor, Matsuoka Central Building, 1-7-1 Nishishinjuku, Shinjuku-ku, Tokyo 160, Japan
CANADA	Pergamon Press Canada Ltd., Suite No. 271, 253 College Street, Toronto, Ontario, Canada M5T 1R5

First edition 1990

Library of Congress Cataloging in Publication Data

International Symposium on Unsteady Aerodynamics and Aeroelasticity of Turbomachines and Propellers (5th: 1989: Beijing, China)

Unsteady aerodynamics and aeroelasticity of turbomachines and propellers: Proceedings of the Fifth International Symposium, September 18-21, 1989, Beijing, China/Edited by Pan Tianmin, Lu Yingjie, Yan Wenxuan, Wu Xiaolu.

p. cm.

1.Turbomachines--Congresses. 2.Propellers--Congresses. 3.Aerodynamics --Congresses. 4.Aeroelasticity--Congresses.

I. Pan, Tianmin. II. Lu, Yingjie. III. Yan, Wenxuan. IV. Wu, Xiaolu. V. Title.

TJ267.157 1989

629.132'3--dc20 89-49007 CIP

British Library Cataloguing in Publication Data

Unsteady aerodynamics and aeroelasticity of turbomachines and Propellers.

(5th: 1989: Beijing, China)

1. Aeroplanes, Aerodynamics

I.Pan, Tianmin. II.Lu, Yingjie. III.Yan, Wenxuan. IV.Wu, Xiaolu.

629.132'3

ISBN 0-08-040187-2 Pergamon Press

ISBN 7-80003-098-9/T.3 International Academic Publishers

Printed in the People's Republic of China

Local Organization Committee

Cao Chuanjun
Zhou Sheng
Lu Yingjie
Hu Zongan
Wu Xiaolu

Academic Committee

Cao Chuanjun
Cui Jiya
Zhou Sheng
Hu Zongan
Wang Zhenhua

Preface

There are 43 papers in the proceedings of the 5th International Symposium on Unsteady Aerodynamics and Aeroelasticity in Turbomachines and Propellers. The decision of holding the Fifth Symposium was made by the Scientific Committee in Cambridge during the Third Symposium in September 1984, and was reaffirmed in Aachen during the Fourth Symposium in September 1987. It is a great pleasure of Beijing University of Aeronautics and Astronautics (BUAA) to host the Fifth Symposium held on the campus of BUAA from 18th to 21st September 1989. The contributors of papers came from eight countries, including England, Federal Republic of Germany, the People's Republic of China, Japan, Pakistan, Poland, the Soviet Union and the United States of America.

For aero-engines, steam turbines, propellers, wind turbines and others, there is always pressure on the designers to make the blades small, light and highly loaded, so as to reduce weight and cost. These effects all tend to make the blades more liable to vibration, and so their is a fundamental vibration limitation on design which will always be paid attention to with us. Therefore, it is our ultimate objective to be able to predict the vibration effects so accurately that very slender and efficient blades can be run in perfect safety. We hope that this symposium will push the branch "Unsteady Aerodynamics and Aeroelasticity of turbomachines and Propellers" forward. We believe that more high level papers will be submitted to the next Symposium.

Cao Chuanjun
Professor, BUAA

Zhou Sheng
Professor, BUAA

CONTENTS

Part One
Unsteady Aerodynamics: Theory and Numerical Simulation

Part Four
Aeroacoustics

Part Five
Steam Turbine

Part Six
Engineering Design

Comment

Part One

Unsteady Aerodynamics: Theory and Numerical Simulation

TWO- AND THREE-DIMENSIONAL PROBLEMS OF UNSTEADY AERODYNAMICS OF LOW LOADED TURBOMACHINERY BLADE ROWS STAGES

K.K. Butenko, A.A. Osipov, K.S. Reyent
(Central Institute of Aviation Motors, Moscow, USSR)

ABSTRACT

Several problems are solved for unsteady ideal gas flow through turbomachinery blade rows and stages consisting of thin low loaded blades. The range of problems under consideration includes the problems of aerodynamic interaction of two plane cascades moving one relative to other, consisting of fixed or oscilating airfoils, the problem of aerodynamic interaction of two annular blade rows, rotating one relative to other, and the problem of unsteady flow through rotating annular row of blades at subsonic, supersonic and transonic (subsonic at the blade root and supersonic at the blade tip) velocities of the basic flow. By means of linear analysis of small disturbances these problems are reduced to the solution of integral equations on surface of one of the blades in every row. Developed calculation programs provide fast estimation of unsteady load on blades of rows and acoustic fields, generated by them.

AERODYNAMIC INTERACTION OF TWO PLANE AIRFOIL CASCADES

1. Interaction of cascades of fixed airfoils

In the first works concerning the aerodynamic interaction of two plane cascades moving one relative to other this problem was solved for incompressible fluid in quasistationary approximation, [1–4] allowing to neglect vortex sheets shedding the trailing edges of airfoils. The solution of the nonlinear problem with regard to vortex sheets was obtained by means of numerical integrating of appropriate unsteady integral equations using the discrete vortex technique. [5,6] The problem of aerodynamic interaction of cascades in compressible gas was solved only by direct numerical integrating of unsteady gas dynamics equations. [7–10] However realization of such an approach demands much computer resource expenditure that brings down considerably its practical usefulness.

In the present work the aerodynamic interaction of cascades of thin low loaded airfoils has been considered. Such a restriction allows to apply the linear analysis of small disturbance and to obtain the solution by means of integral equations technique. Application of this approach does not require much computer time and storage that is important for extensive parameter investigation.

In linear model of small perturbations the unsteady flow field under consideration is the disturbance of uniform steady flow through a cascade of straight line airfoils directed along the stream line in cascade–fixed coordinates. Scheme of such a flow is shown in Fig.1 in coordinate system connected with the front cascade. The subscripts 1 and 2 are ascribed to parameters of front and back cascades respectively, V – the velocity of running flow, W –the velocity of movement of the second cascade relative to the first cascade.

Small bend or turn of all airfoils of one or both cascades results in unsteady periodic in time t aerodynamic interaction of them with period $T_1 = S_2 / W$ and $T_2 = S_1 / W$ in coordinate systems connected with the first and second cascades respectively. The equality $Y = N_1 \leftarrow S_1 = N_2 S_2$ is satisfied for gaps S_1 and S_2 of these cascades, where N_1 and N_2 –numbers of the airfoils of the first and second cascades contained in their minimal common period y in direction y.

Generelized spatial–temporal periodicity of flow parameters takes place in unsteady process under consideration. For pressure $P_v (v = 1.2)$ this property is expressed in the equalities

$$P_1(x_1, y_1, t) = P_1(x_1, y_1 + s_1, t + s_1 / W) \tag{1}$$

$$P_2(x_2, y_2, t) = P_2(x_2, y_2 + s_2, t - s_2 / W) \tag{2}$$

with regard to periodicity of P_v in y with period y and in t with period T_v and using (1) and (2) one can obtain formulaes

$$P_v = \sum_{j=-\infty}^{\infty} \sum_{n=-\infty}^{\infty} P_{vjn}(x) exp[i(\beta_{vin} y_v - \omega_{vj} t)] \tag{3}$$

$$\beta_{1jn} \equiv \frac{2\pi}{S_1}\left(n + j\frac{N_2}{N_1}\right), \quad \beta_{2jn} \equiv \frac{2\pi}{S_2}\left(n - j\frac{N_1}{N_2}\right) \tag{4}$$

$$\omega_{1j} \equiv \frac{2\pi}{S_2} j W, \quad \omega_{2j} \equiv \frac{2\pi}{S_1} j W \tag{5}$$

According to (3) oscillations of any gas parameter,including the oscillation of unknown aerodynamic load on airfoils, may be expressed as an infinite set of harmonic components with known values of frequency ω_{vj}. Within the bounds of linear theory of small disturbance the influence of airfoils on the flow is a superposition of influences of different load harmonic components, which may be discribed by dipole distribution along the airfoil. acoustic field and vortex wakes behind the airfoils induced by every load harmonic component ΔP_{11vj} of all cascade airfoils may be determined using known technique. Then using the tangency condition on both cascades the following infinite system of integral equations for ΔP_{vj} may be obtained

$$\int_0^{L_1} K_{1j}(\xi_1 - \xi_{10}) \Delta P_{ij}(\xi_{10}) d\xi_{10} + \int_0^{L_2} \sum_{n=-\infty}^{\infty} Q_{2nj}(\xi_2, \xi_{20}) \Delta P_{2n}(\xi_{20}) d\xi_{20} = \delta_{jo} \frac{dh_1}{d\xi_1} \tag{6}$$

$$\int_0^{L_2} K_{2j}(\xi_2 - \xi_{20}) \Delta P_{2j}(\xi_{20}) d\xi_{20} + \int_0^{L_1} \sum_{n=-\infty}^{\infty} Q_{1n-j}(\xi_1, \xi_{10}) \Delta P_{1n}(\xi_{10}) d\xi_{10} = \delta_{jo} \frac{dh_2}{d\xi_2} \tag{7}$$

$$j = 0, \pm 1, \pm 2, \cdots$$

Equation (6) corresponds to tangency condition on the first cascade and equation (7) – on the second cascade; L_v – v– cascade airfoil length, ξ_v – coordinate along airfoil; δ_{jo} – Kronecker's symbol. Given function $n_v = h_v(\xi_v)$ describes airfoil deformation of v – cascade in direction of normal relative to its undisturbed surface.

Each kernel function K_{vj} in equations (6) and (7) coincides with kernel function of integral equation corresponding to (6), (7), which arises in the problem of unsteady flow through separate cascade of plates, oscillating at frequency ω_{vj} and fixed phase lag σ_{vj}, where $\sigma_{1j} = 2\pi j N_2 / N_1, \sigma_{2j} = -2\pi j N_1 / N_2$. The functions Q_{vnj} correspond to n– component in time harmonic expansion and j – component in y – expansion of acoustic field, induced by ν– cascade in appropiate coordinate system, and describe influence of it on the other cascade.

The dependence of lift coefficient $C_{L\nu}$ on time at one period of oscillations obtained by means of (6), (7) for the first and second cascades is shown in Fig.2 and 3 in the case $S_1 = 1.0$, $L_2 = 1.0$, $N_2 = 10$, $N_2 = 9$, $\alpha_1 = 0.0$, $M_1 = 0.01$, $W = -1.0$, $\Delta = 0.5$. Here M_1, α_1 – Mach number and angle between velocity vector and x – axis in basic flow relative to the first cascade, Δ – distance between cascades. All the values with dimension of lenght relate to L_1, velocity – to velocity V of basic flow relative to the first cascade. Unsteady aerodynamic interaction of cascades is induced by little change of α_1 relative to undisturbed value. Comparison of results obtained with consideration of 9, 13 and 17 temporal harmonic components of oscilations are presented. There is a coincedence of data obtained for the first cascade with consideration of 13 and 17 harmonic components in Fig.2, which are presented by solid line. More irregular dependece $C_{L2}(t)$ in comparison with $C_{L1}(t)$ is connected with unsteady vortex sheets, shedding the trailing edges of airfoils of the firs cascade, which influence on the second cascade.

2. Interaction of cascades of oscillating airfoils

In flutter investigations it is necessary to know how the interaction of cascades in turbine stage influence on the unsteady aerodynamic characteristics of cascade of oscillating airfoils. Any theoretical investigation of this question in literature are unknown for authors. The complexity of analytical description of such unsteady flow is caused by presence of two different time scales: period T of cascade oscillation and period of cascades moving one relative to other. The presence of two scales, one undivisible by other in general case, don't permit one to seek the solution by means of usual time Fourier decomposition because the flow as a whole isn't periodic.

We shall confine ourselves to the case of oscillation of airfoils of the first cascade at frequency ω and phase lag σ between the oscillations of adjacent airfoils. Displacement of any airfoil surface in direction of its normal may be described by function

$$\eta_1 = h'_1(\xi_1, y_1, t) = h(\xi_1) exp\left\{i\left[\frac{\sigma}{S_1}(y_1 - \xi_1 sin\alpha_1) - \omega t + \varphi_0\right]\right\}$$

Here φ_0 – oscillation phase of initial airfoil at the moment t=0 Let's present the disturbances of all flow parameters and in particular the pressure P'_ν and the velocity component normal to airfoil surface v'_ν by the expression

$$P'_\nu(x_\nu, y_\nu, t) = P_\nu(x_\nu, y_\nu, t) exp\left[i\left(\frac{\sigma}{S_1} - \omega t + \varphi_0\right)\right]$$

$$V'_\nu(x_\nu, y_\nu, t) = U_\nu(x_\nu, y_\nu, t) exp\left[i\left(\frac{\sigma}{S_1} - \omega t + \varphi_0\right)\right]$$

Then the boundary conditions on the airfoils, which are necessary for formulation of boundary problem for $P_\nu(x_\nu, y_\nu, t)$, reduce to relationships

$$V_1(\xi_1, y_1 = x_1 tg\alpha_1, t) = (-i\omega h_1 + \frac{dh_1}{d\xi_1}) e^{-i\frac{\sigma}{S_1}\xi_1 sin\alpha_1} \tag{8}$$

$$V_2(\xi_2, y_2 = x_2 tg\alpha_2, t) = 0 \tag{9}$$

Thus the boundary problem for $P_\nu(x_\nu, y_\nu, t)$is formulated as a problem of stationary disturbance of velocity component normal to airfoil surface which is independent of the number of airfoils and determined by (8), (9). Repeating the previous analysis, we obtain, that P'_i may be expressed in a form(3), where values of β'_{vjn} and ω'_{vj} should be substituted instead of values β_{vjn} and ω_{ij} from (4) and (5), as follows

$$\beta'_{1jn} = \frac{2\pi}{S_1}\left(n + j\frac{N_2}{N_1}\right) + \frac{\sigma}{S_1}, \quad \beta'_{2jn} = \frac{2\pi}{S_2}\left(n - j\frac{N_2}{N_1}\right) + \frac{\sigma}{S_2} \tag{10}$$

$$\omega'_{1j} = \omega + \frac{2\pi}{S_2} jW, \quad \omega'_{2j} = \frac{2\pi}{S_1} jW + \omega - \frac{\sigma}{S_1} W \tag{11}$$

Integral equations obtained from the tangency condition in this case have the from of (6), (7), where the right–hand side would be equal $\delta_{jo}(-i\omega h_1 + dh_1 / d\xi_1)$ for the first equation and zero – for the second equation.

The presence of the second cascade results in scattering the The presence of the second cascade results in scattering the oscillation energy among the whole range of frequencies (11), that differs from the case of separate cascade when there is a single component at the frequency of given airfoil oscillations.

The numerical data illustrating the influence of the presence of the second cascade moving relative to the first cascade upon the unsteady aerodynamic characteristics of oscillating first cascade are shown in Fig.4. The values of casceds and flow parameters are as follows; $S_1 = S_2 = L_2 = 1.0, \Delta = 0.05$, $M_1 = 0.5, \alpha_1 = 45°$, $M_w = 0.42$. Here M_w –the Mach number corresponding to velocity W. The unsteady flow through cascades has been considered which is induced by bending oscillation of the first cascade blades ($h_1(\xi_1) = 1.0$) at frequency $\omega = 2.0$. The main harmonic component amplitude of lift coefficient oscillation C_L versus phase lag σ is shown in Fig.4. The multiple load increase is observed near $\sigma = 300$, that apparently points out the resonant character of cascades interaction in this case.

AERODYNAMIC INTERACTION OF TWO ANNULAR BLADE ROWS

At present the problem of aerodynamic interaction of two annular blade rows has been solved only by the direct numerical methods.[2,13] At the same time the approach developed above for evaluation of two plane cascades allows to solve this problem also in the case of two annular blade rows rotating one relative to other. The method of integral equation which is applicable to this case was developed by Namba.[14]

The flow scheme is presented in Fig.5. The annular rows consist of thin low loaded blades, which within bounds of linear approximation may be considered as lying on the stream surfaces formed by radial raies in coordinate system rotating with the row. We shall confine ourselves to case when the running flow is uniform and directed along the chanal axis in the stationary coordinats system and the flow is subsonic relative to each row.

The periodicity conditions relative to t and to azimuthal direction and the generalized spatial–temporal periodicity of flow parameters in annular blade rows are analogous to the same conditions in plane airfoil cascades. So the pressure perturbation for instance, may be expressed as a double Fourier expension

$$P_v(Z_v, r_v, \theta_v, t) = \sum_{j=-\infty}^{\infty} \sum_{n=-\infty}^{\infty} P_{vjn}(Z_v, r_v) exp[i(\beta_{vjn}\theta_v - \omega_{vj} t)]$$

$$\beta_{1jn} = nN_1 + jN_2, \quad \beta_{2jn} = nN_2 - jN_1$$

$$\omega_{1j} = jN_2 W, \quad \omega_{2j} = -jN_1 W, \quad W = \omega_{T1} - \omega_{T2} \tag{12}$$

Here Z_v, δ_v, θ_v – the axes of cilindrical coordinate system related to v row.

The oscillations of unknown aerodynamic load on blades may be expressed as infinite set of harmonic components at frequencies (12). Using Green formula the influence of each harmonic component of load upon the flow is presented as influence of dipoles distributed along the blade surfaces. Providing the total velocity field induced on blade surfaces satisfy the tangency condition, one can obtain the following infinite system of integral equations for load

$$\sum_{K=0}^{L-1}\int_{-C_\nu/2}^{C_\nu/2} \Delta P_{\nu jk}(Z_{0\nu}) KT_{\nu jk}(Z_\nu - Z_{o\nu}, r_\nu) dZ_{o\nu} + \sum_{n=-\infty}^{\infty}\sum_{K=0}^{L-1}\int_{-C_{\bar\nu}/2}^{C_{\bar\nu}/2} \Delta P_{\nu jk}(Z_{0\bar\nu}) Q_{\bar\nu n-jk}(Z_\nu, Z_{o\bar\nu}, \delta_\nu) dZ_{o\bar\nu}$$

$$= -\delta_{jo} V_\nu \qquad (13)$$

$$\nu = 1.2, \qquad j = 0, \pm 1, \pm 2, \cdots\cdots$$

Here $\bar\nu = 3 - \nu$, C_ν – projection of blade profile on Z – axis, V_ν – normal velocity component, induced by basic flow on blade sruface because of its displacement, with ν_2 – is equal to zero. The functions $\Delta P_{\nu jk}$ are the coefficients of series

$$\Delta P_{\nu j}(Z, r) = \sum_{k=0}^{L-1} \Delta P_{\nu jk}(Z) R_k^{(0)}(\delta) \qquad (14)$$

where $R_k^{(0)}(\delta)$ – is a combination of Bessel functions of first and second kinds of zero order, which has a zero derivative on the ends of interval [h ,1], where h – is hub radius of rotor, and $R_k^{(0)}(r)$ is a k – eigen function of appropritc boundary problem. The kernel function $KT_{\nu jk}$ coincides with kernel function of corresponding problem for a separate row, [14] and the kernel function $Q_{\nu njk}$ describes the influence of one row on other.

Interaction of annular blade rows has a singularity which is absent in the case of plane cascades. The initial reason for interaction is the disturbance of flow through a separate row. Inhomogeneous steady flow behind the plane cascade contains only potential distirbances quickly decreasing in downstream direction. This results in quick decreasing of interaction of plane cascades due to increasing of distance between them.

There are trailing edge vortex wakes side by side with potential distrubance in inhomogeneous flow behind the annular blade row. The vortex wakes contain disturbances of radial and azimuthal velocity components also as a discontinuity of radial velocity components which don't disappear moving off the row in the downstream direction. For this reason the interaction of annular blade rows is preserved at any distance between them.

The numerical results obtained using (13) give evidence that when relative contribution of steady vortex heterogeneity is small the strip theory describes satisfactory the aerodynamic interaction of rows. If the vortex component dominates in the inhomomgeneous flow influencing the second row, application of three–dimensional theory is necessary for true description of rows interaction.

Comparison of three–dimensional and strip–theory results is shown in Fig.6. The curves of lift coefficient dependence C_L on t are presented which was obtained at root, middle and tip of blades of both rows. The case is considered: h = 0.5, $c_1 = c_2 = 0.314$, $N_1 = 20$, $N_2 = 18$ $\omega_{T1} = 0$ $\omega_{T2} = 0.01$, $\Delta = 0.0157$, $M_a = 0.7$. Here $\omega_{T\nu}$ – angular velocity of ν – row rotation, *Ma*– Mach naumber corresponding to axial velocity of basic flow. All the values witn dimension of length relate to radius of outward wall the values with dimension of length relate to radius of outward wall of annular chanal and the values of velocity dimension – to velocity V of basic axial flow. The initial disturbance of normal velocity induced by blades deflection in both rows with attack angle constant along radius. The curves of strip–theory results are not ploted if the value of C_L turned out too small.

UNSTEADY AERODYNAMICS OF ROTATING ANNULAR BLADE ROW IN TRANSONIC FLOW REGIME

While using the method of integral equation for calculation of unsteady aerodynamic characteristics the serious difficulties arise concerning the regularization of appropriate integral equation kernel function. The infinite series in kernel function expression contain Bessel functions which have no asymptotic evaluation required for regularization. The solution of this problem was proposed by Namba, [15] who used the approximate kernel function found by Galerkin method. This approach was applied succesfully for calculation of subsonic unsteady flow through blade row. [14] However using the method in the case of supersonic or transonic (subsonic at the blade root and supersonic at the blade tip) flow essential errors are revealed in the solution obtained by this method which are results of approximate description of singular properties of integral equation kernel function[16].

In the present work a new technique for regularization of integral equation has been proposed which is based on using Green function which is fundamental solution of the problem in some asymptotic sense.

This asymptotic Green function as distinct from the Green function found by Namba, [14-16] reflects the ellipticity or hyperbolicity of the problem in the subsonic or hypersonic region of flow respectively. Such an approach permits one to develop robust computation procedure and provides acceptable convergence of results.

The flow scheme for separate blade row is presented by part of Fig.5, including only one of two rows shown there. In the present work we shall confine ourselves by the case of subsonic axial flow ($M_a < 1$), while in rotating coordinate system the flow may be both subsonic or supersonic. The harmonic flow oscillations at frequency are considered, which are the result of small blades oscillations or influence of week vortex heterogeneities of running flow.

In concluding the integral equation the expression for disturbed pressure p at any point $\bar{r}$ inside chanal is used as follows

$$P((\bar{r})\int_{S_0}\int \Delta P(Z_0,r_0)\frac{\partial}{\partial n_0}G_N(Z-Z_0,r,r_0,\theta-\theta_0)dS_0 \qquad (15)$$

Here S_0 – surface of and selected blade, n_0 – coordinate noraml to blade surface and ΔP – pressure jump across the blade. Green function G_N describes the pressure field of annular row of poles oscillating at frequency ω, placed at points $\{Z_0, r_0, \theta_0 + 2\pi k / N\}$ $k=0,\cdots\cdots,N-1$, and determined by expression[14].

$$G_N \equiv -\frac{N}{4\pi\beta^2}\sum_{\nu=-\infty}^{\infty}\sum_{L=L_0}^{\infty}\frac{1}{\Omega_L^{(n)}}R_L^{(n)}(r)R_L^{(n)}(r_0)Q_L^{(n)}(Z-Z_0,\theta-\theta_0) \qquad (16)$$

$$Q_L^{(n)}(Z-Z_0,\theta-\theta_0) \equiv exp[in(\theta-\theta_0)+i\omega_n\frac{M_a^2}{\beta^2}(Z-Z_0)-\Omega_L^{(n)}|Z-Z_0|]$$

$$\Omega_L^{(n)2} \equiv \frac{1}{\beta^2}(\lambda_L^{(n)2}-\omega_n^2\frac{M_a^2}{\beta^2}), \quad \omega_n \equiv \omega + n\omega_T, \quad \beta^2 \equiv 1-M_a^2$$

$$n=\nu N+6, \qquad \sigma = 0,\pm 1,\pm 2,\cdots\cdots$$

Here $R_L^{(n)}(r)$ – normalized solutions of Bessel equation, which have zero derivatives at the ends of interval [h,1] and are combinations of Bessel functions of first kind $J_n(\lambda_L^{(n)}r)$ and second kind $Y_n(\lambda_L^{(n)}r)$, where $\lambda_L^{(n)}$ eigen values of boundary problem for $R_L^{(n)}$, $L_0 = 1$ if $n=0$ and if $n \neq 0$; N – number of blades in row, ω_T – angular velocity of its rotating, and $2\pi\sigma / N$ – phase lag between the oscillations of adjacent blades. A signum of $\Omega_L^{(n)}$ is selected according to radiation condition.

The presence of $R_l^{(n)}$(r) in (16) makes difficult the regulatization of integral (15) beintegral (15) because asymptotic evaluation for great n of $R_L^{(n)}$ and $\lambda_L^{(n)}$ required for it is absent. In order to solve this problem we shall consider the function $f(Z,Z_0,r,\theta,\theta_0)$ determined by integral transformation

$$f=\int_h^1 \Delta P(Z_0,r_0)G_N(Z-Z_0,r,r_0\theta-\theta_0)r_0dr_0 \tag{17}$$

Using (17) investigation of singular integual (15) has reduced to investigation of singular properties of f. Let's consider n– component of Fourier series for f relative to $\theta-\theta_0$. It may be proved that the function $F^{(n)}(Z_0,r)$ which is a Fourier transformation of $f^{(n)}$ relative z – coordinate satisfies the equation

$$\left[\frac{1}{r}\frac{\partial}{\partial r}(r\frac{\partial}{\partial r})+(\xi^2-\frac{n^2}{r^2})\right]F^{(n)}=\frac{N}{4\pi^2}\Delta P(Z_0,r)e^{-iaz_0} \tag{18}$$

$$F^{(n)}\equiv\frac{1}{2\pi}\int_{-\infty}^{\infty}f^{(n)}e^{-iaZ}dZ,\quad \xi^2\equiv-\alpha^2(1-M_a^2)+2M_a^2\alpha\omega_n+M_a^2\omega_n^2$$

An intricate asymptotic behaviour of eigen function $R_L^{(n)}$ of operator in left–hand side of (18) makes the investigation of asymptotic properties of $F^{(n)}$ for $n\to\infty$ difficult. An approach proposed by Namba for regularization of integral (15),[14–16] may be justified as follows. Let's consider finits system of L approximate eigen functions $F_k^{(n)}$(r) of operator in left–hand side of (18) which by means of Galerkin method we shall express in form of finite series

$$F_k^{(n)}(r)=\sum_{L=0}^{L-1}F_{kl}^{(n)}R_l^{(0)}(r),\qquad k=0,1,\cdots,L-1 \tag{19}$$

The eigen values of ξ in (18) correspond to the eigen function $F_k^{(n)}(r)$. If Δp is a sufficiently smooth function of r then $F^{(n)}$ may be expressed with and accuracy in terms of $F_k^{(n)}$, which satisfies the relationships

$$F_k^{(n)}=F_k^{(\infty)}+O(n^{-2}),\qquad \xi_k^{(n)}=\xi_k^{(\infty)}+O(n^{-2}) \tag{20}$$

In the case of subsonic flow ΔP is a differentiable function of r, which may be presented with any accuracy in a form of finite series of eigen functions $F_k^{(n)}(r)$. The asymptotic behaviour (20) of finite basic $F_k^{(n)}$ allows one to accomplish the regularization of f.

In the case of supersonic flow the pressure field described by integral (15) contains discontinuities, issuing from leading and trailing edges of blades. In this case in approximate solution of Namba the pressure discontinuities of exact solution splits into series of discontinuities produced by radial raies. Such a splitting reflects the complex spatial behaviour of shock surface in exact solution and is its approximation in known sense.

At regime of transonic flow through row the Cauchy's singularity corresponding to subsonic flow and a singularity of δ – function corresponding to supersonic flow present in kernel function of integral equation simultaneously. In this case the pressure discontinuities of approximate solution are formed generaly both in supersonic and subsonic parts of flow. The convergence of numerical solution of such equation near discontinuities becomes worse greatly which is a result of breach of assumption of smooth dependence ΔP on r.

Another method for regularization of f is proposed in the present work. For this purpose the asymptotic solution of equation (18) is used which for subsonic flow looks as follows

$$F^{(n)} = \frac{N}{4\pi^2} \frac{\Delta p e^{-i\alpha z_0}}{\xi^2 - n^2/r^2} [1 + O(n^{-2})] \tag{21}$$

The relationship (21) allows one to choose the singular part of f, which for fixed coinsides with singular part of kernel function of integral (15) in the case of plane flow corresponding to gas flow parameters at given r. Using (21) tangency condition on the blade surface is reduced to integral equation

$$\int_{-c/2}^{c/2} \left\{ \sum_{\nu=-N_0}^{N_0} \left[\int_h^1 \sum_{L=L_0}^{L-1} K_{1l}^{(n)}(Z - Z_0, r, r_0) \Delta P(Z_0, r_0) dr_0 + k_2^{(n)}(Z - Z_0, r) \Delta P(Z_0, r) \right] + K_s(Z - Z_0) \Delta P(Z_0 r) \right\} dz_0$$
$$= V(Z,r) \tag{22}$$

Here v – given oscillation amplitude of gas velocity normal to blade surface, N_0 – number of azimuthal modes under consideration. The kernel function $K_{1l}^{(n)}$ is obtained from the corresponding component of G_N in (16),the kernel function $K_2^{(n)}$ is a principal value of n– component of kernel functions in corresponding plane flow problem, and both this kernel functions form the regular part of integral equation. The kernel function K_s coinsides with the singular part of kernel function in corresponding plane flow problem.

In the case of supersonic flow one fails in obtaining relationship (21). Nevertheless the calculations accomplished by means of (22), display the acceptable convergence of numerical results. Analysing the equation (22) it has been found that the errors of this solution are concentraited near pressure discontinuities, which on surface r =const are described approximately by characteristics of corresponding plane flow. It may be expected that if the parameters of flow and row are such that discontinuities don't incide on adjacent blades or with such incidence the deflection of the line of discontinuity in plane z=const relative to radial ray direction is small the region of considerable errors of ΔP(z,r) remains sufficiently thin and do not influence considerably on the calculation accuracy of row unsteady aerodynamic characteristics.

The convergence of calculation results relative to N_0 is shown in Fig.7. The blade row has been considered at N=20, c=0.05, h=0.6, in axial uniform flow at $M_a = 0.8$ and rotating at $\omega_T = 0.833$. Unsteady loads induced on blades oscillating at frequency $\omega = 20.0$ and at phase lag $\sigma = 5$. The blade oscillation form H(z,r), determining the amplitude of surface displacement in direction normal to its undisturbed position is given as $H = (r-h)^2/(1-h)^2$. Dependence of modulus of lift coefficient C_L on r is presented. The interrupted vertical line corresponds to the value where the relative flow velocity is equal to the sound velocity. In Fig.8 the distribution of modulus of lift coefficient C_p along the blade profile is shown at different sections r=const. The case has been considered corresponding to N=40, C=0.05, h=0.6, $M_a = 0.8$, $\omega_T = 1.0$, $\omega = 2.0$, $\sigma = 5$, which correspods to transonic flow.

It should be noted that as it has been shown by calculations some critical value n_* depended of h exists so that for $n > n_*$ approximating properties of basis $R_L^{(n)}(r)$ becomes worse greatly which results in pronounced worsening of results convergence relative to L. The value n_* is decreasing if h decreases. However this feature $K_{1l}^{(n)}$ of the method proposed here may be eliminated if for approximation of regular kernel function in (22) the basis consisting of eigen functions is used which determined by *Namba*[15] according to (19).

REFERENCES

1. Feindt E–G, Ingr–Arch, 31(4)(1962), 258.
2. Kazimierski Z, Arch Budowy Maszyn, 13(2)(1966).
3. Saren VE, Izv AN SSSR, MJG, (4)(1971), 75.
4. Saren VE, Judin VA, Probl prochnosty, (9)(1977), 104.
5. Kuljaev RL, Jurn Prikl Meh & Tehn Fis, (4)(1976), 61.
6. Belotserkovskij SM et al, Matematicheskoe modelirovanie ploskoparallelnogo otrivnogo obtekanija tel, Nauka, 1988.
7. Erdos JI, Alzner E, McNally W, AIAA Journal, 15(11)(1977), 1559.
8. Gnesin VI, Izv AN SSSR, Energetika & Transport, (1)(1982), 102.
9. Rai MM, AIAA Paper, (361)(1988).
10. Lewis JP, Delaney RA, Hall EJ, AIAA Paper, (2149)(1987).
11. Ni R–h, Trans ASME, J Eng Power 101(3)(1979).
12. Gnesin VI, Izv AN SSSR, MJG, (6)(1982), 138.
13. oya M, Kotake S, Trans ASME, J Eng Gas Turbins Power, 107(4)(1985).
14. Namba M, J Sound & Vibration, 50(4)(1977), 479.
15. Namba M, Aeronautical Research Council, R & M, 3740(1972).
16. Namba M, Ishikawa A, Trans ASME, J Eng Power, 105(1)(1983).

Fig:

Fig.1 Scheme of flow through two airfoils cascades moving one felative to other.

Fig.2 Dependence of lift coefficient of the first cascade on time; ---- - 13 and 17 harmonic components, ---- - in - compressible flow method.[4]

Fig.3 Dependence of lift coefficient of the second cascade on time;..... - 9 harmonic components, ---- - 17 harmonic components, ---- - incompressible flow method.

Fig.4 Dependence of real and imagine parts of the amplitude of main harmonic component of lift oscillation on σ for first cascade; ---- - with cascades interaction, ---- - without cascades interaction.

Fig.5 Scheme of flow through two annular rows of blades rotating one relative to other.

Fig.6 Dependence of lift coefficient of the first and of the second blade rows on time at three fixed values of radius $r = h$, $r_{min} = 0.5(h + 1)$, 1; ---- - three-dimensional theory, ---- - quasi-three-dimensional theory.

Fig.7 convergence of modulus of lift coefficient dependence on r relative to number of considered azimuthal modes in regular part of integral equations kernel at mixed flow regime; □ - N_0=0, 0 - N_0 =1, 0 - N_0 =2.

Fig.8 Distribution of modulus of pressure coefficient along the blade profile at different values of r .

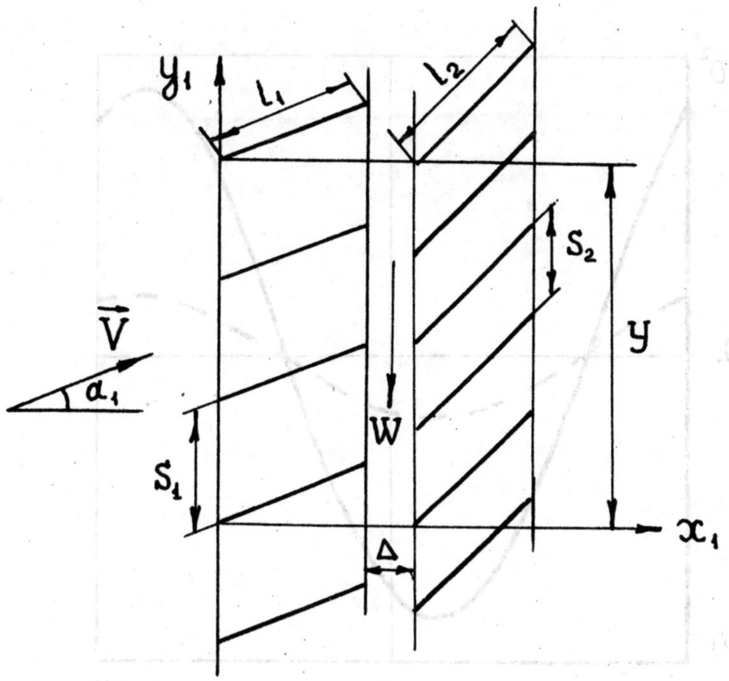

Fig.1 Scheme of flow through two airfoils cascades moving one relative to other.

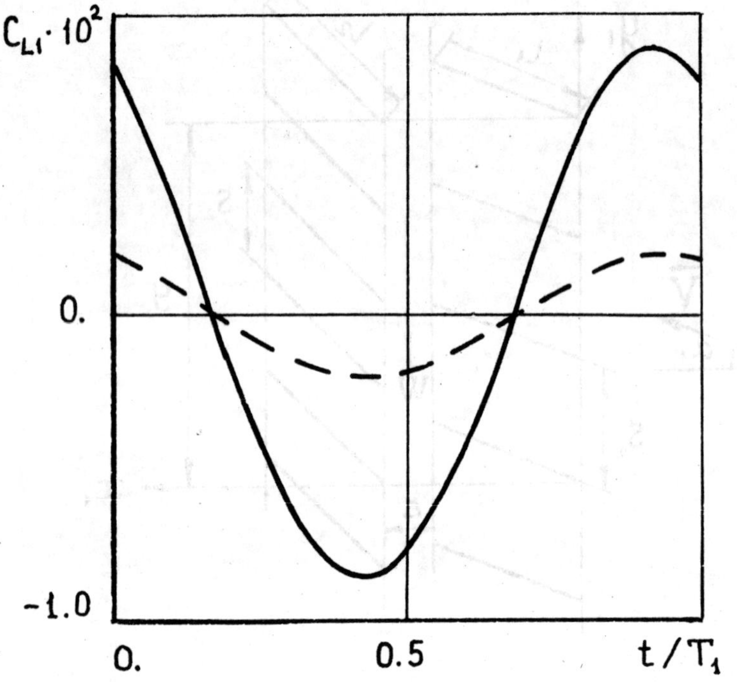

Fig.2 Dependence of lift coefficient of the first cascade on time;---- - 13 and 17 harmonic components, ---- - in - compressible flow method.[4]

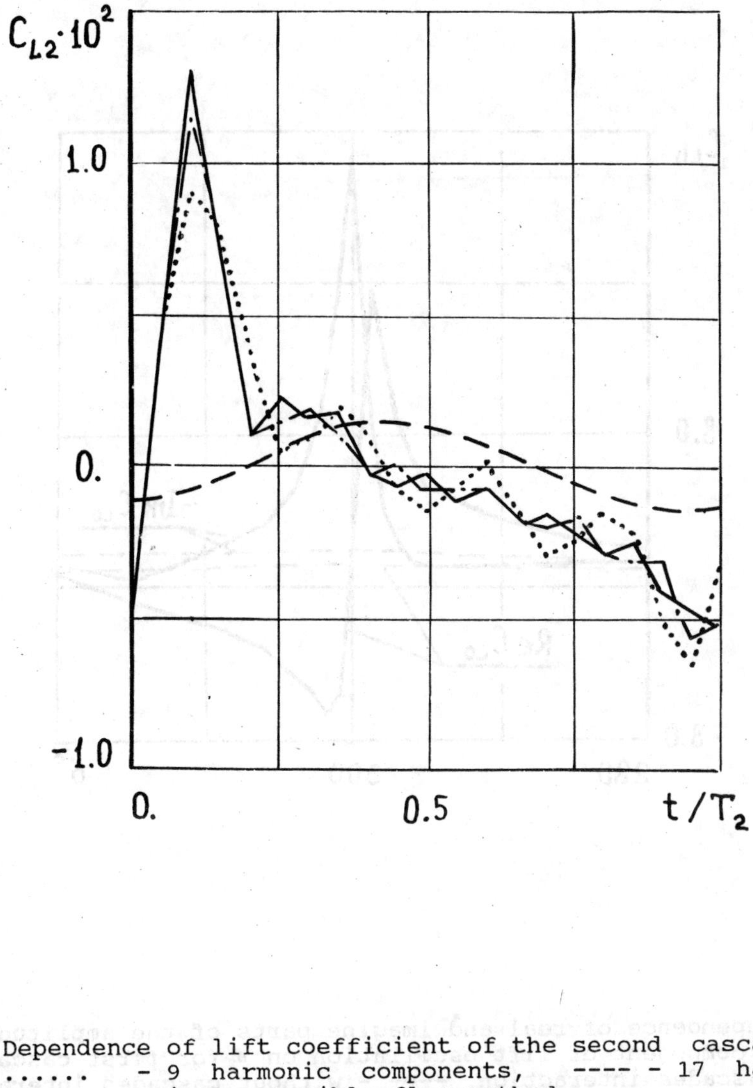

Fig.3 Dependence of lift coefficient of the second cascade on time; ...... - 9 harmonic components, ---- - 17 harmonic components, ---- - incompressible flow method.

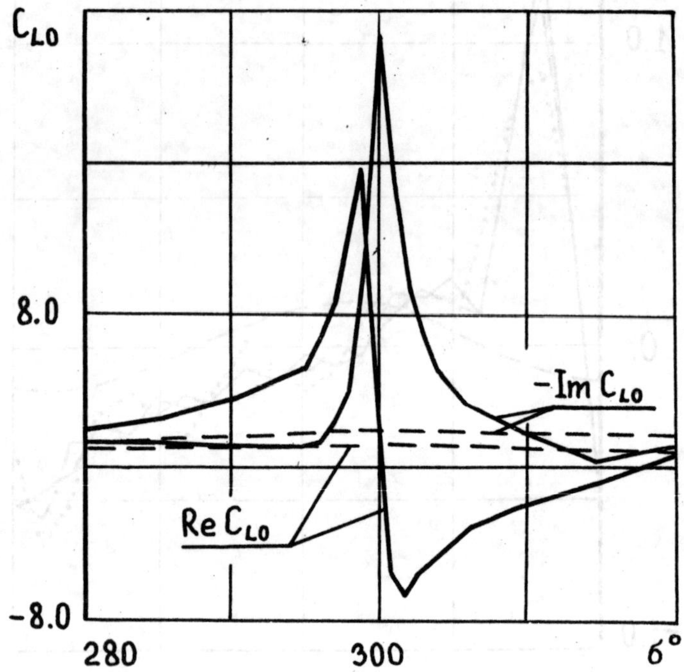

Fig.4 Dependence of real and imagine parts of the amplitude of main harmonic component of lift oscillation on σ for first cascade; ----- with cascades interaction, ---- - without cascades interaction.

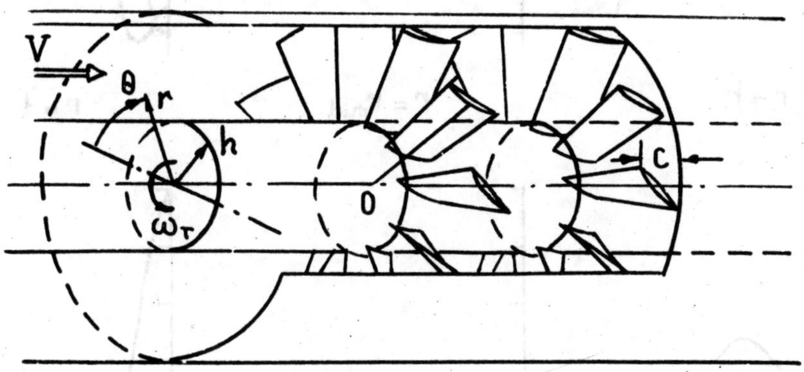

Fig.5 Scheme of flow through two annular rows of blades rotating one relative to other.

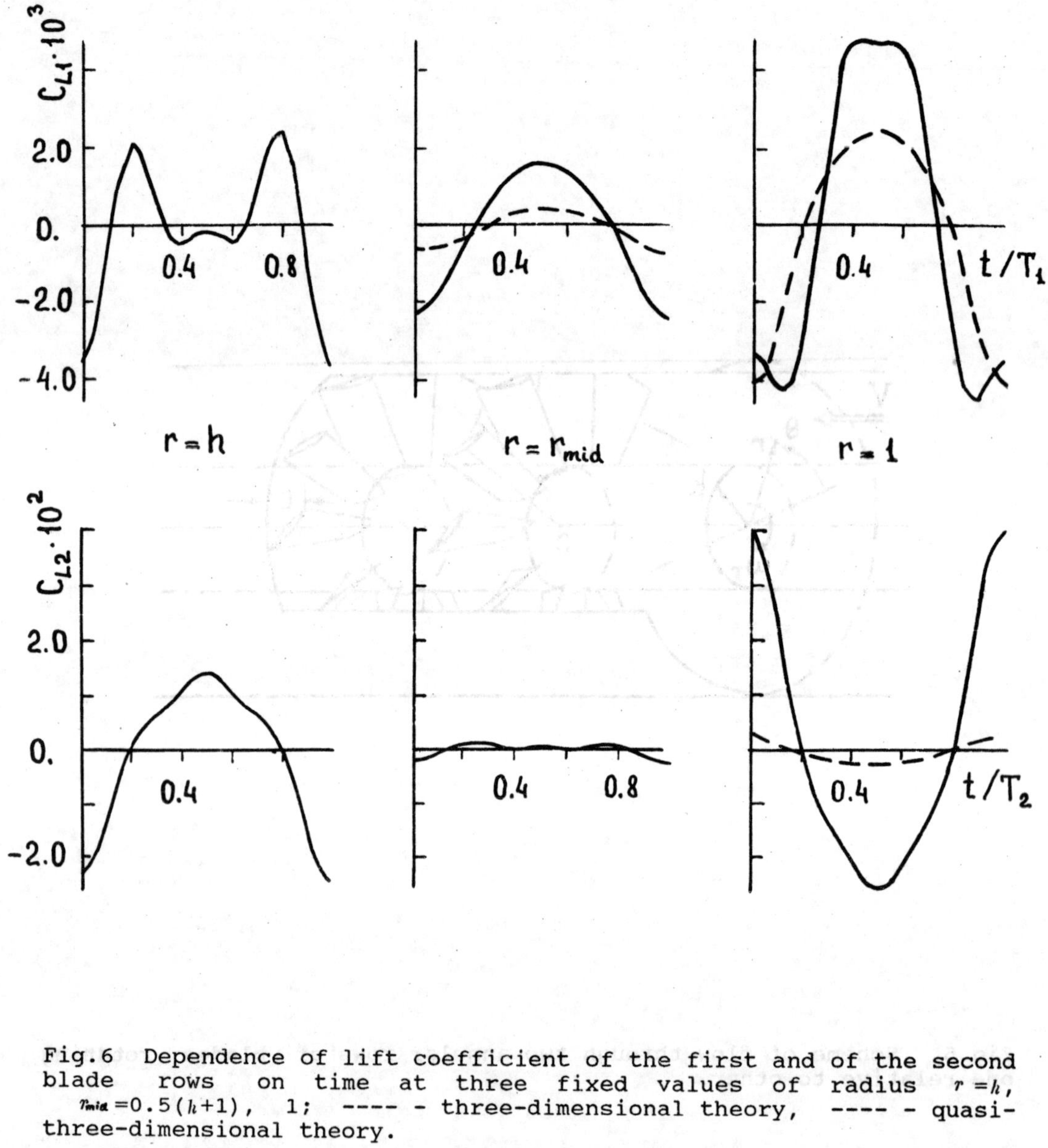

Fig.6 Dependence of lift coefficient of the first and of the second blade rows on time at three fixed values of radius $r = h$, $r_{mid} = 0.5(h+1)$, 1; ---- - three-dimensional theory, ---- - quasi-three-dimensional theory.

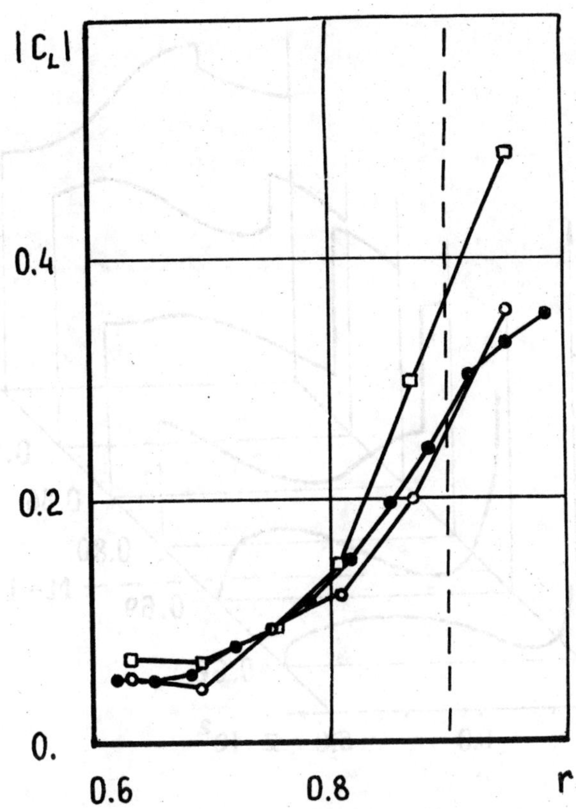

Fig.7 Convergence of modulus of lift coefficient dependence on r relative to number of considered azimuthal modes in regular part of integral equations kernel at mixed flow regime;
□ - $N_0=0$, 0 - $N_0=1$, ● - $N_0=2$.

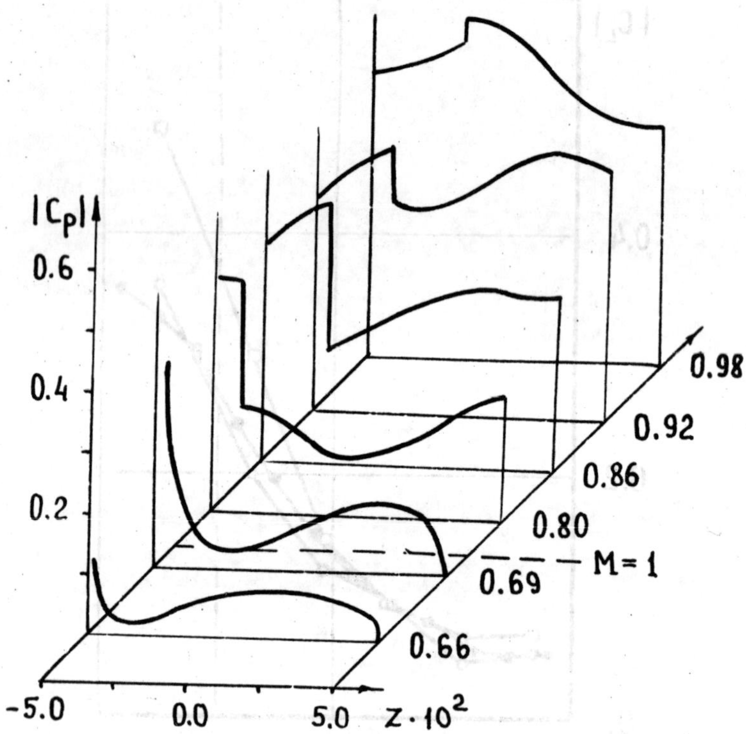

Fig.8 Distribution of modulus of pressure coefficient along the blade profile at different values of r.

MEAN LOADING EFFECTS ON FLUTTER OF SUBSONIC ROTATING ANNULAR CASCADE

Masanobu NAMBA, Kazuhiko TOSHIMITSU and Ping LI
(Kyushu University, Fukuoka, Japan)

ABSTRACT

A three-dimensional theory for unsteady subsonic annular cascade of blades oscillating with nonzero steady loading is developed. The model adopted in this theory is an isolated rotor with *N* vibrating blades rotating at a constant angular velocity in an infinitely long annular duct. It is assumed that the annular cascade blades operate with small mean angle of attack, camber and thickness, and vibrate with infinitesimal displacement amplitude. The steady and unsteady components of disturbance are treated on the basis of the double linearization theory so that both of steady and unsteady flow fields are governed by linear differential equations of constant coefficients, which are solved by using the Green function method.

Extensive parametric studies of the effects of the mean angle of attack, camber and thickness upon the unsteady aerodynamic forces and the flutter boundaries have been conducted. Numerical results of the present three-dimensional theory are compared with the results of the strip theory approximation to demonstrate some features of three-dimensional effects. It is found that the three-dimensional effect is essentially important. The strip theory approximation overestimates the spanwise variation of aerodynamic forces on vibrating blades and in some cases underestimates the contribution of the mean loading upon the overall aerodynamic work.

INTRODUCTION

In the past various theoretical methods to predict unsteady flows in axial turbomachines have been developed, and the flutter characteristics of two-dimensional cascade blades without steady loading have been fairly well understood. However, the actual annular cascade blades have three-dimensional change of profile and operate with steady loading which is known to have important effects on cascade flutter. Therefore, it is necessary to take both of the three-dimensional effects and steady loading effects into account so as to predict the actual behavior of flutter more correctly.

In recent years, several analytical studies dealing with the steady loading effects on cascade flutter have appeared in the literature and useful results have been reported. However, since most of the studies are based on simplified two-dimensional linear cascade models, there is a need to investigate how accurately these theories can predict actual flutter of a rotating annular cascade.

Namba[1] has developed the double linearization theory which can deal with steady loading effects under certain conditions. In this paper, the authors extend the double linearization theory to three-dimensional subsonic annular cascade blades to study the mean loading effects on flutter of a rotating annular cascade.

ANNULAR CASCADE MODEL

As shown in Fig.1, the model considered herein is an isolated rotor composed of *N* blades rotating at a constant angular velocity ω^* in an infinitely long annular duct. The undisturbed fluid flow is of a constant uniform subsonic axial velocity W_a^* and a constant uniform density ρ_o^*. In the following discussion we use dimensionless quantities without asterisk. Lengths are scaled with respect to the blade tip radius r_t^*, time with respect to r_t^*/W_a^*, and density with respect to ρ_o^*. The reference quantities for velocity, pressure and angular frequency are W_a^*, $\rho_o^* W_a^{*2}$, and W_a^*/r_t^* respectively.

We take a cylindrical coordinate system (r,θ,z) which is fixed to the rotor. The hub to tip ratio and dimensionless axial chord length are denoted by h and C_a respectively. Here the axial chord length C_a is assumed to be constant along the span to make the model simplified. Let the mean angle of attack, camber and thickness of blades are given by $\alpha_o\alpha(r)$, $\mu_o\mu(r)f(z)$ and $\tau_o\tau(r)g(z)$ respectively. Then the time mean circumferential distance from the helical streamsurfaces $\theta-\omega z=2\pi m/N$, $(m=1,\cdots,N)$ to the position of blade surfaces can be described by

$$rb_\eta(r,z)=\{-\alpha_o\alpha(r)z+\mu_o\mu(r)f(z)\pm\tau_o\tau(r)\frac{g(z)}{2}\}(1+\omega^2 r^2). \qquad (1)$$

Here symbol ± denotes upper and lower surfaces of the blades.

We assume the blades are undergoing identical harmonic vibration at reduced circular frequency λ and constant interblade phase angle $2\pi\sigma/N$. Therefore both steady and unsteady aerodynamic forces exist on the blades. Hereafter the symbols with overbar denote the steady components of disturbance and the symbols with tilde ~ times $e^{i\lambda t}$ denote the unsteady components of disturbance. We further assume the following:

— The flow is inviscid and isentropic and the fluid is a perfect gas.
— The relative fluid velocity with respect to the blades is subsonic along the whole span.
— The effect of chordwise displacement of blades is negligible.
— Steady disturbance is small compared with the undisturbed uniform flow quantities, and unsteady disturbance is far smaller than the steady disturbance, i.e., the flow field can be treated on the basis of the double linearization theory.

MATHEMATICAL FORMULATION

The governing equation of the flow field is a linearized acoustic wave equation. On the basis of the double linearization theory, we can split it into two linearized equations which govern the steady and unsteady disturbances respectively. Let the steady and unsteady pressure jumps across the blade surface be denoted by $\Delta\bar{p}$ and $\Delta\tilde{p}$, the steady and unsteady mass sources by $\bar{Q}$ and $\tilde{Q}$, and the spanwise unsteady force by $\tilde{D}$. Then the equations governing the steady and unsteady disturbance pressures $\bar{p}$ and $\tilde{p}$ can be expressed in the form

$$[\nabla^2-M_a^2(1+\omega^2r^2)\frac{\partial^2}{\partial s^2}]\bar{p}(r,\theta,z)=\int_h^1 dr_o\int_{-C_a/2}^{C_a/2}\sqrt{1+\omega^2 r_o^2}\sum_{m=0}^{N-1}\{-\Delta\bar{p}(r_o,z_o)\frac{\partial}{\partial n}$$

$$-\bar{Q}(r_o,z_o)\sqrt{1+\omega^2r^2}\frac{\partial}{\partial s}\}\frac{1}{r}\delta(r-r_o)\delta(z-z_o)\delta(\theta-\omega z_o-\frac{2\pi m}{N})dz_o\,, \qquad (2)$$

$$[\nabla^2-M_a^2\{i\lambda+\sqrt{1+\omega^2r^2}\frac{\partial}{\partial s}\}^2]\tilde{p}(r,\theta,z)=\int_h^1 dr_o\int_{-C_a/2}^{C_a/2}\sqrt{1+\omega^2 r_o^2}\sum_{m=0}^{N-1}e^{im2\pi\sigma/N}[-\Delta\tilde{p}(r_o,z_o)\frac{\partial}{\partial n}$$

$$-\Delta\bar{p}(r_o,z_o)a_n(r_o,z_o)\{\frac{i\lambda}{\sqrt{1+\omega^2r_o^2}}\cdot\frac{\partial}{\partial s}-\frac{\partial^2}{\partial n^2}\}-\tilde{D}(r_o,z_o)\frac{\partial}{\partial r}-\tilde{Q}(r_o,z_o)\{i\lambda+\sqrt{1+\omega^2r^2}\frac{\partial}{\partial s}\}$$

$$+\bar{Q}(r_o,z_o)a_n(r_o,z_o)\frac{\partial}{\partial n}\{i\lambda+\sqrt{1+\omega^2r^2}\frac{\partial}{\partial s}\}]\frac{1}{r}\delta(r-r_o)\delta(z-z_o)\delta(\theta-\omega z_o-\frac{2\pi m}{N})dz_o\,, \qquad (3)$$

where r_o and z_o denote the radial and axial coordinates of a disturbance source on the time mean blade surface, $a_n(r,z)e^{i\lambda t}$ denotes the n-wise displacement of blade from the time mean blade surface, while s and n represent the undisturbed streamline direction and the normal direction to the undisturbed streamsurface $\theta-\omega z$=constant. The (r,n,s) coordinate system can be related to the (r,θ,z) coordinate system by

$$\frac{\partial}{\partial n}=\frac{1}{\sqrt{1+\omega^2 r^2}}\left(\frac{1}{r}\cdot\frac{\partial}{\partial\theta}-\omega r\frac{\partial}{\partial z}\right) , \quad (4) \qquad \frac{\partial}{\partial s}=\frac{1}{\sqrt{1+\omega^2 r^2}}\left(\frac{\partial}{\partial z}+\omega\frac{\partial}{\partial\theta}\right) . \quad (5)$$

Equations (2) and (3) imply that the steady disturbance is independent of the unsteady disturbance, while the unsteady disturbance is influenced by the steady disturbance. It is thus necessary to determine the steady flow before solving the unsteady equation. At this stage $\Delta\bar{p}$, $\Delta\tilde{p}$, $\bar{Q}$, $\tilde{Q}$, and $\tilde{D}$ are unknown functions, which can be determined so that the boundary conditions on blades are satisfied.

Using the Green function method, equations (2) and (3) can be solved, and the disturbance pressures can be expressed as follows:

$$\bar{p}(r,\theta,z)=\int_h^1 \mathrm{d}r_\circ \int_{-C_\circ/2}^{C_\circ/2} \{\Delta\bar{p}(r_\circ,z_\circ)\bar{K}_{pF}(r,\eta,z-z_\circ\,|\,r_\circ)$$
$$+\bar{Q}(r_\circ,z_\circ)\bar{K}_{pQ}(r,\eta,z-z_\circ\,|\,r_\circ)\}\mathrm{d}z_\circ , \qquad (6)$$

$$\tilde{p}(r,\theta,z)=\int_h^1 \mathrm{d}r_\circ \int_{-C_\circ/2}^{C_\circ/2} \{\Delta\tilde{p}(r_\circ,z_\circ)K_{pA}(r,\eta,z-z_\circ\,|\,r_\circ)$$
$$+\Delta\bar{p}(r_\circ,z_\circ)a_n(r_\circ,z_\circ)K_{pL}(r,\eta,z-z_\circ\,|\,r_\circ)+\bar{Q}(r_\circ,z_\circ)a_n(r_\circ,z_\circ)K_{pQ}(r,\eta,z-z_\circ\,|\,r_\circ)$$
$$+\tilde{Q}(r_\circ,z_\circ)K_{pS}(r,\eta,z-z_\circ\,|\,r_\circ)-\tilde{D}(r_\circ,z_\circ)K_{pD}(r,\eta,z-z_\circ\,|\,r_\circ)\}\mathrm{d}z_\circ . \qquad (7)$$

Here $\eta=\theta-\omega z$, and the pressure kernel functions are given by

$$\begin{pmatrix} \bar{K}_{pF}(r,\eta,z-z_\circ\,|\,r_\circ) \\ \bar{K}_{pQ}(r,\eta,z-z_\circ\,|\,r_\circ) \end{pmatrix} = \sqrt{1+\omega^2 r_\circ^2} \left(\begin{pmatrix} \frac{\partial}{\partial n_\circ} \\ \sqrt{1+\omega^2 r_\circ^2}\,\frac{\partial}{\partial s_\circ} \end{pmatrix} \bar{G}(r,\theta,z\,|\,r_\circ,\theta_\circ,z_\circ) \right)_{\theta_\circ=\omega z_\circ} , \qquad (8)$$

$$\begin{pmatrix} K_{pA}(r,\eta,z-z_\circ\,|\,r_\circ) \\ K_{pL}(r,\eta,z-z_\circ\,|\,r_\circ) \\ K_{pQ}(r,\eta,z-z_\circ\,|\,r_\circ) \\ K_{pD}(r,\eta,z-z_\circ\,|\,r_\circ) \\ K_{pS}(r,\eta,z-z_\circ\,|\,r_\circ) \end{pmatrix} = \sqrt{1+\omega^2 r_\circ^2} \left(\begin{pmatrix} \frac{\partial}{\partial n_\circ} \\ \frac{\partial^2}{\partial n_\circ^2}+\frac{i\lambda}{\sqrt{1+\omega^2 r_\circ^2}}+\frac{\partial}{\partial s_\circ} \\ \frac{\partial}{\partial n}\{i\lambda-\sqrt{1+\omega^2 r_\circ^2}\}\frac{\partial}{\partial s_\circ} \\ -\frac{\partial}{\partial r_\circ} \\ -i\lambda+\sqrt{1+\omega^2 r_\circ^2}\frac{\partial}{\partial s_\circ} \end{pmatrix} \tilde{G}(r,\theta,z\,|\,r_\circ,\theta_\circ,z_\circ) \right)_{\theta_\circ=\omega z_\circ} , \qquad (9)$$

where Green's functions $\bar{G}$ and $\tilde{G}$ are the solutions to the following equations

$$\{\nabla^2-M_a^2(1+\omega^2 r^2)\frac{\partial^2}{\partial s^2}\}\bar{G}(r,\theta,z\,|\,r_\circ,\theta_\circ,z_\circ)=\sum_{m=0}^{N-1}\frac{1}{r}\delta(r-r_\circ)\delta(z-z_\circ)\delta(\theta-\theta_\circ-\frac{2\pi m}{N}) , \qquad (10)$$

$$[\nabla^2-M_a^2\{i\lambda+\sqrt{1+\omega^2 r^2}\frac{\partial}{\partial s}\}^2]\tilde{G}(r,\theta,z\,|\,r_\circ,\theta_\circ,z_\circ)$$
$$=\sum_{m=0}^{N-1} e^{im2\pi\sigma/N}\cdot\frac{1}{r}\delta(r-r_\circ)\delta(z-z_\circ)\delta(\theta-\theta_\circ-\frac{2\pi m}{N}) . \qquad (11)$$

The expressions for $\bar{G}$ and $\tilde{G}$ in terms of a finite series approximation have been given by Namba[2].

By integrating the linearized Euler's equation of motion, the steady and unsteady velocities $\bar{q}[=(\bar{q}_r, \bar{q}_n, \bar{q}_s)]$ and $\tilde{q}[=(\tilde{q}_r, \tilde{q}_n, \tilde{q}_s)]$ are obtained in the form

$$\bar{q}(r,\theta,z)=\int_h^1 dr_0 \int_{-C_0/2}^{C_0/2} \Delta\bar{p}(r_0, z_0)\bar{K}_{qF}(r,\eta,z-z_0 \mid r_0)$$

$$+\bar{Q}(r_0, z_0)\bar{K}_{qQ}(r,\eta,z-z_0 \mid r_0)\}dz_0 , \tag{12}$$

$$\tilde{q}(r,\theta,z)-\int_h^1 dr_0 \int_{-C_0/2}^{C_0/2} \{\Delta\tilde{p}(r_0, z_0)K_{qA}(r,\eta,z-z_0 \mid r_0)$$

$$+\Delta\bar{p}(r_0, z_0)a_n(r_0, z_0)K_{qL}(r,\eta,z-z_0 \mid r_0)+\bar{Q}(r_0, z_0)a_n(r_0, z_0)K_{qQ}(r,\eta,z-z_0 \mid r_0)$$

$$+\tilde{Q}(r_0, z_0)K_{qS}(r,\eta,z-z_0 \mid r_0)-\tilde{D}(r_0, z_0)K_{qD}(r,\eta,z-z_0 \mid r_0)\}dz_0 , \tag{13}$$

where $\bar{K}_{qF}, \bar{K}_{qQ}, K_{qA}, K_{qL}, K_{qQ}, K_{qS}$, and K_{qD} denote the velocity kernel functions.

The boundary condition to be satisfied is that the flow must be tangent to the blade surfaces. On the basis of the double linearization theory, the condition of flow tangency can be split into a steady part

$$\bar{q}_n(r,\pm 0,z)=\{-\alpha_0\alpha(r)+\mu_0\mu(r)f'(z)\pm\tau_0\tau(r)\frac{g'(z)}{2}\}\sqrt{1+\omega^2 r^2} , \tag{14}$$

and an unsteady part

$$\tilde{q}_n(r,\pm 0,z)=i\lambda a_n(r,z)+\frac{\partial}{\partial z}a_n(r,z)+\bar{q}_r(r,\pm 0,z)\frac{r}{\sqrt{1+\omega^2 r^2}}\frac{\partial}{\partial r}\{\frac{\sqrt{1+\omega^2 r^2}}{r}a_n(r,z)\}$$

$$-a_n(r,z)\frac{\partial}{\partial n}\bar{q}_n(r,\pm 0,z)+\bar{q}_s(r,\pm 0,z)\frac{1}{\sqrt{1+\omega^2 r^2}}\frac{\partial}{\partial z}a_n(r,z)\} . \tag{15}$$

Substituting Eqs.(12) and (13) into (14) and (15), we obtain integral equations for steady pressure difference distribution $\Delta\bar{p}$ and unsteady pressure difference distribution $\Delta\tilde{p}$ as follows

$$\int_h^1 dr_0 \int_{-C_0/2}^{C_0/2} \Delta\bar{p}(r_0, z_0)\bar{K}_{q_n F}dz_0 =\{-\alpha_0\alpha(r)+\mu_0\mu(r)f'(z)\pm\tau_0\tau(r)g'(z)/2\}\sqrt{1+\omega^2 r^2}$$

$$-\int_h^1 dr_0 \int_{-C_0/2}^{C_0/2} \bar{Q}(r_0, z_0)\bar{K}_{q_n Q}dz_0 , \tag{16}$$

$$\int_h^1 dr_0 \int_{-C_0/2}^{C_0/2} \Delta\tilde{p}(r_0, z_0)K_{q_n A}dz_0 = i\lambda a_n(r,z)+\frac{\partial}{\partial z}a_n(r,z)$$

$$+\int_h^1 dr_0 \int_{-C_0/2}^{C_0/2} \Delta\bar{p}(r_0, z_0)[-a_n(r_0, z_0)K_{q_n L}-a_n(r,z)\frac{\partial}{\partial n}\bar{K}_{q_n F}$$

$$+\frac{1}{\sqrt{1+\omega^2 r^2}}\frac{\partial}{\partial z}a_n(r,z)\bar{K}_{q_s F}+\frac{r}{\sqrt{1+\omega^2 r^2}}\frac{\partial}{\partial r}\{\frac{\sqrt{1+\omega^2 r^2}}{r}a_n(r,z)\}\bar{K}_{q_r F}]dz_0$$

$$+\int_h^1 dr_0 \int_{-C_0/2}^{C_0/2} \bar{Q}(r_0, z_0)[-a_n(r_0, z_0)K_{q_n Q}-a_n(r,z)\frac{\partial}{\partial n}\bar{K}_{q_n Q}$$

$$+\frac{1}{\sqrt{1+\omega^2 r^2}}\frac{\partial}{\partial z}a_n(r,z)\bar{K}_{q_s Q}+\frac{r}{\sqrt{1+\omega^2 r^2}}\frac{\partial}{\partial r}\{\frac{\sqrt{1+\omega^2 r^2}}{r}a_n(r,z)\}\bar{K}_{q_r Q}]dz_0$$

$$+\int_h^1 dr_0 \int_{-C_a/2}^{C_a/2} \{-\tilde{Q}(r_0, z_0) K_{q_n S} + \tilde{D}(r_0, z_0) K_{q_n D}\} dz_0 \ . \tag{17}$$

It should be noted that there exist jumps across the helical streamsurface $\eta=\theta-\omega z=0$ in kernel functions $\bar{K}_{q_n Q}$, $K_{q_n L}$, $K_{q_n D}$, $K_{q_n S}$ etc. as $\eta \to 0$. Making the discontinuities at $\eta=\pm 0$ on both sides of Eqs.(16) and (17) be exactly balanced, we can determine $\bar{Q}$, $\tilde{Q}$ and $\tilde{D}$ as follows:

$$\bar{Q}(r,z)=\tau_0 \tau(r) g'(z)\sqrt{1+\omega^2 r^2} \ , \tag{18}$$

$$\tilde{Q}(r,z)=M_a^2 \Delta\tilde{p}(r,z)(i\lambda+\frac{\partial}{\partial z}) a_n(r,z) \ , \tag{19}$$

$$\tilde{D}(r,z)=-\Delta\tilde{p}(r,z)\frac{\partial}{\partial r} a_n(r,z)+\{(i\lambda+\frac{\partial}{\partial z}) a_n(r,z)\}\int_{-C_a/2}^{z} \frac{\partial}{\partial r}\Delta\tilde{p}(r,z) dz \ . \tag{20}$$

Consequently Eqs.(16) and (17) become integral equations for $\Delta\bar{p}$ and $\Delta\tilde{p}$ respectively. As for the physical meanings of $\bar{Q}$, $\tilde{Q}$ and $\tilde{D}$ the reader should refer to Ref.[3].

AERODYNAMIC COEFFICIENTS

We limit our discussions in the present paper to pure bending vibration. The normal displacement of a blade is given by

$$a_n(r,z)=H\cdot C_a\, a_j^{(r\theta)}(r)\sqrt{1+\omega^2 r^2} \ , \tag{21}$$

where H denotes the displacement amplitude at the blade tip, and $a_j^{(r,\theta)}(r)$ denotes the j-th mode circumferential displacement of the reference blade, where the modal shape of a uniform cantilever is assumed. The local aerodynamic force coefficient for bending vibration is defined by

$$C_L^{(b)}(r)=\frac{1}{i\pi\lambda C_a^2\sqrt{1+\omega^2 r^2}}\int_{-C_a/2}^{C_a/2} \Delta\tilde{p}^{(b)}(r,z)dz \ , \tag{22}$$

where $\Delta\tilde{p}^{(b)}$ denotes the pressure difference for vibration of unit amplitude(H=1). The local aerodynamic work per cycle is given by

$$W^*(r)=\oint dt\int_{-C_a^*/2}^{C_a^*/2} \mathrm{Re}[\Delta\tilde{p}^*(r,z)e^{i\lambda t}]\cdot \mathrm{Re}[r_t^* \frac{d}{dt}\{a_n(r)e^{i\lambda t}\}]\sqrt{1+\omega^2 r^2}dz^* . \tag{23}$$

Scaling $W^*(r)$ with respect to $\pi\rho_0^* W_a^{*2} C_a^{*2} r_t^* H^2$, we obtain dimensionless local aerodynamic work

$$W^{(b)}(r)=\pi\lambda C_a\, a_j^{(r\theta)}(r)\cdot \mathrm{Re}[C_L^{(b)}(r)]\sqrt{1+\omega^2 r^2} \ . \tag{24}$$

Then the total dimensionless aerodynamic work per cycle is given by

$$\overline{W}^{(b)}=\int_h^1 W^{(b)}(r)dr \ . \tag{25}$$

The flutter condition can be investigated on the basis of aerodynamic work. When $\overline{W}^{(b)}$ is positive, the vibrating blades get energy from the flow, and flutter will occur in the case of zero mechanical damping.

NUMERICAL RESULTS AND DISCUSSIONS

In the following we conduct parametric study to demonstrate the mean loading effects. Fig.2 shows the variation of the flow conditions along the span for the prescribed ω, NC_0 and h.

To solve Eqs.(16) and (17) we expand both $\Delta\bar{p}$ and $\Delta\tilde{p}$ into a double series in terms of 1the radial eigenfunctions and the Glauert series (see Ref.[1]), so that Kutta condition is automatically satisfied. Hence Eqs.(16) and (17) are reduced to algebraic equations for coefficients of the double series. Making the algebraic equations satisfied at a finite number of control points, we can determine $\Delta\bar{p}$ and $\Delta\tilde{p}$.

Parabolic camber and thickness profiles are assumed, i.e., $f(z)=g(z)=1-4(z/C_0)^2$. Besides the spanwise distributions of the mean angle of attack, camber and thickness are assumed to be given by

$$\alpha(r)=[1+A_1\cdot\cos\{(r-h)\pi/(1-h)\}] \ , \quad (26)$$

$$\mu(r)=[1+M_1\cdot\cos\{(r-h)\pi/(1-h)\}]C_0 \ , \quad (27)$$

$$\tau(r)=[1+T_1\cdot\cos\{(r-h)\pi/(1-h)\}]C_0 \ . \quad (28)$$

The linearity of the present theory allows us to express $\Delta\bar{p}$ in terms of components of the mean angle of attack, camber and thickness as follows

$$\Delta\bar{p}(r,z)=\alpha_0\Delta\bar{p}_\alpha(r,z)+\mu_0\Delta\bar{p}_\mu(r,z)+\tau_0\Delta\bar{p}_\tau(r,z) \ . \quad (29)$$

As for $\Delta\tilde{p}$, we can express it by

$$\Delta\tilde{p}(r,z)=\Delta\tilde{p}_0(r,z)+\alpha_0\Delta\tilde{p}_\alpha(r,z)+\mu_0\Delta\tilde{p}_\mu(r,z)+\tau_0\Delta\tilde{p}_\tau(r,z) \ , \quad (30)$$

where $\Delta\tilde{p}_0$ denotes the contribution of blade motion, i.e. the unsteady loading due to vibration without steady loading, while $\Delta\tilde{p}_\alpha$, $\Delta\tilde{p}_\mu$ and $\Delta\tilde{p}_\tau$ denote the contributions of the mean angle of attack, camber and thickness respectively. The unsteady lift and work coefficients can also be expressed in the form as above.

In Figs.3-12, the real part and imaginary part are denoted by Re[ ] and Im[ ] respectively. The results of the strip theory which assumes a local two-dimensional flow on each cylindrical streamsurface are shown by dashed lines, and the results of the present full three-dimensional theory are shown by solid lines.

Figs.3 and 4 show the pressure difference distribution due to the first order bending vibration. Except at the hub where the strip theory gives zero pressure difference, both the three-dimensional theory and the strip theory give similar chordwise shape at each span. In general the spanwise variation predicted by the strip theory is steeper than that predicted by the present theory. Therefore we can state that the three-dimensional effect reduces the spanwise gradient of aerodynamic force.

In Fig.5, the spanwise distributions of the unsteady lift coefficient for N=30, N=45 and N=60 are illustrated. In this case, the increase of the number of blades N is equivalent to the increase of the blade aspect ratio. It can be seen that the three-dimensional effect, i.e., the difference between the present theory and the strip theory becomes smaller as the number of blades increases.

The effect of Mach number on unsteady aerodynamic force is indicated in Fig.6 where the axial Mach numbers are prescribed at 0.2 and 0.3. In the two cases, the relative Mach numbers at the blade tip are nearly 0.5 and 0.8 respectively. We see that the aerodynamic force becomes greater as the Mach number increases. As a result of quasi-two-dimensional treatment of local Mach number effect, the strip theory gives too steep variation of the unsteady aerodynamic force along the span. In other words the three-dimensional effect is especially important at high axial Mach numbers.

Fig.7 shows the effect of the spanwise distribution of camber. It should be mentioned that according to Eqs.(26), (27) and (28), the steady loading will increase at the tip and decrease at the hub as the parameters A_1, M_1 and T_1 change from 0.5 to -0.5. The result indicates that three-dimensional effect is relatively large when the steady loading increases from hub to tip. Fig.8 shows the unsteady lift coefficients for various reduced frequencies. The three dimensional effect decreases with increasing frequencies.

As shown in Fig.9, the aerodynamic force and three-dimensional effect are heavily dependent on the interblade phase angle. For example, increase of the interblade phase angle from zero to $\pi(\sigma/N=0.5)$ gives a significant change of the lift coefficient.

Figs.10 and 11 show the variation of the unsteady aerodynamic work components of the angle of attack and camber with change of the interblade phase angle respectively. It should be noted that the three-dimensional theory and the strip theory can give opposite predictions about instability of the blade vibration in some cases, e.g. $A_1=-0.5$, $\sigma/N=0.2$, in Fig.10. There are substantial changes in the three-dimensional effect as the steady loading distribution is changed. For example, as shown in Fig.11, the strip theory overestimates $|\overline{W}_u^{(t)}|$ when the camber at the tip is small ($M_1=0.5$), and underestimates $|\overline{W}_u^{(b)}|$ when the camber at the tip is large ($M_1=0.5$). Furthermore, by comparing Fig.12 with Figs. 10 and 11, we see that the contribution of thickness is relatively smaller than the contributions of the angle of attack and camber.

CONCLUSIONS

A three-dimensional unsteady cascade theory has been developed on the basis of the double linearization theory to investigate the mean loading effects on flutter of annular cascades. Numerical results of the present theory are compared with the results of the strip theory approximation. The main conclusions that can be drawn from this study are summarized as follows:

— The three-dimensional effect is essentially important in the case of high Mach number.
— The three-dimensional effect increases as the number of blades and the vibration frequency are decreased.
— The spanwise distribution of steady loading has an important effect on the unsteady aerodynamic force. In general, the three-dimensional effect is large when the steady loadings increase from hub to tip.
— The strip theory approximation fails to predict the instability of vibrating annular cascade blades correctly in some instances.

REFERENCES

1. Namba, M., AIAA Journal, Vol.13(5),(1975),586.
2. Namba, M., R and M, No.3740, British Aeronautical Research Council, (1972).
3. Namba, M. and Toshimitsu, K., Proceedings of the 4th Symposium on Unsteady Aerodynamics and Aeroelasticity of Turbomachines and Propellers, (1987), 453.

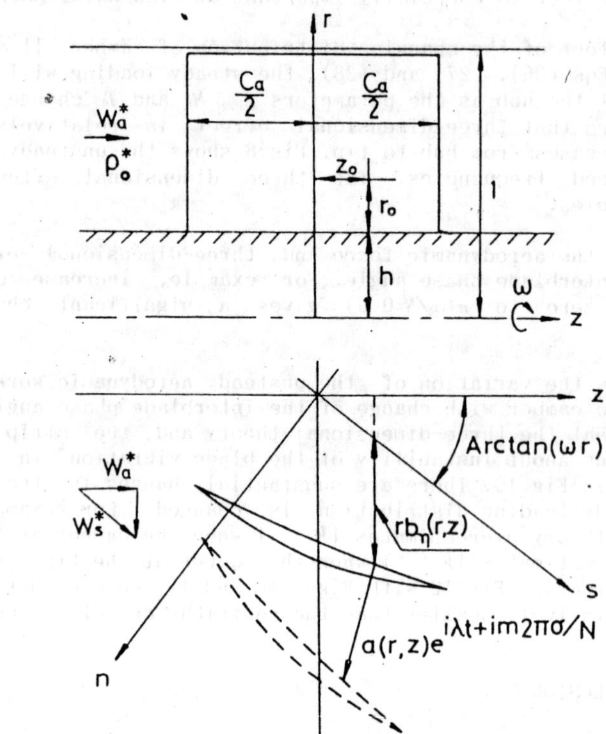

Fig.1 The annular cascade model

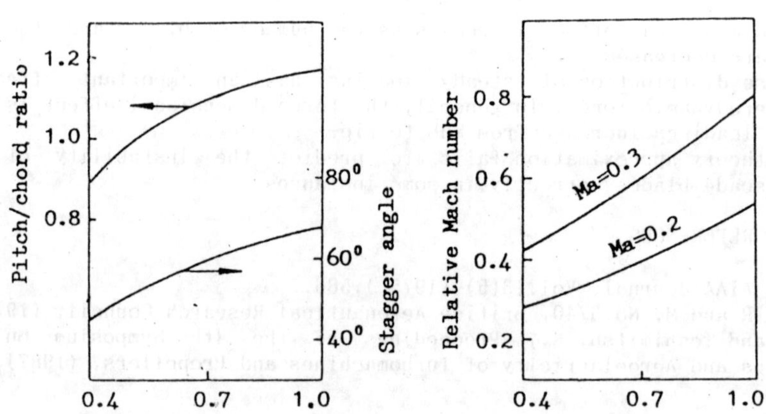

(ω=2.4744, NC_a=2.0, h=0.4)

Fig.2 Distribution of local pitch-chord ratio, stagger angle and relative Mach number along the span for prescribed cascade condition.

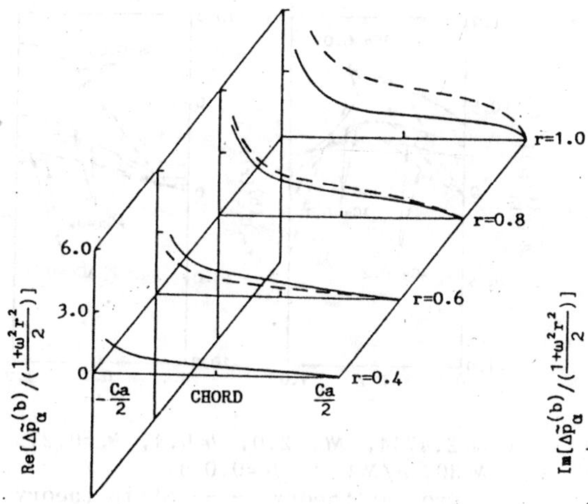

(ω=2.4744, NC_a=2.0, h=0.4, M_a=0.2, N=30, λC_a=0.2, σ/N=0.2, A_1=0.0)
—— Present theory, -----Strip theory

Fig.3 Distribution of the real part of unsteady pressure difference (contribution of the angle of attack)

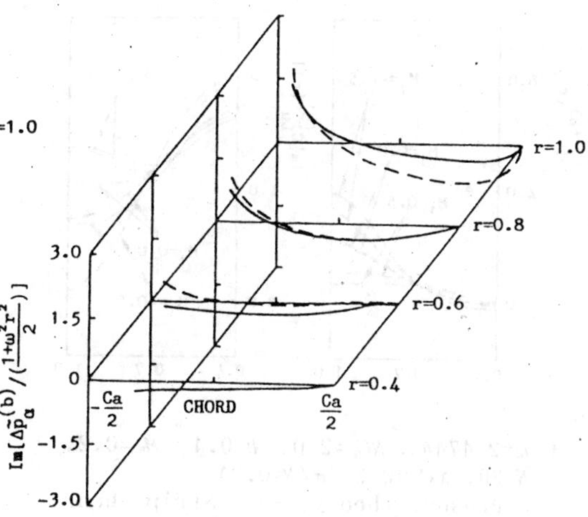

(ω=2.4744, NC_a=2.0, h=0.4, M_a=0.2, N=30, λC_a=0.2, σ/N=0.2, A_1=0.0)
—— Present theory, -----Strip theory

Fig.4 Distribution of the imaginary part of unsteady pressure difference (contribution of the angle of attack)

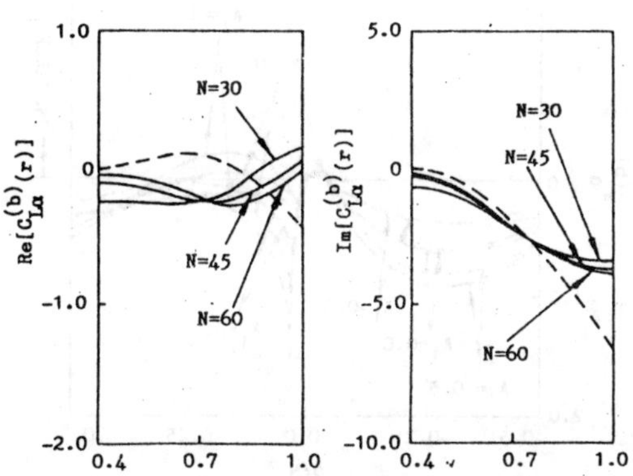

(ω=2.4744, NC_a=2.0, h=0.4, M_a=0.2, λC_a=0.2, σ/N=0.2, A_1=0.0)
—— Present theory, -----Strip theory

Fig.5 Effect of the number of blades on the spanwise distribution of the unsteady aerodynamic force. (contribution of the angle of attack)

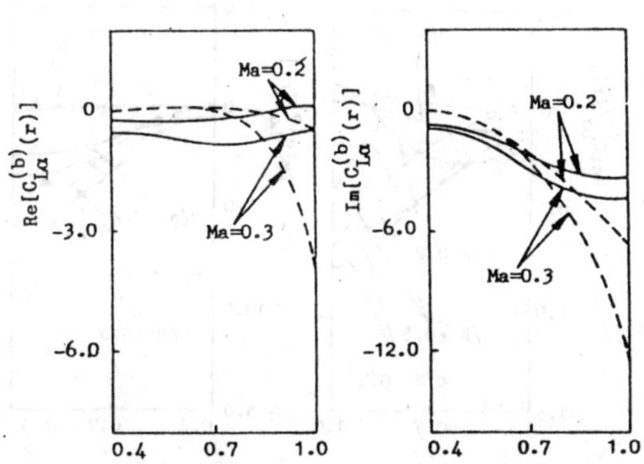

(ω=2.4744, NC_a=2.0, h=0.4, N=30, λC_a=0.2, σ/N=0.2, A_1=0.0)
—— Present theory, -----Strip theory

Fig.6 Effect of Mach number on the spanwise distribution of the unsteady aerodynamic force. (contribution of the angle of attack)

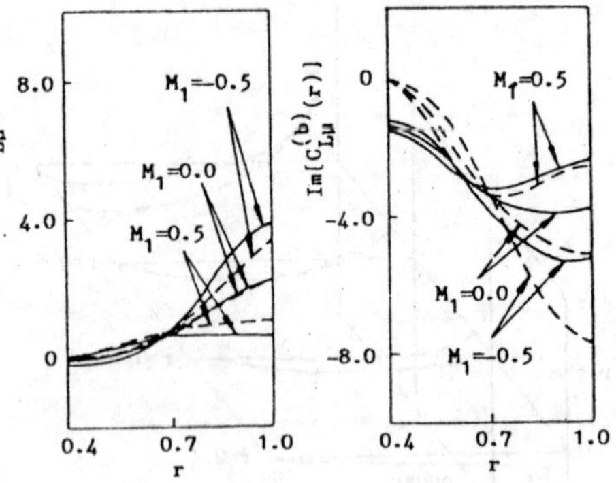

(ω=2.4744, NC_a=2.0, h=0.4, M_a=0.2, N=30, λC_a=0.2, σ/N=0.2)
—— Present theory, -----Strip theory

Fig.7 Influence of the distribution of mean loading on the unsteady aerodynamic force. (contribution of the camber)

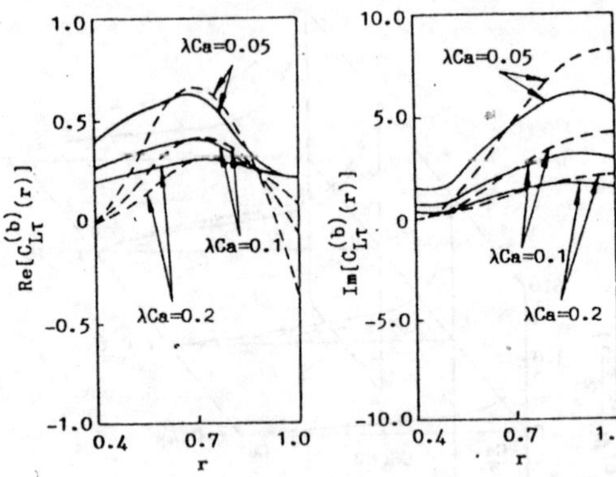

(ω=2.4744, NC_a=2.0, h=0.4, M_a=0.2, N=30, σ/N=0.2, T_1=0.0)
—— Present theory, -----Strip theory

Fig.8 Influence of the reduced frequency on the spanwise distribution of the unsteady aerodynamic force. (contribution of the thickness)

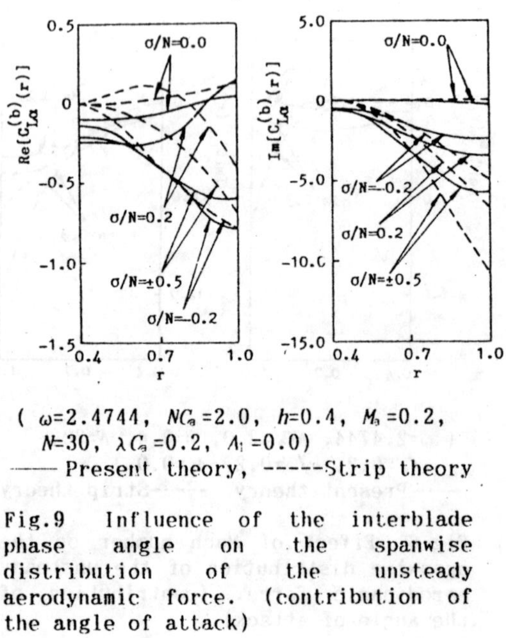

(ω=2.4744, NC_a=2.0, h=0.4, M_a=0.2, N=30, λC_a=0.2, A_1=0.0)
—— Present theory, -----Strip theory

Fig.9 Influence of the interblade phase angle on the spanwise distribution of the unsteady aerodynamic force. (contribution of the angle of attack)

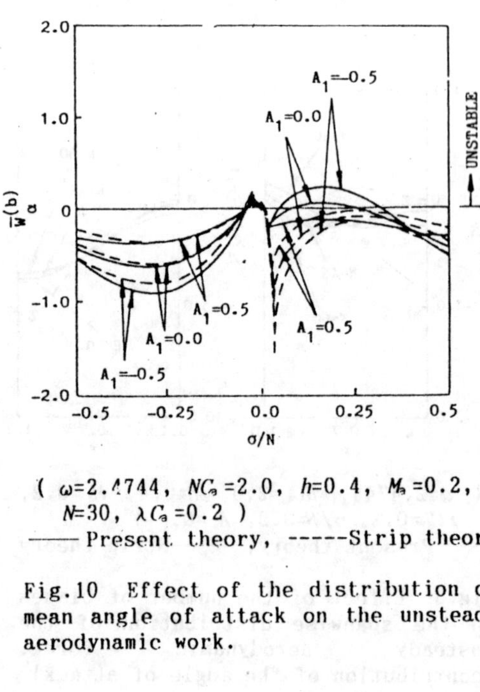

(ω=2.4744, NC_a=2.0, h=0.4, M_a=0.2, N=30, λC_a=0.2)
—— Present theory, -----Strip theory

Fig.10 Effect of the distribution of mean angle of attack on the unsteady aerodynamic work.

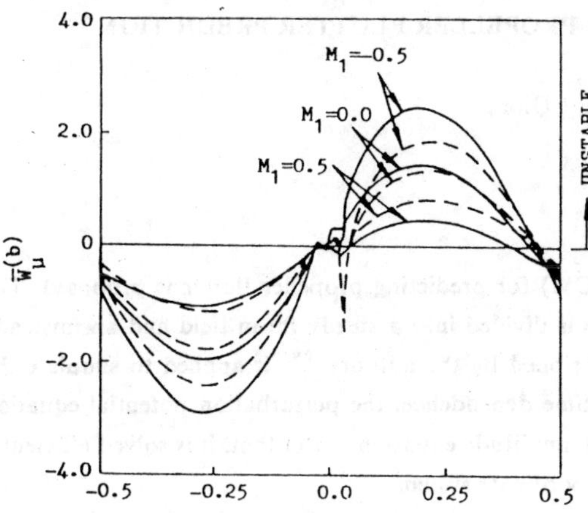

(ω=2.4744, NC_a=2.0, h=0.4, M_a=0.2, N=30, λC_a=0.2,)
—— Present theory, ----- Strip theory

Fig.11 Effect of the camber distribution on the unsteady aerodynamic work.

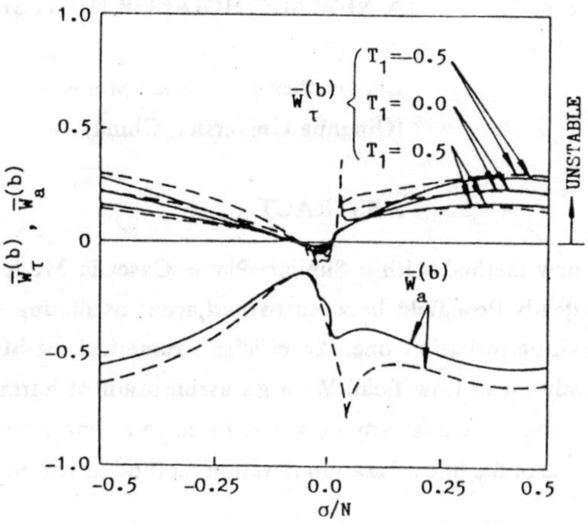

(ω=2.4744, NC_a=2.0, h=0.4, M_a=0.2, N=30, λC_a=0.2,)
—— Present theory, ----- Strip theory

Fig.12 Effect of the thickness distribution and vibration velocity on the unsteady aerodynamic work.

A NEW METHOD FOR HIGH SPEED PROPELLER FLUTTER PREDICTION

Liu Qiusheng Shen Mengyu Zhang Qiong
(Qinghua University, China)

ABSTRACT

A new method with a Similar–Plane–Cascade Model (SPCM) for predicting propeller flutter is proposed. The unsteady flow field between two adjacent oscillating blades is divided into a steady mean field and an unsteady small perturbation one. An efficient numerical method developed by the authors [9] is applied to simulate the steady mean flow field. With an assumption of harmonic time dependence, the perturbation potential equation governing the unsteady flow is reduced to a simple potential amplitude equation. After that, it is solved efficiently by use of the line relaxation iteration method in von Misses coordinate system.

The propeller flutter is studied by means of the "energy method". Predictions have been performed on two propellers. The results agree well with available flight tests and show that the modified Sisto model is applicable to transonic cascade flows. The present method is advantageous in its mathematical simplicity and fast convergence rate. It brings about significant savings in CPU time while providing satisfactory accuracy.

KEY WORDS: Propeller flutter, Flutter prediction, Aeroelastic stability of propellers, Transonic flow.

INTRODUCTION

With high speed and high aerodynamic performance recent propellers are threatened by flutter because high aerodynamic load and light material construction reduce aeroelastic stability of propellers. The flutter trouble should be eliminated in the design stage of propellers to ensure the safety of an aircraft. Of the two kinds of flutter, classical and stall flutter, the stall flutter occurs more frequently for high speed and thin blades. Thus the prediction of stall flutter is especially important. So far, however, it depends mainly on experiments and experience. [1,2,3,4] Theoretical methods have not been fully developed. Ref.4 used a modified theory of an airfoil flutter [5] for the classical flutter of propellers and obtained some results which agree with experiments approximately. But it is unfortunate that neither the two–dimensional airfoil theory [6] nor its modified theory can predict a stall flutter of propellers. In the last decade theoretical analysis and prediction of flutter (including the stall flutter) of turbomachine blades have been developed rapidly. [7] In China many excellent works have been done by Prof. Zhou Sheng and Lin Baozheng. They are of significance for predicting the stall flutter of propellers. Nevertheless, a special study of propeller flutter is still necessary because it is different from the turbomachines. In this paper our attention will be focused on the stall flutter.

As is known to all, stall flutter is in company with large scale separation of flow and interaction between shocks and boundary layer, and the flow is always unsteady. Hence at present it is not practical in engineering application to study the stall flutter theoretically. For this reason, some practical methods developed for predicting stall flutter are based on certain simplifying assumptions.

Although the flow between two blades of a propeller has a typical three–dimensional feature, many experiments have proved that the plane flow on a certain cylindrical section near the tips of propeller blades, which is called "characteristic section", can characterize well the flutter including the stall flutter. It is on this fact that the plane isolated airfoil model (PIAM) is based. It is obvious that the PIAM is reasonable when there are few blades, (e.g.2). However, the majority of propellers of interest to recent turboprob have more blades and the interaction between blades become too strong to neglect. In view of this fact this paper proposes a similar–plane cascade model (SPCM). In this model we take a series of blade sections on the characteristic section as a plane cascade and use it to analyse the aeroelastic stability of propellers.

In the present paper the "energy method" [7] is used for determining the aeroelastic stability.

For simplicity, the fluid is assumed inviscid. In order to consider the effects of separation flow on the stall flutter, the Helmholtz flow model suggested by Sisto and Perumal for an isolated plate airfoil [8] will be extended to the propeller case in this paper.

BASIC EQUATIONS

We study the 2–dimensional flow of a inviscid, perfect gas in the similar–plane–cascade, relative to the reference frame fixed with the propeller rotating at a constant angle speed ω, assuming that the absolute motion of the gas is homo–energetic and irrotational. The governing equations can be written as

$$\left.\begin{aligned} &\frac{\partial \rho}{\partial t} + \nabla \cdot (\rho \vec{q}) = 0 \\ &\frac{\partial \vec{q}}{\partial t} + \nabla\left(\frac{q^2}{2} + \frac{a^2}{\gamma - 1}\right) = 0 \\ &\nabla \times \vec{q} = -2\vec{\omega} \\ &p = \rho^{\gamma}, \quad a^2 = \rho^{\gamma - 1} \end{aligned}\right\} \tag{1}$$

where $\vec{q}$ is the velocity,γ–the ratio of specific heat, ρ,p and a are the density, the pressure and the sound speed respectively. The above quantities, the length and the time t are all non–dimensional quantities, divided by the corresponding reference quantities: a^*,b/a^*,ρ^*,p^*,b and b/a^* (the superscript "*" denotes the critical state,b–the chord length in the x direction).

As the energy method is used to predict flutter, only a time dependent small perturbation of flow should be taken into account. In this case, each of flow variables can be divided into two parts, a steady mean part and an unsteady small perturbation one. That is,

$$\vec{q} = \vec{q}_o + \vec{q}', \qquad p = p_o + p', \qquad \rho = \rho_o + \rho' \tag{2}$$

where the subscript "o" denotes the steady mean part, the superscript "′" indicates the perturbation part dependent on time. At the same time the following relations are assumed to hold:

$$|\vec{q}'| \ll |\vec{q}_o|, \qquad |p'| \ll p_o, \qquad |\rho'| \ll \rho_o$$

Thus

$$q^2 \approx q_o^2 + 2\vec{q}_o \cdot \vec{q}' \tag{3}$$

Substituting the relation (2) into the equation (lc), we can obtain

$$\nabla \times \vec{q}' = 0$$

So a perturbation velocity potential can be introduced as follows

$$\vec{q}' = \nabla \varphi \tag{4}$$

Substituting relations (2) and (4) into the equation (1) and omitting the infinite small quantities of second and higher order, we can divide it into two systems of equations. One is independent of time, the other dependent on time. They are

$$
\left.\begin{aligned}
&\nabla\cdot(\rho_o\vec{q}_o)=0\\
&\nabla\times\vec{q}_o=-2\vec{\omega}\\
&\frac{q_o^2}{2}+\frac{a_o^2}{\gamma-1}=\frac{\gamma+1}{2(\gamma-1)}\\
&p_o=\rho_o^\gamma
\end{aligned}\right\} \qquad (5)
$$

and

$$
\frac{\partial\varphi}{\partial t}+q^2/2\quad+a^2/(\gamma-1)\quad=f(t)=\left(\frac{q_o^2}{2}+\frac{a_o^2}{\gamma-1}\right)_{inlet} \qquad (6)
$$

Notice that the condition that the far upstream inlet boundary values are independent of time has been used in (5c) and (6). The flow governed by equation(5) is called the steady mean flow, and that governed by equation (6) the perturbation flow. Besides, an expression of sound speed a can be derived from equations (3),(4) and (6), and it is

$$
a^2=a_o^2-(\gamma-1)(\varphi_t+\vec{q}_o\cdot\nabla\varphi) \qquad (7)
$$

By taking partial derivative of equation (6) with respect to t and using the relation $a^2=\rho^{\gamma-1}$, we obtain

$$
\frac{\partial^2\varphi}{\partial t^2}+\frac{\partial}{\partial t}\left(\frac{q^2}{2}\right)+\rho^{\gamma-2}\frac{\partial\rho}{\partial t}=0 \qquad (8)
$$

Taking an inner product of equation (lb) with $\vec{q}$ and using the equations (la) and (8), we can make partial derivatives of ρ in (lb) disappear. Then substituting formulas (3), (4), (7) into it, the perturbation potential equation can be finally obtained

$$
\begin{aligned}
&a_o^2\nabla^2\varphi-(\gamma-1)(\nabla\cdot\vec{q}_o)(\vec{q}_o\cdot\nabla\varphi)-\nabla\frac{q_o^2}{2}\cdot\nabla\varphi-\vec{q}_o\cdot\nabla(\vec{q}_o\cdot\nabla\varphi)\\
&=\varphi_{tt}+2\vec{q}_o\cdot\nabla\varphi_t+(\gamma-1)(\nabla\cdot\vec{q}_o)\varphi_t
\end{aligned} \qquad (9)
$$

THE SOLUTION OF THE STEADY MEAN FLOW FIELD

To our knowledge, among all the numerical methods for calculating transonic flow in plane cascades, the method proposed in Ref.9 converges the fastest and is of highest accuracy. So it is used to solve the steady mean flow.

1. Streamline Governing Equation And Its Boundary Conditions

On the characteristic section we choose circular cylindrical surface coordinate system (x,y) as a reference frame, where the x–axis is parelled to the rotating axis, y = rε, ε and r are the peripheral angle and the radius of the characteristic surface respectively (See Fig.1). According to the equation (5a) a stream function ψ is defined as follows

$$
\left.\begin{aligned}
\rho_o q_o\cos\theta&=\frac{\partial\psi}{\partial y}\\
\rho_o q_o\sin\theta&=-\frac{\partial\psi}{\partial x}
\end{aligned}\right\} \qquad (10)
$$

where θ is the flow angle.

Introducing von Misses transformation

$$
\left.\begin{aligned}
\xi&=x\\
\psi&=\psi(x,y)
\end{aligned}\right\} \qquad (11)
$$

the equation (5b) and (10) can be transformed into

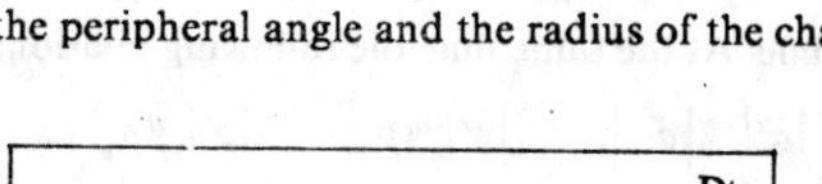

Fig.1 Solving domain in characteristic plane

$$\left.\begin{aligned}\frac{\partial y}{\partial \psi} &= \frac{1}{\rho_o q_o \cos\theta} \\ \frac{\partial y}{\partial \xi} &= tg\theta\end{aligned}\right\} \tag{12}$$

$$\rho_o q_o \frac{\partial q_o}{\partial \psi} = \partial(q_o \sin\theta) / \partial\xi \tag{13}$$

After eliminating all partial derivatives of q_o by use of equations (5c),(10),(12) and (13), we get the streamline governing equation

$$G y_{\xi\xi} + E y_{\xi\psi} + D y_{\psi\psi} = 0 \tag{14a}$$

where

$$\left.\begin{aligned}G &= 1 - \frac{2}{\gamma+1} q_o^2 / (1 + y_\xi^2)(1 - \frac{\gamma-1}{\gamma+1} q_o^2) \\ E &= -2y_\xi / y_\psi \\ D &= (1 + y_\xi^2) / y_\psi^2\end{aligned}\right\} \tag{14b}$$

The discriminant of its type is

$$\triangle = E^2 - 4AD = 4(q_o^2 - 1) / y_\psi^2 (1 - \frac{\gamma-1}{\gamma+1} q_o^2)$$

It can be seen that it is elliptic when $q_o < 1$, parabolic when $q_o = 1$, and hyperbolic when $q_o > 1$.
Besides, from equations (5c), (5d), (12) and (14b) we obtain

$$\rho_o = (\frac{\gamma+1}{2})^{\frac{1}{\gamma-1}} [1 - \frac{\gamma-1}{\gamma+1} q_o^2]^{\frac{1}{\gamma-1}} \tag{15}$$

$$\rho_o q_o = \sqrt{D} \tag{16}$$

The equation (13) also can be rewritten as follows

$$\frac{\partial q_o}{\partial \psi} = \frac{1}{\sqrt{D}} \frac{\partial}{\partial \xi} (q_o \frac{y_\xi}{\sqrt{1 + y_\xi^2}}) \tag{17}$$

As a result, equations (14) and (17), or (15) and (16) instead of (17), form the system of basic equations for calculating the steady mean flow field.

Its boundary conditions, as shown in Fig.1, are as follows

A. At the far upstream inlet boundary the following conditions are specified:

$$q_o = (q_o)_1 = constant$$

$$\theta = \theta_1 = constant$$

B. Conditions on blade surfaces are

$$\left.\begin{aligned}y(\xi,0) &= y_s(\xi) \\ y(\xi,Q) &= y_p(\xi) + T\end{aligned}\right\}$$

where Q is the mass flow rate between two adjacent blades, $y_p(\xi)$ and $y_s(\xi)$ are ordinates of the pressure and suction surface of the blade respectively, T is the pitch.

C. The periodicity condition along cyclic boundaries is

$$y(\xi,Q) = y(\xi,0) + T$$

Besides, in the case of $(q_o)_1 < 1$, Kutta condition must be added to determine the far downstream outflow angle because in SPCM it is unknown. This is unlike the situation in turbine cascades.

2. Difference Equations and Their Solution Method

The Computing domain in the plane (ξ,ψ), as shown in Fig.2, is divided into many rectanglar meshes of unequal size. Discretizing the equation (14), we use the type dependent difference scheme. That is, in the supersonic region an upwind difference scheme of first order accuracy is used in ξ–direction so as to capture a shock automatically, and in other cases a central difference scheme is used.

As a result, the difference equations have following two forms along a relaxation line $\xi = \xi_i$.

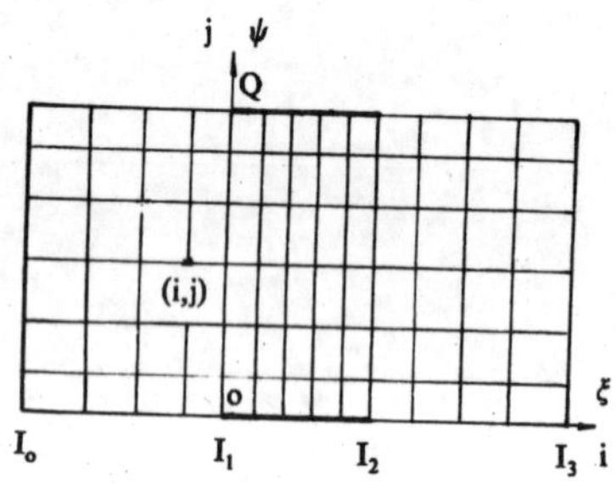

Fig.2 Calculating domain is (ξ,ψ) plane

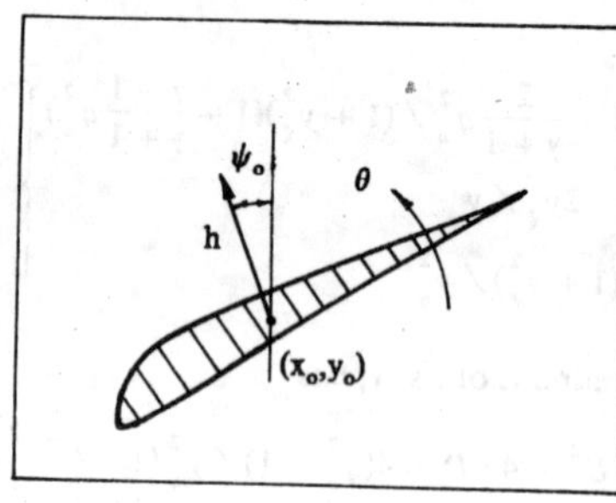

Fig.3 The model of vibration

A. In subsonic region

$$A_{1j}^{(n-1)} y_{i,j-1}^{(n)} + B_{1j}^{(n-1)} y_{i,j}^{(n)} + C_{1j}^{(n-1)} y_{i,j+1}^{(n)} = F_{1j}^{(n-1)} \tag{18a}$$

where

$$A_{1j}^{(n-1)} = -\frac{2c_1 \alpha D_{ij}^{(n-1)}}{c_1 + c_2} + \frac{(H_2 - H_1)c_2}{H_2(c_1 + c_2)} E_{ij}^{(n-1)}$$

$$B_{1j}^{(n-1)} = 2(\alpha \frac{c_1}{c_2} D_{ij}^{(n-1)} + \frac{H_1}{\alpha H_2} G_{ij}^{(n-1)}) - \frac{(H_2 - H_1)(c_2 - c_1)}{H_2 c_2} E_{ij}^{(n-1)}$$

$$C_{1j}^{(n-1)} = -2(\alpha D_{ij}^{(n-1)} + \frac{H_2 - H_1}{2H_2} E_{ij}^{(n-1)}) \frac{c_1^2}{c_2(c_1 + c_2)}$$

$$\begin{aligned} F_{1j}^{(n-1)} = E_{ij}^{(n-1)} [& \frac{H_1^2}{H_2(H_1 + H_2)} (\frac{c_1^2}{c_2(c_1 + c_2)} y_{i+1,j+1}^{(n-1)} + \frac{c_2 - c_1}{c_2} y_{i+1,j}^{(n-1)} \\ & - \frac{c_2}{c_1 + c_2} y_{i+1,j-1}^{(n-1)}) - \frac{H_2}{H_1 + H_2} (\frac{c_1^2}{c_2(c_1 + c_2)} y_{i-1,j+1}^{(n-1)} + \frac{c_2 - c_1}{c_2} y_{i-1,j}^{(n-1)} \\ & - \frac{c_2}{c_1 + c_2} y_{i-1,j-1}^{(n-1)})] + \frac{2}{\alpha} G_{ij}^{(n-1)} [\frac{H_1^2}{H_2(H_1 + H_2)} y_{i+1,j}^{(n-1)} + \frac{H_1}{H_1 + H_2} y_{i-1,j}^{(n-1)}] \end{aligned} \tag{18b}$$

$$H_1 = \xi_i - \xi_{i-1}, H_2 = \xi_{i+1} - \xi_i, c_1 = \psi_j - \psi_{j-1}, c_2 = \psi_{j+1} - \psi_j, \alpha = H_1 / c_1$$

B. In supersonic region

$$A_{2j}^{(n-1)} y_{i,j-1}^{(n)} + B_{2j}^{(n-1)} y_{i,j}^{(n)} + C_{2j}^{(n-1)} y_{i,j+1}^{(n)} = F_{2j}^{(n-1)} \tag{19a}$$

where

$$
\left.\begin{aligned}
A_{2j}^{(n-1)} &= \frac{1}{c_1+c_2}(-2\alpha c_1 D_{ij}^{(n-1)} + c_2 E_{ij}^{(n-1)}) \\
B_{2j}^{(n-1)} &= 2(\alpha\frac{c_1}{c_2} D_{ij}^{(n-1)} - \frac{H_2}{\alpha(H_1+H_2)} G_{ij}^{(n-1)}) - \frac{c_2-c_1}{c_2} E_{ij}^{(n-1)} \\
C_{2j}^{(n-1)} &= -\frac{c_1^2}{c_2(c_1+c_2)}(2\alpha D_{ij}^{(n-1)} + E_{ij}^{(n-1)}) \\
F_{2j}^{(n-1)} &= \frac{1}{c_2(c_1+c_2)} E_{ij}^{(n-1)}[-c_1^2 y_{i-1,j+1}^{(n-1)} - (c_2^2-c_1^2) y_{i-1,j}^{(n-1)} + c_2^2 y_{i-1,j-1}^{(n-1)}] \\
&\quad - \frac{2c_1}{H_1+H_2} G_{ij}^{(n-1)}[y_{i-1,j}^{(n-1)} + \frac{H_2}{H_1}(y_{i-1,j}^{(n-1)} - y_{i-2,j}^{(n-1)})] \\
H_1 &= \xi_{i-1}-\xi_{i-2}, H_2 = \xi_i - \xi_{i-1}, c_1 = \psi_j - \psi_{j-1}, c_2 = \psi_{j+1} - \psi_j, \alpha = H_2/c_1
\end{aligned}\right\} \qquad (19b)
$$

In the above two systems of equations, the superscript (n) denotes the n−th iteration, and G_{ij}, E_{ij} and D_{ij} are the values of G,E and D at the point (i,j) respectively.

After the difference representation of the boundary conditions (they are quite simple, thus have been omitted) are added, a system of equations with a cyclic tridiagonal coefficient matrix can be obtained in both the periodic region and the blade region, along a relaxation line. It has the following form:

$$
\begin{bmatrix} b_1 & c_1 & & & a_1 \\ a_2 & b_2 & c_2 & & \\ & a_3 & b_3 & c_3 & \\ & & & & \\ c_n & & & a_n & b_n \end{bmatrix} \cdot \begin{bmatrix} y_1 \\ y_2 \\ \cdots \\ \cdots \\ y_n \end{bmatrix} = \begin{bmatrix} f_1 \\ f_2 \\ \cdots \\ \cdots \\ f_n \end{bmatrix} \qquad (20)
$$

and can be solved efficiently by use of modified Thomas algorithm. The steps of solution of difference equations are as follows:

A. Assign initial values to y_{ij}, ρ_{ij}, $(q_o)_{ij}$.

B. Calculate $G_{ij}^{(n-1)}$, $E_{ij}^{(n-1)}$ and $D_{ij}^{(n-1)}$, using the formula (14b) and the known $y_{ij}^{(n-1)}$. In them $(q_o)_{ij}^{(n-1)}$ can be found by using formulas (15), (16) or (17). About its details and the treatment of double value problem, see also Ref. [9].

C. Iterate by use of line relaxation method, marching downwind. That is, to solve the equation (20) on each line $\xi=\xi_i$ so as to obtain a new $y_{ij}^{(n)}$.

D. Return to step B.

We repeat the process until the following criterions are satisfied.

$$
\left.\begin{aligned}
max\left|(q_o)_{ij}^{(n)} - (q_o)_{ij}^{(n-1)}\right| &< \varepsilon_1 \\
max\left|y_{ij}^{(n)} - y_{ij}^{(n-1)}\right| &< \varepsilon_2
\end{aligned}\right\} \qquad (21)
$$

where ε_1 and ε_2 are given permissible errors.

3. On Stability

It is very difficult to prove strictly the stability of a method for solving a nonlinear partial differential equation. The following discussion only involves two sub−procedures of the present method.

A. It has been proved that the calculating procedure for solving the equation (20) is stable when the angle θ is so large that the diagonal elements of the coefficient matrix of the equation (20) satisfy the conditions $|b_j|>|a_j|+|c_j|$ for all j [10]. In the Similar−Plane−Cascade of propeller the θ is usually large enough to ensure that the above stability condition holds.

B. Integrating the equation (17) over ψ by means of the Lax–Wendroff scheme,a criterion of stability is

$$\frac{\Delta\psi}{(\sqrt{D})_{min}\cdot\Delta\xi}\leqslant 1 \tag{22}$$

SOLUTION OF THE UNSTEADY PERTURBATION FLOWFIELD

1. Perturbation Potential and Its Amplitude Equations

In von Misses cordinate system the perturbation potential equation (9) can be changed into

$$G\varphi_{\xi\xi}+E\varphi_{\xi\psi}+D\varphi_{\psi\psi}=B_o[\varphi_{tt}+2B_1\varphi_{t\xi}+(\gamma-1)B_2\varphi_t+B_3\varphi_\xi+B_4\varphi_\psi] \tag{22a}$$

where

$$\left.\begin{aligned}
B_o&=\frac{2}{(\gamma+1)-(\gamma-1)q_o^2}\\
B_1&=\frac{q_o}{\sqrt{1+y_\xi^2}}\\
B_2&=\frac{1}{\sqrt{1+y_\xi^2}}[q_{o\xi}+\frac{q_o}{1+y_\xi^2}(Dy_{\xi\psi}-y_\xi y_{\xi\psi})]\\
B_3&=\frac{\gamma q_o}{1+y_\xi^2}(q_{o\xi}+\frac{q_o}{1+y_\xi^2}\cdot Dy_{\xi\psi})+q_o(q_{o\xi}+\frac{1}{2}Eq_{o\psi})\\
B_4&=q_o(\frac{1}{2}Eq_{o\xi}+Dq_{o\psi})+\frac{q_o}{(1+y_\xi^2)y_\psi}y_{\xi\xi}
\end{aligned}\right\} \tag{22b}$$

Assuming that the vibration of blades is of harmonic time dependence, the perturbation potential can be expressed

$$\varphi(\xi,\psi,t)=\bar{\varphi}(\xi,\psi)e^{ikt} \tag{23}$$

where $k(=\frac{2\pi fb}{\overset{\circ}{a}})$ is the reduced frequence, $\bar{\varphi}$ the amplitude of φ, and $i=\sqrt{-1}$.

Substituting the formula (23) into (22), an amplitude equation of the perturbation potential can be obtained

$$G\bar{\varphi}_{\xi\xi}+E\bar{\varphi}_{\xi\psi}+D\bar{\varphi}_{\psi\psi}=C_o\bar{\varphi}+C_1\bar{\varphi}_\xi+C_2\bar{\varphi}_\psi \tag{24a}$$

where

$$\left.\begin{aligned}
C_o&=[-k^2+(\gamma-1)ikB_2]B_o\\
C_1&=(2kiB_1+B_3)B_o\\
C_2&=B_4B_o
\end{aligned}\right\} \tag{24b}$$

Notice that the type of equation (24) is the same as the equation (14).

2. Perturbation Pressure

Using the formulas (7) and (2) as well as $a^2=\gamma p^{\frac{\gamma-1}{\gamma}}$ one can obtain

$$p'=-\frac{\gamma p_o}{a_o^2}(\varphi_t+\vec{q}_o\cdot\nabla\varphi)$$

Letting $p'=\bar{p}e^{ikt}$ and substituting the formula (23) into the above formula, we can obtain the expression of the amplitude of the perturbation pressure

$$\bar{p} = -\frac{\gamma p_o}{a_o^2}(ik\bar{\varphi} + \frac{q_o}{\sqrt{1+y_\xi^2}}\bar{\varphi}_\xi) \tag{25}$$

3. Boundary Conditions

Assume that the model of vibration of blades is a combined bending and torsional one, and the two kinds of vibration have the same frequence and phase, and that the bending direction makes an angle of ψ_0 with the positive y–axis and the torsional center is located at (x_o, y_o). See Fig.3.

So we have expresses

$$\left.\begin{aligned} h &= h_R e^{ikt} \\ \theta &= \theta_R e^{ikt} \end{aligned}\right\}$$

where h_R, and θ_R denote the bending and torsional amplitude respectively, they are all real and small. Besides, we denote the phase difference between the vibration in the blade below and that in the blade above in Fig.1 with β.

For this model of vibration the boundary conditions are as follows.

A. Far field condition:

$\bar{\varphi} = 0$ at far upstream inlet boundary

$\bar{p} = 0$ i.e. $ik\bar{\varphi} + \frac{q_o}{\sqrt{1+y_\xi^2}}\bar{\varphi}_\xi = 0$ at far downstream outlet boundary.

B. The periodicity condition with a shift of phase between the cyclic boundaries

$$\bar{\varphi}_\psi(\xi, 0)e^{i\beta} = \bar{\varphi}_\psi(\xi, Q)$$

C. Solid surface conditions without separation of flow.

The unpermeable condition gives

$$\left.\begin{aligned} &E\bar{\varphi}_\xi + 2D\bar{\varphi}_\psi = ik\{-[h_R \sin\psi_o + \theta_R(y-y_o)]E + 2[h_R\cos\psi_o + \theta_R(x-x_o)]/y_\psi\}, \\ &\qquad \textit{on the suction surface} \\ &E\bar{\varphi}_\xi + 2D\bar{\varphi}_\psi = ik\{-[h_R \sin\psi_o + \theta_R(y-y_o-T)]E + 2[h_R\cos\psi_o + \theta_R(x-x_o)]/y_\psi\}e^{i\beta}, \\ &\qquad \textit{on the pressure surface.} \end{aligned}\right\}$$

D. The conditions on boundaries with separation of flow.

In order to consider the effects of the separated flow on flutter, the present method extends Sisto's idea [8] and assume that the separation bubbles have no thickness and stretch to downstream infinity along the null and maximal flux streamline, with a separation point located at the shock or at the point where the steady mean speed has attained maximum without a shock, and that inside the bubbles the unsteady perturbation pressure becomes zero, i.e.

$$ik\bar{\varphi} + \frac{q_o}{\sqrt{1+y_\xi^2}}\bar{\varphi}_\xi = 0$$

4. Difference Equation and Its Solution

The equation (24) and its boundary conditions can be approximated by difference equations and solved by using a method similar to that described in the previous section.

ANALYSIS OF AEROELASTIC STABILITY

In order to investigate the onset of flutter, it is necessary to calculate the work done by the perturbation airflow on a blade in one period of vibration. This work consists of that done by the perturbation aerodynamic moment when the blade torques, W_M, and that done by the perturbation aerodynamic force when the blade bends, W_L. The steady mean aerodynamic moment and force have no contribution to this work.

Let Lx,Ly denote the components of the unsteady perturbation aerodynamic composite force L in x–and y–directions respectively, they can be expressed as

$$\left.\begin{aligned} L_x &= \bar{L}_x e^{ikt} \\ L_y &= \bar{L}_y e^{ikt} \end{aligned}\right\}$$

So, according to the definition, one can obtain

$$\left.\begin{aligned} \bar{L}_x &= -\oint_{lb} \bar{p} dy \\ \bar{L}_y &= \oint_{lb} \bar{p} dx \end{aligned}\right\}$$

where lb denotes the contour of blade. The instantaneous real displacements of the blade contour in x– and y– directions are

$$\left.\begin{aligned} x_{TR} &= -h_R \sin\psi_o \sin kt \\ y_{TR} &= h_R \cos\psi_o \cos kt \end{aligned}\right\}$$

thus the work of unsteady perturbation aerodynamic force in one period of vibration is

$$W_L = \oint L_{yR} dy_{TR} + L_{xR} dx_{TR} = \pi h_R [\bar{L}_{yI} \cos\psi_o - \bar{L}_{xI} \sin\psi_o]$$

where the subscripts R,I denote the real and imaginary parts of a complex.
If unsteady perturbation aerodynamic moment is written as $M = \bar{M} e^{ikt}$, then its work in one period of vibraton is

$$W_M = \oint_{lb} dw_M = \oint_{lb} M_R d\theta_{TR} = \pi \bar{M}_I \theta_R$$

where

$$\bar{M} = \oint_{lb} \bar{p}[(x - x_o)dx + (y - y_o)dy]$$

Consequently we can obtain the total work

$$W = W_L + W_M$$

According to the "energy method", if the total work is negative then the onset of flutter can not occur, otherwise the onset of flutter depends on a mechanical damping of the blades.

NUMERICAL RESULTS

Numerical results have been obtained for five conditions of climbing and cruising flight of the propellers J16–G10A and Y7, examples 1–5. Their main parameters are listed in Table 1.

Table 1. Main parametres of five examples

No.	V(m / s)	M_1	$\theta_1(deg)$	α(deg)	k	Model of vibration	Propeller
1	150	0.822	57.57	2	0.0645	bending and torquing	J16–G10A
2	50	0.709	78.04	2	0.0275		
3	50	0.709	78.04	8	0.0352		
4	50	0.631	76.52	12	0.359	pure torquing	
5	50	0.625	76.39	12	0.370		Y7

V–Advance speed, M_1–Inlet Mach Number, α–Angle of attack

Numerical results are shown in Fig.4~Fig.6.

Fig.4 shows the distributions of the steady mean pressure on the blade surface. It can be seen that a greater part of aerodynamic load distributes on the front when the angle of attack is large. This differs from that of an isolated airfoil.

Fig.5 shows the effect of phase difference β on the distributions of the real and imaginary parts of the perturbation pressure difference between the pressure surface and suction surface of cascade passage. It is obvious that in general there is a phase shift between the perturbation pressure and vibration of blade.

Fig.6 indicates the relationship of perturbation aerodynamic work with the phase angle β. It can be seen that the work is always negative in all cases, so the onset of flutter cannot occur. These results agree well with experiment and actual flight.

CONCLUDING REMARKS

A. Numerical results have proved that SPCM with the extended Sisto's model is available for prediction of both classical and stall flutter. SPCM agree better with the actual situation than the isolated airfoil one.

B. The streamline–governing equation and perturbation potential equation written in von Misses coordinate system can be quite efficiently solved by use of the line–relaxation iteration method, and quite exact results can be obtained with considerably less computing time.

However, a smaller number of blades and a larger propeller diameter will cause very large cascade–spacing and computational domain so that the efficiency of this method would reduce.

C. The present work is only at the initial stage. A great deal of work remains to be done.

ACKNOWLEDGEMENTS

Acknowledgements are due to Prof. Zhou Sheng, Beijing University of Aeronautics and Astronautics, for his suggestions and support.

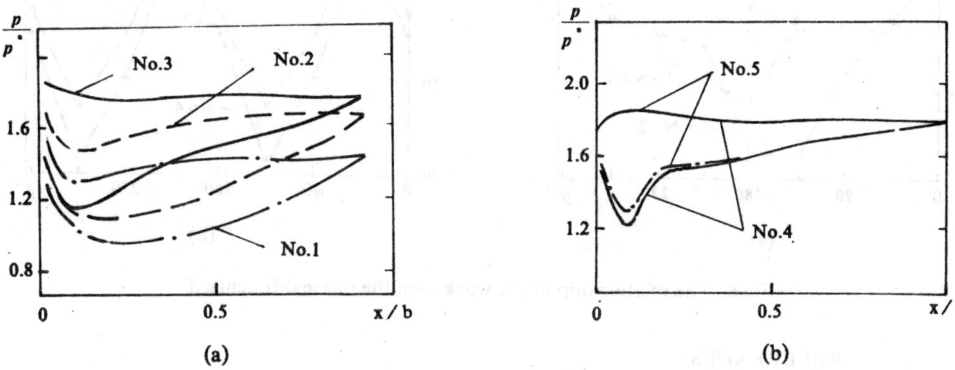

Fig.4 The distributions of steady mean pressure on the surfaces of blade

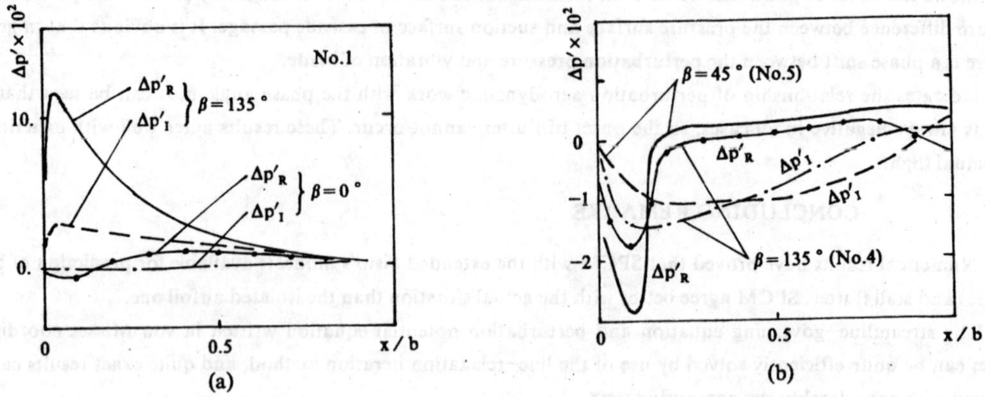

Fig.5 The distributions of perturbation pressure differences between the pressure and suction surfaces for typical phase angles

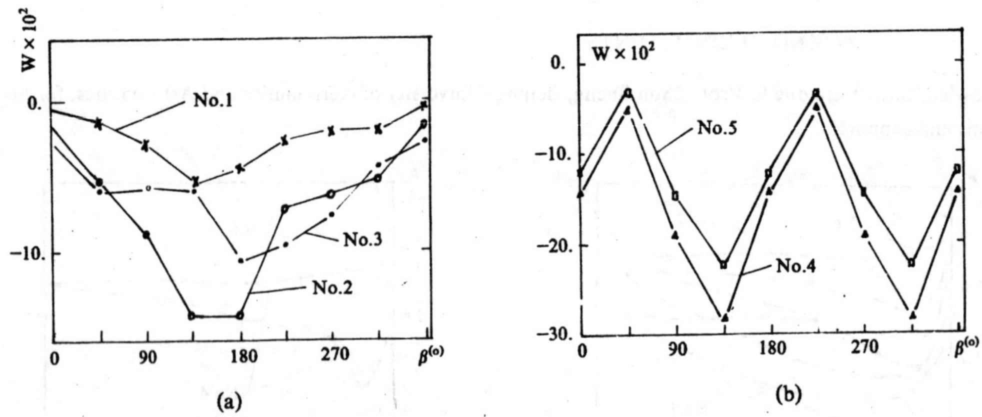

Fig.6 The relationship of the work with the phase difference β

REFERENCES

[1]. H.H.Hubbard, M.F.Burgess &. M.A.Sylvester, NACA TN–3707, 1956.

[2]. J.E.Baker &. R.S.Paulnock, NACA TN–1966, 1949.

[3]. H.G.Ewing, B.Sc et al., A.R.C. Technical Report, R&.M. No.3634 1948.

[4]. J.E.Baker, NACA TN–3357, 1950.

[5]. J.G.Barmby et al., NACA Rep.1014 1951.

[6]. Theodorsen et al., NACA Rep. 685, 1940.

[7]. Shen Mengyu, Zhou Sheng &. Lin Baozheng, Transonic Flow in Turbomachinery, Academic Press, 1988.

[8]. P.V.K.Perumal, F.Sisto, Trans. ASME, Journal of Engineering for Power, 96, 372–382, 1974.

[9]. Liu Qiusheng, Shen Mengyu, ACTA MECHANICA SINICA, Vol. 20, No.4 1988.

[10]. Feng Kang, Numerical methods, National Defence Industry Press, 1978.

A Method for Predicting Stall Flutter under Variable Interblade Phase Angle along Rotating Direction

Yang.X.D Tao.D.P Zhou.S

Beijing University of Aeronautics & Astronautics

ABSTRACT:

The prediction of stall flutter at turbomachinery always needs the assumption of constant interblade phase angle, in which all interblade phase angles of each blades at a rotor are the constant. But many experiments for practical turbomachine have shown that the interblade phase angle is actually variable. This phenomenon shows that the basic assumption of aeroelasticity in turbomachinery is not correct. In this paper, the basic principle of deforming actuator disk method is used to develop a new predicting method by solving linearized Euler's equation.The method can predict the stall flutter under variable interblade phase angle along the rotating direction. The comparison between the computation and the experiment shows that the most unstable blade predicted by the method is very close to the measured one. This proves the predicting method to be effective for practical usage in engineering.

SYMBOLS:

h	amplitude of blade
i	$\sqrt{-1}$ or vector unit
j	vecort vector unit
k_2	wave number in tangential direction
k	reduced frequency
M	Mach number
P	pressure
q	velocity vector of flow
s	vector unit normal to element
t	time
u	velocity component of coordinate x
v	velocity component of coordinate y
α	incidence angle
β	flow angle
γ	ratio of specific heats
δ	deviation angle
θ	stagger angle
σ	solidity
x	total pressure loss coefficient
ρ	density
φ	oscillating phase angle
Subscript:	
n	normal components relative to actuator disk

−	the leading edge of deforming disk
+	the trailing edge of deforming disk
Superscript:	
−	steady component
~	unsteady component
≃	amplitude of unsteady componen'
rel	relative to deforming disk
−∞	upstream
+∞	downstream

INTRODUCTION:

The aeroelasticity of turbomachinery is a branch of aeroelasticity by which the fluid induced vibration is studied in turbomachinery. This branch has been widely used in the aeroengine design, especially in the prediction of compressor blade flutter. Since this branch was founded, some basic assumptions have been used such as the assumption that the interblade phase angle at tangential direction is constant.This assumption simplifies the computation of unsteady flow field around the oscillating cascade, because the flow throgh a cascade channel can represent the whole channels in the rotor.

But recently some experiments about the stall flutter of compressor showed that the interblade phase angles at rotating direction are not constant, and its variation has been measured (Ref.1).Fig.1 and Fig.2 are the flutter amplitude and flutter phase angle of each blades for rotor TS22 in Ref.1. From the figures it can be seen that the interblade phase angle is not constant at rotating direction.The computations of Ref.1 and Ref.2 show that the value of interblade phase angle has very large effect on aerodynamic damping. Maybe the variation of interblade phase angle of each blade at a rotor is why there exists a peak of flutter amplitude at the rotor (Fig.1). But by the assumption of constant interblade phase angle, predicted aerodynamic dampings of each blade are the same. So the basic assumption of constant interblade phase angle may not be correct.

The purpose of this paper is studing the stall flutter under variable interblade phase angle at rotating direction. The unsteady flow under the variable interblade phase angle is simulated numerically by a numerical actuator disk method which was developed by authors.

Since the assumption of constant interblade phase angle is canceled by above−mentioned phenomenon, unsteady flow field through all cascade channels at a rotor must be calculated to predict flutter of all blades. So if the traditional computation aerodynamics of the cascade is used, the computational quantities will be much more.

The actuator disk method (Ref.2 and Ref.3) is a semi−empirical method which have much smaller computational quantities. It can solve unsteady flow under constant interblade phase angle. In the paper this method is combined with computational aerodynamics to get a numerical actuator disk method which can treat variable interblade phase angle at rotating direction. In the characteristic section the method can numerically simulate whole two−dimensional unsteady flow in front of the oscillating blade row and behind it. The flow inside the channels is not considered and the abrupt−jumping conditions across the blade row are used to determine the flow parameters across the blade row. To solve the unsteady flow, the linearized Euler's Equation is used as basic equations. Fig.3 is the sketch map of computation zone.

By using finite difference the convergent solution to the computational example in Fig.1 and Fig.2 is got.The computation shows that the peak of blade aerodynamic damping appears and its position is very near to the measured peak of blade amplitude. So it is obvious that the blade amplitude peak is caused by blade aerodynamic damping peak under the variable interblade phase angle.

In order to study the effect of flutter phase angle on aerodynamic damping, two kinds of phase angle distributions are given. The computation shows that the abrupt variation (rectangle distribution) of interblade phase angle will cause larger negative aerodynamic damping which means more violent blade flutter than harmonic distribution. So the abrupt variation of interblade phase angle should be avoided.

BASIC MODEL:

According to the introduction, the actuator disk method can be used for stall flutter prediction under variable interblade phase angle along rotating direction. But the actuator disk method at Ref.2 and Ref.3 has a assumption: the aerodynamic parameters along Y direction are harmonic:

$$\tilde{f}(x, y, t) = \tilde{\bar{f}}(x)e^{ik_2 y + ikt} \tag{1}$$

By this assumption, the analytical solution can be yielded.

Under the variable interblade phase angle, this assumption is not correct. So the analytic solution for linearized Euler's equation can't be yielded. The authors intend to solve numerically the equations and retain the assumption of harmonic variation with time:

$$\tilde{f}(x, y, t) = \tilde{\bar{f}}(x, y)e^{ikt} \tag{2}$$

According to the principle of actuator disk method, the mathematics–physics model suitable to variable interblade phase angle is as follows:

1) Two–dimensional unsteady flow field at the characteristic section is simulated. Fig.3 represents the computational zone in which line AC and line BD conform to periodic condition.

2) Suppose blade chord is zero. So the blade row becomes a actuator disk across which the aerodynamic parameters will have abrupt jump.

3) Suppose the flow in front of and behind the blade row is inviscid.

4) There are four abrupt–jumping conditions across the actuator disk, which are jump of flow loss, flow angle and conservation of flow rate and energy.

5) The motion of actuator disk is determined by the vibration of blades. Suppose blade vibration is in bending mode. The vibrating phase angles of each blades are determined by measured ones.

$$\begin{cases} \tilde{h}_{x,j} = \bar{\tilde{h}}_{x,j} e^{i\varphi_j + ikt} \\ \tilde{h}_{y,j} = \bar{\tilde{h}}_{y,j} e^{i\varphi_j + ikt} \end{cases} \tag{3}$$

6) Unsteady flow is a small perturbation to the steady one. So every parameters can be divided into steady and unsteady components:

$$f = \bar{f} + \tilde{f}$$
$$\left|\frac{\tilde{f}}{\bar{f}}\right| < < 1.0 \tag{4}$$

7) Suppose unsteady components are harmonic with time(Formula(2)).

8) The steady flow field are uniform in front of and behind the actuator disk.

THE BASIC EQUATIONS & BOUNDARY CONDITIONS:

From the basic assumptions of the problem, Euler's Equations can be used as the basic equations. By using assumption 6), the equations can be linearized as follows:

$$\frac{\partial \tilde{\rho}}{\partial t} + \bar{u}\frac{\partial \tilde{\rho}}{\partial x} + \bar{V}\frac{\partial \tilde{\rho}}{\partial y} + \bar{\rho}\left(\frac{\partial \tilde{u}}{\partial x} + \frac{\partial \tilde{V}}{\partial y}\right) = 0 \tag{5}$$

$$\frac{\partial \tilde{u}}{\partial t} + \bar{u}\frac{\partial \tilde{u}}{\partial x} + \bar{V}\frac{\partial \tilde{u}}{\partial y} + \frac{1}{\bar{\rho}}\frac{\partial \tilde{P}}{\partial x} = 0 \tag{6}$$

$$\frac{\partial \tilde{V}}{\partial t} + \bar{u}\frac{\partial \tilde{V}}{\partial x} + \bar{V}\frac{\partial \tilde{V}}{\partial y} + \frac{1}{\bar{\rho}}\frac{\partial \tilde{P}}{\partial y} = 0 \tag{7}$$

$$\frac{\partial \tilde{P}}{\partial t} + \bar{u}\frac{\partial \tilde{P}}{\partial x} + \bar{V}\frac{\partial \tilde{P}}{\partial y} - \bar{a}^2\left(\frac{\partial \tilde{\rho}}{\partial t} + \bar{u}\frac{\partial \tilde{\rho}}{\partial x} + \bar{V}\frac{\partial \tilde{\rho}}{\partial y}\right) = 0 \tag{8}$$

These equations are unsteady linearized Euler's equation which have three independent variables.From the computational aerodynamics, it is obvious that such equations need much more computational quantities to solve. To reduce the quantities, basic assumption 6) and 8) is used. So the equations become:

$$\bar{u}\frac{\partial \tilde{\bar{\bar{\rho}}}}{\partial x} + \bar{V}\frac{\partial \tilde{\bar{\bar{\rho}}}}{\partial y} + \bar{\rho}\left(\frac{\partial \tilde{\bar{\bar{u}}}}{\partial x} + \frac{\partial \tilde{\bar{V}}}{\partial y}\right) = 0 \tag{9}$$

$$\bar{u}\frac{\partial \tilde{\bar{\bar{u}}}}{\partial x} + \bar{V}\frac{\partial \tilde{\bar{\bar{u}}}}{\partial y} + \frac{1}{\bar{\bar{\rho}}}\frac{\partial \tilde{\bar{P}}}{\partial x} = 0 \tag{10}$$

$$\bar{u}\frac{\partial \tilde{\bar{V}}}{\partial x} + \bar{V}\frac{\partial \tilde{\bar{V}}}{\partial y} + \frac{1}{\bar{\bar{\rho}}}\frac{\partial \tilde{\bar{P}}}{\partial y} = 0 \tag{11}$$

$$\bar{u}\frac{\partial \tilde{\bar{P}}}{\partial x} + \bar{V}\frac{\partial \tilde{\bar{P}}}{\partial y} - \bar{a}^2\left(\bar{u}\frac{\partial \tilde{\bar{\bar{\rho}}}}{\partial x} + \bar{V}\frac{\partial \tilde{\bar{\bar{\rho}}}}{\partial y}\right) = 0 \tag{12}$$

These equations have two independent variables, so the computational quantity is reduced greatly.But by aerodynamics, the type of these equations is very complex in transonic flow field, so it is very difficult to solve it with finite difference. Based on time marching method of computational aerodynamics, it can be supposed that the amplitudes of aerodynamic parameters tend to steady values with the march of time. This assumption can be represented as this:

$$\tilde{f}(x,\ y,\ t) = \tilde{\bar{f}}(x,\ y,\ t)e^{ikt} \tag{13}$$

With the march of time, f(x, y, t) will tend to a steady value f(x, y). By the formula, linearized equations with two independent variables are changed as :

$$\frac{\partial \tilde{\bar{\bar{\rho}}}}{\partial t} + \bar{u}\frac{\partial \tilde{\bar{\bar{\rho}}}}{\partial x} + \bar{V}\frac{\partial \tilde{\bar{\bar{\rho}}}}{\partial y} + \bar{\rho}\left(\frac{\partial \tilde{\bar{\bar{u}}}}{\partial x} + \frac{\partial \tilde{\bar{V}}}{\partial y}\right) = -ik\tilde{\bar{\bar{\rho}}} \tag{14}$$

$$\frac{\partial \tilde{u}}{\partial t} + \bar{u}\frac{\partial \tilde{u}}{\partial x} + \bar{V}\frac{\partial \tilde{\bar{\bar{u}}}}{\partial y} + \frac{1}{\bar{\bar{\rho}}}\frac{\partial \tilde{\bar{P}}}{\partial x} = -ik\tilde{\bar{\bar{u}}} \tag{15}$$

$$\frac{\partial \tilde{\bar{V}}}{\partial t} + \bar{V}\frac{\partial \tilde{\bar{V}}}{\partial x} + \bar{V}\frac{\partial \tilde{\bar{V}}}{\partial y} + \frac{1}{\bar{\bar{\rho}}}\frac{\partial \tilde{\bar{P}}}{\partial y} = -ik\tilde{\bar{V}} \tag{16}$$

$$\frac{\partial \tilde{\bar{P}}}{\partial t} + \bar{u}\frac{\partial \tilde{\bar{P}}}{\partial x} + \bar{V}\frac{\partial \tilde{\bar{P}}}{\partial y} - \bar{a}^2\left(\frac{\partial \tilde{\bar{\bar{\rho}}}}{\partial t} + \bar{u}\frac{\partial \tilde{\bar{\bar{\rho}}}}{\partial x} + \bar{V}\frac{\partial \tilde{\bar{\bar{\rho}}}}{\partial y}\right) = -ik(\tilde{\bar{P}} - \bar{a}_2\tilde{\bar{\bar{\rho}}}) \tag{17}$$

Equations (14)–(17) are the basic equations in the method.

The boundary conditions of computational zone (Fig.3) are:

1) Inlet boundary(line AB):

According to aerodynamics a characteristic condition can be used and three conditions must be given:

$$\tilde{\rho}^{\bullet}_{-\infty} = \tilde{P}^{\bullet}_{-\infty} = 0 \tag{18}$$

$$\tilde{u}_{-\infty} = \tilde{V}_{-\infty} tan(\beta_{-\infty}) \tag{19}$$

2) Exit boundary(line CD):

Three characteristic conditions is used to determine the parameters at the boundary and a condition must be given:

$$\tilde{P}_{\infty} = 0 \tag{20}$$

3) Periodic boundary(line AC line BD):

The corresponding parameters in two boundaries are equal to each other.

4) Deforming actuator disk (line EF)

There are four abrupt–jumping conditions across the disk:

a) The continuity of flow rate across the disk:

$$(\bar{\rho}_{-} + \tilde{\rho}_{-}) \cdot u^{rel}_{n,-} = (\bar{\rho}_{+} + \tilde{\rho}_{+}) \cdot u^{rel}_{n,+} \tag{21}$$

b) The conservation of total enthalpy across the disk:

$$\frac{\gamma}{\gamma-1}\frac{P_{-}}{\rho_{-}} + \frac{(q^{rel}_{-})^2}{2} = \frac{\gamma}{\gamma-1}\frac{P_{+}}{\rho_{-}} + \frac{(q^{rel}_{-})^2}{2} \tag{22}$$

c) The abrupt jump of flow angle across the disk:

As usually, the quasi–steady assumption is used to determine unsteady variation by steady empirical relation.The empirical steady relation of flow angle at the trailing edge is:

$$\delta = g(\alpha_{-}, M_{-}, \sigma, \theta) \tag{23}$$

So the unsteady flow angle at the trailing edge can be determined by relation (23).

d) The abrupt jump of flow loss across the disk:

The quasi–steady assumption is also used to solve the unsteady flow loss across the disk. The empirical steady relation of flow loss is:

$$x = x(\alpha_{-}, M_{-}, \sigma, \theta) \tag{24}$$

Then four abrupt–jumping cinditions across the actuator disk have been got.

The initial flow field is supposed to be zero. The computation in this paper proves that such a initial flow field does not increase computational quantities obviously.

THE FINITE DIFFERENCE SCHEME:

In order to solve the equations with finite difference, MacCormack finite volume diffrence scheme is developed for the basic equations. The equations are yielded on the formula (14)–(17):

$$\frac{\partial}{\partial t}\int_{vol} \overline{U} d(vol) + \int_{s} \overline{H} \cdot \vec{n} ds = -\int_{vol} ik\overline{U} d(vol) \tag{25}$$

$$\overline{\mathbf{U}} = \begin{bmatrix} \bar{\tilde{\rho}} \\ \bar{\tilde{u}} \\ \tilde{\bar{V}} \\ \tilde{\bar{P}} - \bar{a}^2 \bar{\tilde{\rho}} \end{bmatrix} \tag{26}$$

$$\overline{\mathbf{H}} = \begin{bmatrix} \bar{i}(\bar{u}\tilde{\rho} + \bar{\rho}\tilde{u}) + \bar{j}(\bar{V}\tilde{\rho} + \tilde{\bar{v}}\rho) \\ \bar{i}(\bar{u}\tilde{u} + \tilde{P}/\bar{\rho}) + \bar{j}(\bar{V}\tilde{u}) \\ \bar{i}(\bar{u}\tilde{V} + \tilde{P}/\bar{\rho}) + \bar{j}(\bar{V}\tilde{V} + \tilde{P}/\bar{\rho}) \\ \bar{i}(\bar{u}\tilde{P} - \bar{a}^2\bar{u}\tilde{\rho}) + \bar{j}(\bar{V}\tilde{P} - \bar{a}^2\tilde{V}\tilde{\rho}) \end{bmatrix} \tag{27}$$

The finite difference scheme is as this:

$$\overline{\mathbf{U}}_{i,\ j}^{n+\frac{1}{2}} = \overline{\mathbf{U}}_{i,\ j}^{n} - \Delta t \cdot k \cdot i \cdot \overline{\mathbf{U}}_{i,\ j}^{n} - \frac{\Delta t}{\Delta_{i,\ j}} \cdot (\overline{\mathbf{H}}_{i,\ j}^{n} \cdot \bar{S}_3 + \overline{\mathbf{H}}_{i,\ j-1}^{n} \cdot \bar{S}_1 + \mathbf{H}_{i-1,j}^{-n} \cdot \vec{\mathbf{S}}_2 + \overline{\mathbf{H}}_{i,\ j}^{n} \cdot \bar{S}_4) \tag{28}$$

$$\overline{\mathbf{U}}_{i,\ j}^{n+1} = \frac{1}{2}(\overline{\mathbf{U}}_{i,\ j}^{n+\frac{1}{2}} + \overline{\mathbf{U}}_{i,\ j}^{n}) + \frac{1}{2}\Delta t \cdot k \cdot i \cdot (\overline{\mathbf{U}}_{i,\ j}^{n+\frac{1}{2}} + \overline{\mathbf{U}}_{i,\ j}^{n}) - \frac{\Delta t}{\Delta i,\ j} \cdot (\overline{\mathbf{H}}_{i,\ j}^{n+\frac{1}{2}} \cdot \bar{S}_2 + \overline{\mathbf{H}}_{i,\ j+1}^{n+\frac{1}{2}} \cdot \bar{S}_3 + \overline{\mathbf{H}}_{i+1,\ j}^{n+\frac{1}{2}} \cdot \bar{S}_4 + \overline{\mathbf{H}}_{i,\ j}^{n+\frac{1}{2}} \cdot \bar{S}_2) \tag{29}$$

The skech map of the element is refered to at Ref. 5.

COMPUTATIONAL ANALYSIS:

The computational example is rotor TS22 which encounters the stall flutter at the experiment (Ref.1 & Ref.4). Fig.1 gives out the distribution of measured dynamic blade stress in each blades. Fig.2 is interblade phase angle which was measured at the test. From the Fig.1 it is obvious that there exists a peak of dynamic stress. The test shows that the blade amplitude is caused by stall flutter.From Fig.2 it can be seen that the each interblade phase angle at the rotor is not constant.

The oscillating phase angle of formula (2) is inputed according to the distribution at Fig.2.Because the rotor has shroud, the characteristic section is located in 95% height of blade.

Fig.4 is the computational aerodynamic damping of each blade. In the figure there is the peak of minus aerodynamic damping and its position is very near to the measured one. The measured result shows that blade No.0–6 have entered the stall flutter zone. The computational result is that blade No. 7–9 have the minus aerodynamic damping.So the computational results is satisfactory to predict the position of flutter peak.

Fig.5 is computational stall flutter boundary under variable interblade phase angle along the rotating direction and the boundary under constant interblade phase angle. From the figure, it can be seen that the boundary under variable interblade phase angle is more satisfactory than constant interblade phase angle, so the assumption of variable interblade phase angle is more reasonable.

To analyze the effect of distribution of interblade phase angle on the aeroelastic stability, two kinds of phase angle distribution are given in Fig.6 and Fig.7. Fig.6 is the sine distribution and Fig.7 is rectangle distribution. Fig 8 is the computational aerodynamic damping of rotor TS22 with two kinds of phase angle distribution. It is shown from the figure that the minus aerodynamic damping under rectangle distribution is larger than sine distribution. And the violent stall flutter always appears under plus interblade phase angle. So this proves that the abrupt large plus interblade phase angle should be avoided as far as possible.

CONCLUDING REMARKS:

1) A lot of testing results show that the interblade phase angle of each balde is not constant in the stall flutter. This paper predict the stall flutter under variable interblade phase angle firstly and the useful computational results is got.

2) From the computation of the paper, the actuator disk method is effective for the prediction of stall flutter with variable interblade phase angle along the rotating direction. It predicts the position of flutter peak in rotor TS22 which is very near to the measured position of blade dynamic stress.

3) The plus abrupt variation of interblade phase angle will cause more violent stall flutter, so such a variation should be avoided as far as possible.

REFERENCE:

1. R.M.Chi and A.V. Srinivasan:
Some Recent Advances in the Understanding and Predicting of Turbomachine Subsonic Stall Flutter.
ASME Paper 84–GT–151

2. J.J.Adamczyk:
Analysis of Supersonic Stall Flutter in Axial–flow Compressor by Actuator Disk Theory.
NASA–TP 1345, 1978

3. Yang Xiaodong; Tang Zhiming and Zhou Sheng:
An Actuator–Disk Model for Predicting Stall Flutter of Steam Turbines.
ASME Paper 85–IGT–99

4. H.Stargardter:
Subsonic / Transonic Stall Flutter Study
NASA CR–165256

5. Yang Xiodong:
The Researsh for the Basic Model of Aeroelasticity in Turbomachinery.
Ph.D Thesis in BUAA, November, 1988

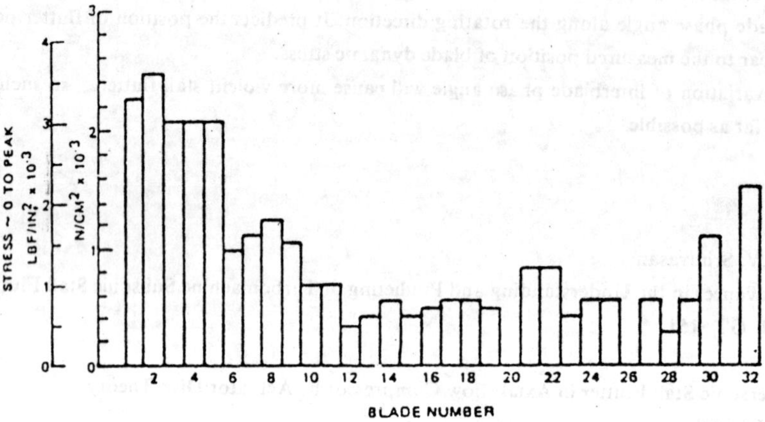
STRESS ~ 0 TO PEAK
LBF/IN² x 10⁻³
N/CM² x 10⁻³
0
1
2
3
4
2
4
6
8
10
12
14
16
18
20
22
24
26
28
30
32
BLADE NUMBER

Fig. 1

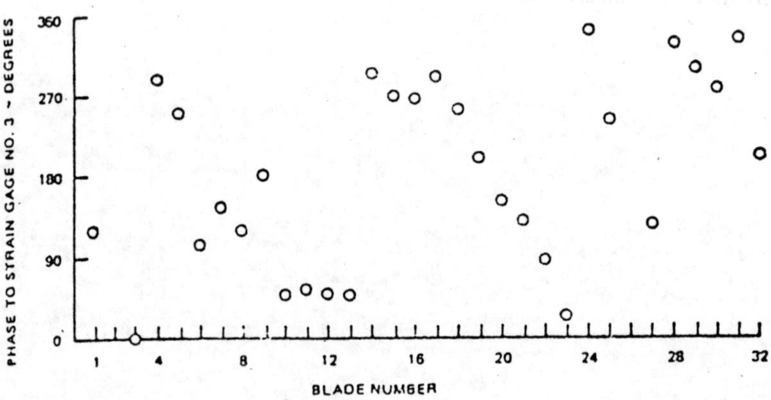
PHASE TO STRAIN GAGE NO. 3 ~ DEGREES
360
270
180
90
0
1
4
8
12
16
20
24
28
32
BLADE NUMBER

Fig.2

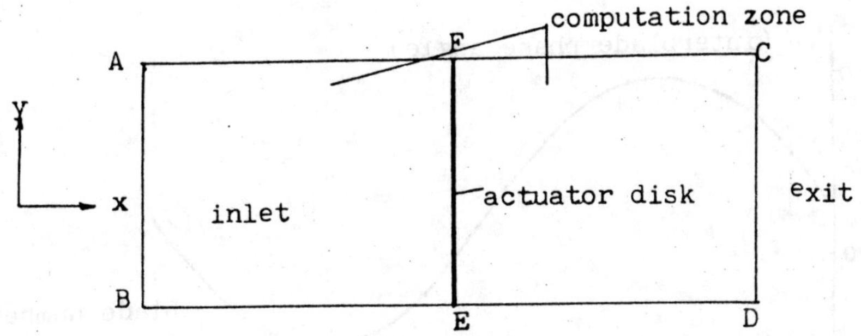

Fig.3

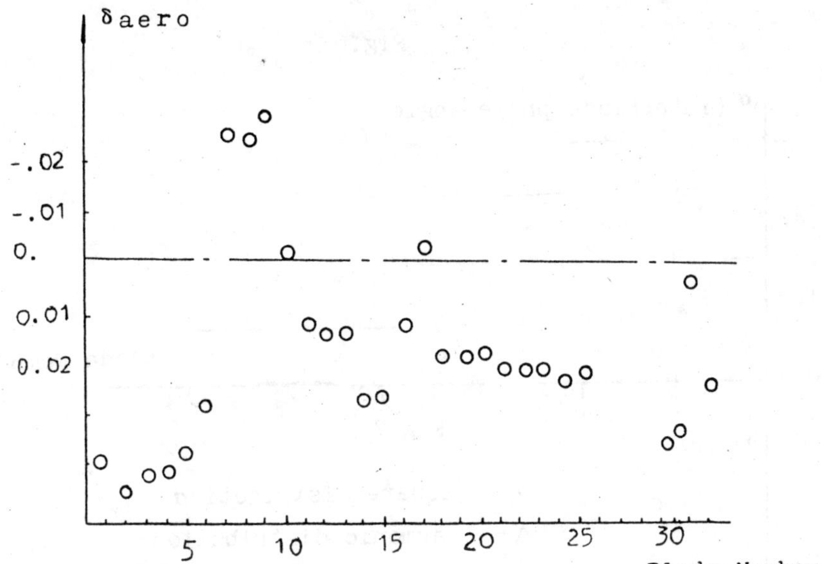

Fig.4

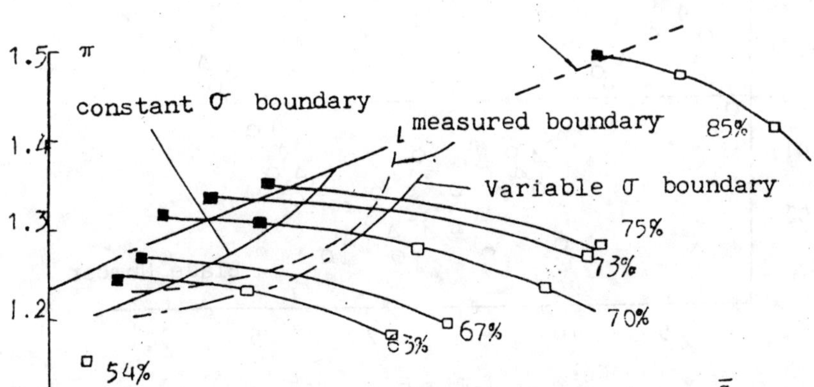

Fig.5

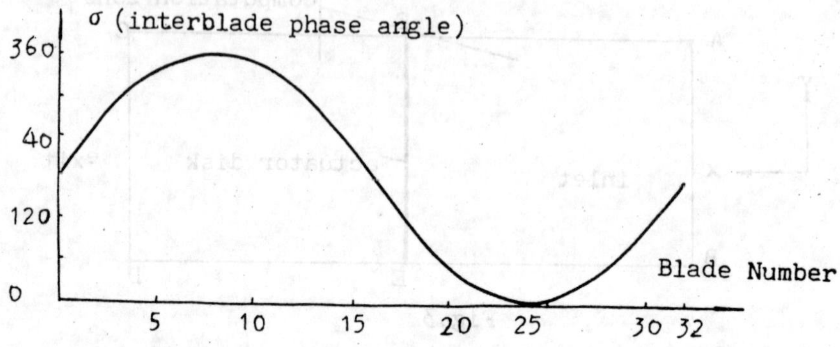

Fig.6

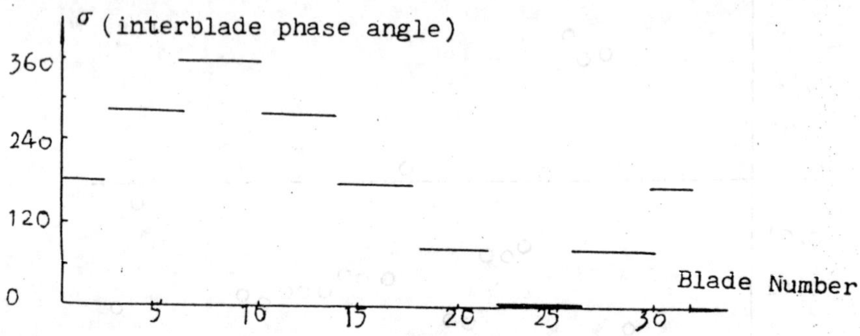

Fig.7

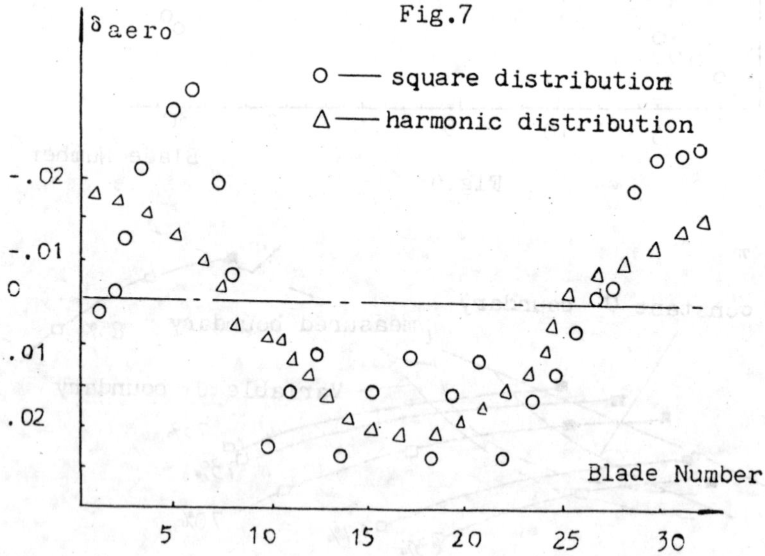

Fig.8

BLADE LOADS DUE TO UNSTEADY FLOW IN TURBOMACHINERY CASCADES

H.E. Gallus and H.-P. Kau
(Institut für Strahlantriebe und Turboarbeitsmaschinen, RWTH Aachen, Templergraben 55, D - 5100 Aachen)
S. Servaty
(MTU München GmbH, Dachauer Str. 665, D - 8000 München)

ABSTRACT

An algorithm for calculating unsteady blade forces and moments due to oscillating cascades and incoming wakes is presented. The flow is regarded as subsonic, two-dimensional and inviscid.

The non-linear Euler-equations in conservative law form are solved by a combined method. MacCormack's explicit predictor-corrector scheme is used at interior points. Time-dependent boundary conditions are formulated by means of a characteristics method.

The algorithm is validated by comparison with linearized theory. Experimental data are presented as well and compared with computational results.

A study is performed showing the influence of important aerodynamic and geometric parameters on the time-dependent forces and moments.

NOMENCLATURE

c	velocity
e	energy
E, F	flux vectors
J	Jacobian
p	pressure
s	entropy
t	time
TG	blade pitch
TD	pitch of the wake
u	velocity in x-direction
U	flow values (ρ, ρu, ρv, e)
v	velocity in y-direction
x, y	coordinates in physical plane
ξ, η	coordinates for transformation
$\vec{\xi}$	local normal vector
α	upstream flow angle
γ	ratio of specific heats
ν	frequency
ρ	density

Subscripts

B	blade
n	numerical
t	total value

Superscripts

n	value at time t
n+1	value at time $t+\Delta t$
$\wedge$	values after the transformation
$\overline{}$	values after the predictor step of the MacCormack scheme
*	value at the boundary after MacCormack iteration

INTRODUCTION

The reliability of modern axial-flow turbomachines mainly depends on blade loads due to flow-induced vibrations. They are caused by unsteady phenomena such as blade row interaction, turbulence, stall, self-excited blade vibrations and inlet distortion.

This paper presents a method to predict the aerodynamic blade loads in idealized two-dimensional, inviscid and subsonic flow taking into account wake influence and blade vibrations. These effects are covered by the time-dependent two-dimensional Euler-equations. They are formulated in finite difference form on a time-dependent grid. Boundary conditions are taken into account by a post-correction method using one-dimensional characteristics /1,2/. For interior points the explicit MacCormack scheme is used /3/. Calculating unsteady phenomena the discretisation area can be reduced to only one channel using a special formulation of a periodicity condition. To show the accuracy of the algorithm in the 2d-area it has first been checked computing the steady flow through compressor and turbine cascades. The prediction capability of the time-dependent computational results is shown by comparison with several experimental data.

GOVERNING EQUATIONS

The unsteady flow is described by the two-dimensional, inviscid and compressible Euler-equations written in conservative form. They are formulated on a cartesian coordinate grid with x being the axial direction:

$$\frac{\partial}{\partial t} U + \frac{\partial}{\partial x} E(U) + \frac{\partial}{\partial y} F(U) = 0 \tag{1}$$

where

$$U = \begin{pmatrix} \rho \\ \rho u \\ \rho v \\ e \end{pmatrix} \qquad E(U) = \begin{pmatrix} \rho u \\ \rho u^2+p \\ \rho uv \\ (e+p)u \end{pmatrix} \qquad F(U) = \begin{pmatrix} \rho v \\ \rho uv \\ \rho v^2+p \\ (e+p)v \end{pmatrix}$$

The system is closed with the assumption of perfect gas.

$$e = \frac{p}{\gamma-1} + \frac{\rho}{2}(u^2 + v^2)$$

MATHEMATICAL TRANSFORMATION

To solve the Euler-equations for a time-dependent problem a moving computational grid (x,y,t) has to be used where at each time step the physical boundaries coincide with grid lines (Fig. 1). To avoid space interpolations on this grid a mathematical model for the formulation of the derivatives $\partial/\partial x$ and $\partial/\partial y$ is used transforming the physical coordinates (x,y) at a fixed time into a new set (ξ,η) /4/. Using this transformation the Euler-equations can be written as

$$\frac{\partial}{\partial t}\hat{U} + \frac{\partial}{\partial \xi}\hat{E} + \frac{\partial}{\partial \eta}\hat{F} = 0 \qquad (2)$$

where

$$\hat{U} = U/J$$
$$\hat{E} = (U\ \xi_t + E\xi_x + F\xi_y)\ /\ J$$
$$\hat{F} = (U\ \eta_t + E\eta_x + F\eta_y)\ /\ J$$

with the Jacobian

$$J = \xi_x\ \eta_y - \xi_y\ \eta_x = \frac{1}{x_\xi\ y_\eta - x_\eta\ y_\xi}$$

Expressing all terms as a function of ξ, η and t the Jacobian can be eliminated from $\hat{E}$ and $\hat{F}$. The equations can be recasted analytically for $\hat{E}$ and $\hat{F}$ giving:

$$\hat{E} = U * U_K + p\begin{pmatrix}0\\ y_\eta\\ -x_\eta\\ uy_\eta - vx_\eta\end{pmatrix}$$

$$\hat{F} = U * V_K + p\begin{pmatrix}0\\ -y_\xi\\ x_\xi\\ vx_\xi - uy_\xi\end{pmatrix} \qquad (3)$$

with the abbreviations

$$U_K = x_\eta\ (y_t - v) + y_\eta\ (u - x_t)$$

$$V_K = x_\xi\ (v - y_t) + y_\xi\ (x_t - u)\ .$$

Eq. 3 shows that $\hat{E}$ contains the factor $1/\Delta\eta$, $\hat{F}$ $1/\Delta\xi$ and $\hat{U}$ $1/\Delta\xi\Delta\eta$, respectively. Thus Eq. 2 is independent from $\Delta\xi$ and $\Delta\eta$ if for a fixed solution point all derivatives in ξ- or η-direction are formulated between the same grid points.

ALGORITHM FOR INTERIOR POINTS

These transformed Euler-equations can now easily be solved at interior points by the explicit predictor/corrector MacCormack scheme /3/

Predictor step:

$$\overline{U_{i,j}^{n+1}} = J_{i,j}^{n+1} \left\{ \frac{U_{i,j}^{n}}{J_{i,j}^{n}} - \frac{\Delta t}{\Delta \xi} \left[\hat{E}_{i+1-\delta,j}^{n} - \hat{E}_{i-\delta,j}^{n} \right] - \frac{\Delta t}{\Delta \eta} \left[\hat{F}_{i,j+1-\varepsilon}^{n} - \hat{F}_{i,j-\varepsilon}^{n} \right] \right\} \tag{4}$$

The predictor step gives a first approximation $\overline{U_{ij}^{n+1}}$ of the values U_{ij}^{n+1} at time t^{n+1}.

The corrector step gives a second approximation and includes averaging yielding the new value at time t^{n+1}

$$U_{i,j}^{n+1} = J_{i,j}^{n+1} * \frac{1}{2} \left\{ \frac{\overline{U_{i,j}^{n+1}}}{\overline{J^{n+1}}} + \frac{U_{i,j}^{n}}{J_{i,j}^{n}} - \frac{\Delta t}{\Delta \xi} \left[\overline{\hat{E}_{i+\delta,j}^{n+1}} - \overline{\hat{E}_{i-1+\delta,j}^{n+1}} \right] - \frac{\Delta t}{\Delta \eta} \left[\overline{\hat{F}_{i,j+\varepsilon}^{n+1}} - \overline{\hat{F}_{i,j-1+\varepsilon}^{n+1}} \right] \right\} \tag{5}$$

The time step Δt is bounded by the CFL criterion /5/.

Due to a proposal of MacCormack /6/ the four possible choices for approximating the derivatives $\partial\hat{E}/\partial\xi$ and $\partial\hat{F}/\partial\eta$ are cycled through during the course of a calculation (Fig. 2) by changing δ and ε from "1" to "0". This rotation introduces a numerical oscillation with the frequency

$$\nu_n = \frac{1}{4 \ \Delta t} \tag{6}$$

To ensure that the physical phenomena to be computed are not influenced by these effects the frequency ν_B of the blade vibration must have a certain distance from ν_n. In practice one period of blade oscillation will be divided into 200 or more time steps depending on the allowed maximum of Δt so that this condition is satisfied automatically since

$$\nu_n \geq 50 \quad \nu_B \quad . \tag{7}$$

BOUNDARY POINTS

The MacCormack algorithm yields only the solution at inner points. At the boundaries of the computational domain (B1 to B8 in Fig. 3) a special treatment is necessary. Following conditions are assumed:

- At B1 the stagnation values p_t, T_t and the flow angle α are prescribed. Wake influence can be introduced by a circumferential distribution of these values.
- At B2 the static pressure p is assumed known.
- At the blade profiles the flow tangency condition is used, i.e.

  $$\vec{c} * \vec{\xi} = \vec{c}_B * \vec{\xi} \tag{8}$$

  where $\vec{c} = (u,v)^t$ denotes the velocity of flow at the solution point, $\vec{c}_B = (u_B,v_B)^t$ describes the local velocity of the blade and $\vec{\xi}$ is the normal vector to the profile.
- Downstream of the blading, the boundaries B5 and B6 are represented by the slipstream leaving the trailing edge of each blade.
- At B7/B8 a special periodicity condition is used which is explained later on.

To calculate the boundary values the MacCormack scheme is used approximating the unknown derivatives normal to the boundaries by one-sided finite differences. These new values do not satisfy the above mentioned boundary conditions. They are corrected by means of one-dimensional characteristics. A detailed description of this method is given by Veuillot and Meauzé /2/, Fourmaux /1/ and Gallus/Kau /7/. The basic algorithm is modified by a special treatment at the solid wall. As the flow is inviscid a particle running along the wall should not change its entropy. The entropy at the position of the particle at time t can be computed by linear interpolation along the wall itself. This value is combined with the pressure p^{n+1} at the new position, and yields the new density by solving the equation

$$\frac{ds}{dt} = 0 \quad . \qquad (9)$$

PERIODICITY CONDITION (B7, B8)

Solving the Euler-equations on the periodic lines B7/B8 with the MacCormack scheme a special assumption has to be made for the derivatives in circumferential direction ($\partial/\partial\eta$). Computing the steady solution the numerical grid can be closed by a simple periodicity condition using the identity of point P1 and P2 in Fig. 3.
On the other hand, computing unsteady phenomena a special treatment has to be found since time-dependent fluctuations on the "periodic" lines must be taken into account. If one is interested in the real physical phenomena during the transition from the initial values to the periodic state, the computational domain cannot be limited to one channel. It must be extended over as many channels as needed due to a geometric similarity that allows to close the computational domain with the simple periodic condition. If only the periodic state has to be computed a distinction can be made in the treatment of the periodic line depending on the vibration mode.

In travelling wave mode all blades vibrate at the same frequency and amplitude but there exists a circumferential constant phase lag between the blade motion, and therefore between the flow situation in each blade passage and its neighbours. Thus in periodic state the missing derivatives or absolute values on the line B7 (B8) can be found at another time especially at a time just computed during the last period - at line B7 (B8) of the neighbouring channel which is identical to B8 (B7) of the channel itself. Therefore the history of the derivatives or absolute values on these lines are stored over one period and restored to complete the actual MacCormack step.

The basic idea of this special treatment of the boundaries B7/B8 has first been introduced by Erdos and Alzner /8/. During the periodic state it reduces the computational area to one single channel. Its physical background is not valid for the transient from an initial state to the periodic situation. There may occur stability problems. Stability can be improved by introduction of an additional damping coefficient at the periodic lines. It is formulated in dependence of the actual values and the values of the last period. At time t^n there are two values for each point of line B7 (B8), one results from the last time step, the other has been stored from line B8 (B7) for the treatment of the periodicity condition. The damping consists of a simple averaging of these two values and drives the algorithm to the periodic state. It vanishes after the periodic state has been reached. The validity of this formulation has been tested and proved for different interblade phase angles by comparison with calculations using more than one channel.

RESULTS

The validity of the presented method is checked by comparison with several experimental data and with numerical results of other computational methods. The spatial (2d) accuracy can be illustrated by steady state results. Examples are presented in Ref. /7/ and /9/.

The prediction capability of the algorithm for time-dependent results will be shown by comparison with both, numerical methods and experimental data.

COMPARISON WITH LINEARIZED METHOD

Fig. 4 shows a comparison between results of the presented method and the linearized algorithm LINSUB from Smith/Whitehead /10/ for a flat plate cascade in bending mode. As expected for this simple geometry the linearization does not produce great deviations and the agreement between both algorithms is fairly well. For all reduced frequencies a stable motion is predicted with a region of less damping near $k = 1.4$. Fig. 5 presents a comparison for a sinusoidal wake at the upstream boundary showing the computed dynamic pressure and phase on the profile's surface. Results are presented for two cases: A: blade pitch TG identical to wake pitch TD and B: TD = 2 * TG. Both algorithms agree. The results of the Euler code for case B were computed using the described periodicity condition. Therefore these results also validate this special treatment including the damping technique.

To demonstrate the necessity for the development of a non-linearized Euler-method Fig. 6 shows the dynamic force coefficient for a variation of the reduced frequency for the complex geometry of the Fourth Standard Configuration /11/ in bending mode. Because of the wide range of the local Mach number ($Ma = 0.28$ to 1.0), these results of the Euler-solver are compared with LINSUB for the Mach numbers 0.3, 0.6 and 0.9. It can be seen that for this thick profile the linearized program cannot predict the real behaviour of the cascade. Best results are given with a mean Mach number of 0.6 but it remains a factor of 5 compared with the non-linearized computational values.

In addition to the investigations presented above Fig. 7 gives a short survey of the influence of important parameters on the dynamic force coefficient for a cascade in bending mode. The dynamic behaviour is mainly influenced by the vibration frequency, the interblade phase angle and also depends on the Mach number. To study the influence of these parameters in detail comprehensive investigations are planned in the near future for different geometries.

COMPARISON WITH EXPERIMENTAL DATA

For the validation of the presented method a lot of experimental investigations have been made at the Institut für Strahlantriebe und Turboarbeitsmaschinen since the last UAATP-meeting 1987 held in Aachen. Fig. 8 shows a selected result for a compressor cascade in bending mode with an interblade phase angle of $\sigma = 180°$, $\nu = 132$ Hz and an upstream Mach number $M_1 = 0.23$.

The unsteady pressure was measured in 5 positions D1 to D5 on the pressure side and in 6 positions S1 to S6 on the suction side. The comparison with computed results shows a very good agreement in amplitude and phase, except for D1. Unfortunately, the pressure transducer S1 was out of order. The acceleration measured in the experiment shows a second, higher frequency corresponding to the first torsional mode. This frequency occurs also in the pressure signals. The difference at point D1 is caused by the relatively coarse grid.

INFLUENCE OF THE WAKE FORM ON THE DYNAMIC PRESSURE DISTRIBUTION

The incoming wake can be described by different models. Fig. 9 shows the computed dynamic pressure distribution when a sinusoidal wake (in circumferential direction) is prescribed at the upstream boundary. On the pressure side of the profile there is a continuous decrease of the dynamic amplitude while on the suction side behind a short decrease the amplitude increases at the second half of the surface.

To simulate the wake influence of a preceeding cascade a gaussian distribution is better suited describing a more or less sharp wake and a region of undisturbed flow. The computed results (Fig. 10) show a change not only in amplitude but in the general distribution of the dynamic pressure on the surface of the profile. Besides, a Fourier-analysis shows that the influence of higher frequencies has increased due to the non-harmonic parts of the incoming wake.

CONCLUSION

The presented method computes the two-dimensional, unsteady, subsonic flow. The non-linear Euler-equations in conservative form are solved at interior points with MacCormack's predictor-corrector scheme. Boundary conditions are handled with a post correction method using one-dimensional characteristics.
Compared with a linearized method (flat plate) good agreement is achieved for time-dependent geometries as well as for incoming wakes. Results for oscillating cascades show good agreement with experimental data.
The aeroelastic behaviour of a turbomachinery cascade mainly depends on the frequency, the interblade phase angle and the Mach number. Important geometric parameters are camber, thickness and stagger. Computing the wake influence of a preceeding cascade the shape of the incoming wake has great influence on the dynamic pressure.

REFERENCES

/1/ Fourmaux, A.: "Unsteady Flow Calculation in Cascades", ASME-Paper 86-GT-178, 1986

/2/ Veuillot, J.P., Meauzé, G.: "A 3d Euler Method for Internal Transonic Flows Computation with a Multi-Domain Approach", AGARD-LS No. 140, 1985

/3/ MacCormack, R.W.: "The Effect of Viscosity in Hypervelocity Impact Cratering", AIAA-Paper No. 69-354, May 1969

/4/ Thompson, J.F., Thames, F.C., Mastin, C.W.: "Boundary-Fitted Curvilinear Coordinate Systems for Solution of Partial Differential Equations on Fields Containing Any Number of Arbitrary Two-Dimensional Bodies", NASA CR-2729, 1977

/5/ Courant, R., Friedrichs, K., Lewy, H.: "Über die partiellen Differentialgleichungen der mathematischen Physik", Math. Ann., Vol. 100, 1928, pp. 32-74

/6/ MacCormack, R.W.: "Current Status of Numerical Solutions of the Navier Stokes Equations", AIAA-Paper No. 85-0032, Jan. 1985

/7/ Kau, H.-P., Gallus, H.E.: "Numerical Investigation of Unsteady Flow in Oscillating Turbine and Compressor Cascades", AGARD-PEP, Paper 74A-2

/8/ Erdos, J.I., Alzner, E.: "Numerical Solution of Periodic Transonic Flow through a Fan Stage", AIAA-Journal, Vol. 15, No. 11, Nov. 1977

/9/ Servaty, S., Gallus, H.E., Kau, H.-P.: "Computation of Aerodynamic Blade Loads Due to Wake Influence and Aerodynamic Damping of Turbine and Compressor Cascades", presented at the Fourth Symposium on Unsteady Aerodynamics and Aeroelasticity of Turbomachines and Propellers, Aachen, West-Germany, September 6-10, 1987

/10/ Smith, S.N.: "Discrete Frequency Sound Generation in Axial Flow Turbomachines", University Engineering Dept., Cambridge, Report and Memoranda No. 3709, 1972

/11/ Bölcs, A., Fransson, T.H. (eds.): "Aeroelasticity in Turbomachines - Comparison of Theoretical and Experimental Cascade Results", Communication du Laboratoire de Thermique Appliquée et de Turbomachines, No. 13, Lausanne, EPFL 1986

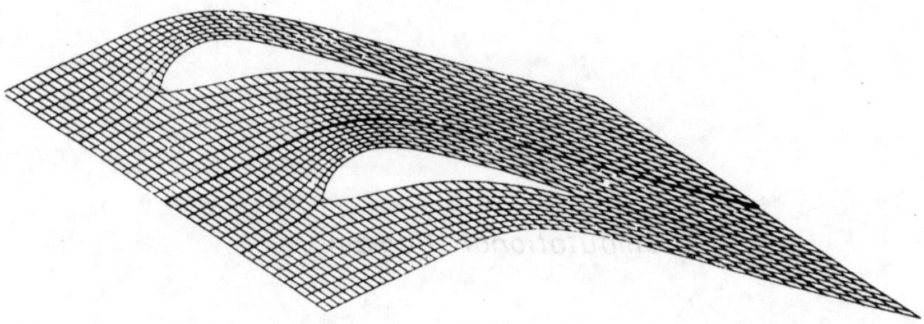

Fig. 1: Typical computational grid for a turbine cascade

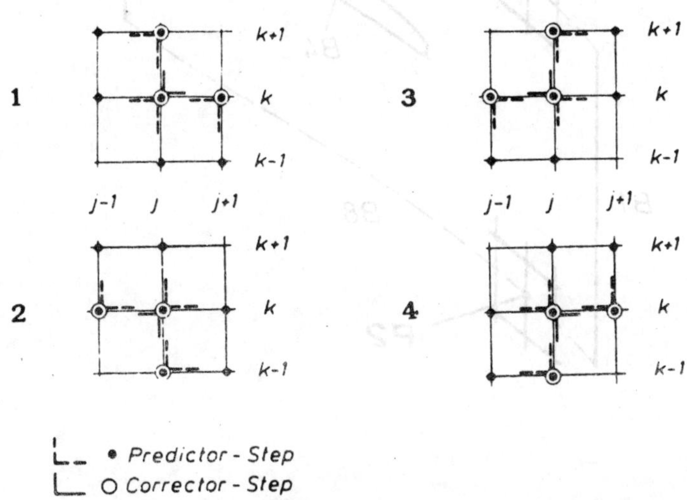

Fig. 2: Discretisation of the derivatives in the MacCormack-scheme

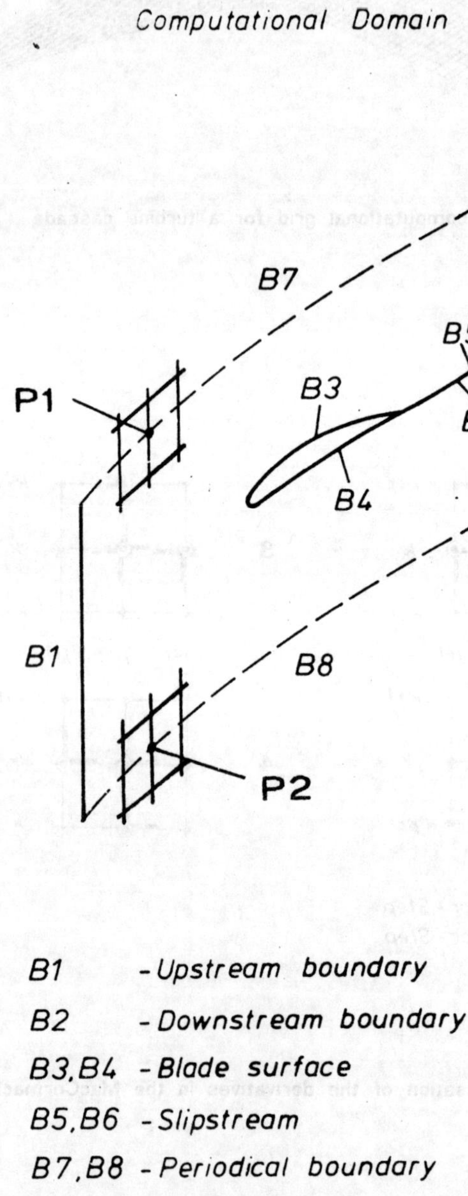

Fig. 3: Location of boundaries

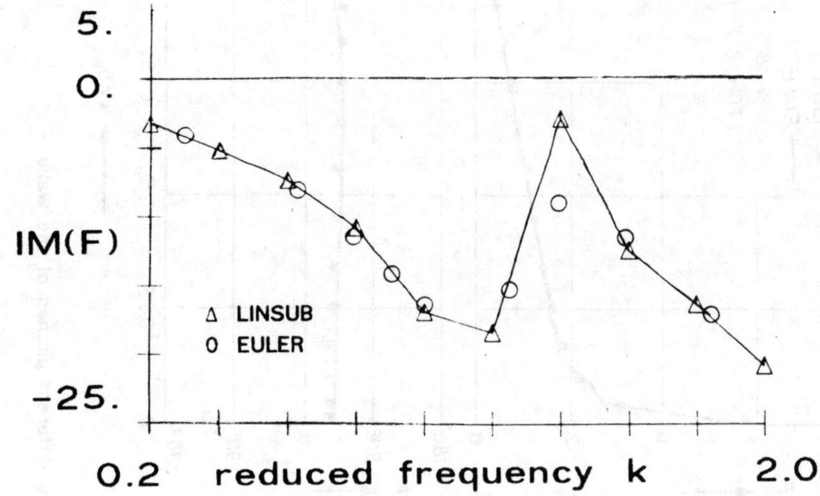

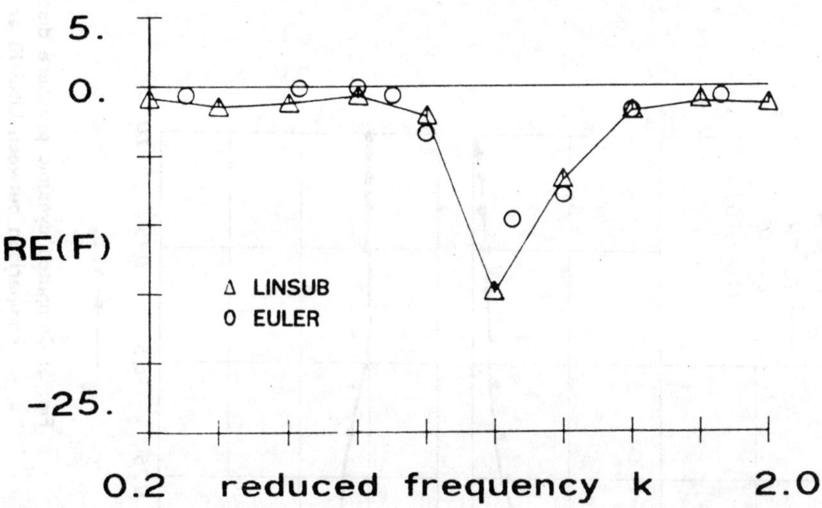

Fig. 4: Real- and imaginary-part of the dynamic force coefficient for a flat plate cascade in bending mode ($s/c = 1.$, $Ma = 0.6$, $\lambda = 30^o$, $\sigma = 180^o$) - comparison between linearized method LINSUB and presented Euler-solver.

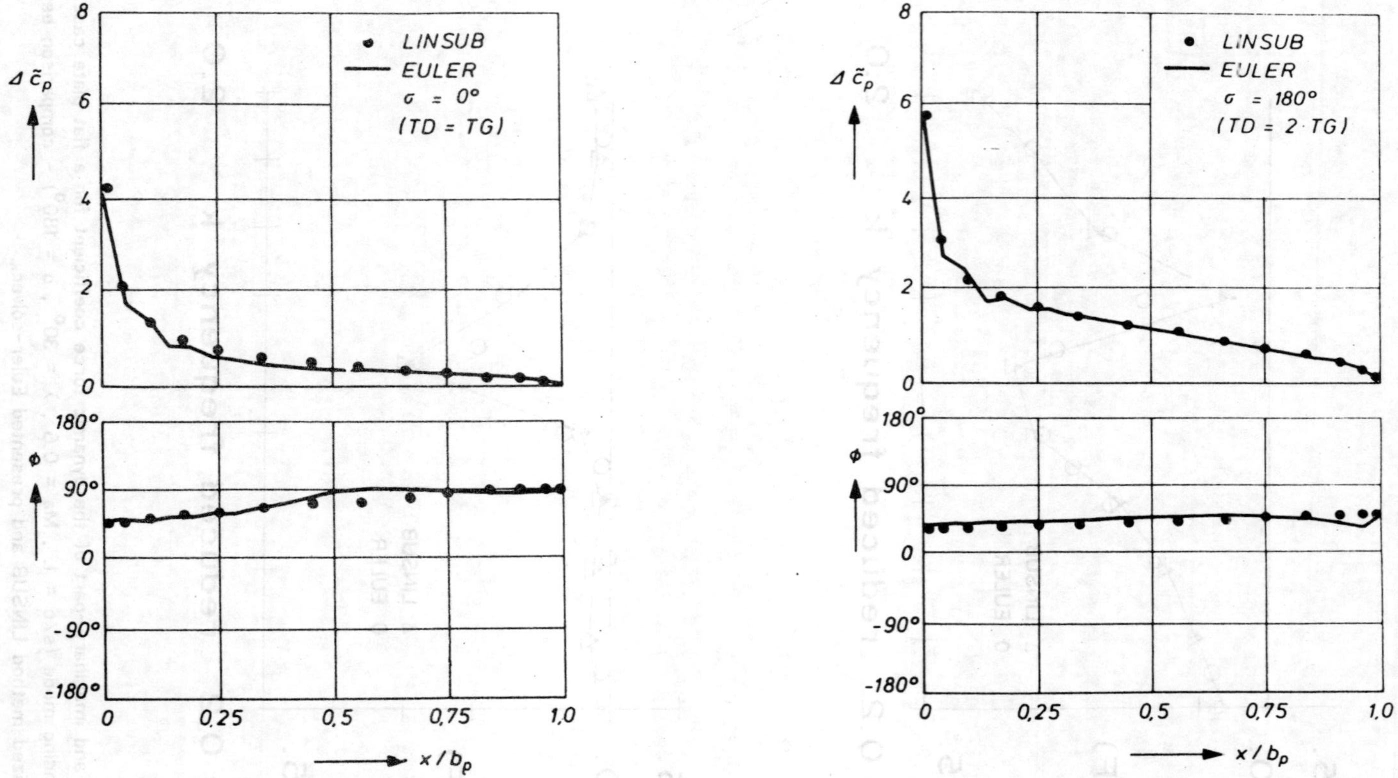

Fig. 5: Computed dynamic pressure distribution for two different pitches of the wake – comparison between LINSUB and EULER

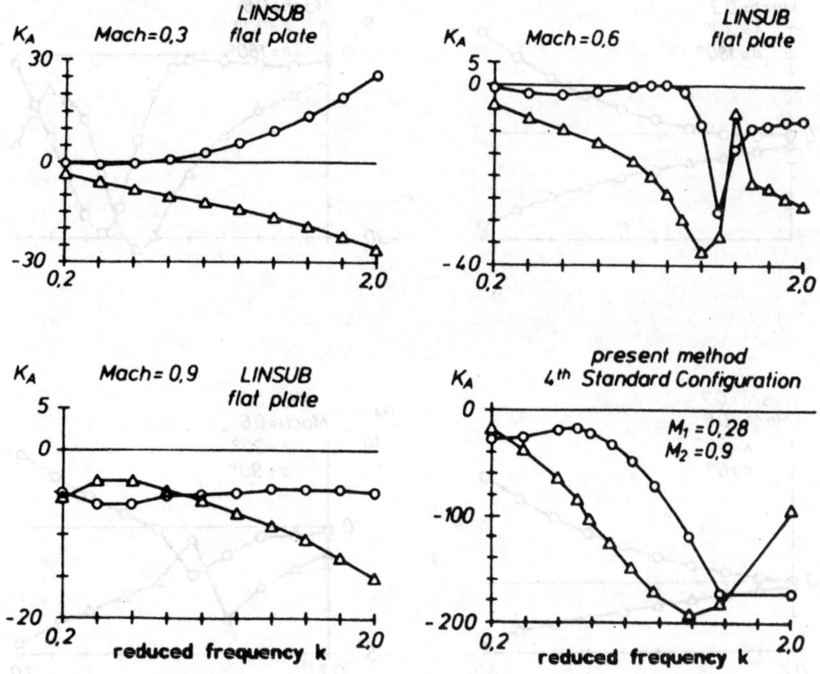

O real part

Δ imaginary part

$$K_A = \frac{\int \tilde{p}(x,t) * (\vec{e}_n * \vec{e}_f) \frac{dx}{c}}{h * (\overline{p}_{t-\infty} - \overline{p}_{-\infty})} \qquad k = \frac{c*\pi*\nu}{2*v_{ref}}$$

flat plate: $\lambda = 60^o$, $s/c = 0.7$, $\sigma = 180^o$

4th Standard Configuration: $\lambda = 56.6^o$, $s/c = 0.76$, $\sigma = 180^o$

Fig. 6: Calculated aerodynamic force coefficient as a function of reduced frequency for the high subsonic "Fourth Standard Configuration" - comparison with the linearized method LINSUB (flat plate)

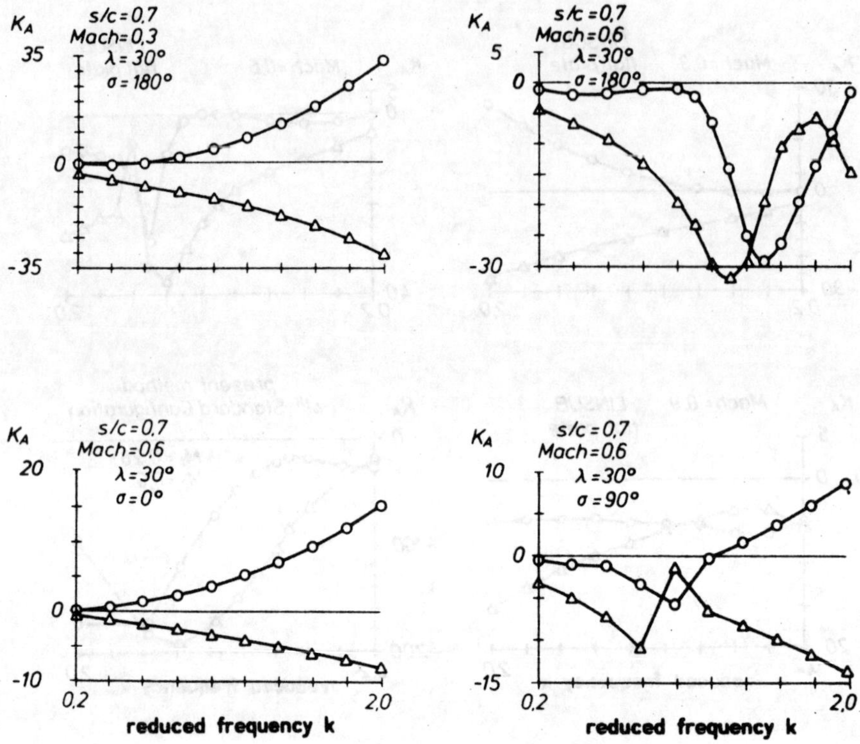

O real part
Δ imaginary part

$$K_A = \frac{\int \tilde{p}(x,t) * (e_n * e_f) \frac{dx}{c}}{h * (\overline{p}_{t-\infty} - \overline{p}_{-\infty})} \qquad k = \frac{c*\pi*\nu}{2*v_{ref}}$$

Fig. 7: Influence of frequency, Mach number and interblade phase angle on the aerodynamic force coefficient of a bending flat plate

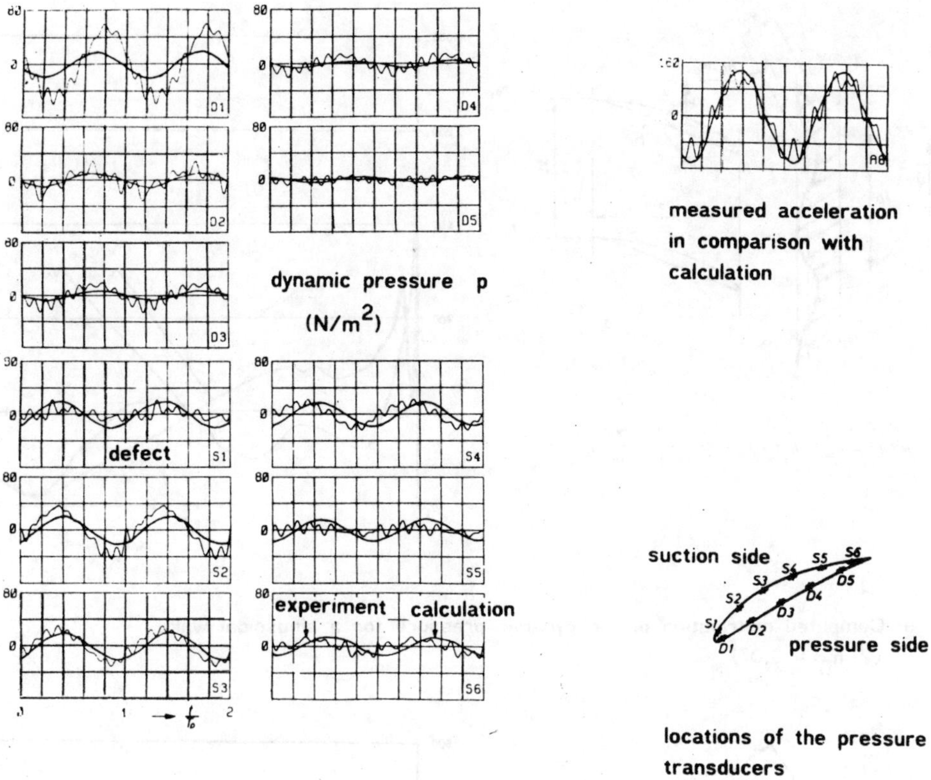

Fig. 8: Dynamic pressure for the compressor cascade bending in TWM; σ = 180°, ν = 132Hz, M_1 = 0.228 - comparison between measured and calculated results

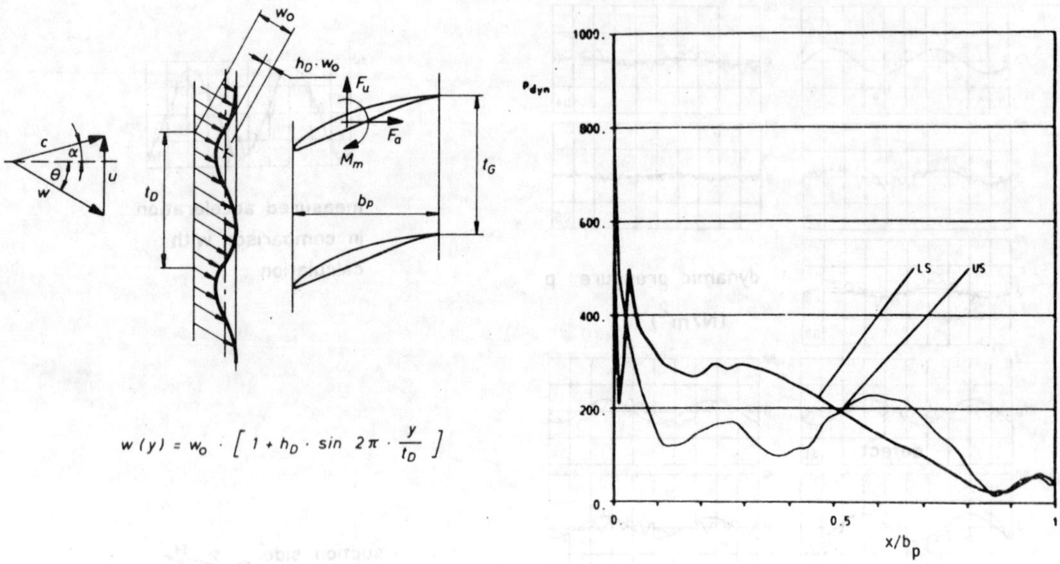

Fig. 9: Computed distribution of the dynamic pressure for a sinusoidal wake
($2*h_D = 0.07$)

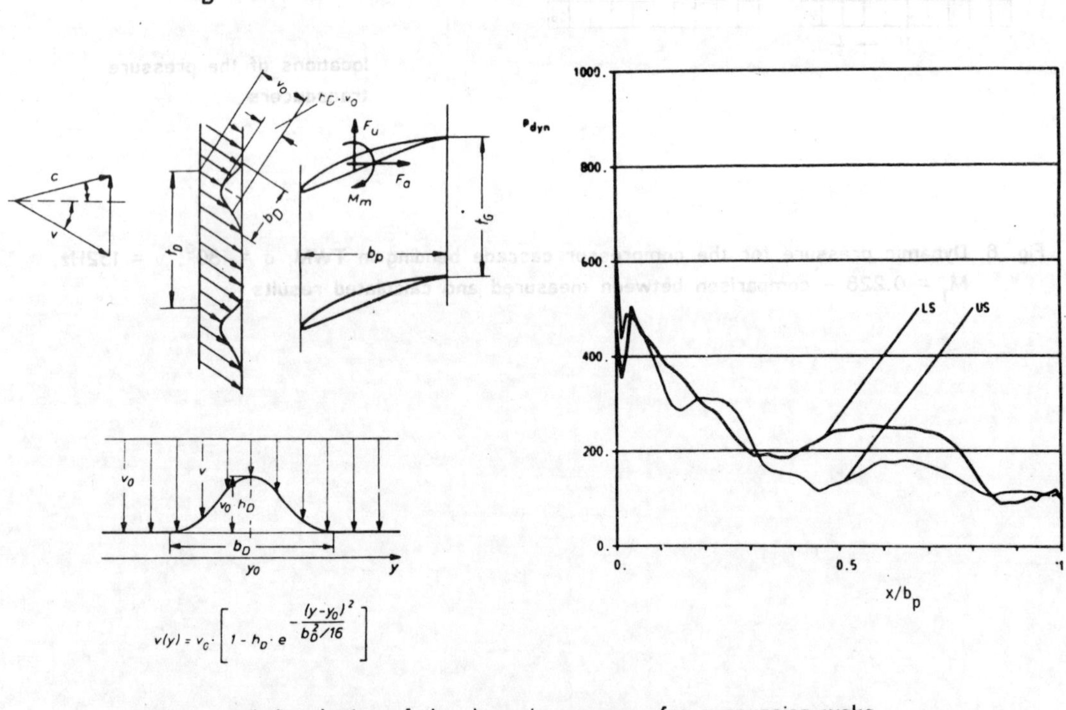

Fig. 10: Computed distribution of the dynamic pressure for a gaussian wake
($h_D = 0.07$, $b_D = 0.4$)

SOLUTION TO OSCILLATING FLOW AROUND A VIBRATING BLADE USING "PARAMETRIC POLYNOMIAL METHOD" AND ITS AEROELASTIC STABILITY ANALYSIS

Chen Zuoyi and Shu Hong

(Tsinghua University, Beijing, China)

ABSTRACT

Formulation and numerical simulation results of oscillating flow around a vibrating blade using "parametric polynomial method" is presented in this paper, including solutions of Euler equation for ideal fluid and Navier–Stokes equation for viscous fluid. As a further application, aeroelasticstability analysis was made for different incidence angles. Corresponding oscillating flow experiments were made to justify the method Agreement of results of experiment and numerical simulation is fair. We believe that the usage of "parametric polynomial method" in solving oscillating flow field around a vibrating blade is straightforward and effective.

INTRODUCTION

The solution to oscillating flow around a vibrating blade is the major problem in aeroelastic stability analysis. Both experimental and numerical approaches are applicable. Due to the complicity involved in measurement of oscillatory flow, especially the difficulty in simulation of the exact operating condition, the experimental approach is not practical.On the contrast, the development of an acurate and simple numerical approach to solve oscillating flow is of great significance. The method by Theodorsen, although widely used and simple in calculation, is limited to flow around flat plate with zero angle of attack. Ref(2) is meant to take into account the effects of incidence angle and aerofoil geometry, but is also limited by irrotational requirement since it solves basically the amplitude eauation of velocity potential. The computation cost is relatively high due to the complicity in treatment of boudary condition.The "parametric polynomial method" presented in this paper is based on the priciple that any flow parameter can be written as polynomial of coordinate variables, and thus the problem of solving a nonlinear governing equation is transformed into a problem of determination of coefficients in a parametric polynomial. Governing equations could be either and Euler equation or a Navier–Stokes equation. According to the priciple of oscillatng fluid mechanics(Ref(3)), the determination of oscillatory flow parameters should be preceded by the determination of steady flow parameters. The "parametric polynomial method" presented here applies to both steady and oscillatory flow field.

The above method has been applied to wind turbine blade calculation under various operating conditions. The test cases include 9 different incidence angles of −20°, −15°, −10°, −5°, 0°, 10°,15°, 20°, different oscillating frequencies and upstream flow velocities. Corresponding measurement of unsteady pressure distribution on the surface was made and work done by aerodynamic force was calculated for same flow conditions to justify the method. Experimental details can be found in Ref(4). The agreement of results of experiment and numerical simulation is fair.

GOVERNING EQUATIONS FOR "PARAMETRIC POLYNOMIAL METHOD"

Navier–Stokes equations for non–uniform ν can be written as following:

$$u\frac{\partial u}{\partial x}+v\frac{\partial u}{\partial y}=-\frac{1}{\rho}\frac{\partial p}{\partial x}+\nu\left[\frac{\partial^2 u}{\partial x^2}+\frac{\partial^2 u}{\partial y^2}\right]+\frac{\partial \nu}{\partial y}\left[\frac{\partial u}{\partial y}+\frac{\partial v}{\partial x}\right]+2\frac{\partial \nu}{\partial x}\frac{\partial u}{\partial x} \tag{1}$$

$$u\frac{\partial v}{\partial x}+v\frac{\partial v}{\partial y}=-\frac{1}{\rho}\frac{\partial p}{\partial y}+\nu\left[\frac{\partial^2 v}{\partial x^2}+\frac{\partial^2 v}{\partial y^2}\right]+\frac{\partial \nu}{\partial x}\left[\frac{\partial u}{\partial y}+\frac{\partial v}{\partial x}\right]+2\frac{\partial \nu}{\partial y}\frac{\partial v}{\partial y} \tag{2}$$

the transformation for a body fitted coordinates ζ and η has the general form:

$$\frac{\partial}{\partial x}=\frac{1}{J}\left[y_\eta\frac{\partial}{\partial \zeta}-y_\zeta\frac{\partial}{\partial \eta}\right] \tag{3}$$

$$\frac{\partial}{\partial y}=\frac{1}{J}\left[-x_\eta\frac{\partial}{\partial \zeta}+x_\zeta\frac{\partial}{\partial \eta}\right] \tag{4}$$

$$\frac{\partial^2}{\partial x^2}=\left[\frac{1}{J^2}y_\eta y_{\eta\zeta}-\frac{1}{J^2}y_\zeta y_{\eta\eta}+\frac{1}{J}y_\eta^2\frac{\partial(\frac{1}{J})}{\partial \zeta}-\frac{1}{J}y_\zeta y_\eta\frac{\partial(\frac{1}{J})}{\partial \eta}\right]\frac{\partial}{\partial \zeta}+\left[\frac{1}{J^2}y_\zeta y_{\zeta\eta}-\frac{1}{J^2}y_\eta y_{\zeta\zeta}\right.$$
$$\left.+\frac{1}{J}y_\zeta^2\frac{\partial(\frac{1}{J})}{\partial \eta}-\frac{1}{J}y_\zeta y_\eta\frac{\partial(\frac{1}{J})}{\partial \zeta}\right]\frac{\partial}{\partial \eta}+\frac{1}{J^2}y_\eta^2\frac{\partial^2}{\partial \zeta^2}+\frac{1}{J^2}y_\zeta^2\frac{\partial^2}{\partial \eta^2}-\frac{2}{J^2}y_\eta y_\zeta\frac{\partial^2}{\partial \zeta\partial \eta} \tag{5}$$

$$\frac{\partial^2}{\partial y^2}=\left[\frac{1}{J^2}x_\eta x_{\eta\zeta}-\frac{1}{J^2}x_\zeta x_{\eta\eta}+\frac{1}{J}x_\eta^2\frac{\partial(\frac{1}{J})}{\partial \zeta}-\frac{1}{J}x_\eta x_\zeta\frac{\partial(\frac{1}{J})}{\partial \eta}\right]\frac{\partial}{\partial \zeta}+\left[\frac{1}{J^2}x_\zeta x_{\zeta\eta}-\frac{1}{J^2}x_\eta x_{\zeta\zeta}\right.$$
$$\left.+\frac{1}{J}x_\zeta^2\frac{\partial(\frac{1}{J})}{\partial \eta}-\frac{1}{J}x_\zeta x_\eta\frac{\partial(\frac{1}{J})}{\partial \zeta}\right]\frac{\partial}{\partial \eta}+\frac{1}{J^2}x_\eta^2\frac{\partial^2}{\partial \zeta^2}+\frac{1}{J^2}x_\zeta^2\frac{\partial^2}{\partial \eta^2}-\frac{2}{J^2}x_\eta x_\zeta\frac{\partial^2}{\partial \zeta\partial \eta} \tag{6}$$

Substitute (3), (4), (5), (6) into (1) and (2), result in Navier–Stokes equations in body fitted coordinates. Flow parameters can be written in terms of ζ and η (as shown in Figure 1):

$$u=A_u\eta^3+B_u\eta^2+C_u\eta+D_u \tag{7}$$

$$v = A_v\eta^3 + B_v\eta^2 + C_v\eta + D_v \tag{8}$$

$$p = A_p\eta^6 + B_p\eta^5 + C_p\eta^4 + D_p\eta^3 + E_p\eta^2 + F_p\eta + G_p \tag{9}$$

$$\nu = A_\nu\eta^3 + B_\nu\eta^2 + C_\nu\eta + D_\nu \tag{10}$$

where coefficients Au, Bu, ···, C_ν, D_ν are functions of ζ. Substitute polynomials (7), (8), (9) and (10) into governing equations in ζ and η, yield equations consisting of all these coefficients and their partial differentials with respect to s . The procedure could be simplified if one solves for partial differentials of $\partial Au / \partial\zeta$, $\partial Av / \partial\zeta$, ... etc.with respect to ζ. We further drop out the terms of second order derivatives with respect to ζ (parabolic assumption). The parameters can be easily found from $\partial / \partial\zeta$ terms at the previous calculation nodes. Equations are discreterized at sufficient number of nodal points at cross–section of constant ζ such that number of equations matches the number of unknown coefficients.

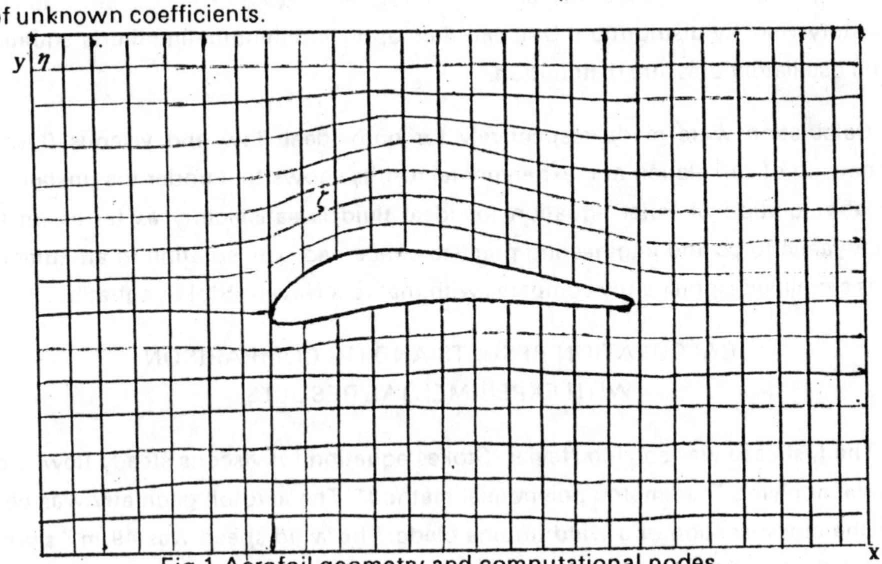

Fig.1 Aerofoil geometry and computational nodes

Boundary conditions are also included.

In the calculation, we neglect the variation of viscosity and combine the momentum equation to eliminate pressure term. In so doing, we reduce the number of unknown coefficients in velocity polynomial, and make it possible to integrate momentum and continuity equations along surface of constant ζ.

The order of velocity polynomial depends on the flow characteristics. The choice of order three is necessary for stalled flow or flow with recirculation zone to guarantee an inflection point. Once the steady flow field is known, we can determine the oscillating flow applying the same "parametric polynomial method". The Navier–Stokes amplitude equation is as following:

$$iw\bar{u} + \bar{u}\frac{\partial \bar{u}}{\partial x} + \bar{u}\frac{\partial \bar{u}}{\partial x} + \bar{v}\frac{\partial \bar{u}}{\partial y} + \bar{v}\frac{\partial \bar{u}}{\partial y} = -\frac{1}{\zeta}\frac{\partial \bar{p}}{\partial x} + \nu[\frac{\partial^2 \bar{u}}{\partial x^2} + \frac{\partial^2 \bar{u}}{\partial y^2}] \quad (11)$$

$$iw\bar{v} + \bar{u}\frac{\partial \bar{v}}{\partial x} + \bar{u}\frac{\partial \bar{v}}{\partial x} + \bar{v}\frac{\partial \bar{v}}{\partial y} + \bar{v}\frac{\partial \bar{v}}{\partial y} = -\frac{1}{\zeta}\frac{\partial \bar{p}}{\partial y} + \nu[\frac{\partial^2 \bar{v}}{\partial x^2} + \frac{\partial^2 \bar{v}}{\partial y^2}] \quad (12)$$

According to the oscillating velocity distribution, its amplitude can be written in polynomial expression:

$$\bar{u} = \bar{A}u\eta^2 + \bar{B}u\eta + \bar{C}u \quad (13)$$

$$\bar{v} = \bar{A}v\eta^2 + Cv \quad (14)$$

Similarly, substitute (13) and (14) into transformed momentum and continuity amplitude equations, and solve for $\partial\bar{A}u / \partial\zeta, \partial\bar{B}u / \partial\zeta$, ..., etc . to obtain oscillatory velocity distribution. One can then apply momentum amplitude equation to obtain oscillatory pressure distribution.

Calculation were made respectively for both ideal flow and viscous flow. The comparison of calculation and experimental results shows that under the unstall condition, the solution of Euler equation for ideal fluid is satisfactory as far as unsteady aerodynamic force and engineering practice concerned. The solution to an Euler equation is simplified significantly compares with that to a Navier–Stokes equation.

CALCULATION RESULTS AND ITS COMPARISON WITH EXPERIMENTAL RESULTS

The first case was solving Navier–Stokes equation for viscous steady flow around a blade, applying "parametric polynomial method". The aerofoil geometry was chosen from the cross–section of a wind turbine blade. The wind speed was 49 m / s with incidence angles of 9°and 29°respectively. Calculated surface pressure distributions can be found in Figure 2 and 3. Experiment was conducted for same aerofoil aerofoil geometry.

The second case was solving the oscillatory flow around an oscillating blade applying "parametric polynomial method". The wind speed ws 55 m / s.Oscillating frequency was 2Hz. Calculations and experiments were made for 9 different incidence angles. Figure 4, 5 and 6 illustrate the oscillatory pressure amplitude on the blade surface for incidence angles of −10°, 0°and 20°respectively. Figure 7 shows the work done by unsteady aerodynamic force with various incidence angles. The comparison of calculationand experimental results are fair.

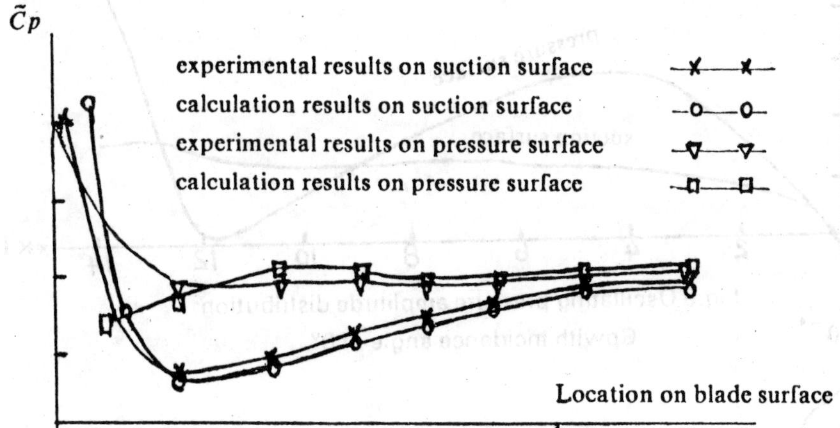

Fig.2 Steady pressure distribution $\tilde{C}p(=\tilde{p}/\frac{1}{2}\rho U_{\infty}^{2})$ with incidence angle of 9°.

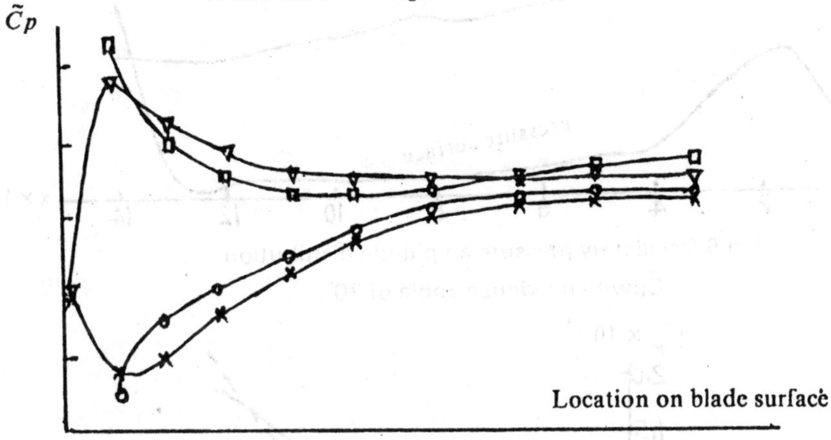

Fig.3 Steady pressure distribution with incidence angle of 29°.

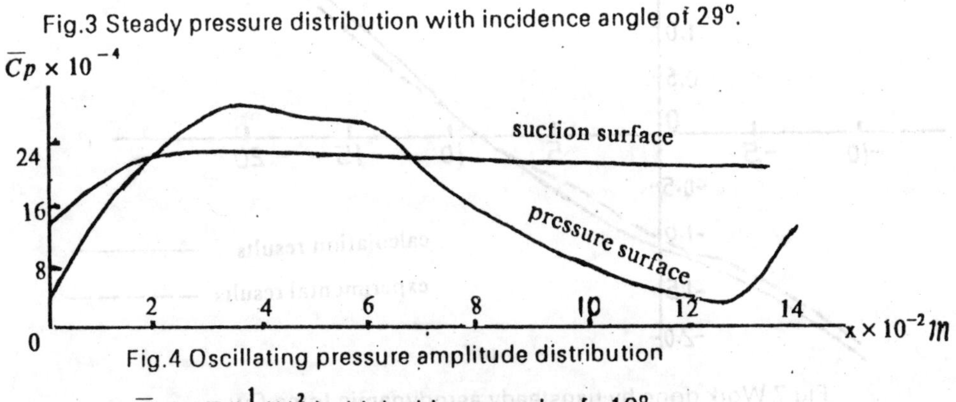

Fig.4 Oscillating pressure amplitude distribution $\overline{C}p(=\overline{p}/\frac{1}{2}\zeta U_{\infty}^{2})$ with incidence angle of $-10°$.

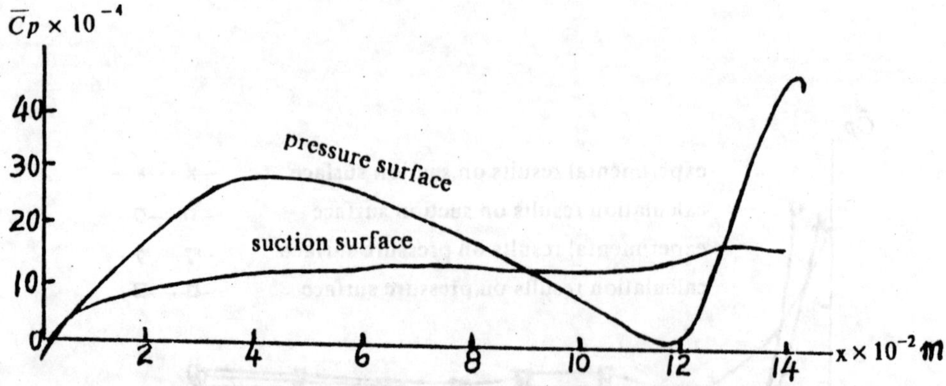

Fig.5 Oscillating pressure amplitude distribution
$\overline{C}p$with incidence angle of 0°.

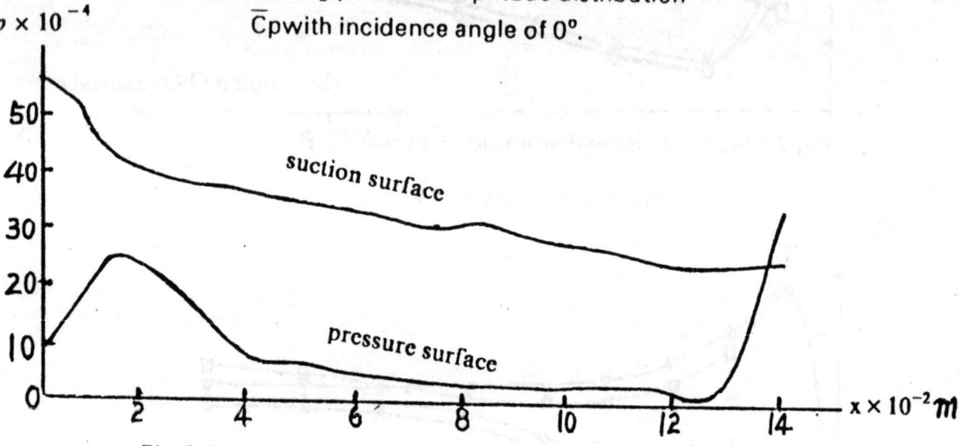

Fig.6 Oscillatory pressure amplitude distribution
$\overline{C}p$with incidence angle of 20°.

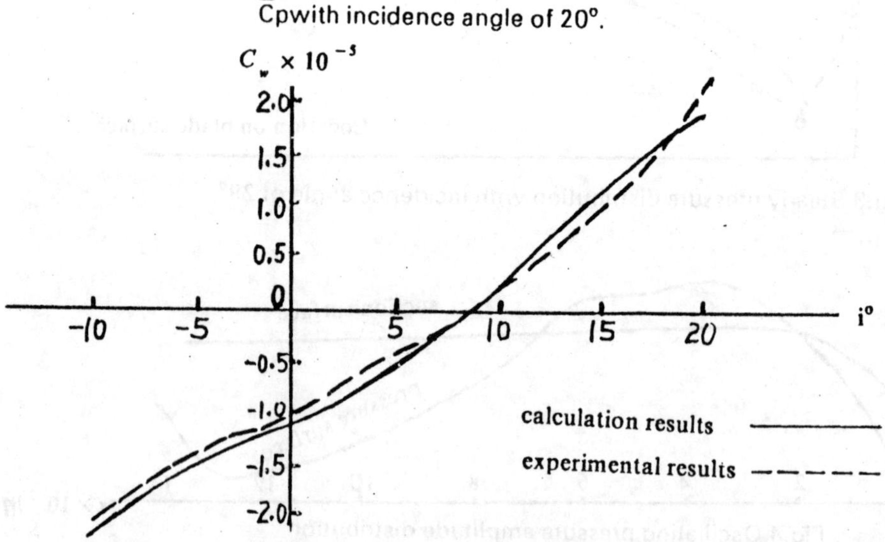

Fig.7 Work done by usnsteady aerodynamic force Cw
for different incidence angles

REFERENCES

1. Theodorsen T., "General Theory of Aerodynamic Instability and the Mechanismof Flutter", NACA Report, 496, 1935.

2. Chen Zuoyi and Wen Feng, "Two Dimensional Unsteady Aerodynamic Forceon an oscillating Blade of Wind Turbine", ACTA Aerodynamic Sinica, Vol. 5, No.1, 1987.

3. Chen Zuoyi, "The Oscillating Fluid mechanics", Hydroelectric Publishing House, 1988

4. Zhong Shan, "The Research on Viscous Oscillating Flow Field of Vibrating Blade", Postgraduate paper, Department of Thermal Engineering, Tsinghua University, Beijing, China, 1988.

VARIATIONAL FORMULATION OF 2-D UNSTEADY TRANSONIC AERODYNAMICS OF OSCILLATING CASCADES

Gao-Lian liu
(Shanghai Institute of Mech. Engrg, Shanghai 200093, China)

ABSTRACT

A family of variational principles for 2-D unsteady transonic flow with shocks around oscillating cascades is presented, in which the theory of Ref. [3] is improved in two aspects. Firstly, the non-reflecting inlet/outlet boundary conditions are incorporated. Secondly, a new unsteady Kutta condition [11] is employed. This theory serves as a sound theoretical basis for the finite element analysis of unsteady flow and is valid for perfect gas as well as for ideal steam [15].

INTRODUCTION

In recent years there has been an ever-increasing interest in the research on unsteady flow in turbomachines, since both the destructive self-excited (flutter) and forced (responce) vibrations of blades are directly related to unsteady flow phenomena [1-7]. On the other hand, the unsteady transonic cascade flow problem is extremely complex and very difficult to solve. There remains much work to be done in order to understand and compute it reliably.

In Ref.[3] this problem has been handled by formulating it in a variational principle (VP) family, in which most of boundary conditions (BC) and matching conditions across unknown oscillating shocks and free trailing vortex sheets have been converted into natural ones via functional variations with variable domain. Ref.[3], however, still involves some deficiencies regarding the inlet/outlet BC and Kutta condition. It is the purpose of this paper to suggest two major improvements on Ref.[3]. Firstly, to describe properly the physical process of wave propagation, the non-reflecting (radiation) inlet/outlet BC are derived [6-9] and incorporated in the VPs. Secondly, a new unsteady Kutta condition suggested recently in Ref.[11] is employed.

BASIC EQUATIONS FOR 2-D UNSTEADY TRANSONIC CASCADE FLOW

The dimensionless governing equations of this problem are [3,6]

$$\frac{\partial\rho}{\partial t} + \nabla\cdot(\rho\vec{\Lambda}) = 0 \tag{1}$$

$$\nabla\Phi = \vec{\Lambda} \tag{2}$$

$$p/\rho + (r-1)\left(\frac{\partial\Phi}{\partial t}\Lambda^2/2\right) = 1 \tag{3}$$

$$p = \rho^r \tag{4}$$

Set the cascade oscillate with a period T and with an interblade phase angle σ, but otherwise arbitrarily. The time delay ΔT between two adjacent blades is related to σ by $\Delta T = T\sigma/2\pi$. To make this unsteady flow behavior in cascade more perceptible, it is expedient, following Ref.[3], to depict it in a 3-D space R^3 directly (Fig.1). This figure may be thought of intentionally as an imaginary steady flow past an imaginary cascade of twisted blades at rest in the 3-D space xyz.

1.Boundary Conditions (BC)

In the region upstream of the cascade two computational boundary surfaces A_{u-} & A_{u+}(Fig.2) are generated by straight lines emanating from the airfoil leading edges (e.g.31, 3'1', 45, 4'2) with a fixed (assigned) angle α . Obviously, the shape of $A_{u\pm}$ and any flow parameter X on them are periodic both in time and in the y-direction with an interblade time delay ΔT , so we have the following periodic BC on $A_{u\pm}$:

$$X(x, y \pm j \cdot \Delta Y, t \pm j \cdot \Delta T \mp kT) = X(x,y,t) \qquad (j,k = 0,1,2,\cdots) \tag{5}$$

If the equations of $A_{u\pm}$ are $y = Y_{\pm}(x,t)$, we can write the periodic BC on them:

$$Y(x,t) = Y(x, t \pm j \cdot \Delta T \mp kT) \mp j \cdot \Delta Y \qquad (j,k = 0,1,2,\cdots) \tag{6}$$

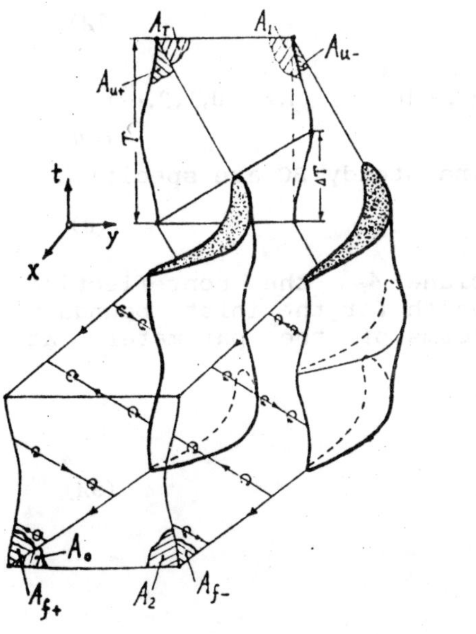

Fig.1 Unsteady flow past 2-D oscillating cascade

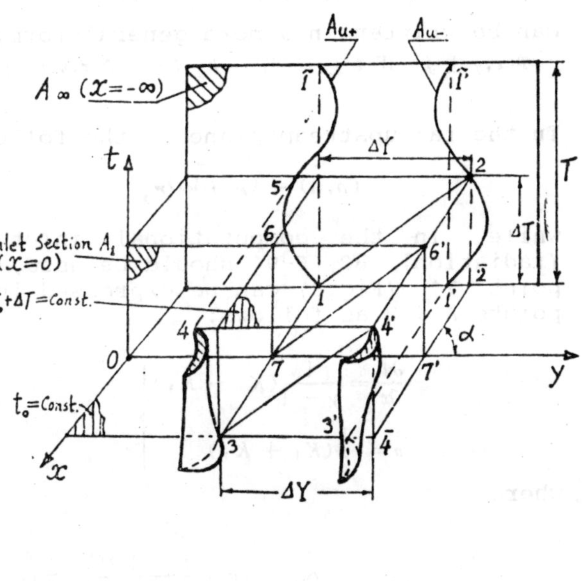

Fig.2 Upstream region of a 2-D oscillating cascade

The flow at upstream infinity (plane A_∞ at $x=-\infty$,Fig.2) can always be assumed to be steady [5]. This can be justified in the following way. At first we consider a steady 2-D cascade flow. Then, let the cascade start to vibrate periodically, the resulting flow disturbances will propagate outwards progressively in different manner, depending on whether the x-component of the far upstream Mach number $M_{\infty,x}$ is greater than 1 or not. If $M_{\infty,x}>1$, the flow upstream of the cascade will never be disturbed and remains to be steady. If, however, $M_{\infty,x}<1$, the disturbances wil propagate upstream and go out through the computational inlet plane A_1 (at $x=0$, Fig.2) without reflection. After a sufficient long but finite period of time t_a (though theoretically $t_a\rightarrow\infty$) a stable periodic unsteady flow will be established at least downstream of A_1. Since t_a is practically finite, the disturbances will never reach the upstream infinity A_∞, therefore in computational practice we can always assume the flow at A_∞ to be steady, so$(\partial\Phi/\partial t)_\infty=0$. Consequently, the potential values at points 2 and $\bar{2}$ (Fig.2) are equal: $\Phi_2=\Phi_{\bar{2}}$. Noting that

$$\Phi_{\bar{2}} - \Phi_1 = \Lambda_{\infty y} \cdot \Delta Y$$

yields

$$\Phi_2 = \Phi_1 + \Lambda_{\infty y} \cdot \Delta Y.$$

Because of the interblade time delay ΔT, all flow parameters along 24′ & 13 must be identical, so $\Phi_{6'} - \Phi_2 = \Phi_7 - \Phi_1$, and we get

$$\Phi_{6'} - \Phi_7 = \Lambda_{\infty y} \cdot \Delta Y,$$

which, together with the periodicity condition

$$\Phi(x,y,t \pm kT) = \Phi(x,y,t), \tag{7A}$$

can be written in a more general form;

$$\Phi(x,y \pm j \cdot \Delta Y, t \pm j \cdot \Delta T \mp k \cdot T) = \Phi(x,y,t) \pm j\Lambda_{\infty y} \cdot \Delta Y, \qquad (j,k = 0,1,2,\cdots) \tag{7B}$$

In the far upstream plane the following steady BC are specified

$$(\rho, \vec{\Lambda}) = (\rho_\infty, \vec{\Lambda}_\infty)_r, \tag{8}$$

while in the computational inlet plane A_∞ the nonreflecting (radiation) BC [6-9] should be used, which for the inlet boundary point 1′ (Fig.3) can be expressed in terms of the parameters at points A & B as follows:

$$\left.\begin{aligned} \frac{\partial\Phi}{\partial x} &= \frac{1}{\gamma-1}(K_{\text{I}} - K_{\text{II}}) \\ a &= \frac{1}{2}(K_{\text{I}} + K_{\text{II}}) \end{aligned}\right\} \tag{9A}$$

where

$$\left.\begin{aligned} K_{\text{I}} &= \left(a + \frac{\gamma-1}{2}\frac{\partial\Phi}{\partial x}\right)_A \cong \left(a + \frac{\gamma-1}{2}\Lambda_{\infty x}\right)_\infty \\ K_{\text{II}} &= \left(a - \frac{\gamma-1}{2}\frac{\partial\Phi}{\partial x}\right)_B \end{aligned}\right\} \tag{9B}$$

In Eq. (9B) the sound speed should be calculated from

$$a^2 = p/\rho = 1 - (\gamma - 1)\left\{\frac{\partial \Phi}{\partial t} + \frac{1}{2}(\nabla \Phi)^2\right\} \tag{9C}$$

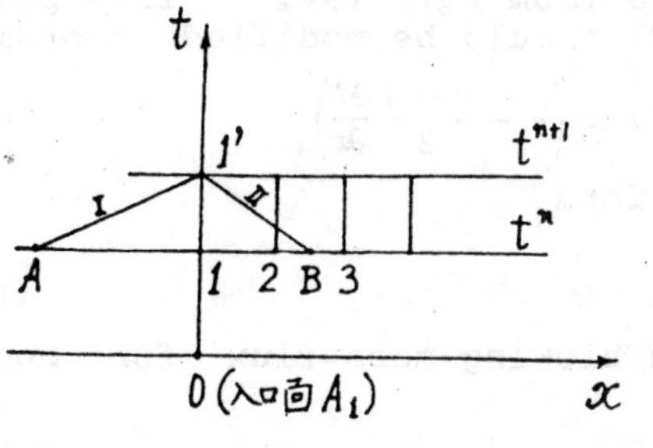

Fig.3 For nonreflecting BC at boundary point 1'

In the downstream region the free trailing vortex sheets $A_{f\pm}$ (Fig.1) are taken to be the periodic boundaries, on which, apart from the geometric periodic conditions Eq. (6), the following matching conditions must also be satisfied

$$p_-(x, t + \Delta T) = p_+(x, t) \qquad A_n = U_f \tag{10}$$

The last equation clearly shows that no fluid passes through the free vortex sheets.
Moreover, at the trailing edge the unsteady Kutta condition must be imposed. The classical unsteady Kutta condition, $p_s = p_p$, used in Ref.[3] and widely in the literature, is questioned by some investigators in recent years [11-13] due to its clear violation at high reduced frequencies and should be replaced by the Giesing's model [11]: As long as no seperation upstream of the trailing edge (TE), the rear stagnation point still coinsides with the TE all the time. Accordingly, following Ref.[4], in taking the variable-domain variation of vortex sheets A we fix the rear stagnation point just at the TE by simply specifying the position vector $\vec{r}_{TE}$ =given, so that

$$(\delta r_{w})_{TE} = 0\,, \tag{11}$$

which is the mathematical (variational) formulation of the Giesing's unsteady Kutta condition. In this way, the unsteady Kutta condition Eq. (11) can be handled in the present variable-domain theory even much more simply than the classical one, which can only be satisfied by successively adjusting the stagnation point.

In the far (but finite distance) downstream outlet plane A_2 (Fig.1), obviously, no BC should be imposed when $M_{2x} > 1$; if, however, $M_{2x} < 1$ the BC can be dealt with in two ways:
a) The pressure at outlet A_2 is specified: Then the A_{2x} at A_2 can be calculated from the relation of the right-running characteristics:

$$\left(\frac{\partial \Phi}{\partial x}\right)_2 = \frac{2}{\gamma - 1}(K_1 - a_2) = \frac{2}{\gamma - 1}(K_1 - p_2^{(\gamma - 1)/2)}) \tag{12A}$$

b) Treated similarly to the inlet plane A_1: Assuming the steady outflow parameters at downstream infinity A'_∞ ($x=+\infty$ Fig.1) to be known, the Λ_{2x} at A_2 can be computed from Eq. (9A). If A_2-plane is placed at $x=0$ in Fig.3, the Eq. (9B) should be modified accordingly:

$$K_1 = \left(a + \frac{\gamma-1}{2}\frac{\partial\Phi}{\partial x}\right)_A, \qquad K_{\mathrm{I}} = \left(a - \frac{\gamma-1}{2}\frac{\partial\Phi}{\partial x}\right)_{+\infty} \tag{12B}$$

The BC on blade surface A_4 has the form

$$\rho(\Lambda_n - U_w) = (q_n)_w, \tag{13}$$

where q_n is the specified suction/blowing mass flux for boundary layer control and/or cooling.

On the unknown shocks A_s the following Rankine-Hugoniot relations [3] should be met:

$$\left.\begin{aligned} &[\rho(\Lambda_n - U_s)] = 0, \\ &[p]/\gamma + \rho(\Lambda_n - U_s)[\Lambda_n] = 0 \end{aligned}\right\} \tag{14}$$

$$[\Lambda_\tau] = 0, \tag{15}$$

$$[H] = (\gamma-1)U_s[\Lambda_n] \tag{16}$$

2. Initial/Final Conditions on A_o^T
From Eq.(5) we have for any flow parameters X on A_o^T :

$$X(x,y,t \pm jT) = X(x,y,t), \qquad (j = 0,1,2,\cdots) \tag{17}$$

and from Eq.(7A)

$$\Phi(x,y,o) = \Phi(x,y,T) \tag{18}$$

VARIATIONAL PRINCIPLES(VPs)

Starting from the above formulation of the unsteady cascade flow problem and following a systematic approach to the derivation and transformation of VPs[16], the following general results can be obtained.

1. Generalized VP I: The solution to the above unsteady transonic flow problem of oscillating cascades makes the following functional J_1 stationary: $\delta J_1 = 0$ with independent variation of Φ, $\vec{\Lambda}$, p, ρ, A_f and A_s ,

$$J_1(\Phi,\vec{\Lambda},p,\rho,A_f,A_s) = \iiint_{(V)} \left\{\rho\left(\frac{\Lambda^2}{2} - \frac{\partial\Phi}{\partial t} - \vec{\Lambda}\cdot\nabla\Phi\right) - \frac{p}{\gamma(\gamma-1)}\left[1 - \ln\left(\frac{p}{\rho^\gamma}\right)\right] + \frac{\rho}{\gamma-1}\right\}dv + L \tag{19}$$

where

$$L = - \iint\limits_{(A_1+A_2)} \left(\rho \frac{\partial \Phi}{\partial x}\right)_{pr} \Phi \cdot dA + \iint\limits_{(A_4)} \frac{q_*}{\sqrt{1+u_{wpr}^2}} \Phi \cdot dA \ ,$$

$$dV = dx \cdot dy \cdot dt \ .$$

By means of the general formula for the functional variation with variable domain [3,16] and using the new Kutta condition Eq.(11) as an essential BC, we can derive from $\delta J_1 = 0$ the following stationary conditions:

Euler's equations: Eqs(1)-(4).

Natural BC:

on A_1 & A_2 : $\rho \Lambda_* = \rho_{pr}(K_{\rm I} - K_{\rm II})/(\gamma - 1)$, [equivalent to Eq.(9A)]

where $\rho_{pr}^{\gamma-1} = a^2 = (K_{\rm I} + K_{\rm II})/4$.

on $A_{*\pm}$: Using Eq.(7B) as essential BC leads to the natural BC

$$[\rho(U_* - \Lambda_*)]_+(x,t) = [\rho(U_* - \Lambda_*)]_-(x,t,\Delta T),$$

which, together with $A_{*+}(x,t) = A_{*-}(x,t+\Delta T)$ and $U_{*+}(x,t) = U_{*-}(x,t+\Delta T)$ derived from Eqs(7B) & (6) respectively, results in the natural BC: $\rho_+(x,t) = \rho_-(x,t,\Delta T)$.

on A_4 : Eq.(13).

on $A_{f\pm}$: Using essential condition $(\delta r_* \cdot)_-(x,t,\Delta T) = (\delta r_* \cdot)_+(x,t)$ leads to the generalized Weierstrass-Erdmann's corner condition Eq.(10).

on A_s : Imposing $\Phi_- = \Phi_+$ and $(\delta r_* \cdot)_- = (\delta r_* \cdot)_+$ as essential conditions, we get the generalized Weierstrass-Erdmann's corner condition Eq.(14). Eq(15) & (16) can be obtained from $\Phi_- = \Phi_+$ and Eq.(3) respectively (vide Ref.[3] for details).

on A_o^T : The BC can be handled in two ways:

a) Imposing Eq.(18) as essential condition yields the natural periodic condition

$$\rho(x,y,o) = \rho(x,y,T) \ , \qquad (20A)$$

which, together with Eqs(3), (4) & (18), leads to

$$X(x,y,o) = X(x,y,T) \ , \qquad (20B)$$

where X stands for $p, \vec{\Lambda}$ & $\partial\Phi/\partial t$.

b) Converting Eq.(18) into a natural condition via augmenting by the following boundary term:

$$\iint\limits_{(A_o)} [\Phi - \Phi(x,y,T)]\rho \cdot dA \ . \qquad (21)$$

This will lead to the natural initial/final conditions Eqs (20A), (18) and (20B).

So far we have shown that this generalized VP really provides us with the solution to the unsteady transonic flow problem under study.

2. Subgeneralized VP II: $\delta J_{II}=0$ subject to constraint Eqs (3) & (4) with independent variation of $\Phi', \vec{A}, A_f$ & A_s .

$$J_{II}(\Phi,\vec{A},A_f,A_s) = -\iiint_{(V)} \left[1-(\gamma-1)\left(\frac{\partial\Phi}{\partial t}+\frac{A^2}{2}\right)\right]^{\frac{1}{\gamma-1}} \{\vec{A}\cdot\nabla\Phi$$

$$+\frac{\gamma-1}{\gamma}\frac{\partial\Phi}{\partial t}-\frac{1}{\gamma}\left(1+\frac{\gamma+1}{2}A^2\right)\}dV + L \quad (22)$$

It is easy to show similarly to VPI that $\delta J_{II}=0$ leads to Euler's Eqs (1) & (2) and the same natural boundary/initial conditions as those of VP I.

3. VP III: $\delta J_{III}=0$ subject to the constraint Eqs(2)-(4) with independent variation of Φ, A_f & A_s .

$$J_{III}(\Phi,A_f,A_s) = \iiint_{(V)} \left(\frac{p}{\gamma}\right)dv + L \quad (23)$$

From $\delta J_{III}=0$ we get the Euler Eq.(1) and the same natural boundary/initial conditions as those in VP I.

4. VP IV: $\delta J_{IV}=0$ with free variation of Φ, A_f & A_s.

$$J_{IV}(\Phi,A_f,A_s) = -\iiint_{(V)} \left\{1-(\gamma-1)\left[\frac{\partial\Phi}{\partial t}+\frac{(\nabla\Phi)^2}{2}\right]\right\}^{\frac{1}{\gamma-1}} \left\{\frac{\gamma-1}{2}(\nabla\Phi)^2+(\gamma-1)\frac{\partial\Phi}{\partial t}-1\right\}\frac{dv}{\gamma}+L \quad (24)$$

Its Euler's equation and natural boundary/initial conditions are the same as those of VP III.

5. A General Family of Generalized VPs: Based on the foregoing VPs a much more general family of generalized VPs containing a number of free parameters Ci or given functions Ci(x,y,t) (i=1,2,...) can be constructed quite straightforward via a linear combination method [3]. Just in this way the following 3-parameter family of generalized VPs is formed: $\delta J_G=0$

$$J_G(\Phi,\vec{A},p,\rho,A_f,A_s,C_i) = C_1J_I + C_2 + J_{II} + C_3J_{III} + \left(1-\sum_{i=1}^{3}C_i\right)J_{IV} \quad (25)$$

We can show that J_0 and J_1 have the same Euler's equations and natural boundary/initial conditions.

CONCLUSIONS

The variational theory of unsteady cascade flow of Ref.[3] is improved herein. It is free from most of the simplifying assumptions adopted usually in the literature. By using the variable-domain variation together with a new finite element with self-adjusting build-in discontinuities [16] and the artificial density concept [14] all unknown discontinuities (shocks and free vortex sheets) could be captured automatically and with mimimal smearing. It is interesting to note that making use of the periodicity of the oscillation we have succeeded in transforming the original initial-boundary value problem into a boundary value problem in R^3, so that it becomes very suitable for taking advantage of the variational finite element method. The theory applies directly not only to perfect gas, but also to ideal steam [15] and can be extended to 3-D unsteady flow in rotors [16].

NOMENCLATURE(Supplementary)

A_1, A_2 computational inlet & outlet boundaries resp. (Fig.1 & 2)

A_4, A_f, A_s airfoil surface, trailing vortex sheets & shocks resp. (Fig.1)

A_∞, A'_∞ boundaries at upstream infinity & downstream infinity resp. (Fig.2)

A_0, A_T solution domain boundaries at t=0 and t=T resp. (Fig.1)

A_t^T $=A_0 \cup A_T$

U_n, U_s, U_f, U_w normal speed of A_n, A_s, A_f and A_4 in R^2 resp.

H dimensionless stagnation enthalpy: $H=p/\rho+(\gamma-1)/2\ \Lambda^2$

γ specific heat ratio.

p, ρ dimensionless pressure & density resp.

q dimensionless mass flux $=\rho\Lambda$

R^2, R^3 2-D space xy and 3-D space xyt resp.

$\vec{r}$ position vector in R^3

V solution domain in R^3

ΔY cascade spacing (pitch)

$\vec{\Lambda}$ nondimensicnal fluid velocity

Φ velocity potential

$[X]$ $=X_+,-X_-$, jump of X across a discontinuity

Subscripts:

n, n' normal components in R and R resp.

pr prescribed

s, p suction & pressure sides of airfoil resp.

$-$, $+$ left & right sides of a discontinuity resp.

τ tangential compopnent

∞ at upstream boundary $A_\infty=(x=-\infty)$

REFERENCES

1. Chan S.T.K. & Chen H.C., AIAA-Paper 77-446 (1977).

2. Gallus H.E. et al., In: Proc. Symp. on Unsteady Aerodynamics of Turbomachines & Propellers, Cambridge, U.K., (1984) 73-92.

3.Liu G.L., AIAA-Paper 87-1426 (1987)
4.Liu G.L., Science in China (Series A), 32(6) (1989), 707-715.
5.Ni R.H. & Sisto F., ASME J. Engrg for Power, 98(2) (1976), 165-170.
6.Shen M.Y., Zhu S. & Lin B.Z., Transonic Flow in Turbomachines, Science Press, Beijing, China (1988).
7.Fransson T.H. & Pandolfi M., ASME-Paper 86-GT-304 (1986).
8.Jameson A. & Schmidt W., Computer Methods in Appl. Mech. & Engrg, 51 (1985), 467-493.
9.Gopalakrishnan S. & Bozzola R., ASME J. Engrg for Power, 95(4) (1973) 340-344.
10.Shapiro A.H., The Dynamics & Thermodynamics of Compressible Fluid Flow, Vol, II, Ronald Press (1954).
11.Poling D.R. & Telinis D.P., AIAA-J. 24(2) (1986), 193-199.
12.Fleeter S., AIAA-J, 18 (5) (1980), 497-503.
13.Sears W.R. AIAA-J, 14(2) (1976), 216-220.
14.Hafez M.M. et al., AIAA-J. 17 (8) (1979) 838-844.
15.Traupel W., Thermische Turbomaschinen, Band 1, 3. Aufl., Springer-Verlag, Berlin (1977).
16.Liu G.L., Variational Principles of 3-D Compreesible Flow in Turbomachines, Lecture notes, Shanghai Institute of Mech. Engrg and Shanghai Society of Engrg Thermpohysics, (1981).
17.Finlayson B.A., The Method of Weighted Residuals & Variational Principles. Acad. Press (1972).

APPENDIX NON-REFLECTING BOUNDARY CONDITIONS FOR UNSTEADY CASCADE FLOW

The full potential equation for 2-D unsteady flow is [6]

$$\frac{\partial^2\Phi}{\partial t^2} - (a^2 - \Lambda_x^2)\frac{\partial^2\Phi}{\partial x^2} + 2\Lambda_x\Lambda_y\frac{\partial^2\Phi}{\partial x\partial y} - (a^2 - \Lambda_y^2)\frac{\partial^2\Phi}{\partial y^2} + \frac{\partial}{\partial t}\left[\left(\frac{\partial\Phi}{\partial x}\right)^2 + \left(\frac{\partial\Phi}{\partial y}\right)^2\right] = 0 \quad (A\text{-}1)$$

with a^2 determined form

$$\frac{\partial\Phi}{\partial t} + \Lambda^2/2 + a^2/(\gamma - 1) = 1$$

In the far field upstream and downstream of the cascade (e.g. at the section A_1 & A_2 , Fig.1) the flow, though disturbed by the oscillating cascade at $M_x<1$, can still be approximately assumed to be uniform in the y-direction [9]. Accordingly, Eq.(A-1) can be simplified to

$$\frac{\partial^2\Phi}{\partial t^2} - (a^2 - \Lambda_x^2)\frac{\partial^2\Phi}{\partial x^2} + 2\Lambda_x\frac{\partial^2\Phi}{\partial t\partial x} \cong 0. \quad (A\text{-}2)$$

It has the following characteristic relations [10]:
characteristic eq.:

$$\frac{\partial t}{\partial x}_{I,II} = (\Lambda_x \pm a)^{-1}, \quad (A\text{-}3)$$

hodograph characteristics:

$$(\partial\Phi_{,t}/\partial\Phi_{,x})_{I,II} = -(\Lambda_x \mp a), \quad (A\text{-}4)$$

compatibility eq.:

$$a \pm \frac{\gamma - 1}{2}\Lambda_x = K_{I,II}, \quad (A\text{-}5)$$

where $\Phi_{,t}$ and $\Phi_{,x}$ denote $\partial\Phi/\partial t$ and $\partial\Phi/\partial x$ respectively, and the subscripts I & II refer to the right- & left-running characteristics respectively. from Eq. (A-5) we obtain

$$\left.\begin{aligned} \Lambda_x = \Phi_{,x} &= (K_{I} - K_{II})/(\gamma - 1) \\ a &= (K_{I} + K_{II})/2 \end{aligned}\right\} \qquad (A\text{-}6)$$

where $K_{I,II}$ are refered to as Riemann's invariants.

FLOW FIELD AROUND AN OSCILLATING CASCADE

Makoto Kobayakawa
(Kyoto University, kyoto 606, Japan)
Masahiko Ogushi
(All Nippon Airways Ltd., Tokyo 100, Japan)

ABSTRACT

An implicit approximate factorization algorithm is applied to an unsteady cascade flow on dynamic grids. The compressible Navier-Stokes equation is used to solve high Reynolds number turbulent flow through heaving cascade. The flow field of the cascade is much different from that of an airfoil. The shock is generated in the rear part of the cascade and is bent as the cascade goes upward. Unfortunately, this computational results cannot be compared with any experiment in this paper.

INTRODUCTION

With the progress of computers, significant advances have been made in the area of computational aerodynamics. We can find several computational results for the flow field which condition is very near to the real flow. Steady and unsteady flows around airfoils are obtained at high Reynolds number, i.e. Navier-Stokes analysis is wide spread[1]-[4]. Even dynamic stall of an airfoil is simulated in computer[5][6]. Furthermore, accuracy is improved by TVD scheme or second and fourth order artificial viscosity technique[7].

Several papers have been published about the flow around three dimensional bodies with complex shape. The authors solved the flow field around propfans with potential[8], Euler and Navier-Stokes analyses[9]. One of the most important problem of propfan aerodynamics is the flutter of the blades. In order to overcome this difficulty, we have to obtain unsteady flow of propfanblades. We can see some papers about the unsteady flow around wings and three dimensional bodies. However, the authors could not find the paper on numerical calculation of the unsteady flow around cascade, i.e. oscillating cascade.

This paper solves the Navier-Stokes equation for an oscillating two dimensional cascade at high Reynolds number. Algorithm used in the present computation is an implicit difference method with approximate factorization. Airfoil section of the cascade is NACA-0012 and the chord line is set parallel to the free stream. Each airfoil oscillates in heaving with same phase. Dynamic grid systems are generated by interpolation technique.

GOVERNING EQUATIONS

The two dimensional Navier-Stokes equation in conservation form is used in the present analysis. The equation expressed in cartesian coordinates is transformed into the equation expressed in non-orthogonal curvilinear coordinatesas follows:

$$\frac{\partial \hat{Q}}{\partial \tau} + \frac{\partial \hat{F}}{\partial \xi} + \frac{\partial \hat{F}}{\partial \eta} = \frac{1}{R_e}\left[\frac{\partial \hat{R}}{\partial \xi} + \frac{\partial \hat{S}}{\partial \eta}\right], \tag{1}$$

where

$$\hat{Q} = \frac{1}{J}\begin{bmatrix}\rho \\ \rho u \\ \rho v \\ e\end{bmatrix}, \hat{E} = \frac{1}{J}\begin{bmatrix}\rho U \\ \rho uU+p\xi_x \\ \rho vU+p\xi_y \\ (e+p)U-p\xi_t\end{bmatrix}, \hat{F} = \frac{1}{J}\begin{bmatrix}\rho V \\ \rho uV+p\eta_x \\ \rho vV+p\eta_y \\ (e+p)V-p\eta_t\end{bmatrix}$$

$$\hat{R} = \frac{1}{J}\begin{bmatrix}0 \\ \xi_x\tau_{xx}+\xi_y\tau_{xy} \\ \xi_x\tau_{xy}+\xi_y\tau_{yy} \\ \xi_x R_4+\xi_y S_4\end{bmatrix}, \hat{S} = \frac{1}{J}\begin{bmatrix}0 \\ \eta_x\tau_{xx}+\eta_y\tau_{xy} \\ \eta_x\tau_{xy}+\eta_y\tau_{yy} \\ \eta_x R_4+\eta_y S_4\end{bmatrix} \tag{2}$$

$$e = \frac{p}{\gamma-1}+\frac{1}{2}\rho(u^2+v^2)$$

$$\tau_{xx} = \lambda(u_x+v_y) + 2\mu u_x$$

$$\tau_{xy} = \mu(u_y+v_x)$$

$$\tau_{yy} = \lambda(u_x+v_y) + 2\mu v_y$$

$$R_4 = u\tau_{xx} + v\tau_{xy} + \frac{\kappa}{(\gamma-1)P_r}\frac{\partial a^2}{\partial x}$$

$$S_4 = u\tau_{xy} + v\tau_{yy} + \frac{\kappa}{(\gamma-1)P_r}\frac{\partial a^2}{\partial y} . \tag{3}$$

In the above equations u , and v are non-dimensionalized by $a_\infty/\sqrt{\gamma}$, and ρ , p and T are non-dimensionalized by ρ_∞ , p_∞ ,and T_∞ , respectively. Furthermore, μ is calculated by the Sutherland's formula

$$\tilde{\mu} = \tilde{\mu}_1\left[\frac{\tilde{T}}{\tilde{T}_1}\right]^{3/2}\frac{\tilde{T}_1+\tilde{S}}{\tilde{T}+\tilde{S}} , \tag{4}$$

where subscript 1 denotes the value at the freestream. The constants γ , P_r , and $\tilde{S}$ are given by 1.4, 0.72 and 110.4K, respectively.

The physical space and computational space are illustrated in Fig.1.

NUMERICAL METHOD

AF scheme

The numerical algorithm used in the present analysis is approximate factorization scheme (AF scheme). This implicit difference method is slightly old but has several superior points in convergency, stability and short CPU time.

Eq.(1) is factorized after applying first order accurate difference to time derivative as follows:

$$(I + \Delta\tau\partial_\xi\hat{A} - \frac{\Delta\tau}{R_e}\partial_\xi\hat{M})(I + \Delta\tau\partial_\eta\hat{B} - \frac{\Delta\tau}{R_e}\partial_\eta\hat{N})\Delta\hat{Q}^n =$$

$$-\Delta\tau\left\{\partial_\xi\hat{E}^n + \partial_\eta\hat{F}^n - \frac{1}{R_e}(\partial_\xi\hat{R}^n + \partial_\eta\hat{S}^n)\right\} , \tag{5}$$

where $\hat{A}$, $\hat{B}$, $\hat{M}$, and $\hat{N}$ are Jacobian matrices, i.e.

$$\hat{A} = \frac{\partial\hat{E}}{\partial\hat{Q}} , B = \partial F , \hat{M} = \frac{\partial\hat{R}}{\partial\hat{Q}} , \text{and } \hat{N} = \frac{\partial\hat{S}}{\partial\hat{Q}} . \tag{6}$$

The detail expressions of these terms are abbreviated. The unknown vector $\Delta\hat{Q}^n = \hat{Q}^{n+1} - \hat{Q}^n$ can be obtained by sequential inversion process as follows:

$$(1 + \vartheta_\xi \hat{A} - \frac{1}{R_e}\vartheta_\xi \hat{M} - \varepsilon_i J^{-1}\nabla_\xi \Delta_\xi J)\Delta \hat{Q}^* =$$

rhs(5) - (mixed second and fourth order dissipation term along ξ -direction)

$$(1 + \vartheta_\eta \hat{B} - \frac{1}{R_e}\vartheta_\eta \hat{N} - \varepsilon_i J^{-1}\nabla_\eta \Delta_\eta J)\Delta \hat{Q}^n =$$

$\Delta \hat{Q}^*$ - (mixed second and fourth order dissipation term along η -direction)

$$\hat{Q}^{n+1} = \hat{Q}^n + \Delta \hat{Q}^n \tag{7}$$

The mixed second and fourth order dissipation terms by Jameson type[7] are used in order for good shock capturing.

Boundary conditions

The boundary conditions are applied explicitly. On the body surface, the contravariant velocities are set to zero while pressure and density are determined by extrapolation from the interior. Upstream line from the leading edge and wake line are periodic boundaries, therefore, all variables corresponding along these lines should be equal. In the farfield, for subsonic flow, the boundary conditions is based on locally one dimensional Riemann invariants. In the case of supersonic outflow, all variables are extrapolated.

For the initial condition, the steady solutions are set to the cascade at the lowest position. The cascade starts to oscillate up and down. During the oscillation, each airfoil does not change its relative position. The displacement $h(t)$ is given by

$$h(t) = \frac{A_m}{2}(1 - \cos\omega t)\ , \tag{8}$$

where A_m is amplitude.

RESULTS AND DISCUSSIONS

Before the Navier-Stokes analysis, we have to generate grid systems. At each time step an appropriate grid system is necessary. This grid is obtained by interpolation technique developed by Chyu et al.[10]. To begin with, H-type grid is generated in the space between two NACA-0012 airfoils as shown in Fig.2. Sorenson's method[11] which solves Poisson's equation in order to generate the grid system is used in the present paper. The grid points are 251×81 and the minimum spacing normal to the body is about 10^{-4} chord. The highest position of the grid is obtained by parallel translation of distance A_m along y -direction.

The steady solution is depicted in Fig.3 which shows density distribution. The flow conditions are M_∞=0.8 , R_e=10^7 and Baldwin- Lomax turbulence model[11] is used for the viscosity coefficient. This figure shows shock near the trailing edge clealy. The pressure distribution and the velocity vectors near trailing edge are illustrated in Figs.4 and 5, respectively. The flow field around an NACA-0012 airfoil included in cascade is considerably different from a single airfoil.

For the next step, let the cascade start to oscillate from the above mentioned steady solution. The unsteady solution is obtained for time step $\Delta\tau$=0.001 , and number of iteration N=10000 , then, the reduced frequency k=0.4 . The results are shown in Figs.6, 7, and 8. Each figure shows every 500 iteration level. In the figures, h shows non-dimensionalized displacement $h(t)/A_m$ and in Fig.8 solid line shows the pressure distribution on the lower surface and dotted line on the upper surface of the airfoil. From Fig.6, we can see that the shock is retarding as the cascade moves upward. Especially, the shock is bent near the center line of two airfoils. From Fig.7, the shock on the airfoil surface retards until near the trailing edge. The induced angle of attack at h=0.5 is −2 degree. Fig.8 depicts the vortex shed from the cascade because of adverse flow as the induced angle of attack is increased. The size of vortex becomes larger until h becomes 0.5 to 0.7 and decreases until h becomes to 1.0. The flow field returns to the initial state through nearly adverse process to h=0 .

The flow field around an oscillating cascade is considerably different from that of a single airfoil, however, comparison of the present analysis with other computation was not possible because the author could not find any paper of similar computation.

CONCLUDING REMARKS

A flow field of an oscillating cascade is solved with an implicit approximate factorization algorithm. The compressible Navier-Stokes equation is used for the present analysis. Dynamic grid is generated with interpolation technique, and number of grid points is 251×81 . The flow conditions are $M_\infty = 0.8$, R_e=10^7 , and the airfoil section is NACA-0012. The cascade oscillates with heaving in same phase.

The results show that the shock is generated near the rear part of the cascade, and is bent as the cascade moves upward. At the same time, a vortex is shed from the trailing edge. Unfortunately, this computational results could not be compared with any other computation.

REFERENCES

[1]T.H. Pulliam, 'Euler and Thin Layer Navier-Stokes Codes : ARC2D, ARC3D', Notes for Computational Fluid Dynamics User's Workshop, 1984.

[2]J. Nakamichi, 'A Verification of Unsteady Navier-Stokes Solutions around Oscillating Airfoils', NASA TM 88341, 1986.

[3]W.K. Anderson et al., 'Extension and Applications of Flux-Vector Splitting to Unsteady Calculations on Dynamic Meshes', AIAA Paper 87-1152-CP, 1987.

[4]C.L. Rumsey and W.K. Anderson, 'Some Numerical and Physical Aspects of Unsteady Navier-Stokes Computations over Airfoils Using Dynamic Meshes', AIAA Paper 88-0329, 1988.

[5]U.B. Mehta, 'Dynamic Stall of an Oscillating Airfoil', AGARD Paper No.23, 1977.

[6]K. Ono, 'Numerical Study on the Dynamic Stall Process of a NACA 0012 Airfoil', AIAA Paper 85-0128, 1985.

[7]T.H. Pulliam and J.L. Steger, 'Recent Improvements in Efficiency, Accuracy, and Convergence for Implicit Approximate Factorization Algorithms', AIAA Paper 85-0360, 1985.

[8]M. Kobayakawa et al., 'Calculations of High Speed Propeller performances Using Finite Difference Methods', ICAS Proceedings, pp.1451-1460, 1986.

[9]M. Kobayakawa and I. Hatano, 'Flow Field Around a Propeller by Navier-Stokes Equation Analysis', AIAA Paper 88-3150, 1988.

[10]W.J. Chyu et al., 'Calculation of Unsteady Transonic Flow over an Airfoil', AIAA Journal, Vol.6, No.6, July, pp.679-686, 1978.

[11]R.L. Sorenson, 'A Computation Program to Generate Two-Dimensional Grids about Airfoils and Other Shapes by Use of Poisson's Equation', NASA TM-81198, 1981.

[12]B.S. Baldwin and H. Lomax, 'Thin Layer Approximation and Algebraic Model for Separated Turbulent Flows', AIAA Paper 78-257, 1978.

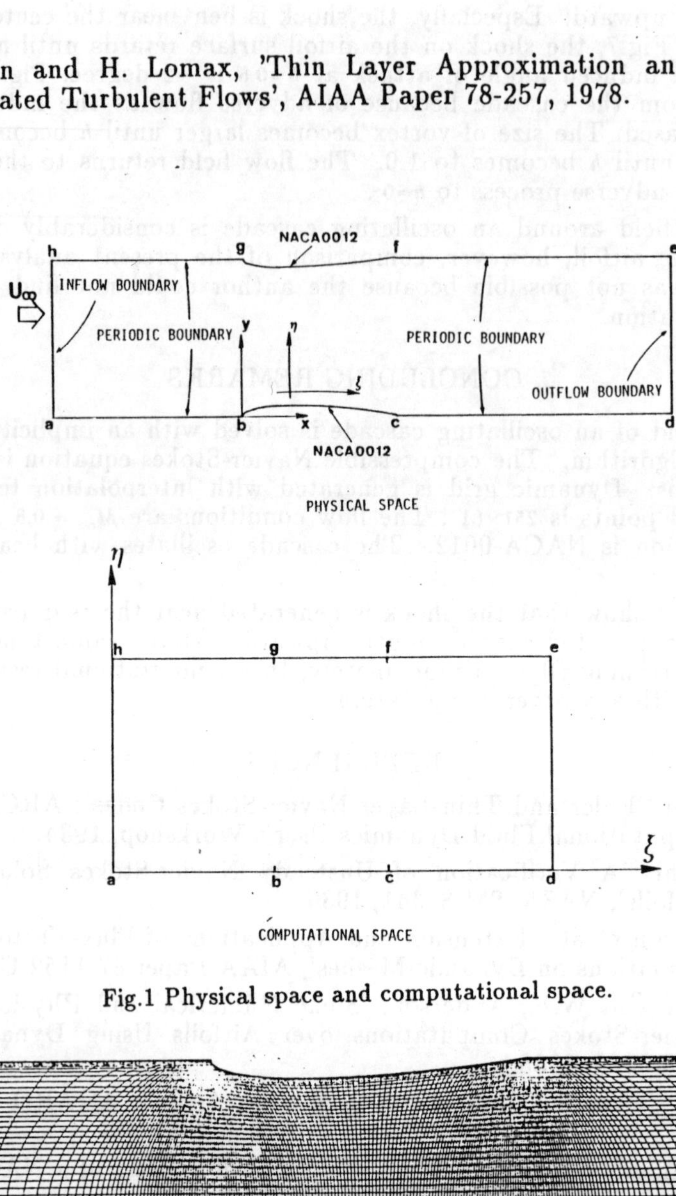

Fig.1 Physical space and computational space.

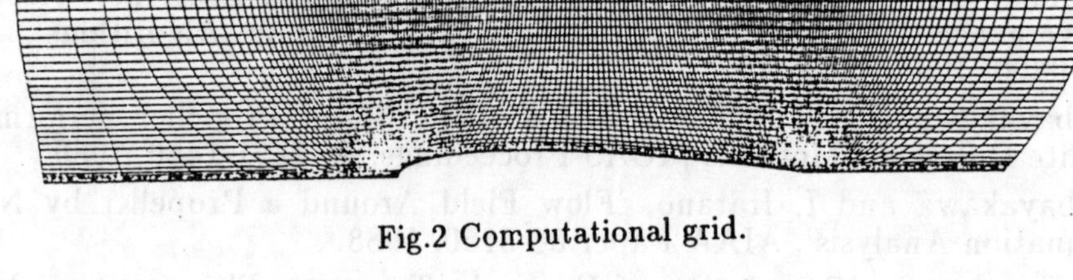

Fig.2 Computational grid.

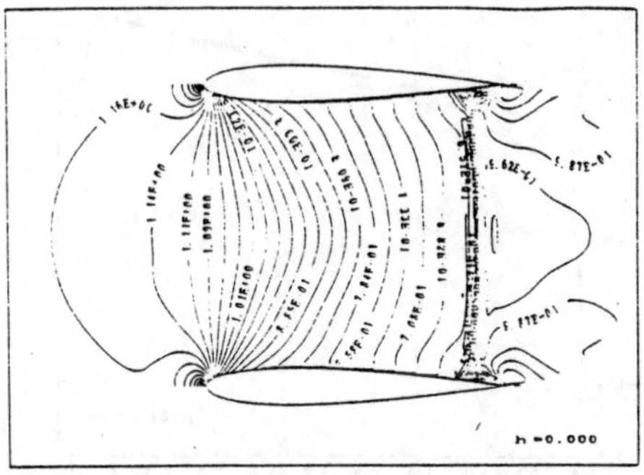

Fig.3 Steady solution (a)-density distribution.

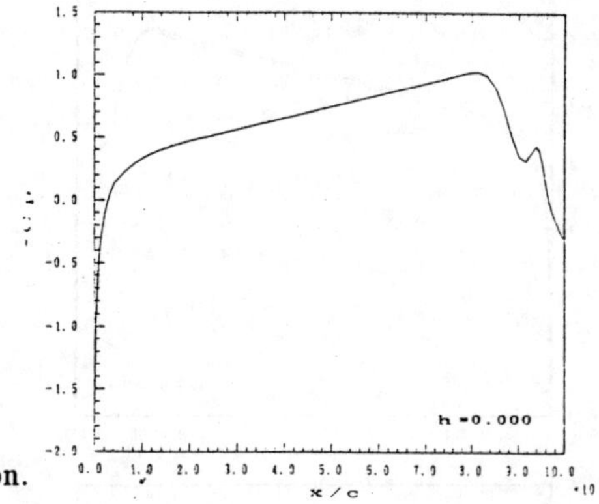

Fig.4 Steady solution (b)-pressure distribution.

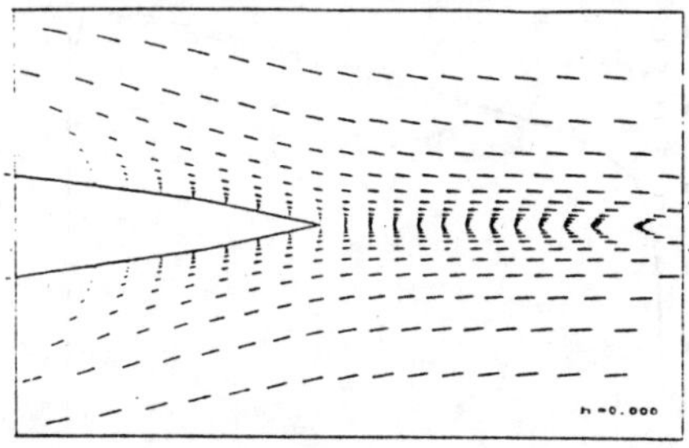

Fig.5 Steady solution (c)-velocity vectors.

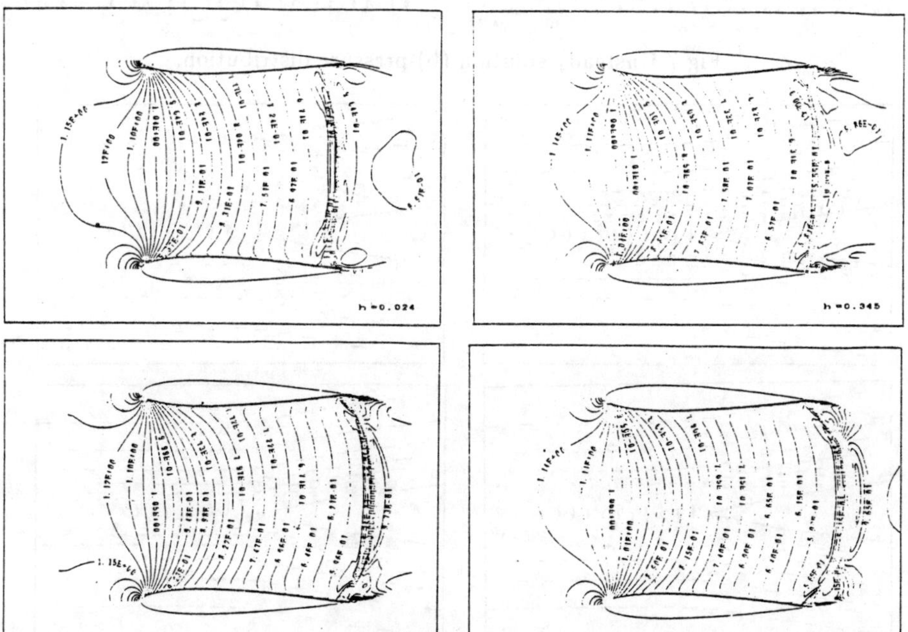

Fig.6 Unsteady solution (a)-density distribution.

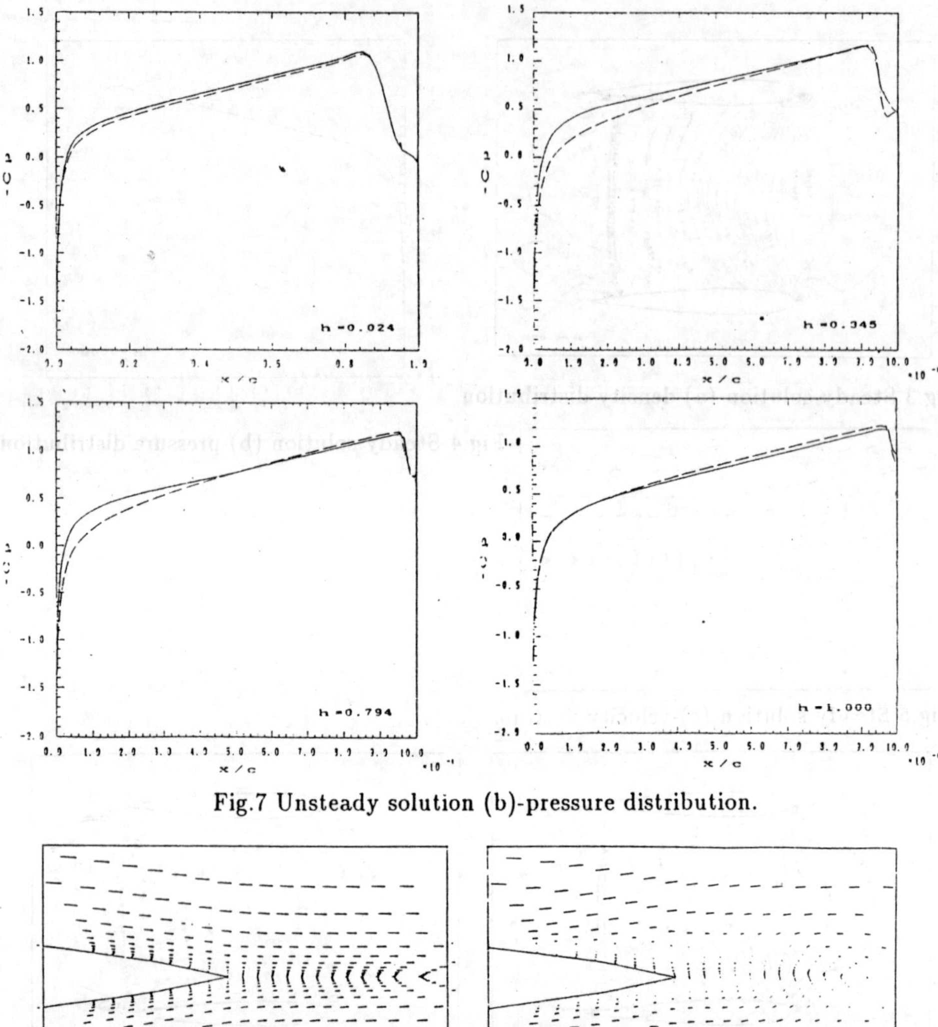

Fig.7 Unsteady solution (b)-pressure distribution.

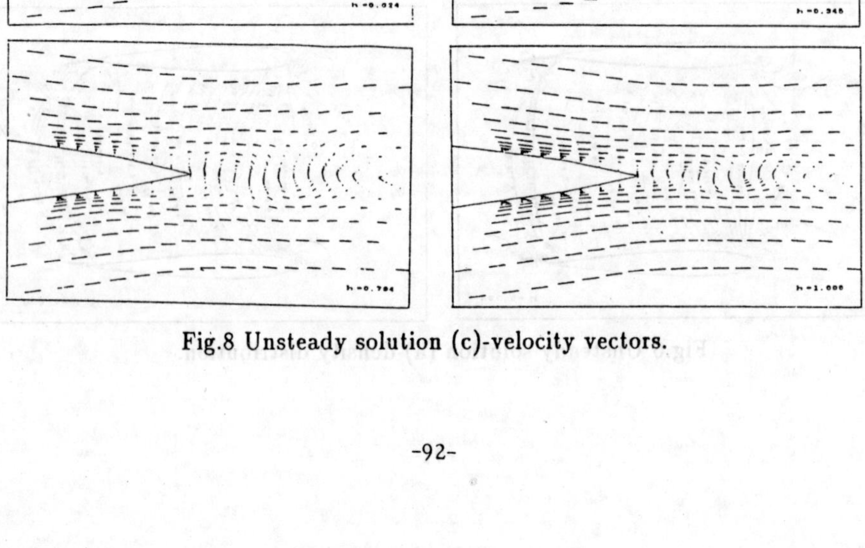

Fig.8 Unsteady solution (c)-velocity vectors.

NUMERICAL SIMULATION OF TRANSONIC FLOW THROUGH OSCILLATING AND MULTI—ROW TWO—DIMENSIONAL AIRFOIL CASCADES

A. B. Arkadyev, V. A. Vanin and S. V. Yershov
(Ukr. SSR Academy of Sciences, Institute for Problems in Machinery, Kharkov, USSR)

ABSTRACT

The problems of aerodynamic interaction of two and three reciprocally mobing blade rows, as well as the unsteady flow about isolated oscillating airfoils in cascade are considered. The solution of the problems under review can be achieved through the set of algorithms which have been devised using the numerical integration of Euler equations by Godunov's difference scheme and its high resolution modifications. A computational analysis of turbine and compressor cascades has been made. An effect of flow regime, as well as mode, frequency, amplitude and phase shift of oscillating airfoils on the performance of cascades air damping has been investigated. The analysis of local parameters has enabled to get a deeper insight into the mechanisms of flutter origin. The influence of Strouhal number, ratio of blade-numbers and axial spacing on the unsteady forces has been examined for the aerodynamic interaction of two blade-rows. Effects of wake segments splitting have been achieved numerically as a result of the wake/rotor interaction. In the case of three blade-rows the influence of axial spacing and moving and static cascades blade-numbers ratios on unsteady loads has been inchanging the relative tangential stator rows shift.

NOMENLATURE

$\overline{A}$	— total work coefficient, Eq. (4)	p	— pressure
C_A	— work coefficient, Eq. (3)	$q=\rho v^2/2$	— dynamic head
$C_P=p/(q_{in}h)$	— pressure coefficient	s	— blade pitch
c	— blade chord	$\bar{s}=s/c$	— pitch-to-chord ration
F	— force	$T=2\pi/\omega$	— period
h	— nondimensional oscillation amplitude	t	— time
$k=\omega c/(2v_{in})$	— reduced frequency	u	— velocity of rotor row movement
M	— Mach number	v	— flow velocity
z	— number of blades	ω	— circular frequency
α	— exit flow angle	**Subscripts**	
β	— stagger angle	o	— cell centre and current time level
γ	— ratio of specific heats	1,2,3	— row number

Δ	— axial spacing	ex, in	— freestream flow parameter at exit or inlet
δ	— aerodynamic logarithmic decrement	tot	— total parameter
θ	— relative tangential shift of two stator rows in three-row cascade	x, y	— axial and tangential directions; also axial and tangential bending
ρ	— density	α	— torsion
σ	— interblade phase angle	η	— chord normal bending
x	— nondimentional chordwise coordinate, positive downstream		

INTRODUCTION

Presently, there is a considerable progress in the development of computational methods for aerodynamic analysis of turbomachine blading. Significant successes have been achieved in two- and threedimentional steady inviscid and viscous flow calculations. The employment of these methods makes it possible to raise the efficiency of turbomachines and to reduce the expenses for their designing. At the same time the results of last studies indicate that unsteady and aeroelastic phenomena are becoming more and more the main factor which influences the reliability of turbomachines and restrains the augmentation of their power. But the theory of unsteady flow phenomena in turbomachines is well developed only for the incompressible fluid approximation.

The potential equation and different types of its linearization are actually used in a number of works for the calculation of unsteady compressible flows in cascades. However, significant difficulties occur here both with the airfoil shape consideration and shocks modelling. As a rule, one has to neglect changes in entropy across the shock. It should be emphasized that the analysis of transonic flows is the most complicated one. For this type of flow the aerodynamic characteristics dependence on the airfoil thickness and camber, the angle of attack and some other geometrical and regime parameters is nonlinear. Small blade shape or flow regime variation can result in the change of the mean position and strength of the shocks and affect the unsteady loads to a greater extent. This is one of the peculiarities, due to which the accounting for the interaction between steady and unsteady disturbances becomes paramountly important in the transonic case.

The steady transonic flows are simulated more accurately within the framework of the approach based on the numerical integration of the full Euler equations. Such an approach is the universal one. It enables the calculation of unsteady flows through real cascades of turbomachines in a wide range of Mach numbers including transonic regimes. Besides, there is a possibility to capture shocks and contact discontinuities numerically in an implicit way and to simulate the jet flows, as well as free borticity. For oscillating airfoils unlike linear theories where the impermeability conditions for solid walls have to be imposed at some fixed surface, one can prescribe them directly at the moving boundaries. As a result, the relative displacement of airfoils in the process of their oscillation can be taken into account automatically. Moreover, the calculation can be extended naturally to the three-dimensional flows and with the equations being supplemented by viscous terms together with a model of turbulence to viscous flows for high Reynolds number.

The Godunov's scheme is practised widely for integration of the Euler equations. This scheme and its different modifications have been successfully used for calculation of internal and external aerodynamic problems[1,2]. The present paper contains the results of this scheme application to the numerical simulation of unsteady transonic flows through cascades of oscillating airfoils and multi-row cascades.

GOVERNING EQUATIONS AND NUMERICAL METHODS

The unsteady cascade flow is described by the Euler equations represented in integral conservation law form:

$$\oiint f dxdy + adydt + bdxdt = 0 \tag{1}$$

where

$$f = \begin{Vmatrix} \rho \\ \rho v_x \\ \rho v_y \\ E \end{Vmatrix}, \quad a = \begin{Vmatrix} \rho v_x \\ \rho v_x^2 + p \\ \rho v_x v_y \\ (E+p)v_x \end{Vmatrix}, \quad b = \begin{Vmatrix} \rho v_y \\ \rho v_x v_y \\ \rho v_y^2 + p \\ (E+p)v_y \end{Vmatrix},$$

$E = p/(\gamma - 1) + \rho v^2/2$ is the total energy.

We consider the isolated blade row for the calculation of the flow through oscillating cascaded airfoils and the set of two or three rows, if the stator/rotor aerodynamic interaction is investigated. In general case it is necessary to include in the computational domain all blade-to-blade passages which the total unsteady flow spatial period contains. However, it is enough to consider only one blade-to-blade passage of each row when the unsteady flow through a two-row cascade or oxcillating airfoils is investigated. The inlet, exit and axial spacing, if necessary, periodicity regions are added to the computational domain. To perform the calculations in relative coordinates this domain is divided into zones. Each of them contains only one row.

The total enthalpy, the entropy function and the flow direction angle are specified at the inlet and the static pressure is prescribed at the exit. It is reasonable to use the non-reflecting boundary conditions for the convergence acceleration. In this case the right Riemann invariant at the inlet and the left one at the exit are fixed instead of the total enthalpy and the static pressure, respectively. The impermeability condition is imposed at the blade surface.

The interpolation procedures and, if necessary, the generalized periodicity condition provide the solution joining at the lines which divide the computational domain into zones. The generalized (or the simple if it is possible) periodicity condition is used at the lines separating the adjacent blade-to-blade passages in the cascade one from another.

The Eqs. (1) are approximated by the finite volume Godunov-type scheme using the one-dimensional Riemann problem solution in order to calculate the fluxes across the cell edges. The initial left and right states for this problem are given by those at the centres of the adjacent cells. Thus, we obtain the first order Godunov scheme. This scheme is used for the unsteady three-row cascade flow calculations. The second-order modified TVD Godunov-Kolgan scheme is constructed similarly to the first-order method. At each cell edge the Riemann problem is solved for some specified pair of left and right states. The extension to second-order is achieved by using extrapolation in space and time to obtain these states:

$$\psi = \psi_o + (x - x_o)\left(\frac{\partial \psi}{\partial x}\right)_o + (y - y_o)\left(\frac{\partial \psi}{\partial y}\right)_o + (t - t_o)\left(\frac{\partial \psi}{\partial t}\right)_o \quad (2)$$

where $\psi = \{\rho, v_x, v_y, P\}$. The spatial derivatives are determined using the two-point formulas and the min-mod principle[3]. The time derivatives are calculated then utilizing the differential nonconservation from of the Eqs. (1) as it has been proposed by Eidelman et al[4]. Such a scheme is second-order accurate on smooth monotonous solutions and the maximum permissible time step corresponds to the CFL number $\gamma = 1$. This scheme is used for the calculation of unsteady flows through cascades of oscillating airfoils and two-row cascades.

FLOW THROUGH OSCILLATING CASCADES

The periodic unsteady flow through a cascade of airfoils which oscillate harmonically in time with the same amplitude and a constant but arbitrary interblade phase angle is considered. The calculations are performed at dynamic grids. The zones of grid change are localized near vibrating airfoils. The work of unsteady aerodynamic forces on various parts of the airfoil surface during each vibratory cycle is estimated by means of the work coefficient distribution

$$C_A(x) = \int_0^T P(x,t) v_n(x,t) dt / (\pi q_{in} h^2 c) \quad (3)$$

where v_n is the velocity of the airfoil movement projected on the inner normal vector of its surface. On the areas where $C_A(x) > 0$ energy is supplied from the flow to the airfoil. The stability of the system as a whole is determined by coefficient of aerodynamic damping

$$W = -\overline{A} = -A/(\pi q_{in} h^2 c^2) \quad (4)$$

Where A is the total work per unit span done by unsteady aerodynamic forces during a cycle of oscillation. Negative value of W indicate system instability if the mechanical damping is neglected.

It has been ascertained that for oscillating subsonic and supersonic flat-plate cascades the results of the present method are on a whole in a good agreement with those obtained by linear theories. In a number of cases, however, there were some distinctions. Two examples are presented in Fig. 1 and 2. Fig. 1 shows the distributions of imaginary (Im) and real (Re) parts of the unsteady pressure difference across the blade due to torsion. the full lines represent the present obtained by Caspar and Verdon[5] and the points represent the present. Good agreement is seen here. Fig. 2 shows the comparison between the present results (points) and the results obtained by Gorelov et al. (full line) for the aerodynamic damping coefficient due to bending vibration as a function of interblade phase angle. There is a considerable discrepancy for the range of phase angles near the resonance point where the aerodynamic damping changes sharply.

Fig. 3 shows the comparison between the experimental[7] and present results for the aerodynamic logarithmic decrement δ_o due to bending motion of one blade in cascade as a function of exit Mach number. The typical impulse turbine cascade has been considered with the following parameters: maximum airfoil thickness to chord ratio $= 0.36$, turning $= 137°$, stagger angle $= -10°$, relative blade spacing $= 0.6$. The frequency of oscillation was 1770 Hz. The present method does not allow to say, to reproduce the conditions of Mach numbers which have been considered up to now the calculations were conducted for six different values of interblade phase angle (Fig. 4). The points in Fig. 3 were obtained then by averaging on the base of the formula[7]

$$\delta_o = \frac{1}{2\pi}\int_o^{2\pi} \delta(\sigma) d\sigma \quad (5)$$

The agreement in Fig. 3 seems to be satisfactory. Rather coarse grid (29x6) was used for these calculations.

The investigation of the aerodynamic damping and the flutter boundaries in transonic cascades has been conducted using the present method. The most detailed investigation was carried out for the cascade which represented the tip section of a modern fan. The steady velocity λ contours and the cascade geometry are shown in Fig. 5. The cascade operated in supersonic relative flow with a subsonic axial component. There was a strong shock in the passage.

Fig. 6 and 7 show the results obtained for the aerodynamic damping coefficient due to bending and torsional oscillations, correspondingly, with different values of the reduced frequency and the interblade phase angle. The pitching axis location was at the midchord. The frequency of the stability loss for bending mode is seen to be significantly lower than the one for torsional mode ($k=0.05$ and $k=0.5$, respectively). The conditions of the acoustic resonance appearance were estimated using the relations from the work[7,8]. The undisturbed flow parameters at upstream infinity were used for this purpose. The following values of the resonance interblade phase angles have been obtained (all of them are expressed as whole numbers): 351° and 357° for $k=0.05$; 324° and 347° for $k=0.2$; 253° and 322° for $k=0.6$; 182° and 296° for $k=1$; 93° and 264° for $k=1.5$. As it is seen in Fig. 6, for bending oscillations with $k=0.6$, 1 and 1.5 the aerodynamic damping falls sharply near one of the two resonance points for a given frequency. For torsional mode this effect is not so pronounced.

The next two figures illustrate the mechanisms of flutter origin. Fig. 8 corresponds to bending and Fig. 9 — to torsional mode, part a — to the suction surface and part b — to the pressure one. Each presented value of the interblade phase angle corresponds to the minimal or close to the minimal aerodynamic damping for a given frequency. For bending mode the loss of stability is connected with energy supply from the flow to the airfoil in the region of the shock on the pressure surface (Fig. 8, b). This result is consistent with the possibility of the shock-induced bending flutter origin at low reduced frequencies in the form of a backward travelling wave which has been noted previously[9,10]. Besides, it is seen in Fig. 8 that for $k=1$ and $\sigma=300°$, that is to say, for the point which is very close to the resonance one the work coefficient is positive along almost the whole pressure surface. As to the torsional instability it is caused mainly by the transmission of energy to the airfoil on the front part of the pressure surface (Fig. 9, b) where the flow is supersonic. This mechanism of the torsional instability origin, apparently, is similar to the mechanism of the purely supersonic unstalled flutter[11]. For the latter one the most important interaction between the neighbouring blades is connected also with the fall and reflection of the disturbances created by the neighbouring lower blade motion on the forward part of the given blade pressure surface[8]. But the strong shock presence in the passage influenced the unsteady aerodynamic performance and caused some distinctions between these mechanisms. The purely supersonic unstalled flutter is characterized by the stabilization effect of the exit pressure increase[11]. However, in our case the effect of the exit pressure increase turned out to be equivocal (Fig. 10). For $k=0.5$ close to the instability boundary the three flow regimes have been considered with different values of the exit pressure. The pressure ratios were 1.67, 1.9 and 2.13. All previous results correspond to the latest of these values. Fig. 10 shows that while the increase of the pressure ratio leads to the monotonous rise of the maximum aerodynamic damping, there is obvious nonmontony in the instability domain. The corresponding changes in the local characteristics are shown in responding changes in the local characteristics are shown in Fig. 11. Evidently, they are caused by the rise of the shock strength and its upstream displacement when the exit pressure increases.

As it is seen in Fig. 10, the differences between the maximum and minimum values of the aerodynamic damping coefficient rise when the pressure ratio increases. This shows that the aerodynamic coupling between the cascaded blades grows. One can see the analogous growth of the coupling when the steady loading increases in Fig. 4.

The investigation of the unsteady aerodynamic performance dependence upon the amplitude of oscillations has been conducted. It has been revealed that this dependence is essentially nonlinear. Alterations in the local characteristics take place mainly near the shocks when the amplitude changes. It is necessary to take into account not only the first but also the second and the third harmonics of the unsteady quantities there. At the same time, however, it has been ascertained that for the investigated range of frequencies ($k \geqslant 0.6$), one may use with a considerable degree of exactness the superposition principle in order to estimate the influence of the aerodynamic responses, caused by different components of the airfoil motion, on the stability of the coupled modes of oscillations.

It should be noted that Fig. 8, 9 and 11 show the simultaneous presence on the airfoil surface of several alternating zones where the energy is supplied to the airfoil and taken away from it to the flow. This observation confirms once more the extreme difficulty of the energy exchange process between the flow and airfoils.

STATOR/ROTOR AERODYNAMIC INTERACTION TAKING INTO ACCOUNT VISCOUS WAKES

The method for the calculation of unsteady flows through two reciprocally moving blade rows has been used in order to study the peculiarities of interacting viscous and potential nonuniformities in an axial space. Herein, the simulation of viscous trailing wakes was based on the dissipative properties of the difference scheme. The total stagnation was specified on the stator blade blunted trailing edge for this. Such a boundary condition provides a flow separation at the trailing edge angle points. As a result, the stagnant flow zone with the increasing entropy level and vorticity generation forms trailing wake. Its intensity is controlled by the trailing edge thickness.

The velocity λ contours for the stator cascade are given in Fig. 12. Parts a and b correspond to the calculations without and with wakes, respectively. Fig. 13 displays the total pressure distribution at the transonic rotor cascade exit for various distances x from the blade edges. Here the lines 1, 2, 3 and 4 correspond to the calculated results for distances $x/s = 0.13, 0.24, 0.32, 0.43$, and the lines 5, 6, 7 and 8 correspond to the experimental data for distances $x/s = 0.14, 0.23, 0.33$ and 0.43, respectively.

The Strouhal number, ratio of blade-numbers and stator-rotor axial spacing influence on the unsteady forces has been examined. It has become clear that the reduced frequency rise results in the increase of unsteady pressure oscillations amplitudes on the rotor blades (in the first place at the leading edge). At the same time it leads to the increase of the phase shift of these oscillations at the neighbouring points of the blade surface.

As a result, as it is shown in Fig. 14, significant reduction of the unsteady circumferential force amplitude ΔF_y is observed after its certain increase.

Fig. 15 illustrates the dependence of the unsteady circumferential force ΔF_y on the blade-numbers ratio Z_1/Z_2. Analogous results which are characterized by the diminution of ΔF_y with Z_1/Z_2 increase, has been obtained for all cascades studied.

The dependence of ΔF_y from the axial space Δ is given in Fig. 16. The model which incorporates trailing wakes (line 1), unlike the model without them (line 2), is in a satisfactory agreement with the LPI experimental data[12] (points). The unsteady load unmonotone behaviour is caused by the interaction of viscous and potential nonuniformities which extend in the axial space with the angle $\pi/2-\alpha_1$ one to another. This effect cannot be obtained if the wakes are ignored.

The effects of wake split have been obtained numerically as a result of the wake/rotor interaction. Lines of entropy function P/ρ^{γ} constant deviation from its time-averaged values are shown in Fig. 17. The wake (that is the entropy high level zones) passing through a vane canal is accompanied by the following effects. The split wake segment turn is due to different flow velocities at the blade sides. The wake backflow from the pressure surface to the suction one leads to wake segment widening at the suction side and to its narrowing at the pressure one. The entropy high level zone is concentrated on the blade suction surface. The split wake segments draw up along the absolute flow direction ($\alpha_2 \approx 78°$) and have a periodical structure.

Pair large-scale vortices are induced in the blade-to-blade passage at the expense of the return flow in a wake. Fig. 18 shows the relative velocity vector increments $\Delta\vec{v}$ defined as follows:

$$\Delta\vec{v} = \vec{v}(t) - \frac{1}{T}\int_0^T \vec{v}(t)dt \qquad (6)$$

The vorticity structure caused by the wakes dissipates significantly as it passes through the vane canal and the unsteady velocity vector distribution at the rotor row exit is comparable with the time-averaged one.

The results represented in Fig. 17 and 18 correspond to the physical notions and agree qualitatively with Hodson's numerical and experimental data[13]. However, it should be noted that the present numerical model of stator/rotor aerodynamic interaction is more complete since both the potential effects and the viscous wakes influence are taken into consideration.

STATOR/ROTOR/STATOR AERODYNAMIC INTERACTION

The influence of two stator rows on the rotor unsteady loads can be estimated using the numerical method of three-row cascade flows simulation. This method mades it possible to investigate the peculiarities of the extension and interaction of unsteady disturbances having different fundamental frequencies. The blade-numbers ratio, stator rows reciprocal position and axial spacing effects can also be simulated.

The typical compressor stage is considered (Fig. 19). It consists of the first stator (S1), a rotor (R) and the second stator (S2). The stage characteristics are as follows: blade-numbers ratio $Z_1 : Z_2 : Z_3 = 5 : 3 : 4$, axial spacings ratio $\Delta_1/\Delta_2 = 2.564$, the initial value of the stator rows reciprocal position $\theta = 0.43$. In this case the unsteady circumferential force on the rotor blades turned out to be modulated as it is shown in Fig. 20. The diminution of the axial space Δ_1 to the Δ_2 value has resulted in the increase of the beats while the time-averaged rotor load remained immutable. At the same time the unsteady load amplitude increased on the S1 row. As to the S2 row the spectrum composition changed here, the multiple harmonics have become commensurable with the basic one.

The rotor blade surface unsteady pressure distribution consideration revealed the possibility of the fluctuation level reduction if $Z_1 = Z_3$ and $\theta = \theta_*(\Delta_1, \Delta_2)$. When the influence of the parameter θ on the unsteady loads is investigated, the range of its change is restricted to the value D_τ. As one can show easily the range D_τ is equal to S_3/Z_1 value when $Z_1 \neq Z_3$ and, otherwise, it is equal to S_1 value.

The time distributions of the unsteady pressures in six points on each side of the rotor blade are presented in Fig. 21. The points were disposed uniformly chordwise with points 1 and 6 situated at the leading and the trailing edges, respectively. The blade-numbers ratio as $Z_1 : Z_2 : Z_3 = 5 : 3 : 5$ in this case. The results correspond to the tangential shift value θ_*, for which the pressure pulsations in the points 1 and 6 are in the counter phase. The optimal value θ_* turned out to be close to 0.5 S_1.

The unsteady components of the circumferential force ΔF_y and the axial one ΔF_x as functions of θ are presented in Fig. 22. The time distributions of the unsteady forces acting on the S1, R and S2 blades are shown in Fig. 23 for the case of $\theta = \theta_*$. Further fluctuation level reduction was achieved by means of the axial spacing Δ_1 decrease to the Δ_2 value with the concerted change of θ.

The calculations for the two two-row cascades have been made to analyse the joint influence of the S1 and R rows (the blade-numbers ratio was 5 : 3 : 0) and in the second case the R and S2 rows (0 : 3 : 5) were considered. The boundary conditions were specified based on the three-row cascade (5 : 3 : 5) calculation. As the calculation results comparison displays (Fig. 23, 24 and 25), not only the adjacent rows interaction takes place but the S1 and S2 row influence on each other. This effect cannot be obtained by means of the superposition of the computational results for the two two-row cascades.

Thus, the reduction of the rotor blade unsteady load ΔF_y is possible not only by the traditional way, that is to say, by the axial spacings increase but also with the aid of the axial spacings diminution with the concerted change of the stator rows reciprocal position. The stator rows with equal blade-numbers are suggested for use.

The authors of this manuscript are research scientists.

REFERENCES

1. Godunov S. K. et al., Numerical Solution of Multidimensional Fluid Dynamics Problems, Nauka (1976), 400.
2. Sokolovsky G. A. and Gnesin V. I., Unsteady Transonic and Viscous Flows, Naukova dumka (1986), 263.
3. Kolgan V. P., Ucheniye zapisky CAGI, 6(1)(1975), 9.
4. Eidelman S, Colella Ph and Shreev RP, AIAA J, 22(11)(1984), 1609.
5. Caspar JR and Verdon JM, AIAA J, 19(12)(1981), 1531.
6. Gorelov D. N., Kurzin V. B. and Saren V. E., Cascaded Airfoils Unsteady Aerodynamic Performance

Album, Nauka(1974), 152.
7. Samoylovich G. S. , Turbomachine Blades Oscillations Exitation, Mashinostroyeniye(1975), 288.
8. Verdon J. M. , Trans ASME, Ser A, 99(4)(1977), 509.
9. Goldstein M. E. , Brawn W and Adamczyk J. J. , J Fluid Mech, 83(3)(1977), 569.
10. Bendiksen O. O. , AIAA J, 24(7)(1986), 1179.
11. Snyder L. E. and Commerford G. L. , Trans ASME, Ser A, 96(4)(1974), 379.
12. Sukhanov A. I. , Turbine Blades Oscillations Aerodynamic Exitation Method Improvement, LPI (1984), 20.
13. Hodson H. P. , Trans ASME, Ser A, 107(2)(1985), 337.

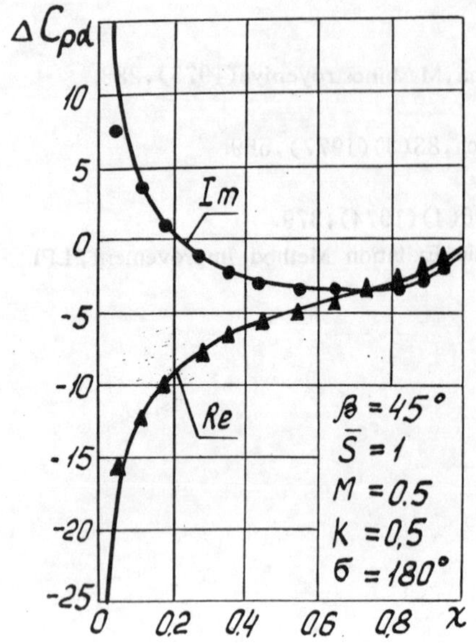

Fig. 1

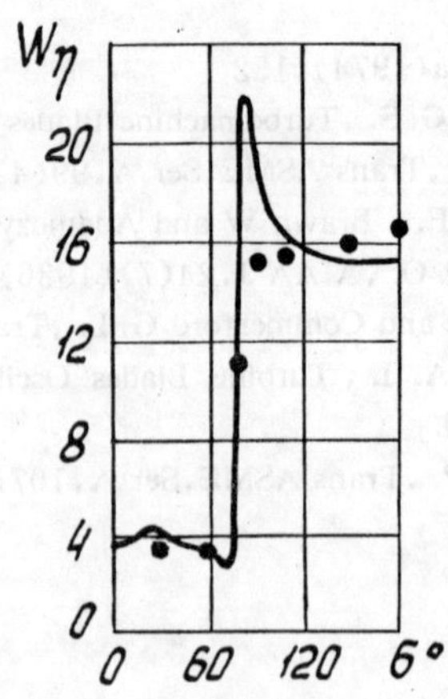

Fig. 2

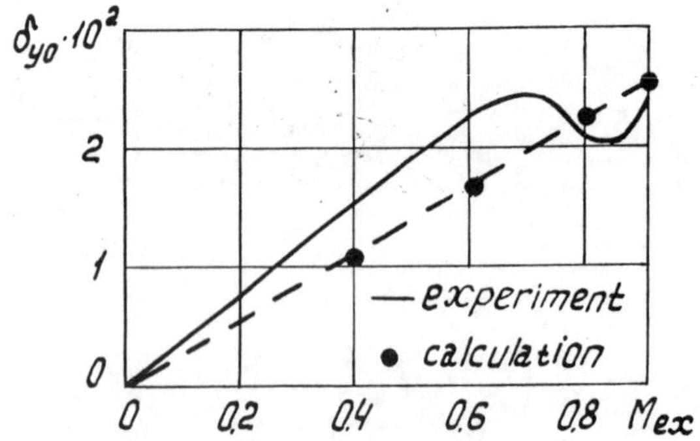

Fig. 3

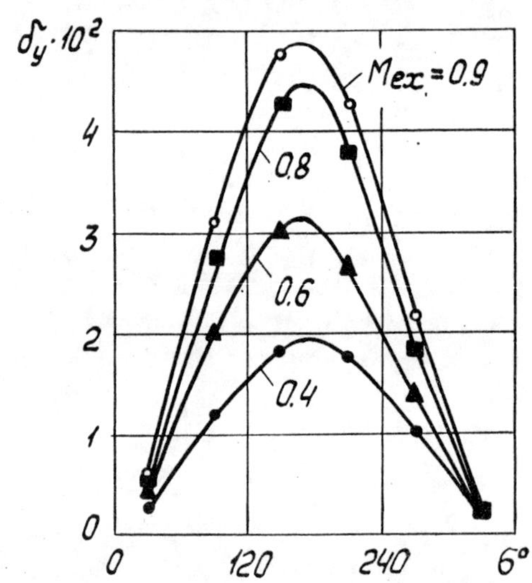

Fig. 4

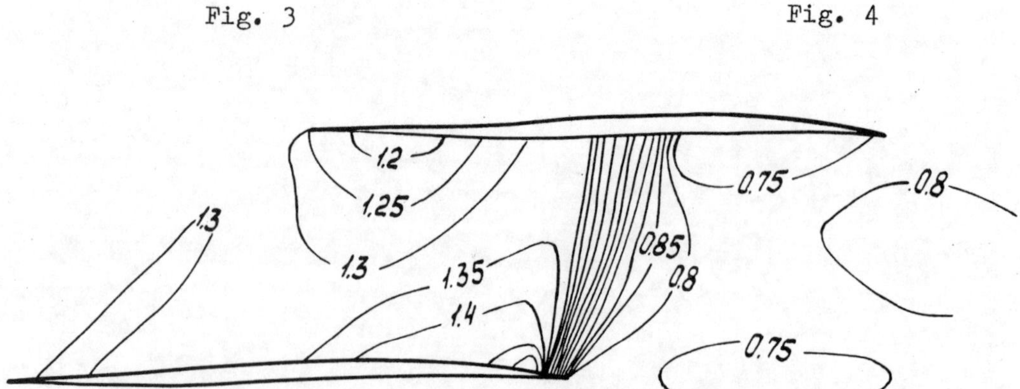

Fig. 5

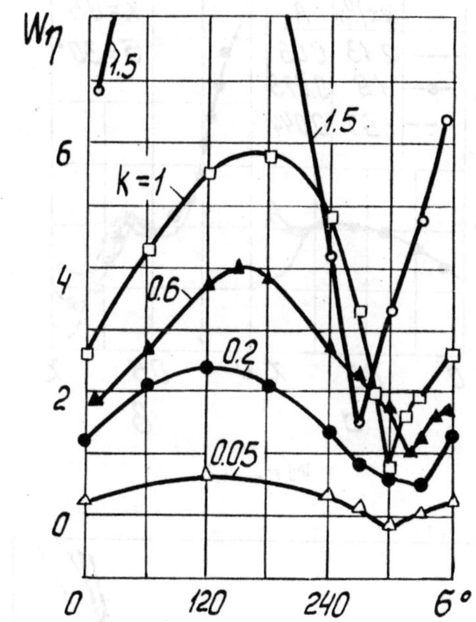

Fig. 6

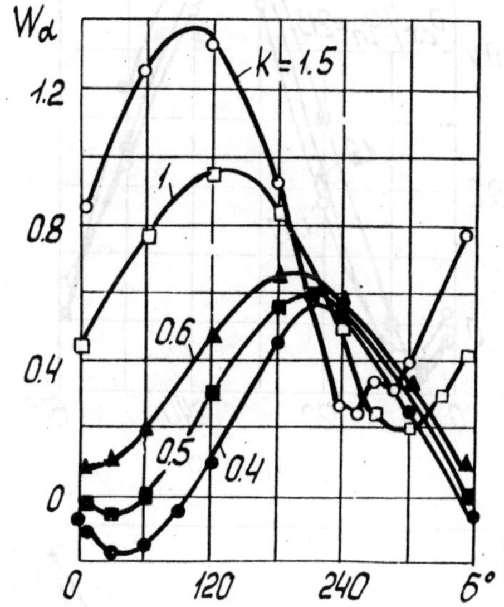

Fig. 7

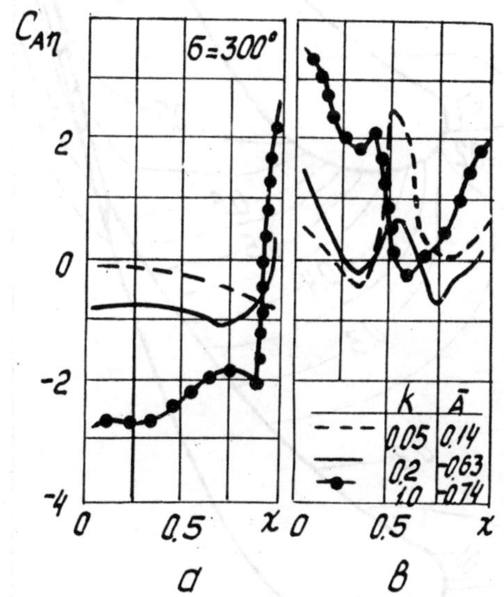

Fig. 8

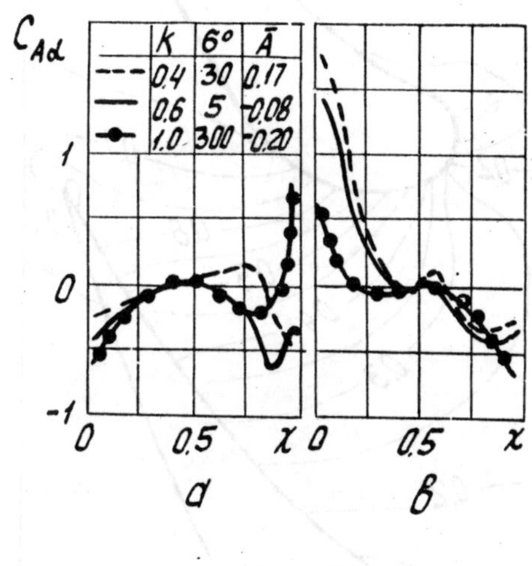

Fig. 9

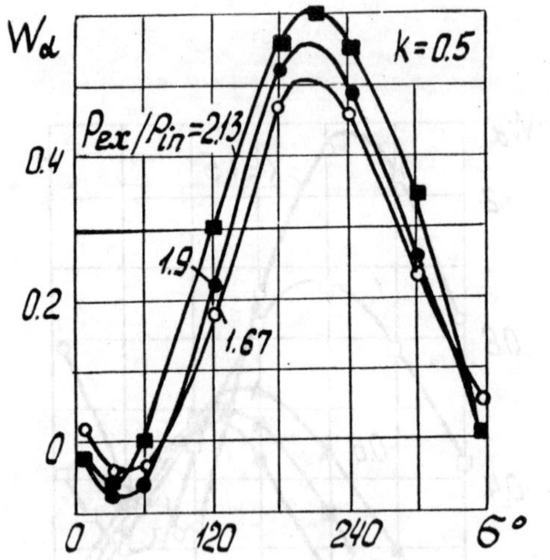

Fig. 10

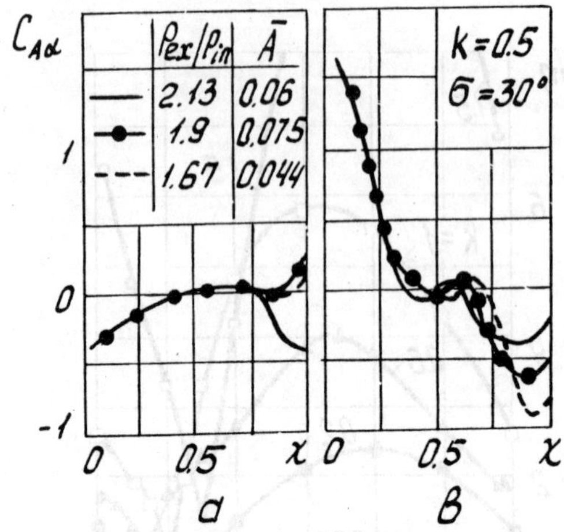

Fig. 11

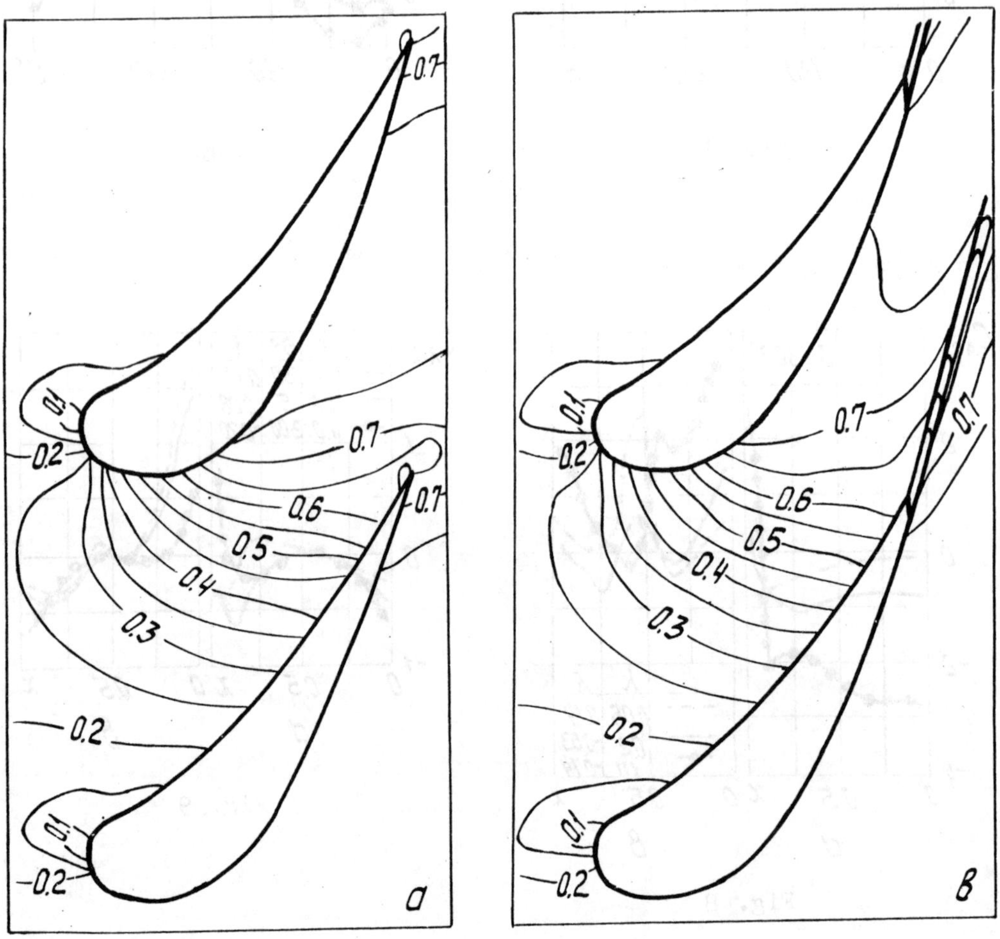

Fig. 12

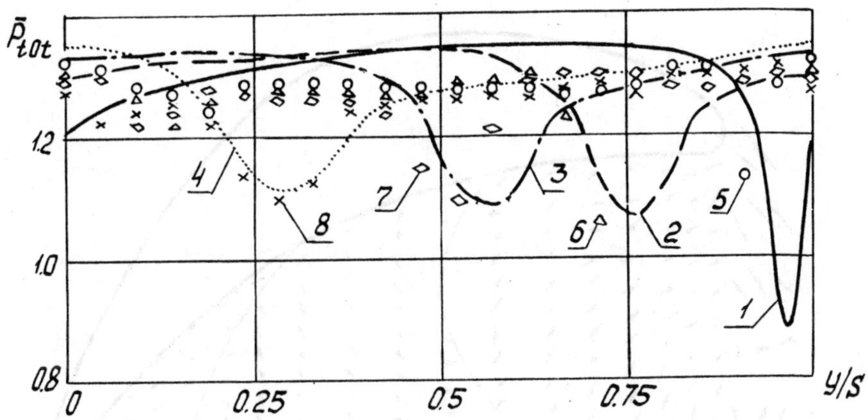

Fig. 13

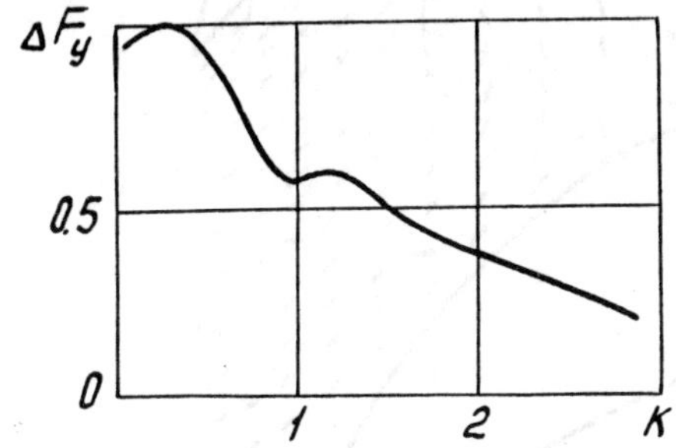

Fig. 14

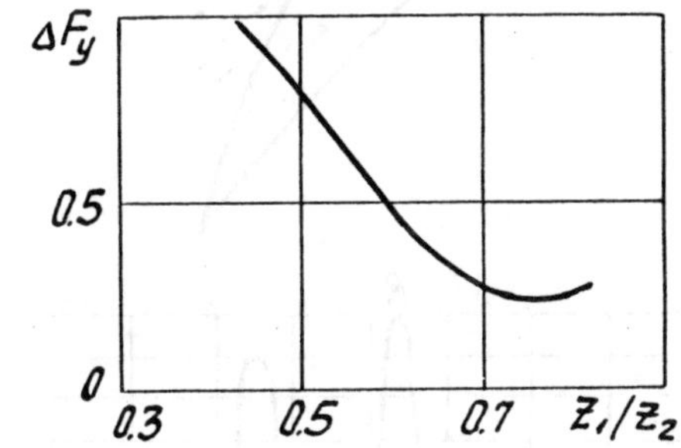

Fig. 15

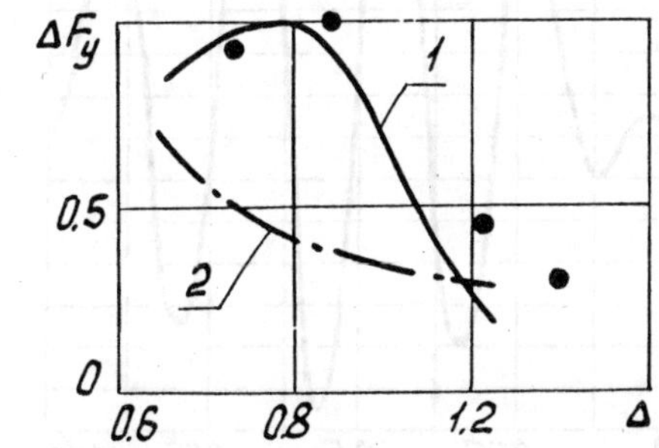

Fig. 16

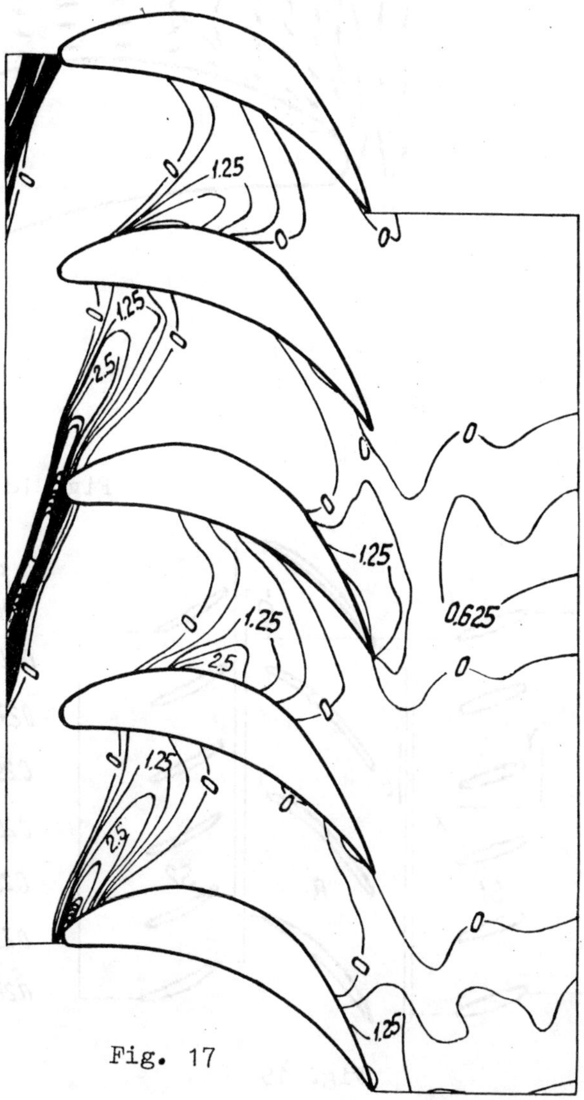

Fig. 17

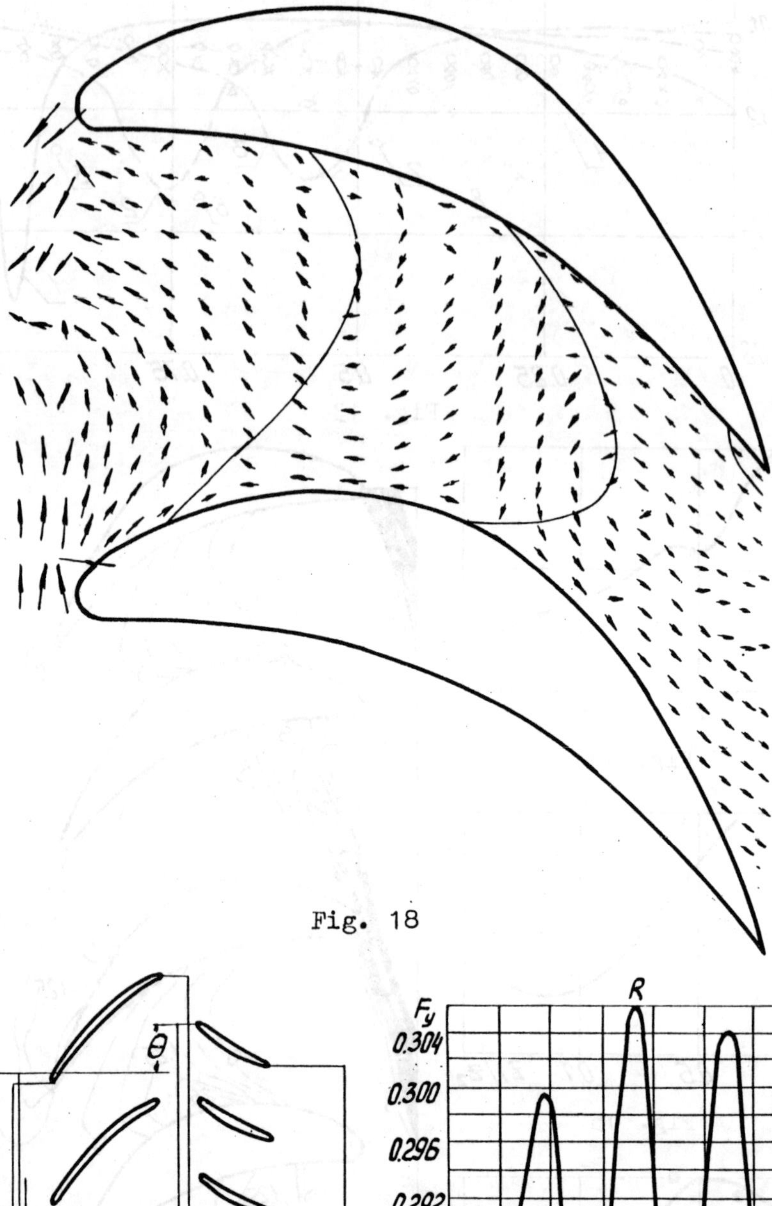

Fig. 18

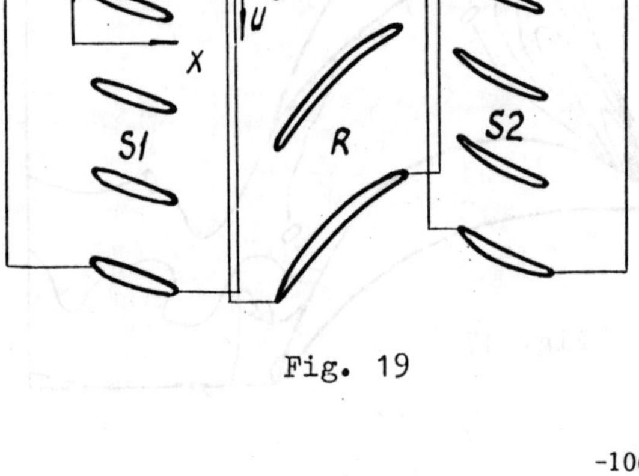

Fig. 19

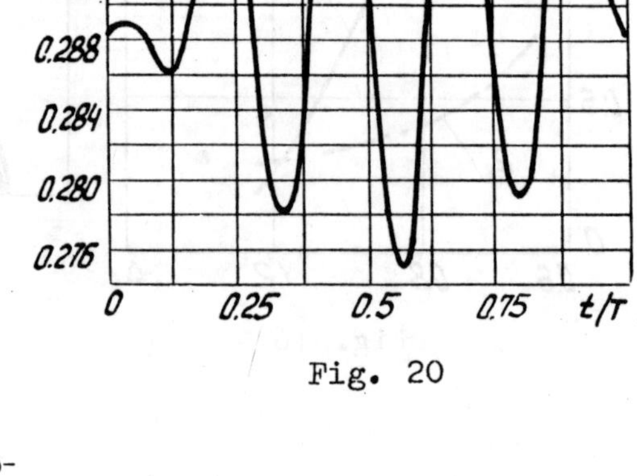

Fig. 20

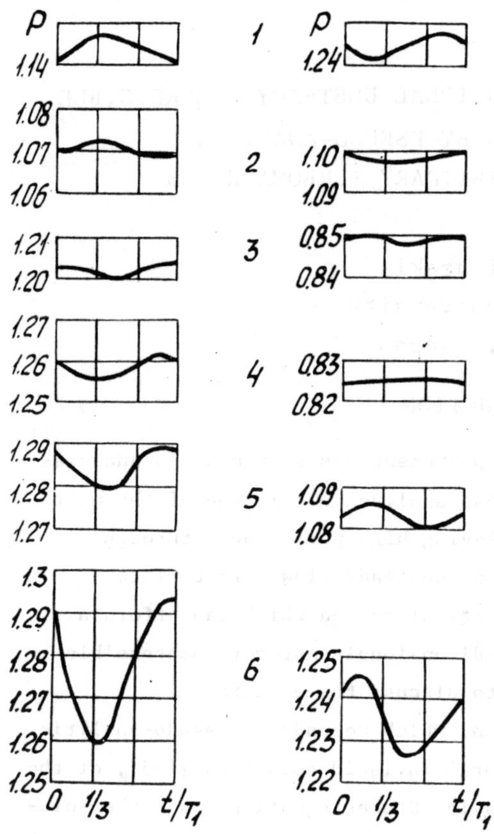

Fig. 21

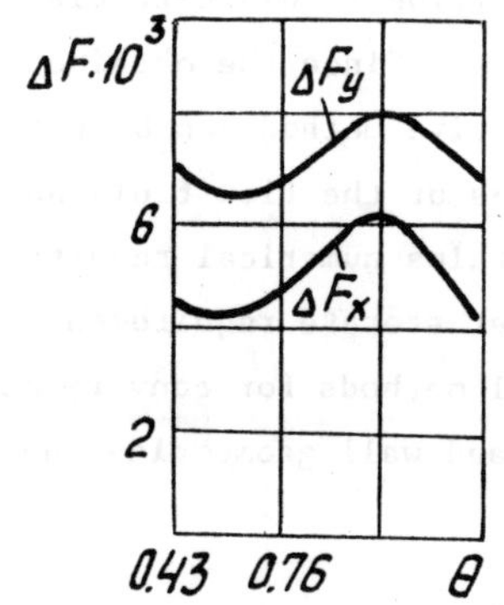

Fig. 22

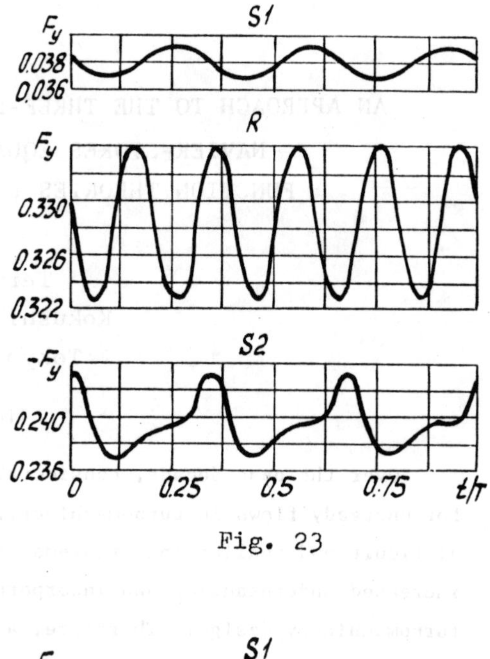

Fig. 23

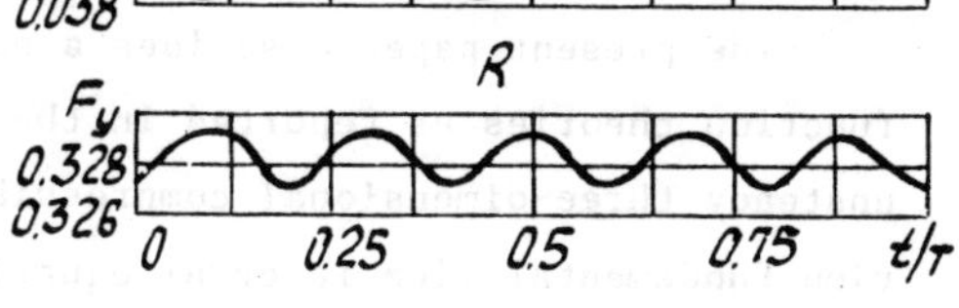

Fig. 24

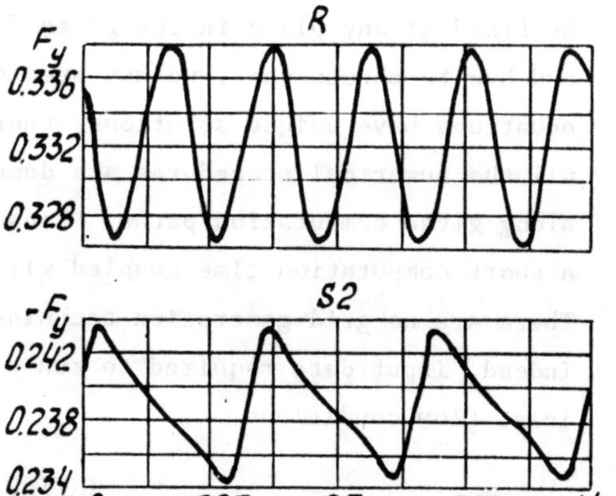

Fig. 25

AN APPROACH TO THE THREE-DIMENSIONAL UNSTEADY COMPRESSIBLE NAVIER-STOKES EQUATION BY PSEUDO-ANALYTIC FUNCTION THEORIES FOR ARBITRARY TURBOMACHINES

Teruo Miyazaki
Kokushikan University
Tokyo 154. Japan

INTRODUCTION

Over the past decade, considerable improvement has been made in analyses for unsteady flows in turbomachinery. These analyses become one of the most difficult and challenging problems in achieving high performance through increased understanding and incorporation of unsteady flow effects into turbomachinery design. Therefore, a theoretical method which can afford a reliable and stable solution of the three-dimensional unsteady compressible Navier-Stokes equation is now inevitable to discuss the effects.

The present paper describes a solution, which depends on pseudo-analytic function theories as reported in the authors' inviscid solutions [1-3], of the unsteady three-dimensional compressible Navier-Stokes equation. For the solution fundamental line integral equations with respect to complex velocities defined on three different kinds of relative stream surfaces, basically owed to Wu's approach [4], are derived from the Navier-Stokes equation and aero-thermodynamic relations. Using cross-sectional stream surfaces [3], which may be fixed at any place in the given flow field, in addition to blade-to-blade and hub-to-shroud ones, we may predict secondary flows. Since the obtained equations have unique solutions, therefore, the iterative method can be employed. All the numerical procedures are done along boundaries of the flow field and along given computation paths. Consequently, one obtains numerical results by a short computation time coupled with minimum computer storage requirements. There are no grid generation techniques nor empirical methods for convergence. Indeed, input data required to run a case are blade and wall geometries and inlet flow conditions.

GOVERNING EQUATIONS

The well-known continuity and momentum equations cast in the blade-relative frame using cylinderical coordinates (x, ϕ, r) are

$$\partial\rho/\partial t + \vec{\nabla}\,(\rho\cdot\vec{W}) = 0 \qquad (1)$$

$$\partial\vec{W}/\partial t - \vec{W}\times(\vec{\nabla}\times\vec{W}) + 2\cdot\omega\times\vec{W}$$
$$= \partial\vec{W}/\partial t - \vec{W}\times(\vec{\nabla}\times\vec{V})$$
$$= -\vec{\nabla}I + T\cdot\vec{\nabla}s + \vec{F}f \qquad (2)$$

For the present flow problem three kinds of relative stream surfaces shown in Fig. 1, for example, are introduced. Also the unit normal vector $\vec{n}$ perpendicular to the relative velocity W satisfies

$$\vec{n}\cdot\vec{W} = 0 \qquad (3)$$

In the numerical computation process we need the energy equation, which defines the pressure from the solution of (2), a state one and boundary conditions.

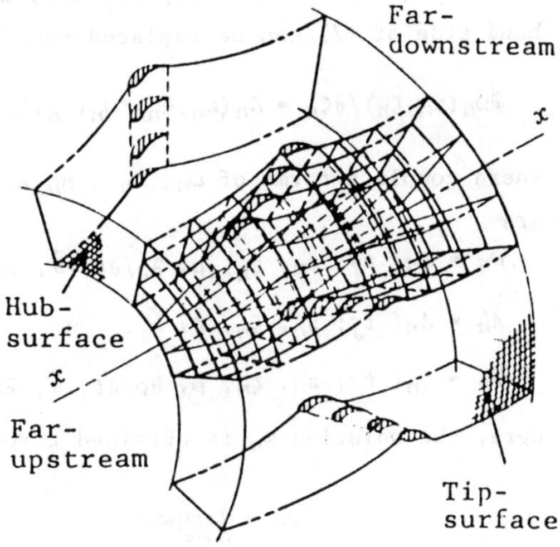

Fig. 1 Three kinds of relative stream surfaces for axial-flow and axial-discharge mixed-flow turbomachines

BASIC RELATIONS

We use complex coordinates and complex velocities on relative stream surfaces to reduce fundamental equations from governing ones [1-3], such that

$$z_n = \xi_n + i\cdot\eta_n \quad , \quad \zeta_n = \xi_n - i\cdot\eta_n \qquad (\ n = 1,2,3\ ) \qquad (4)$$
$$w_n = W_p - i\cdot W_q, \quad (\ p \neq q,\ p,q = x,\phi,r\ ) \qquad (5)$$

No essential difference exists between both methods for axial-flow and axial-discharge mixed-flow turbomachines and for radial-flow and radial-discharge mixed-flow ones. Therefore, we limit our description only on procedures for the former turbomachines without loss of generality hereafter.

FUNDAMENTAL DIFFERENTIAL EQUATIONS

We write the formal derivatives with respect to z_n and ζ_n such that

$$\partial/\partial z_n = 1/2\cdot(\ \tilde{\partial}/\tilde{\partial}\xi_n - i\cdot\tilde{\partial}/\tilde{\partial}\eta_n\ )\ ,\quad \partial/\partial\zeta_n = 1/2\cdot(\ \tilde{\partial}/\tilde{\partial}\xi_n + i\cdot\tilde{\partial}/\tilde{\partial}\eta_n\ ) \qquad (6)$$

Then we have the formal derivative of w_n with respect to ζ_n as

$$\partial w_n/\partial\zeta_n = 1/2\cdot(\ \tilde{\partial}W_p/\tilde{\partial}\xi_n + \tilde{\partial}W_q/\tilde{\partial}\eta_n\ ) - i/2\ (\ \tilde{\partial}W_q/\tilde{\partial}\xi_n - \tilde{\partial}W_p/\tilde{\partial}\eta_n\ ) \qquad (7)$$

Using (1) and (2) respectively, the real and imaginary parts of the right hand side of (7) can be replaced such that [3-5]

$$\partial w_n(z_n,\zeta_n)/\partial\zeta_n = A_n(z_n,\zeta_n)\cdot w_n(z_n,\zeta_n) + B_n(z_n,\zeta_n)\cdot\overline{w_n(z_n,\zeta_n)} + C_n(z_n,\zeta_n) \quad (8)$$

where complete forms of w_n, A_n , B_n and C_n at arbitrary time instant $t = t_j$ are

$$w_n = w_n(\ t_j;\ z_n,\ \zeta_n,\ \rho,\ \partial\rho/\partial t,\ \vec{W},\ \partial\vec{W}/\partial t,\ \vec{F}_f\ ) \quad (9)$$

$$A_n = A_n(\ t_j;\ z_n,\ \zeta_n,\ \rho,\ \vec{W}\ ) \quad (10),(11)$$

$$C_n = C_n(\ t_j;\ z_n,\ \zeta_n,\ \rho,\ \partial\rho/\partial t,\ \vec{W},\ \partial\vec{W}/\partial t,\ \vec{F}_f\ ) \quad (12)$$

Here, the solution w_n is obtained through the iteration such that

$$w_n = \lim_{N\to\infty}[w_n]_N \quad (13)$$

FUNDAMENTAL INTEGRAL EQUATIONS

From the pseudo-analytic function theory [5] every function w_n of z_n and and ζ_n given by (8) satisfies the following integral equation

$$w_n(z_n,\zeta_n) = W_{n,i}(z_n) - \frac{1}{\pi}\cdot\iint_{D_n}\frac{\partial w_n/\partial\zeta_n'}{z_n'-z_n}\,d\xi_n'd\eta_n' \quad (14)$$

where $W_{n,i}$ is an arbitrary holomorphic function. On the other hand, by forming the Cauchy integral of both sides of (14), we obtain [5]

$$W_{n,i}(z_n) = \frac{1}{2\cdot\pi\cdot i}\cdot\oint_{\Gamma_n}\frac{w_n(z_n',\zeta_n')}{z_n'-z_n}\,dz_n' \quad (15)$$

where Γ_n is the boundary of flow field D_n.

Now the pseudo-analytic function w_n on the left hand side of (8) can be replaced with the other one Φ_n, which has a difference of an arbitrary holomorphic function and satisfies.

$$\partial\Phi_n(z_n,\zeta_n)/\partial\zeta_n = \partial w_n(z_n,\zeta_n)/\partial\zeta_n \quad (16)$$

Hence we may assume the function w_n on the respective side of (8) as two different ones hereafter.

For the convenience of numerical computation, then, we reduce (14) into a line integral equation such that [1-3]

$$w_n(z_n,\zeta_n) = W_{n,i}(z_n) + \Phi_n(z_n,\zeta_n)$$
$$-1/(2\cdot\pi\cdot i)\cdot\oint_{\Gamma_n}[\Phi_n(z_n',\zeta_n')/(z_n'-z_n)]dz_n' \quad (17)$$

Finally the fundamental equations in the present paper are (8), (15) and (17).

NUMERICAL PROCEDURES

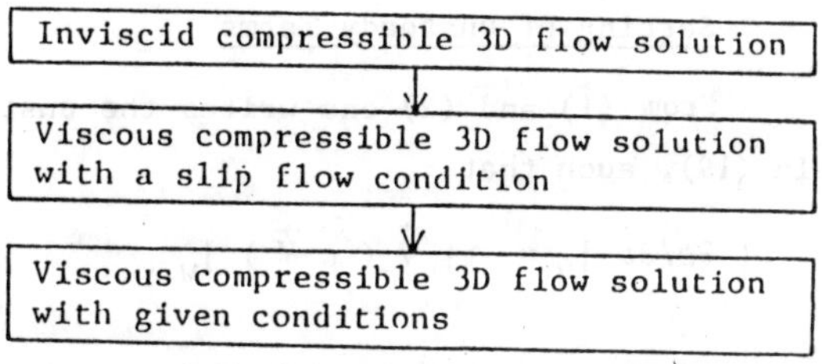

Fig. 2 Scheme of viscous compressible 3D flow computation

The whole directional components of the continuity equation (1) and the momentum one (2) and the other relations of thermodynamics are to be satisfied through the iterative process just in the same way as in [1-3]. Fig. 2 shows the outlines of the present numerical method. We will describe numerical procedures of viscous and unsteady terms belong with boundary conditions below.

Computation of Viscosity Term

The formal derivative of the left hand side of (7) can be written as

$$4\cdot\partial(\,\partial w_n/\partial\zeta_n\,)/\partial z_n = \tilde{\partial}^2 W_p/\tilde{\partial}\xi_n^2 + \tilde{\partial}^2 W_p/\tilde{\partial}\eta_n^2 - i\cdot(\,\tilde{\partial}^2 W_q/\tilde{\partial}\xi_n^2 + \tilde{\partial}^2 W_q/\tilde{\partial}\eta_n^2\,) \quad (18)$$

Using (18), in the iteration process at the instant $t = t_j$, we have the formal derivative (7) and the viscosity term $\vec{F}_f$ as follows. In the first place we set

$$[\,\partial[w_n]_{N-1}/\partial\zeta_n\,]_N^* = [A_n]_N\cdot[w_n]_{N-1} + [B_n]_N\cdot\overline{[w_n]_{N-1}} + [\,C_n(\,t_j;\,z_n,\,\zeta_n,\,[\rho]_{N-1},\,[\partial\rho/\partial t]_{t=t_{j,M}},\,[\vec{W}]_{N-1},\,[\partial\vec{W}/\partial t]_{t=t_{j,M}},\,[\vec{F}_f]_{N-1}\,)\,]_N^* \quad (19)$$

where M shows a repetition number with respect to time instant.

Then the formal derivative shown in (19) can be numerically obtained. Now the components of viscosity term is expressed as

$$[\vec{F}_f]_N = [\,\vec{F}_f(\,t_j;\,z_n,\,\zeta_n,\,[\vec{W}]_{N-1},[\,\partial[w_n]_{N-1}/\partial\zeta_n\,]_N^*,\,\partial[\,\partial[w_n]_{N-1}/\partial\zeta_n\,]_N^*/\partial z_n\,)\,]_N \quad (20)$$

Hence we obtain the formal derivative of w_n with respect to ζ_n in the iteration process such that

$$[\,\partial[w_n]_{N-1}/\partial\zeta_n\,]_N = [A_n]_N\cdot[w_n]_{N-1} + [B_n]_N\cdot\overline{[w_n]_{N-1}} + [\,C_n(\,t_j;\,z_n,\,\zeta_n,\,[\rho]_{N-1},[\partial\rho/\partial t]_{t=t_{j,M}},\,[\vec{W}]_{N-1},\,[\partial\vec{W}/\partial t]_{t=t_{j,M}},\,[\vec{F}_f]_N\,)\,]_N \quad (21)$$

We assume the first approximation of $\vec{F}_f$ as $[\,\vec{F}_f\,]_0 = 0$

Setting of Unsteady Terms

From (1) and (2) one writes the unsteady terms $[\partial\rho/\partial t]_M$ and $[\partial W/\partial t]_M$ shown in (19), such that

$$[\ \partial\rho/\partial t\ ]_M = -\ [\ \vec{\nabla}\ (\ \rho\cdot\vec{W}\ )\ ]_{M-1} \tag{22}$$

$$[\ \partial\vec{W}/\partial t\ ]_M = [\ -\ \vec{\nabla}I + T\cdot\vec{\nabla}s + \vec{F}_f\ ]_{M-1} + [\ \vec{W}\times(\ \vec{\nabla}\times\vec{W}\ ) - 2\cdot\omega\times\vec{W}\ ]_{M-1} \tag{23.a}$$

or

$$= [\ -\ \vec{\nabla}I + T\cdot\vec{\nabla}s + \vec{F}_f\ ]_{M-1} + [\ \vec{W}\times(\ \vec{\nabla}\times\vec{V}\ )\ ]_{M-1} \tag{23.b}$$

Here functions on the right hand sides of (22) and (23) are known with respect to each time step and they also include the unsteady terms obtained in the preceding step. We, therefore, give the terms through the following two schemes.

For starting several time steps we compute the viscous compressible three-dimensional flow at each instant t_j in (t_o, t_f), where t_o stands for the initial and t_f the final instant. Using a well-known series expansion rule, for example, we assume $[\rho]_M$ and $[\vec{W}]_M$ calculated discretely, as being continuous. Then the unsteady terms, the derivatives of $[\rho]_M$ and $[W]_M$, are fixed numerically. After the several time steps we repeat the procedures (22) and (23) until the convergence will be attained.

A NUMERICAL EXAMPLE

The code simplified with the assumption of a slip flow condition to consider the role of unsteady terms, $\partial\rho/\partial t$ and $\partial\vec{W}/\partial t$, in the Navier-Stokes equation, is applied to the study of the transonic axial flow through the overhung rotor without inlet guide vanes, with the cylindrical casing [6,7] were obtained with a tip relative Mach number of 1.3, radially constant stagnation temperature rise and uniform inlet flow condition. For the given flow we employed a simplified mathematical model as a flow through a single blade row in an annular duct with uniform axial inlet flow and pitchwise symmetricity assumptions. The boundary condition at far-upstream and -downstream was

$$V_r = 0,\quad \partial V/\partial\phi = 0 \tag{24}$$

Further simplifications as Prandtl number of 1.0 and adiabatic walls were also made and the tip clearance effect was omitted.

Now the observed Mach number distributions [7] show pitchwise asymmetricity. The presence of possible shock wave-boundary layer interaction effects in the blade passages [6] might be the main cause of the asymmetricity. However, the weak oscillation of inlet flow around a mean value was also reported [6]. Therefore, to illustrate the unsteadiness effect of inlet axial Mach number on the flow, we assumed the inlet Mach number to vary sinusoidally around the given value with an amplitude of 0.01, as an example.

Calculations were carried out using six S1, two S2 and an S3 surface, where the experimental data were measured [6]. Sinusoidal oscillation of inlet axial Mach number was expressed with 9 instants in each time step. The reference points of 60 stood for the given blade section in respective S1 surface. Also the boundaries of S2 and S3 surfaces were represented by 150 and 104 reference points, respectively. The code is written in FORTRAN for an M680H; it requires a memory of 2MB, as working volume. The total calculation converged after 10 iterations of the time steps, satisfying a convergence criterion of $\varepsilon = 0.5 \times 10^{-3}$. The number of repetitions N, needed to obtain converged viscous compressible flow at each instant was gradually reduced as the scheme advanced. Moreover, the stability in the whole computation procedures was kept very well. The CPU time was about 3 hours.

Fig. 3 shows computed and measured outlet air angles. The mean values of computed angles show good agreement with experimental values. Fig. 4 shows Mach number distributions at 0.1-axial-chord-down-stream of the rotor. Due to the unsteadiness of inlet flow condition, calculated Mach number distributions fluctuated and then are shown in the figure by banded regions enclosed by separated solid lines, which stand for converged values. Also broken lines show the first approximations, which satisfy the steady Navier-Stokes equation for each instant. Each banded regions are composed of nine lines obtained at different time instants. The good coincidence between computed and measured values is attained in average. Comparing the first approximations and converged values with the experimental results, the effect of unsteady terms, even so weak, may not be negligiable to consider the asymmetry.

CONCLUSIONS

A theoretical solution of the unsteady compressible Navier-Stokes equations in turbomachinery geometries

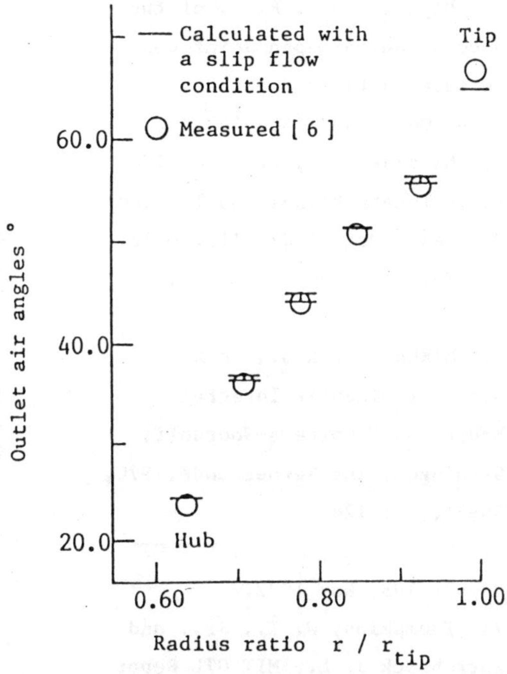

Fig. 3 Outlet air angles at different radius ratios

has been developed using the pseudo-analytic function theories. Its numerical code was applied to the study of the unsteady flow phenomenon in a simplified model. Despite the employed assumptions, a good example to explain a significant role of the unsteady terms in the Navier-Stokes equation is obtained.

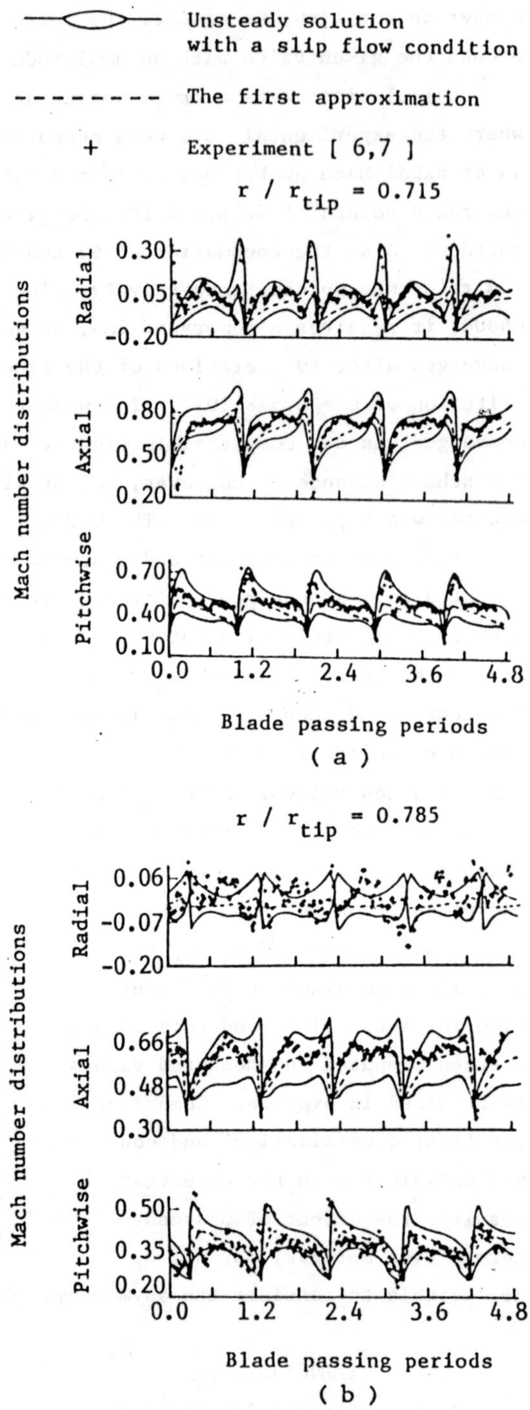

Fig. 4 Mach number distributions at 0.1-axial-chord-downstream of rotor

REFERENCES

1. Miyazaki, T., and Hirayama, N., ASME Paper 86-GT-111, June 1986.
2. Miyazaki, T., Proc. of the Second China-Japan Joint Conference on Fluid Machinery, Xian, Oct. 1987, 456-463.
3. Miyazaki, T., Proc. of 1987 Tokyo International Gas Turbine Congress, Oct. 1987, II265-272.
4. Wu, C-H., NACA TN 2604, 1952.
5. Mikhailov, L.G., "A New Class of Singular Integral Equations," Wolters-Noordoff, Groningen, The Netherlands, 1970, 66-71, 121-124.
6. Kerrebrock, J. L., MIT GTL Report 108, May 1972.
7. Thompkins, W. T., Jr., and Kerrebrock J. L., MIT GTL Report 123, Sept. 1975.

RESEARCH ON AXIAL FLOW COMPRESSOR RESPONSE TO INLET PERIODICAL DYNAMIC DISTORTION

Zhang Hui Min, Feng Qi
(Nanjing Aeronautical Institute,Nanjing,China)

ABSTRACT

In this paper, a theoretical model is presented to predict the response of an axial–flow compressor to periodic dynamic distortion.A new flow field model is established by dividing a blading channel into three regions for the purpose of increasing the maximum response–frequeucy of the blading.Numerical calculations are satisfied with the experimental data taken from a two–stage axial–flow compressor with five perturbation frequencies.As shown both in calculations and experiments,the axial compressor responds differently to different frequeney,but there is a certain frequeucy to which the tested compressor is very sensitive and its performance may be deteriorated rapidly under this unique frequency which we call "sensitive frequeney".

INTRODUCTION

The inlet distortion for the compressor is an important and difficult problem, the solution of which has been still far from satisfaction even after several decades' research.In the past (even nowadays),most researchers concentrated their attentions to the steady distortion (especially circumferential distortions) with respect to absolute coordinate system 1,2,3.Those theoretical and experimental results achieved so far are conducive to understanding the mechanism of unsteady flow field and its influences on performance of the compressor due to the appearance of inlet distortion.On the other hand, the unsteady (or dynamic) distortion which is more complex both in theory and experiment is relatively less explored in comparison with steady distortion.As well known,there are two categories of unsteady distortion with regard to absolute coordinate system.One is called "random distortion" and the other belongs to that having some definite laws such as sinusoidal (or periodic sinusoidal),square wave disturbances and so on.For the time being,the aim of this paper is solely concerned with establishing unsteady flow model (especially that of periodic dynamic distortion) and corresponding numerical method to solve the latter.

FLOW FIELD MODEL

Under the action of inlet distortion,the flow field in compressor is generally unsteady viscous three dimensional and is too complex to be simulated both numerically and experimentally unless some simplifying assumptions are made.
According to the " traces of substantial surface" concept first proposed in 5,an unchanged (time

indepedent)trace of certain   tial surface could be found, the flow on which is unsteady two dimensional and could be simulated approximately the unsteady three dimensional flow field in case of inlet dynamic distortion.In our case when the axial compressor is a two–stage,low pressure and constant blade height fan,the unchanged trace of substantial surface is a cylindrical surface at r=0.215m.For those axial compressors with varying annular area along the flow path and large amplitude variation of fluid flow encountered by high pressure compressors,the trace of substantial surface is likewise according to the "trace theory"but is some what complicated to be determined.

For cylindrical trace of substantial surface,the flow field is unsteady two dimensional and each stage of the compressor is composed by bladed region (including rotor and stator)and blade–free region (axial space between rotor and stator and other parts without blades).

For the purpose of increasing the maximum response frequeucy of the blading,the bladed channel is further divided by three regions as shown in Fig.1 and are defined as follows:

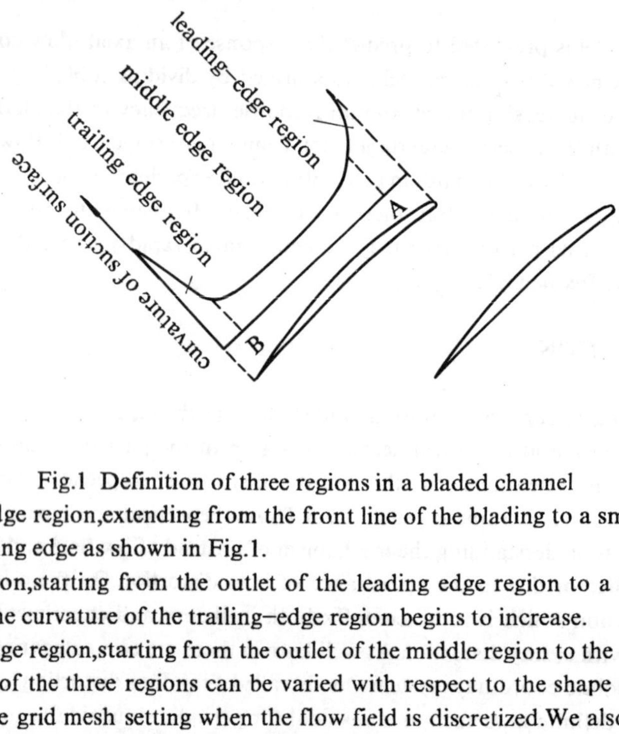

Fig.1 Definition of three regions in a bladed channel

A.Leading–edge region,extending from the front line of the blading to a smaller curvature point (point A) at the leading edge as shown in Fig.1.

B.Middle region,starting from the outlet of the leading edge region to a point (point B as shown in Fig.1)where the curvature of the trailing–edge region begins to increase.

C.Trailing–edge region,starting from the outlet of the middle region to the outlet of the trailing–edge.

Demarcation of the three regions can be varied with respect to the shape of the blade profile and adjusted with the grid mesh setting when the flow field is discretized.We also assume that all losses and angle turning of the blading are concentrated in leading–edge and trailing–edge regions only.In middle region,the flow is assumed one dimensional nonviscous with somewhat variation in channel area.

From the above–mentioned, this flow model belongs also to "semi–actuator disc" but with minute flow description of the channel.

GOVERNING EQUATIONS IN DIFFERENT FLOW REGIONS

Generally speaking,the unsteady viscous 3–D flow field in turbomachinery is governed by unsteady Navier–Stokes equations and may be simplified and take different forms which are suitable for different regions in our flow model (i.e. unsteady 2–D viscous flow).

1. BLADE–FREE REGION

In this region, as we neglect viscosity of the fluid, the N–S equations are transformed into unsteady 2–D Euler equations in conservative form as follows:

$$\frac{\partial \vec{U}}{\partial t}+\frac{\partial \vec{F}}{\partial x}+\frac{\partial \vec{G}}{\partial y}=0 \tag{1}$$

$$U=\begin{pmatrix}\rho\\ \rho u\\ \rho v\\ e\end{pmatrix} \qquad F=\begin{pmatrix}\rho u\\ p+\rho u^2\\ \rho uv\\ u(p+e)\end{pmatrix} \qquad G=\begin{pmatrix}\rho v\\ \rho uv\\ p+\rho v^2\\ v(p+e)\end{pmatrix}$$

where $e=p/(k-1)+\rho/2(u^2+v^2)$; x,y are axial and circumferential coordinates and u,v are velocity components along x,y axis respcetively.

2.MIDDLE REGION

In this region, we assume the flow is unsteady, nonviscous, adiabatic one–dimensional but with some flow area variation in flow direction. Its governing equations are:

$$\frac{\partial \vec{U}}{\partial t}+\frac{\partial \vec{F}}{\partial l}=0 \tag{2}$$

$$U=\begin{pmatrix}\rho A\\ \rho A w_l\\ e\end{pmatrix} \qquad F=\begin{pmatrix}\rho A w_l\\ pA+\rho A w_l^2\\ w_l\,(Ap+e)\end{pmatrix}$$

where $e=pA/(k-1)+w_l^2\rho A/2$. l, $w_l$are coordinate and velocity along flow direction respectively.

3. LEADING–EDGE AND TRAILING–EDGE REGIONS

In these two regions, we express the law of conservation of mass, momentum and energy in integral forms. Since the length(L) of these two regions are relative short in comparison with the blade chord, variations of the flow parameters from inlet to outlet in these short regions may be taken approximately as linear. Assumming also in these short regions the flow is incompressible and neglecting further the heat transfer and the viscous work along the boundary, the equations of continuity, momentum and energy may be respectively written in the following form after integration:

$$\frac{L}{2}\left(\frac{\partial \rho_1}{\partial t}+\frac{\partial \rho_2}{\partial t}\right)+\rho_2 u_2-\rho_1 u_1=0 \tag{3}$$

$$\frac{L}{2}\left[\frac{\partial}{\partial t}\left(\frac{\rho_1 w_1^{\,2}}{2}\right)+\frac{\partial}{\partial t}\left(\frac{\rho_2 w_2^{\,2}}{2}\right)\right]+u_2\left(p_2+\frac{\rho_2 w_2^{\,2}}{2}\right)-u_1\left(p_1+\frac{\rho_1 w_1^{\,2}}{2}\right)$$

$$-\frac{1}{2}(p_1+p_2)(u_2-u_1)=-\frac{1}{8}\rho_{10} w_{10}^{\,2}(u_1+u_2)\omega \tag{4}$$

$$\frac{L}{2}\frac{1}{k-1}\left(\frac{\partial p_1}{\partial t}+\frac{\partial p_2}{\partial t}\right)+\frac{1}{2}(p_1+p_2)+\frac{1}{k-1}(p_2u_2-p_1u_1)$$
$$=\frac{1}{8}\rho_{10}w_{10}{}^2(u_1+u_2)\omega \qquad (5)$$

where Subscript "10" is cascade inlet; 1 and 2 are inlet and outlet of that region; L is the length of that region.

As mentioned above, all losses in the blade channel are assumed to be concentrated in these two regions. If ω is the total loss coefficient of the total pressure loss in the channel and equally distributed in these regions, then the loss coefficient ω' in each region is $\omega'=\omega/2$. These unsteady ω may be obtained through the correction of its steady value ω_s by the generally used equation:

$$\omega=\omega_s-\tau_\omega d\omega/dt \qquad (6)$$

where, ω_s is the steady value of ω obtained from experiment or known cascade characteristics; τ_ω is time lag coefficient. This value may be obtained through unsteady flow experiment or predicted by empirical equation such as recommended in 4.

As regards the outlet flow angle β, the relation between its unsteady and steady value has the same form as equation(5), i.e.

$$\beta=\beta_s-\tau_\beta d\beta/dt \qquad (7)$$

where β_s—steady outlet flow angle obtained from steady experiment or known cascade characteristics; τ_β—time lag of outlet flow angle, also may be taken from experiment or calculated by a empirical relation 4.

In addition to the above six equations connecting inlet and outlet parameters, there is another relation between outlet velocities v_1 and u_2:

$$v_2=u_2 tg\beta \qquad (8)$$

NUMERICAL METHOD AND BOUNDARTY CONDITIONS

1.NUMERICAL METHOD

Mac Cormack's predictor–corrector two step difference method was used to solve flow equations in blade–free region and middle region in blading.

For model equation

$$\frac{\partial U}{\partial t}+\frac{\partial F}{\partial x}+\frac{\partial G}{\partial y}=0$$

$$U_{i,j}^{\overline{n+1}}=U_{i,j}^{n}-S_x(F_{i,j}^{n}-F_{i-1,j}^{n})-S_y(G_{i,j}^{n}-G_{i,j-1}^{n})$$

$$U_{i,j}^{n+1}=[U_{i,j}^{n}+U_{i+1,j}^{\overline{n+1}}-S_x(F_{i+1,j}^{\overline{n+1}}-F_{i,j}^{\overline{n+1}})-S_y(G_{i,j+1}^{\overline{n+1}}-G_{i,j}^{\overline{n+1}})]/2$$

where $S_x=\Delta t/\Delta x, S_y=\Delta t/\Delta y, \Delta t$ should satisfy CFL condition.

2.BOUNDARY CONDITIONS

In this research,conditions given at compressor inlet and outlet are:inlet total pressure p^*,temperature T^*,inlet flow angle and outlet static pressure.

Conditions needed in the numerical procedure are determined by differential relations along characteristics.

A general program which can be used to the simulation of various types of periodic dynamic distortion flow in axial compressors has been compiled according to the preceding flow model and numerical method.

CALCULATING EXAMPLE

Computations and comparisons with experiments were made on the cylindrical stream surface at r=0.215m in a two–stage fan.Normal cascade characteristics (at r=0.215 m) of the compressor rotors and stators are shown in Fig.2. Rectangular grid mesh was used with 55 grid points equally spaced along circumference.The compressor flow fields were simulated by 5 perturbation frequencies of

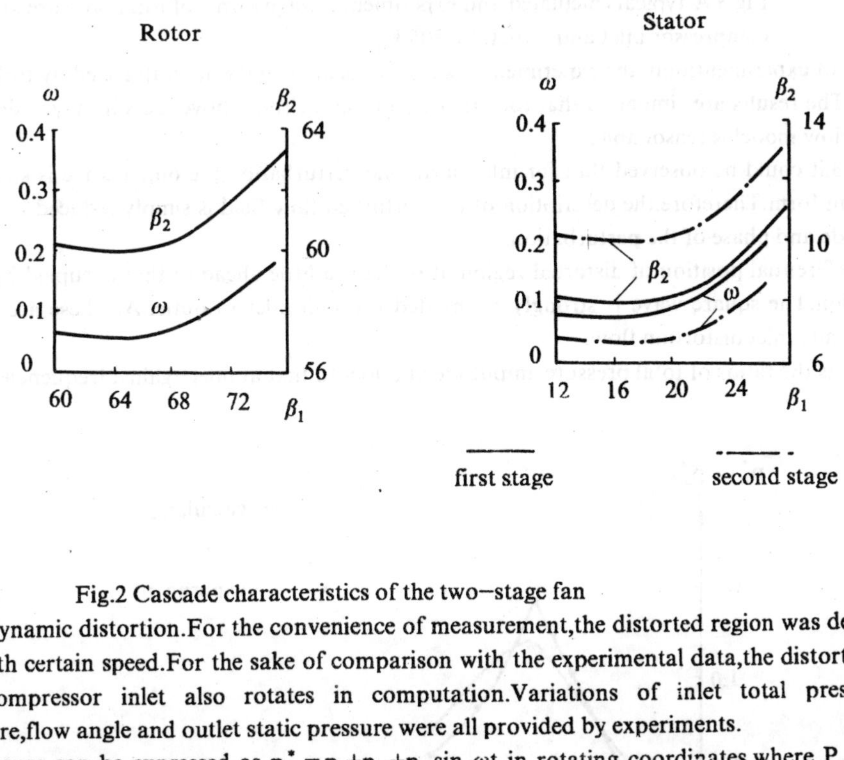

Fig.2 Cascade characteristics of the two–stage fan

periodic dynamic distortion.For the convenience of measurement,the distorted region was designed to rotates with certain speed.For the sake of comparison with the experimental data,the distorted region at the compressor inlet also rotates in computation.Variations of inlet total pressure,total temperature,flow angle and outlet static pressure were all provided by experiments.

Total pressure can be expressed as $p^* = p_o + p_{sq} + p_{ac}\sin \omega t$ in rotating coordinates,where $P_{sq}$is square wave and $\omega = 2\pi f$;five frequencies f=280,305,320,345,365 $H_z$were used in computation.Inlet total temperature,flow angle and outlet static pressure are all kept constant.The results of computation are summarized as follows.

A typical calculated and experimental wave forms of total pressure at compressor inlet and cutlet are shown in Fig.3. It can be seen that the calculated amplitude and phase of the perturbation agree well

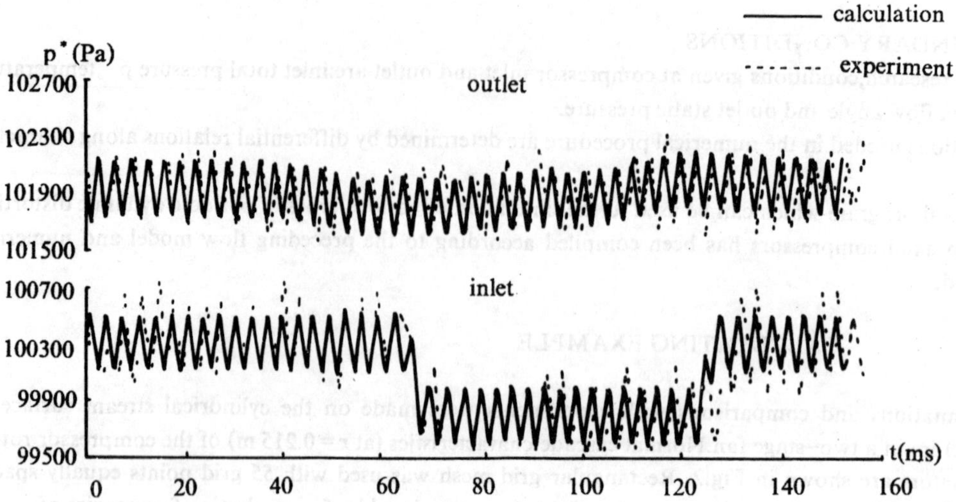

Fig.3 A typical calculated and experimental wave forms of total pressure at compressor inlet and outlet, f = 305 H_z.

with those of experiment;only the experimental wave forms are some what influenced by turbulent disturbance. The results are similar to that for other frequencies.From above facts it maybe deduced the proposed flow model is reasonable.

From Fig.3,it could be observed that for inlet harmonic perturbation,the outlet wave is still basically harmonic in form.Therefore,the description of the perturbed flow field is simply reduced tc variations of amplitude and phase of the perturbation.

The circumferential position of distorted region at outlet is a little ahead of that occupied by inlet distorted region.The square wave is strongly attenuated through inlet to outlet.All these are similar to those of steady inlet distortion flow.

2. Fig.4 plots the ratios of total pressure amplitude at outlet to that at inlet against frequencies of inlet

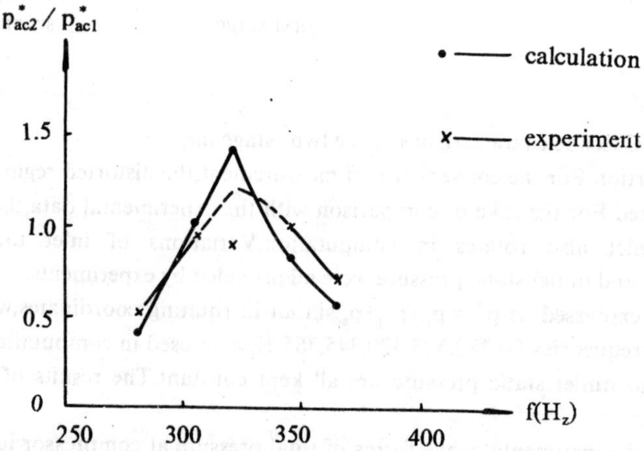

Fig.4 Variation of the ratio of amplitude of outle total pressure perturbation to that of static pressure with frequencies.

perturbation.It can be seen that the theoretical and experimental results agrees well except at f=320 H_z(it may be a strange point with respect to experimental data).

The variations of amplitude and phase of total pressure in high pressure region and low pressure region through the compressor are shown in Fig.5 at f=305 H_z.It shows that the amplitudes and phases are approximetely equal in these two regions, which suggests that the inlet plane waves of perturbation

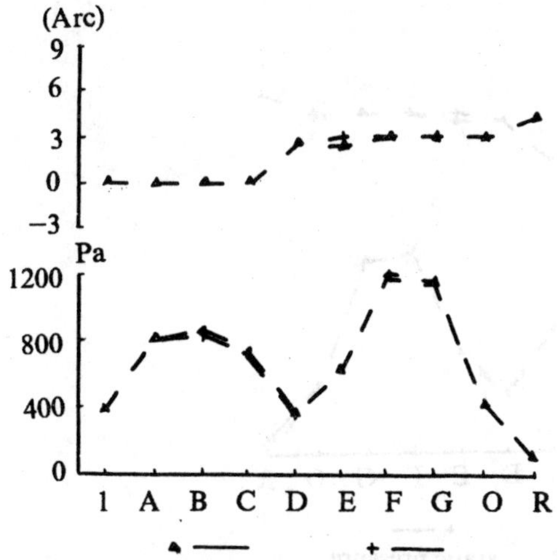

Fig.5 Varations of amplitade and phase of total pressuse preturbation in high and low pressure regions.

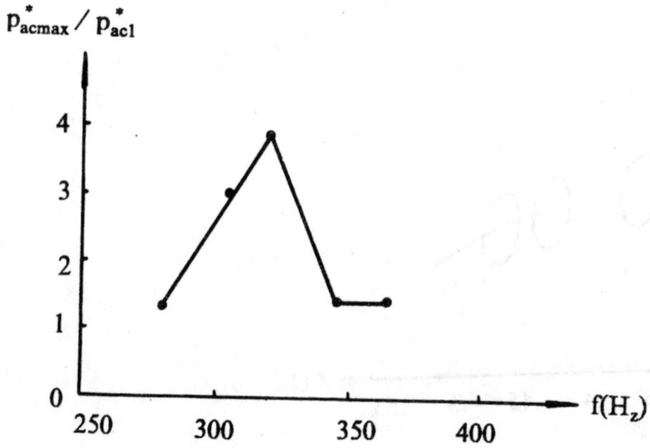

Fig.6 Variation of maximum amplitude to inlet amplitude ratio frequency

propagate with plane waves in the distorted flow field.The maximum amplitude of perturbation occures in the second stator at F,G in the figure.The variation of ratio of outlet maximum amplitude to inlet amplitude with frequencies is illustrated in Fig.6. It can be seen that the maximum amplitude is more than three times of inlet amplitude at f = 320 H_z.

Fig.4 and Fig.6 all suggest that the compressor is sensitive to perturbation with frequency 320 H_z.

3.Fig.7 shows the variations of amplitudes and pheses of total pressure and static pressure in high pressure region at f = 305 H_z.It shows that the amplitude and phese of total pressure is near to those

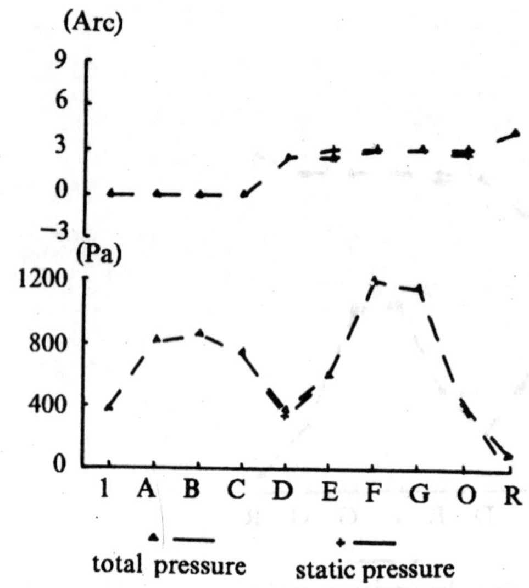

Fig.7 Variations of amplitude and phase of total pressure and static pressure perturbation in high pressure region at f = 305 H_z

of static pressure.The results are similar in low pressure region and / or at other frequencies.

4. Fig.8 shows dynamic characteristics of the compressor at f = 305 H_z.The circumferential averaged curve illustrated is between those of high and low pressure regions.It can be seen that instability of flow

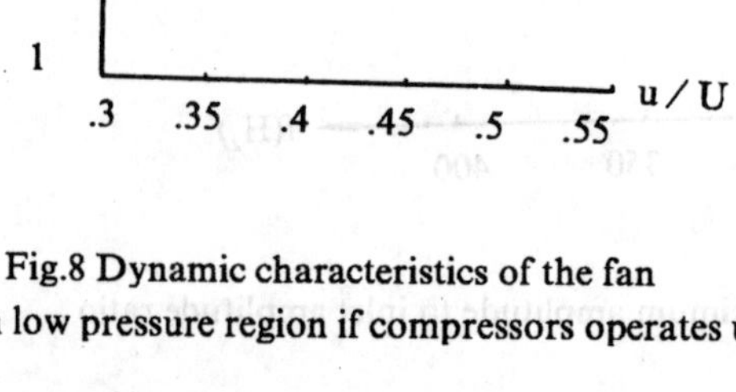

Fig.8 Dynamic characteristics of the fan

may first occur in low pressure region if compressors operates under inlet periodic dynamic distortion.

CONCLUSION

In this paper,systematic research on flow model and corresponding numerical simulations for inlet periodic dynamic distortion in axial compressor have been undertaken. A more reasonable flow model have been founded for inlet periodic dynamic distortion flow field. Numerical experiments were compared with physical experiments at 5 frequencies in an actual compressor.Results of computations are obtained as follows.

Outlet waves are near harmonic when inlet waves are harmonic.The amplitude and phase of perturbation in high pressure region is nearly equal to those in low pressure region.The square wave distortion is attenuted through compressor inlet to outlet.The response of axial compressor to periodic dynamic distortion is different for different frequencies.The investigated compressor is very sensitive to perturbation with frequency of 320 H_z. Given the frequency of perturbation,instability of flow may first occure in low pressure region if compressors operates at periodic dynamic distortion inflow.

REFERENCES

1. Adamczyk,J,J. AIAA paper N0.74–49.
2. Colpin,J. ASME paper N0.78–GT–34.
3. Billet,G.et al; IMech E C75 / 84.
4. Sharma,P.B.,Railly,J.W. Journal Mechanical Eugineering science,22(6) (1980).
5. Feng Qi,Doctoral thesis,Power Engineering Deparment, Nanjing Aeronautical Institute,China.

AN INVESTIGATION OF DYNAMIC BREAKING IN THE COMPRESSOR STALL / SURGE LIMIT

Guo Cai Tang
(Power Engineering Department, Nanjing Aeronautical Institute, Nanjing 210016, The People's Republic of China)

ABSTRACT

This paper presents a new phenomenon called dynamic breaking of the operating point in the compressor stall / surge limit, while the throttle setting is in a region near to the stall / surge limit on the unstalled side. This region is of importance due to higher pressure rise and higher efficiency. The investigation reveals that the dynamic breaking may result either in a steady flow condition with the compressor working at a point on the unstalled part , although the operating point first enters the limit;or in an unsteady flow —fully developed rotating stall with the compressor working at another point on the rotating stall curve ,or surge ,although the throttle setting is somewhat away from the limit position.

Two critical values of the parameter B are identified:Br,at which the system behaviour will change from unstalled flow to rotating stall ;Bs,at which the behaviour will transit from rotating stall to surge .Their variation with changes in parameters are also studied.

NOMENCLATURE

a sound speed
A flow through area
B non–dimensional stability parameter

$$B = \frac{U}{2a}\sqrt{\frac{V_p}{A_c L_c}}$$

C compressor pressure rise
C_{ss} steady–state pressure rise
C_x axial velocity
DT time spent by the valve for one stroke
F throttle pressure drop
G nondimensional number ,defined as

$$G = \frac{L_T A_c}{L_c A_T}$$

L effective length of equivalent duct

$\dot{m}$ mass flow

P pressure

ΔP plenum pressure rise ,($P_{plenum} - P_{ambient}$)

t time

U mean rotor velocity

V_P plenum volume

ρ density of air

τ compressor flow field time constant

Ψ $\Delta P / \frac{1}{2}\rho U^2$

Φ $\frac{C_x}{U}$

Subscripts

p plenum

c compressor

T throttle

Superscripts

~ nondimensionalized variable

INTRODUCTION

The effect of unsteady flows on compressor performance and life expectancy has been of considerable interest and importance in the application of axial –flow compressor to engines. Perhaps the most significant flow fluctuations are those associated with rotating stall and surge , which are the two types of instability modes.Therefore, a fairly sizable work has been devoted to the investigation of these two unsteady phenomena.

Over the most part of the range of possible compressor speeds there is a minimum flow point for each speed where the compressor will either surge or rotating stall, depending upon the geometry of the compressor receiver ,or more generally ,depending upon the B parameter[1].These surge or stall points define the so–called surge or stall line , which separates the whole compressor performance map into the two areas:the steady flow area and the unsteady flow area with the stall line or surge line being the instability limit.

When the flow rate is increased from some point in the unsteady flow area by opening the throttle valve,then the operating point of the compression system will go along the in–rotating–stall characteristic,and will recover from rotating stall at a flow rate ,usually greater than the one determined by the stall line which is obtained while the flow rate is decreased by closing the throttle valve .Therefore ,there is a hysteresis loop for a compressor performance characteristic at a constant speed,and this loop may be quite large as shown in Figure 1.

For a given compression system ,there must be a correspondence between the flow rate or operating point and the throttle setting,as far as the steady flow is concerned .For example,if one closes the throttle from its fully open position M towards the position L (instability limit) very slowly,then the operating point of the compression system will go along the characteristic ml , in particular ,if the setting is at B, then the operating point must be at point b,meaning the pressure rise and the flow rate should be $\triangle P_b$ and Φ_b respectively.

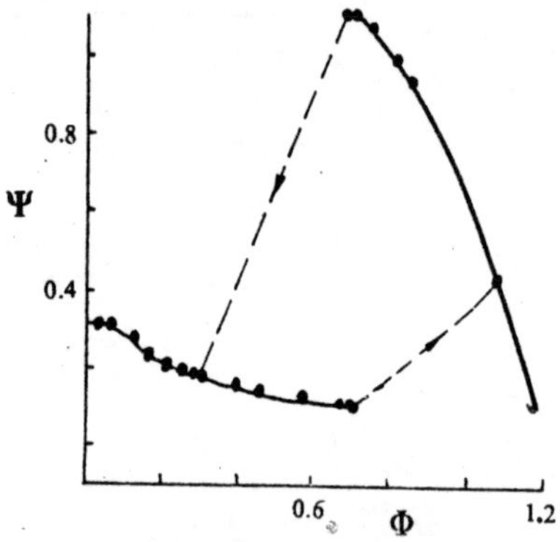

Fig.1 Hysteresis loop (from reference 5)

The part lr of the characteristic is vary important in the application because on this part the pressure rise is greater and the efficiency is higher.Hence, it is very likely for the operating point of an engine to be chosen on this part .Moreover ,for a modern advanced engine with good maneuverability it is desirable to change condition quickly ,so that the operating point of the compression system will not chang quasi –steadily.Recent experimental result on a low speed compressor facility at Nanjing Aeronautical Institute,revealed that the operating point will go along a curve which is higher than steady –state characteristic when the valve is quickly closed ,so that the flow rate is very possible to become less than that of point l,when the final setting of the throttle valve is at some point near to the limit point l on the part lr ,Figure 2.This is an interesting situation because the operating point has already broken in the stall / surge point while the throttle valve setting is actually fixed at a position which is opener than the limiting position L.Will the compression system run into rotating stall or surge because of the entering of the operating point into the unsteady flow area ? or will the compressor system finally work at a point b on the steady–state characteristic ml due to the throttle setting is on the steady flow area side ? and how is the behaviour of the compression system affected by the various factors? These are what the present investigation attempts to attack.

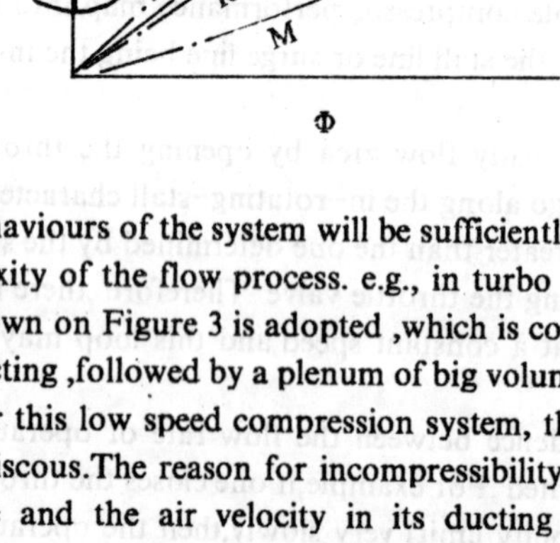

Fig.2 Throttle setting and operating point, dashed line:for quick valve motion

FLUID DYNAMIC MODEL

To answer the abovementioned questions,it is now necessary to examine the unsteady flow process in the compression system.For the present investigation it is appropriate to choose such a compression system as the fluid model that the typical transient behaviours of the system will be sufficiently revealed yet the flow is as simple as possible so that the complexity of the flow process. e.g., in turbo –engines,can be avoided .Therefore a compression system as shown on Figure 3 is adopted ,which is composed of a low speed axial compressor and its inlet / outlet ducting ,followed by a plenum of big volume ,connected to an exhaust pipe with an exit throttling valve. For this low speed compression system, the flow is assumed to be incompressible,one dimensional and inviscous.The reason for incompressibility assumption is quite clear ,because the system is a low speed one and the air velocity in its ducting is also low enough.The assumption that the flow is one dimensional needs some explanation .Obviously ,the detailed flow field in stalled compressor.blade rows is an unsteady ,three dimensional ,viscous one. However ,what we are concerned about in the present investigation is the system transient behaviour,which means how the system operating point moves,and what the final condition is as the result of the changes in some factors.In other words what is needed is the global behaviours of the system as a whole rather than the minute details of the flow field in the compressor blade rows. The influence of the viscous and three dimensionaltiy is taken into account all in the characteristic of the compressor which is represented as an actuator disk across which there is some jump in the properties according to the characteristic . The viscous effect in the system ducting is neglected here only for clarity, no loss in generality of the qualitative results will be introduced.For this fluid model with these assumptions the following governing equations may be obtained:

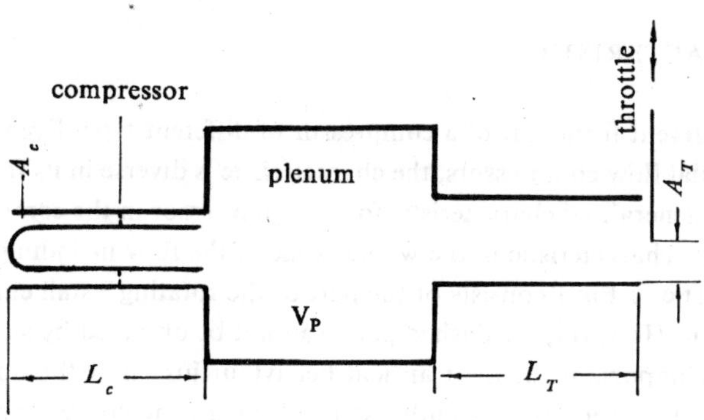

Fig.3 Fluid dynamic model

From the unsteady momentum equation for the compressor and its ducting ,one gets

$$\frac{d\tilde{\dot{m}}_C}{d\tilde{t}} = B(\tilde{C} - \triangle\tilde{P}) \tag{1}$$

for the outlet exhaust pipe ,

$$\frac{d\tilde{\dot{m}}_T}{d\tilde{t}} = (\frac{B}{G})(\triangle\tilde{P} - \tilde{F}) \tag{2}$$

Continuity of mass in the plenum leads to

$$\frac{d\triangle\tilde{P}}{d\tilde{t}} = (\frac{1}{B})(\tilde{\dot{m}}_c - \tilde{\dot{m}}_T) \tag{3}$$

For unsteady flow,the compressor will not respond in a manner as for quasi–steady flow,a first order transient response model to simulate the time lag in compressor response is adopted here ,which was proved to be quite reasonable for the present purpose[1]

$$\frac{d\tilde{C}}{d\tilde{t}} = \frac{1}{\tilde{\tau}}(\tilde{C}_{ss} - \tilde{C}) \tag{4}$$

To complete the description of the problem ,we select to close the valve as a way to initiate the dynamic flow process,pushing the operating point to move in the required manners.Thus

$$\tilde{F}=(\frac{A_c}{A_T})^2\dot{m}_T^{\,2} \tag{5}$$

This set of differential equations (Eqs.1to 5) are to be solved to obtain the trasient behaviours and the outcome.

COMPRESSOR CHARACTERISTIC

The characteristic of a compressor is different from that of a compressor of different type .Even for the compressors of the same type ,e.g.,for axial flow compressors, the characteristic is diverse in its form and magnitude.Hence it is good to employ a generalized characteristic for the compressor in the compression system used here.We need a steady –state characteristic in the whole range of the flow including unstalled part and stalled part .The characteristic of Fig.1 consists of the part of the rotating –stall curve and the unstalled part ,obtained by experiment.However, the dashed parts can not be obtained by simple experiments, and these two parts are vary important for the transient behaviour just after the operating point getting across the stall / surge point. Recently some authors[2,3]put forward the conception of the axisymmetric compressor characteristic.

The axisymmetric compressor characteristic is usually represented by a cubic curve :

$$\Psi = \Psi_o + H[1+\frac{3}{2}(\frac{\Phi}{W}-1)-\frac{1}{2}(\frac{\Phi}{W}-1)^3] \tag{6}$$

based on which and using Moore's rotating stall flow model and Galerkin method ,the in–rotating stall characteristic is also calculated ,being (H:the semi–height;W:the semi–width)

$$\Psi = \Psi_o + H[1-\frac{3}{2}(\frac{\Phi}{W}-1)+\frac{5}{2}(\frac{\Phi}{W}-1)^3] \tag{7}$$

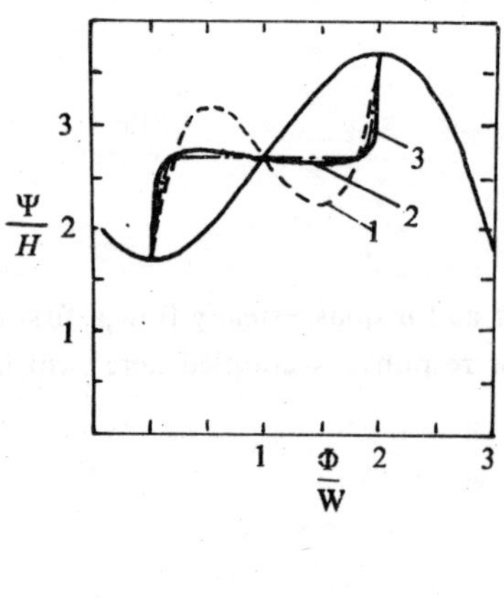

Fig.4 Axisymmetric characteristic and rotating stall curves .

1.Galerkin solution. 2.exact solution 3.Equation 8.

Both the axisymmetric and rotating stall characteristics are shown in Fig.4 It can be seen easily that the system hysteresis appears naturally in the theory since it is set by the stability of the system at the various intersection points of the throttle and compressor curves .When closing the throttle (into stall) the compressor curve is the axisymmetric one ,when opening the throttle from operation in fully developed rotating stall,the compressor curve is the rotating stall one.There can thus be a difference in the throttle settings at which transition from unstall to stall and from stall to unstall occur.

The Galerkin solution was used for the rotating stall curve in references 2 and 3 ,because it can reflect some important aspects.However, it is an approximate one as compared with the exact solution (also in Fig.4),which is a numerical solution and can not be expressed in an analytical function .In the present work,a relation between $\Phi$and $\Psi$is adopted,(Φ_{oo} is the Φ at $\Phi = W$)

$$\Psi - \Psi_{oo} = D(\Phi - \Phi_{oo})^{P}$$

with D and p being constants .Change in p may modify the trend of the curve to approach the exact solution as closely as possible .In summary , we will use a generalized compressor characteristic which consists of an axisymmetric one of cubic function (Equation 6) as unstalled part and a relation (Equation 8) as rotating stall curve which is very near to the exact solution and which naturally provides hysteresis effect for the system behaviour.

CALCULATION PROCEDURE

The method used here is basically a forth order predictor–corrector.At first,the throttle setting is set at the fully open position ,say, at the point with $\Phi = 0.7$,then the throttle is closed to a final setting,say, to the setting B,within a time interval DT.The calcalation is carried out step by step until a final condition is reached,which may be a steady–state operating point, fully developed rotating stall or surge.

DT is used to designate the closing rate of the throttle valve,small DT means quick motion of the valve. The time lag $\tilde{\tau}$ is different for the different ranges of the characteristic; for the onset and cessation of rotating stall,the value of $\tilde{\tau}$ was taken according to the experimental data[1] and for the unstalled part , $\tilde{\tau}$ is very small. The final valve setting is represented by the coefficient of the throttle line ,FKE, the throttle line being:

$$\tilde{F} = FKE\dot{\tilde{m}}_T^{\,2}$$

for the characteristic adopted here,the coefficient of the throttle line which passes through the peak point (stall point)is equal to 5.5556.The coefficient of the throttle line which is tangent to the rotating stall curve equals 4.88,therefore the coefficients FKE,$4.88 < FKE < 5.5556$,are the range we are interested in .

TYPICAL BEHAVIOURS

First let us examine the typical solutions to the set of the differential equations.The parameters are:FKE = 5.35,GN2 = 2,DT = 2 and GN1 = 0.1. (GN2 and GN1 are the lag constants for stalled and unstalled part respectively) The transient behaviours are shown on Fig.5 to Fig.8, corresponding to different values of the parameter B in an increasing sequence.

1.When B is rather small,for example,$B < 0.2$, the operating point moves along a curve,which never surpasses the limit point ,$\Phi = 0.48$, and after a process of spiral,

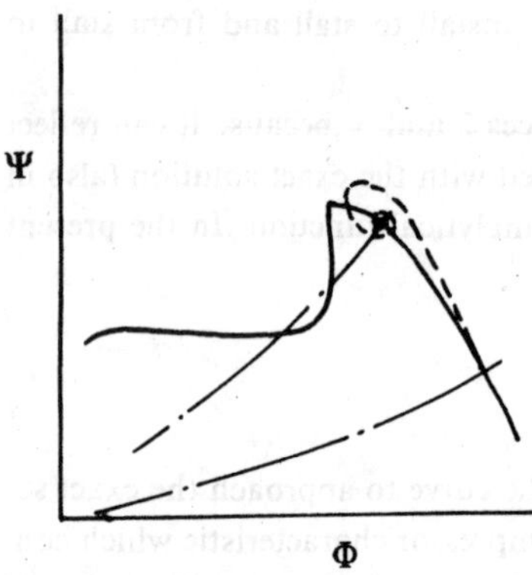

Fig.5 Spiral process,finally working on unstalled part.

at last settles down at the point where the throttle line meets the axisymmetric curve and,which is on the unstalled part.This is the situation usually expected,due to the fact that the throttle setting is on the steady flow area side as compared to the limiting position(Fig.5).

2.Increase in the value of B results in a wider spiral process,and when B is greater than some value, the spiral begins to enter the unsteady flow area.The spiral first breaks in the instability limit,then comes back across the limit,finally settles down at the point which is exactly the same as in case 1(Fig.6).

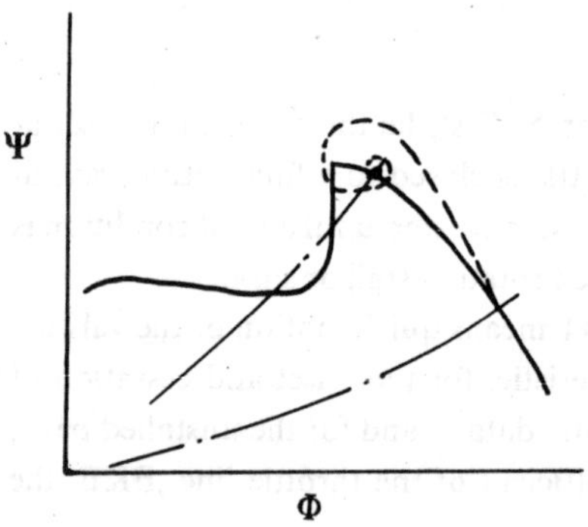

Fig.6 Spiral process,first entering stall limit finally working on unstalled part

3.Further increase in the value of B makes the operating point go farther into the unsteady flow area.In this case, the operating point does not come back across the limit, instead,it goes down to a new steady state point after another spiral process, where the throttle line again meets the compressor characteristic,but now being the fully developed rotating stall chatacteristic rather than the axisymmetric one as in case 2(Fig.7).

4.For still greater values of B,the partern of the transient behaviour changes once more.The operating point in this situation will not settle down at somepoint, just continues its motion, forming a surge cycle(Fig.8).

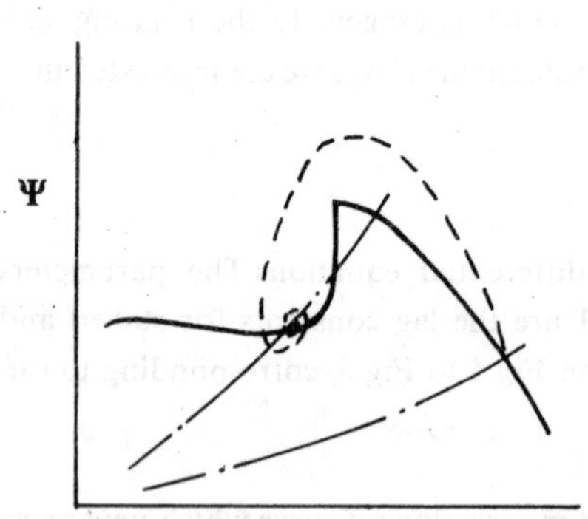

Fig.7 Working on rotating stall part

Thus there are five typical transient behaviours for a throttle setting between l and r, if the quasi steady operation is added to the above four cases . A quasi steady operation is nothing but the case in which DT is sifficiently great for the valve to close from its fully open position to the setting concerned, and the operating point goes up just along the unstalled part until the intersection point is reached.

To the author's knowledge, this is the first time to recognize that for a throttle setting in the hysteresis loop and a valve motion from wide open position to a more closed position (the flow rate changes from great to small), there are three possible final paterns: operation with neither rotating stall nor surge on unstalled part; operation in fully developed rotating stall mode on rotating stall part; operation with surge. Moreover, it is also recognized that the system may operate on the unstalled part with no rotating stall and surge even though the operating point may first break in the unsteady flow limit; on the other hand , the system may present rotating stall or surge although the throttle setting is more open than its limit position.

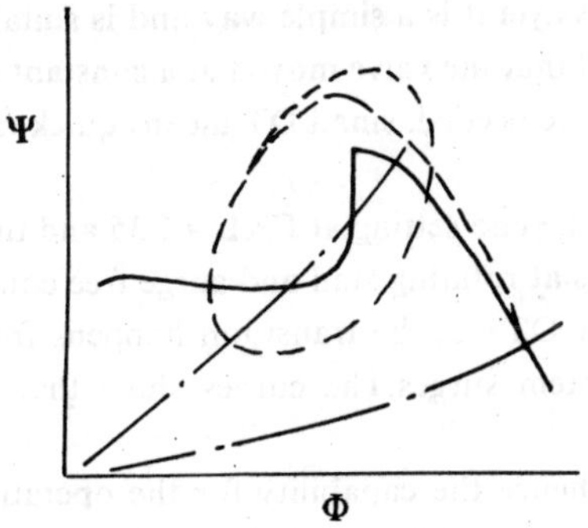

Fig.8 Surge formed

While explaining why this is so, it is interesting to note that there is some important difference between quasi steady operations and dynamic operations.First in a quasi steady operation, closing the valve from its fully open setting to the setting B only moves the operating point just to the point b, with the operating point going closely along the steady state characteristic.In a dynamic process(closing the valve quickly), the operating point is driven beyond the point b due to the unsteady flow effect, thus may be pushed into the unsteady flow area.

Once the operating opint gets into this area,the transient behaviovr will be determined by the solution to the set of the equations, with the position of the throttle setting being an important factor. For the setting L, the throttle line is just through the limit point, and all part of the rotating stall curve from s1 to 1 is under the throttle line, the operating point thus will go directly towards point sl,while for the setting B, there is some part of the rotating stall curve which is above the throttle line. This makes it possible for the operating point to get back across the instability limit,and finally settles down at point b.

NUMERICAL RESULTS

The typical bebaviours of the comoression system have been shown just for the specific situations and specific system parameters. However, in applications,it is of importance to know how the behaviour changes with the factors which may affect the dynamic process, in particular, when the transitions of the behaviour types will happen(e.g, from operation at unstalled part to rotating stall and from rotating stall to surge) and how they vary with the associated parameters. we will choose some important facters and examine their influence on the transitions.

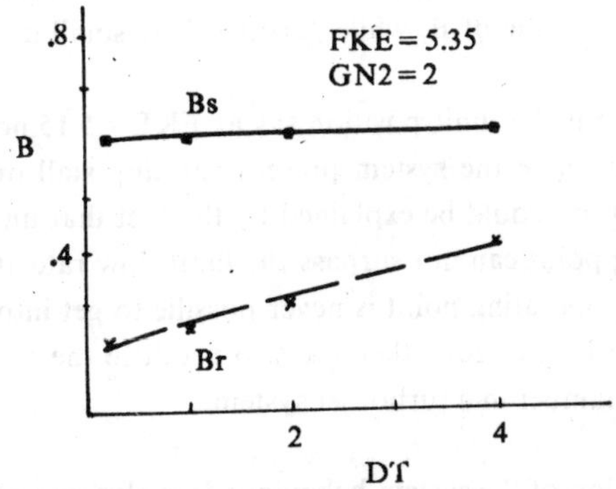

Fig.9 Influence of DT

As is well known, this set of equations are strongly nonlinear, and an analytical solution to the equations has been considered impossible up to now. Therefore, numerical results are presented here to give a description of the transitions of the system behaviour. The transitions are described here,as most anthors did, by the B parameter. Here, we have two critical values of B, namely Br, at which the transition occurs from unstalled operation to rotating stall, and Bs , at which the transition from rotating stall to surge occurs. In what follows we will present and discuss the change in both Br and Bs while some parameters are changed.

1.Variation of the System Behaviour with Valve Moving Rate

As was metioned before, one of the features of the problem concerned in this study is the dynamic process, which is quite different from the quasi steady process. In real application there are many ways to produce dynamic process.The valve motion is one of possible ways,yet it is a simple way and is suitable to show the unsteady flow effect. In the calculation,it was assumed that the valve moved at a constant velocity from its fully open setting to an end setting, DT being the time needed. small DT means quick motion.

The transition values of B are shown in Figure 9. This case is for the end setting at FKE = 5.35 and time lag equal to 2(GN2) and 0.1(GN1) respectively .The system works at rotating stall and surge free condition at rather small value of B. At a value of B, e.g, B = 0.31 for DT = 2, the transition happens from unstalled operation to ratating stall mode. At B = 0.62, the system surges.The curves show that Bs changes little with DT, while Br increases with increase in DT.

A greater value of DT means rather slow motion of the valve, hence the capability for the operating point to rush into the instability limit becomes weak,and a greater value of B is needed to push the operating point into the unsteady flow area with certain deepness. It could be expected that at a very great value of DT Br and Bs should become infinite,because the operation becomes quasi–steady, and the operating point will not get into the unsteady flow area.

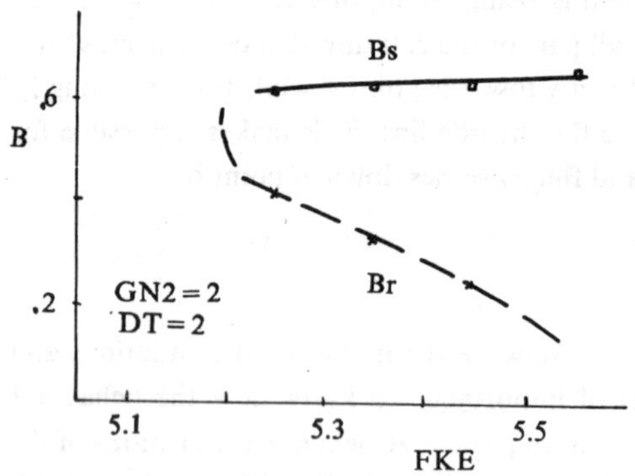

Fig.10 Influence of FKE

2.Variation of the System Behaviour with the End Setting Position.

Fig.10 demonstrates the results for different end setting positions for GN2 = 2, GN1 = 0.1, DT = 2. The setting for the limit point is at FKE = 5.5556, and that for the recovery from rotating stall is at FKE = 4.88. When the throttle setting is put nearer to limit setting, the value of Bs remains basically the same,while the valus of Br gets small.It is easy to understand that if the setting is far away from the limit setting,the operating point has to run a longer distance before it could rush into the instability limit,which means a greater value of Br needed.In fact,when the setting is put at the limit position,the operating point will fall down to a position where fully developed rotating stall prevails at any small value of B, while perturbed by small disturbance.

On the other hand, when the setting is far enough away from the limit position,say,at FKE = 5.15,no value of B,i,e,no matter how great the value of B is, could make the system present rotating stall or surge, it just works free of rotating stall and surge.This situation could be explained by the fact that under the condition (GN2 = 2,GN1 = 0.1,DT = 2),the operating point can not surpass the limit flow rate at the value of B concerned.However,it does not mean that the operating point is never possille to get into the limit under other conditions,e.g., for other valve motion law, or for other ways to produce the dynamic process, especially by abrupt burning in combustion chamber in a turbo–jet system.

3.Variation of the System Behaviour with the Time Lag.

The time lag constant plays an important role for determination of the system behaviour.It is also one of the parameters which could only be determined by experience or experimental data at present time.

However, there are very limited experimentally obtained data published,e.g. the data from Reference 1, which is very useful for the situations under the similar conditions.Nevertheless, it is not clear to what extent they could be used in somewhat different cases.Hence it is of significance to know how the time lag constant affects the system behaviour,particularly, the transition values of the parameter B.

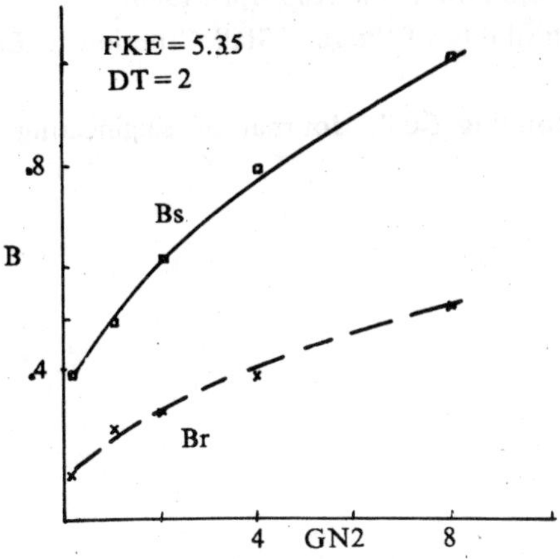

Fig.11 Influence of time lag

Fig.11 shows the trends of Bs and Br for FKE=5.35 and DT=2, GN1=0.1.With the increace in the time lag constant GN2,both Br and Bs increase.Reference 4 gives the result for the critical value of B at increasing values of GN for the transition from rotating stall to surge when the initial point just at the peak point.The trends are very similar.Physically,when the time lag constant becomes greater,the rotating–stall curve from l to s is made less steep as far as its actual role is concerned, resulting in the increased values of Br and Bs.

CONCLUSIONS

1.For the unstalled part of the compressor characteristic there is a correspondence between the throttle setting and the operating point in quasi–steady operations,whereas in quick or dynamic operations the correspondence includes much more content.

2.For the throttle setting in the region of the hysteresis loop,there are five possible typical system behaviours when closing the valve:a. the operating point goes along the steady state characteristic and stops at the intersection of the throttle line and the characteristic;b. the operating point surpasses the intersection but then rests at the point; c. the operating point breaks into the instability limit then comes back to the intersection point. d. the operating point first runs into the instability limit then goes down and finally settles down at a new point: the intersection of the throttle line and the rotating stall curve with the compressor work in a fully–developed rotating stall mode;e. the operating point first breaks into the instability limit and then develops into surge.

3.Dynamic breaking into the instability limit may result either in steady–flow with the operating point on the unstalled part of the characteristic due to the throttle being at a opener setting than the limit position,or nonsteady flow (rotating stall or surge) due to the operating point entering into the unsteady flow region and the throttle is at a closer setting than that for recovery from rotating stall,depending on the parameters:DT,FKE,GN2.

4.Numerical results show that while the critical value Bs,for the transition from rotating stall to surge,is approximately constant,the critical value Br, for the transition from unstalled operation to rotating stall,changes with FKE and DT.

5.The time lag constant is an important parameter,which will lead quite large change in Br and Bs. How to obtain an appropriate value for the time lag is an issue deseving further investigation.

REFERENCES

1.Greitzer, E,M."Surge and Rotating Stall in Axial Flow Compressors, part1:Theoretical Compression System Model,"Journal of Engineering for Power,vol.98,No.2 April,1976

2.Moore,F,k,and Greitzer,E,M," A Theory of Post–stall Transients in Axial Compression systems:Part1—Development of Equations," Journal of Engineering for Gas Turbines and Power.vo1,108,April 1986
3.Greitzer,E,M,and Moore,F,k,"A Theory of Post–stall Transients in Axial Compression Systems :"Part 11—Application", Journal of Engineering for Gas Tarbines and Power ,Vol .108,April 1986.
4.Tang ,G,C and Zhang,W.M ," An Investigation of Applicability of Surge / Stall Criterion to Engineering Problems".ASME paper 85–IGT–22.
5.Day,I.J." Prediction of Compressor Performance in Rotating Stall." Journal of Engineering for Power,Vol.100,January 1978.

NUMERICAL SOLUTION OF 2-D TRANSONIC FLOW THROUGH AN AXIAL TURBINE STAGE

Tadeusz Chmielniak, Włodzimierz Wrøblewski
(Silesian Technical University, ul. Konarskiego 18
44-100 Gliwice, Poland)

ABSTRACT

This paper presents the solution of the steady und unsteady two-dimensional transonic flow through an axial turbine stage. The rotational, adiabatic, inviscid flow of the perfect gas is considered. The analysis of the problem has been based on the time dependent Euler equations. In order to solve the problem, the finite volume method and a time-marching method have been applied. The results of numerical calculations for a chosen geometry of an axial turbine stage are presented.

INTRODUCTION

In modern heavy load stages of turbomachines, transonic and supersonic flows are often met. In recent years a great progress in computing transonic flow is noticeable. Particularly methods based on the time dependent equations for the conservation of mass, energy, and momentum have been developed. This lets maintain the hiperbolic type of differential equations in the whole computational domain for the whole velocity range. The main attraction of the time dependent Euler equations is the ability to compute mixed subsonic-supersonic flows with automatic capturing of shock waves. These equations are solved by using a time-marching technique. One of the first applications of the explicit time marching methods were done by Godunov [1] and McDonald [2]. These methods have found wide applications and are constantly being improved.

In the calculations of transonic flows, also the Denton method is successfully applied. This method has been widely used for solutions of the Euler equations through blade rows e.g. for flow in the meridional plane by Spurr [3], for unsteady flow by Mitchell [4], for wet steam flow by Bakhtar et al.[5] and for three-dimensional flow by Sarathy [6]. Another commonly used method is the MacCormack method . It is used for steady and unsteady (e.g. Ispas et al.[7]) flow calculations either in finite difference form in a rectangular computational domain obtained after a coordinate transformation or directly in finite volume formulation in the physical domain.

The corrected viscosity scheme, applied in this paper, is derived from the original Lax scheme. Some modifications of the Lax scheme, wich were introduced by Couston [8] and recently by Van Hove [9] and Arts [10], have caused that this method has become attractive to calculate two- and three-dimensional problems. The corrected viscosity scheme is a one-step, first order accuracy in time, explicit scheme.

Particular attention should be given to the unsteady effects of blade row interactions. It is important to predict these effects, especially when the blade row of interest is a rotor since measurement difficulties in rotating frame of references.

One of the first attempts at predicting unsteady effects with using a time-marching method was made by Mitchell [4]. He considered a single blade passage for which he used in calculations the Denton scheme with the assumed unsteady condition at the inlet.

Ispas et al.[7] considered the similar problem and employed the MacCormack scheme for a single blade passage. They avoided the problem by solving the flow throughout the number of blade passages for which tne flow repeated. They applied the calculation scheme with the second order accuracy in time but assumed the isotropic flow, which is guite problematic.

Hodson [11], on the basis of the Denton scheme, did the calculations of the flow through the blade passage with the application of modified boundary conditions, what lets consider a single blade passage. In comparision with the experiment, the applied method gives good results.

A similar problem was considered by Sokolowsky et al.[12]. They took into account the two- and three-dimensional unsteady flow through a turbine stage using the Godunov scheme, and obtained interesting results.

It is worth noting that in the present considerations the models of inviscid, unsteady flows are used. It results from difficulty in the unsteady flow calculations which practically limited now the use of viscid models. The conformity of the results obtained from the inviscid flow calculations with the experiment is often good [11], which proves that the viscosity effects are not always dominant.

MATHEMATICAL FORMULATION OF THE PROBLEM

The problem is further considered with the following simplifications:

(a) flow is 2-D, plane,
(b) flow is inviscid and adiabatic
(c) gas is perfect
(d) no blade vibration

Flows through the blade passage are governed by mass, momentum and energy conservation principles. The system of the conservation equations for volume V can be written in the integral form for the considered flow model as follows:

$$\frac{\partial}{\partial \tau} \int_V U \, dV + \int_V F \, dV = 0 \qquad (1)$$

where:

$$U = \begin{Bmatrix} \rho \\ \rho \bar{w} \\ \rho e \end{Bmatrix} \qquad F = \begin{Bmatrix} \rho \overline{wn} \\ \rho \bar{w}(\overline{wn}) + \bar{n}p \\ \rho (e+p/\rho)\overline{wn} \end{Bmatrix}$$

Using the perfect gas law and the definition of total absolute or relative energy, (1) is completed by the following relation:

$$e = \frac{1}{\gamma-1}\frac{p}{\rho} + \frac{w^2}{2} + \frac{u^2}{2} \qquad (2)$$

The problem to be solved is an initial boundary value problem where an initial value distribution has to be know with all boundary conditions.

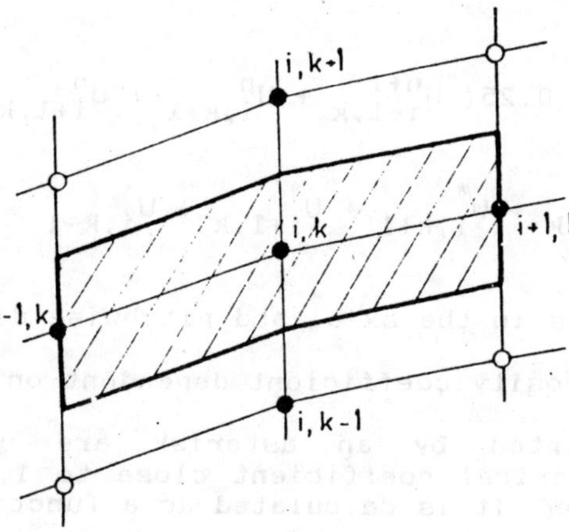

Fig.1.Bitrapezoidal element

DISCRETIZATION OF THE FLOW DOMAIN

The calculation domains of the stator and rotor are discretized by a numerical grid. The numerical grid is composed of quasi-streamwise lines and pitchwise lines. The quasi-stream lines are equally spaced in the pitchwise direction. The pitchwise lines are variably spaced; they diminish from the inlet to the leading edge and increase from the trailing edge to the outlet. Within the blade passage the pitchwise lines are variably spaced, depending on the required precision and the expected density gradients. On the numerical grid a bitrapezoidal element is defined according to Arts [10]. The element is formed by the same pitchwise lines and quasi-stream lines (Fig.1). The numerical domain for the steady flow calculation is shown in fig.2.a, while fig.2.b shows the domain for the unsteady flow calculation.

NUMERICAL PROCEDURE

Let us consider a small volume V over which the system of equations (1) is integrated. By the use of the Gauss-Ostrogradzky divergence theorem and averaged quantities (1) can be written:

$$\hat{U}^{n+1} = \hat{U}^{n} - \int_{\tau^n}^{\tau^{n+1}} \frac{1}{\Delta V} \sum_{m=1}^{M} (\hat{F}\,\Delta S)_m d\tau \tag{3}$$

where:

M - number of surfaces which limit a volume V
^ - averaged quantities

Assuming the finite volume method to determine a subintegral function $\hat{F}$ and the corrected viscosity scheme to integrate in the time partition $\Delta\tau$, equation (3) can be formulated as follows:

$$U_{i,k}^{n+1} = \frac{\Delta\tau}{V}\sum_{m=1}^{M}(\hat{F}\Delta S) + 0.25(\,U_{i-1,k}^{n+1} + U_{i,k+1}^{n} + U_{i+1,k}^{n} + U_{i,k-1}^{n}\,) +$$

$$- 0.25\alpha(\,U_{i-1,k}^{*} + U_{i,k+1}^{*} + U_{i+1,k}^{*} + U_{i,k-1}^{*} - 4U_{i,k}^{*}\,) \qquad (4)$$

where:
- i,k - spatial indices in the axial and pitchwise directions
- n - index of time
- α - numerical viscosity coefficient dependent on density gradient

The terms superscripted by an asterisk are updated every Nv iterations. α is a numerical coefficient close to 1, used to provide the effective correction. It is calculated as a function of the local

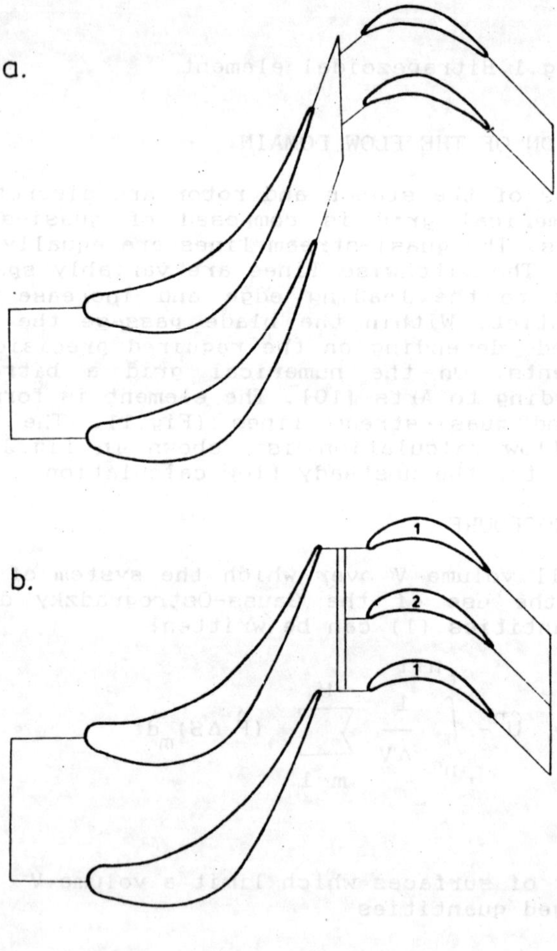

Fig.2.Calculation domain for steady (a) and unsteady (b) flow problem

density gradient in order to ensure the necessary damping near the flow discontinuity.

Corrected viscosity scheme (4) is stable for the Courant- Friedrichs-Lewy condition, which limits the time step $\Delta\tau$.

The detailed description of this method can be found in [10].

BOUNDARY CONDITIONS

In order to have a well-posed problem, (1) must be completed with a set of boundary conditions.

The missing calculation points outside the computational domain are replaced by the corresponding points at the other periodic boundary. Along a solid walls an impermeability condition must be fulfilled. As the solution procedure explicitely uses the transport terms, this condition is satisfied by setting to zero the mass and energy fluxes and by considering only the static pressure acting on the suction and pressure side in the calculation of the momentum fluxes.

The conditions applied on the inlet and outlet plane of a stage are the same as in most other time marching methods (where the axial velocity is subsonic). On the upstream boundary, the absolute stagnation density and stagnation pressure are specified together with the absolute flow direction. On the downstream boundary the static pressure is specified and held constant (the so called non-reflecting condition can also be used).

In calculating the boundary conditions on the outlet from the stator domain and on the inlet to the rotor domain are not set. Depending whether the problem of a steady or unsteady flow is considered, parameters on the connection of the stator and rotor numerical grids are updated in a two different ways.

1.Steady flow

In the problem of the steady flow through a stage, the connection of the stator and rotor numerical grids is realized by assuming an averaged parameters in the pitchwise direction on the lines IA=MX-1 of stator and IA=2 of rotor (Fig.3.a). The problem is considered in the two systems of reference, absolute for the stator and relative for the rotor. From the averaging parameters notated in the relative system of reference, values on line IA=MX of the stator (which overlaps with line IA=1 of the rotor) can be calculated using the one-dimensionsl corrected viscosity scheme By updating the values from the absolute system of reference to the relative one, static parameters are determined due to the following equations:

$$H_a - c_y u = H_w - u^2/2 = \text{const}$$

$$\frac{T_{ow}}{T_{oa}} = 1 - \frac{u^2}{2H_a}\left(2c_y/u - 1\right) \tag{5}$$

$$\frac{p_{ow}}{p_{oa}} = \left(\frac{T_{ow}}{T_{oa}}\right)^{\frac{\gamma}{\gamma-1}} ; \quad \frac{p_{ow}}{p_{oa}} = \left(\frac{\rho_{ow}}{\rho_{oa}}\right)^{\gamma}$$

$$c_x = w_x \quad ; \quad c_y = w_y + u$$

where:

H - enthalpy; subscripts: o - stagnation, a - absolute, w - relative.

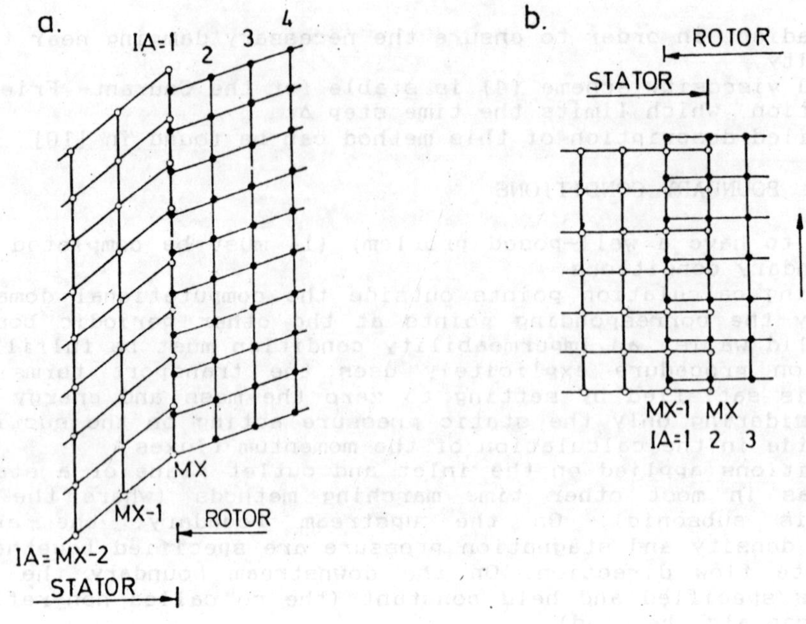

Fig.3.Stator and rotor grid connection for steady (a) and unsteady (b) flow problem

2.Unsteady flow

The numerical grids, that discretize the computational domain of the stator and rotor, have a common field (Fig.3.b). Line IA=MX-1 of the stator overlaps line IA=1 of the rotor and line IA=MX of the stator overlaps line IA=2 of the rotor.

The solution procedure is based on a repeated interaction between two blade row calculations. The flow parameters on line IA=MX-1 of the stator are updated from the absolute system of reference into the relative one due to equations (5) at each node. It permits to determine parameters on line IA=1 of the rotor taking into account an interaction of blade rows at a given moment of time. In the same way the values from the nodes on line IA=2 of the rotor are updated into the values on the nodes on line IA=MX of the stator.

STEADY FLOW PREDICTION

Calculations are performed on the hub section of a last stage of the large output steam turbine. The blade passages of a turbine stage are discretized by the numerical grid with 21x67 grid points on the rotor and 21x60 grid points on the stator (Fig.2.a). The described method lets perform calculations for any stage blade ratio.

The rotor blade row moves, relatively to the stator blade row, at velocity u. The interrow spacing / stator pitch ratio is equal 0.37. Flow parameters are:

$$\rho_{o\alpha} = 0.102 \quad [kg/m^3]$$
$$p_{o\alpha} = 0.0144 \quad [MPa]$$
$$\alpha_o = 0 \quad [^o]$$
$$p_2 = 0.005 \quad [MPa]$$

The calculations are performed simultaneously for the stator and rotor. As the initial data, the stagnation parameters distribution in the whole computational domain of the stage is assumed. To eliminate numerical instability, the selection of the suitable reduction of the outlet static pressure and increase of angular velocity of the rotor blade row has been made. During the first 800 iterations, the static pressure at the outlet from the stage is droped till demanded value p_2 is reached. From 500th to about 1500th iteration the angular velocity of the rotor blade row is gradually increased. The calculations are carried on until the steady-state is reached in the whole computational domain of the stage. Changes of the relative internal energy $\Delta e/e$ was less then $2x10^{-5}$ and fluctuations of the outlet angle β_2 less then 0.02°. Using a IBM PC 386 computer, the necessary CPU time per grid point and per time step was $2.85x10^{-4}$s. 8000 time steps were performed.

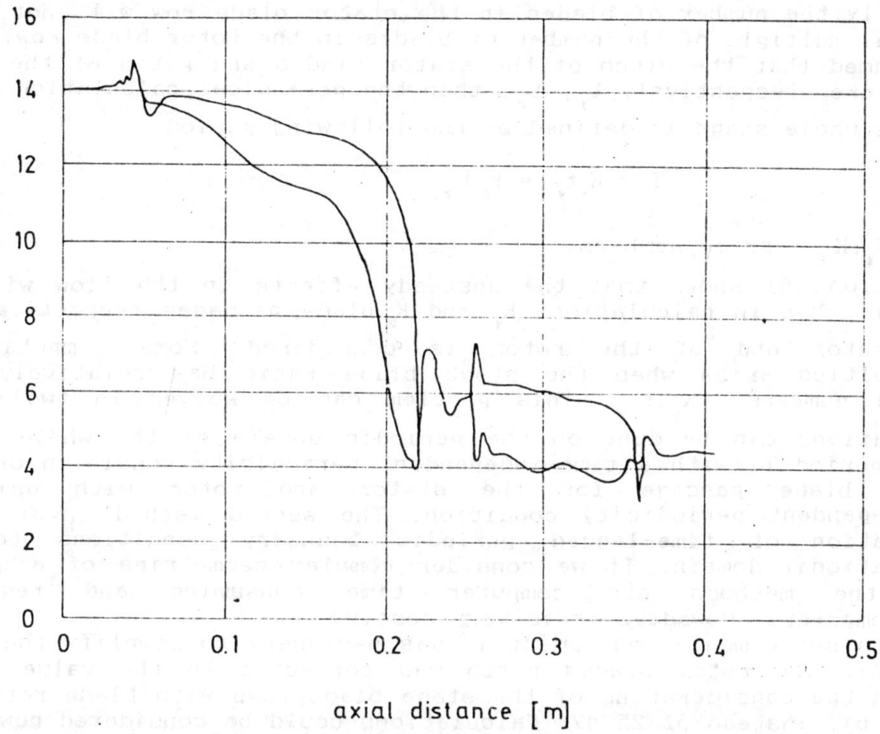

Fig.4.Blade surface static pressure distribution for steady flow calculations

The average in time parameters distribution in the stage domain is the result of the calculations. Fig.4 shows the static pressure distribution on the stator and rotor blade surfaces. In the stator and rotor blade passages are supersonic areas. In the stator maximal isentropic Mach number is equal 1.6 and in the rotor it is 1.4. Chosen average flow parameters are:

$$\alpha_1 = 68.1\ [^\circ]$$
$$\beta_1 = 50.5\ [^\circ]$$
$$p_1 = 5484\ [Pa]$$
$$\beta_2 = 51.6\ [^\circ]$$
$$p_{ov} = 8145\ [Pa]$$

where: α, β -flow angles (absolute, relative), subscripts: 1 - inlet of rotor, 2 - outlet of stage

In order to determine the variation range of the flow parameters in the stage, it is necessary to solve the problem of the unsteady flow through the stage.

UNSTEADY FLOW PREDICTION

Modelling unsteady effects in the blade passages of a stage is based on determining the interaction between the stator and rotor blade rows. The unsteady effects are only due to the rotation.

Usually,the number of blades in the stator blade row will not be an integral multiple of the number of blades in the rotor blade row. If it is assumed that the pitch of the stator blades and pitch of the rotor blades are, respectively t_1, t_2, then the periodical calculation domain for the whole stage is defined by the following period:

$$T = K_1 t_1 = K_2 t_2 \tag{6}$$

where K_1, K_2 - prime numbers.

Equation (6) shows that the unsteady effects in the flow will be periodic when in calculations K_1 and K_2 blade passages respectively of the stator and of the rotor is considered. Some computational difficulties arise when the stage blade ratio has relatively big natural numbers K_1, K_2. This problem can be solved in two ways. Calculations can be done on the periodic domain of the whole stage (with period T) with a time-independent periodicity condition or on a single blade passage for the stator and rotor with using a time-dependent periodicity condition. The second method involves an application of time-lagged periodic boundary conditions to the computational domain. If we consider complex geometries of a stage, both the methods are computer time consuming and required supercomputers. Nowadays it is very costly.

Due to our computer equipment it was necessary to simplify the stage geometry. The rotor blades pitch was corrected to the value which enabled the consideratins of the stage blade rows with blade ratio 1:2 (Fig.2.b), instead of 25:47. Calculations could be considered now for a one stator blade passage and two rotor blade passages.

The computational domain of the stator is discretized by the numerical grid with 25x82 points, while the rotor domain, consisting of two passages, is discretized by the grid with 26x74 points. In the calculations of the unsteady flow through a stage, the initial condition is assumed from the solution of the steady flow for each of blade passages separately. This kind of an initial condition does not require initial calculations which are done in the steady flow case.

Using a IBM PC 386 computer, 15000 time steps were performed. The calculations are carried on till the periodic changes of parameters in each grid point of the stator and rotor are achieved. The periodicity condition was controled by the calculation for each point the relative parameters fluctuation in each time step. The maximal periodicity error was 0.25%.

Fig.5.a shows the example of the course of pressure fluctuations in one of the numerical grid points of the stator; fig.5.b shows the course of these fluctuations in one of the grid points of the rotor. Because of the blades ratio 1:2, the period of the parameters changes in the stator domain is twice longer than the period in the rotor domain.

Fig.6 shows the static pressure distribution on the stator blade surfaces for example for three different positions of the blade rows. The first position corresponds to the position shown in fig. (3/3T). The second and third positions result from the displacement of the blade rows, respectively of 1/3T and 2/3T (T is in this case a stator blades pitch). According to fig.6, the pressure changes in the

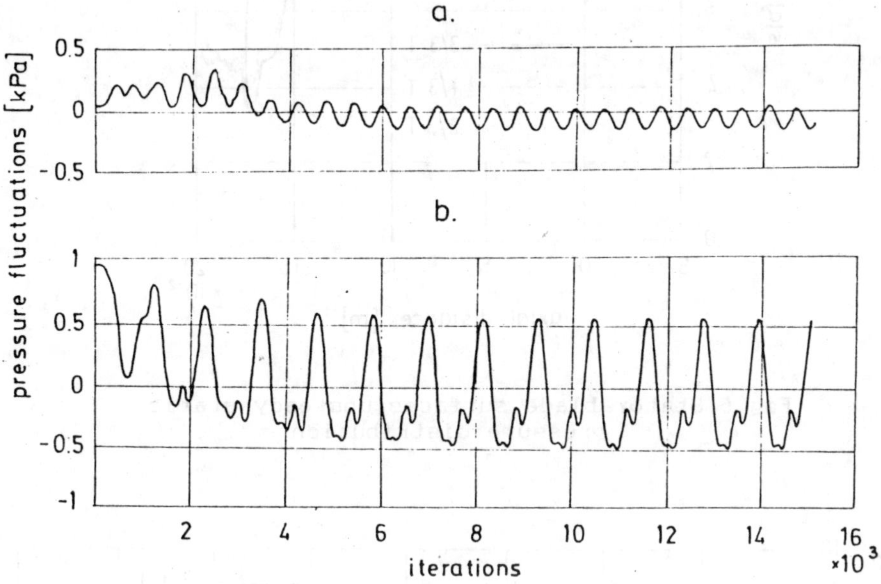

Fig.5.Convergence of the solution to a periodic state in stator (a) and rotor (b)

stator, except a small outlet region at the suction side, are not significant. With respect to the transonic flow in the stator, upstream propagation of the flow perturbations, arised during the blade rows interaction, in the stator blade passage are not observed. At a given time in the rotor row, flows are different in different blade-to-blade passages. Fig.7 shows the static pressure distribution on the surfaces of rotor blade No.1, and fig.8 on the surfaces of rotor blade No.2.

Pressure changes on rotor blades are significant. Particularly big changes occur in the inlet part of the rotor blade passage. Chosen average flow parameters changed within the range:

$$\alpha_1 = 66.1 - 69.2 \quad [^\circ]$$

$$\beta_1 = 48.5 - 52.8 \quad [^\circ]$$

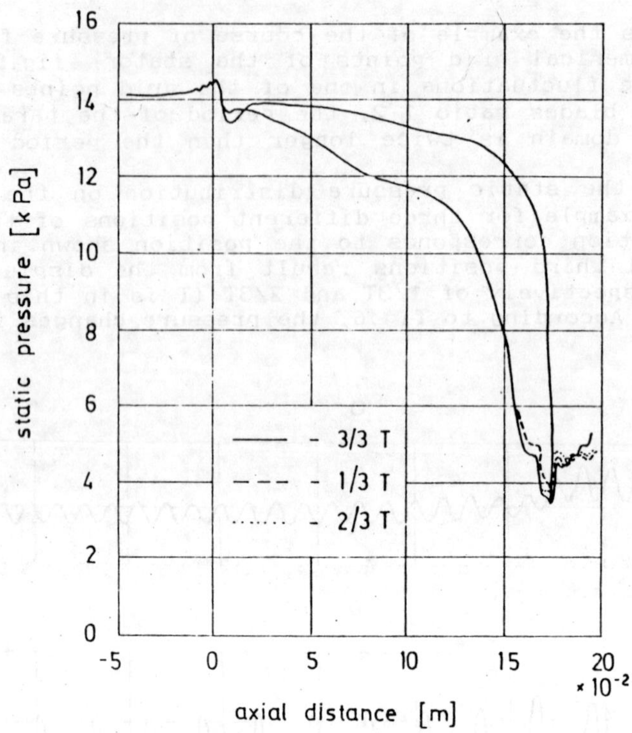

Fig.6.Stator blade surfaces unsteady static pressure distribution

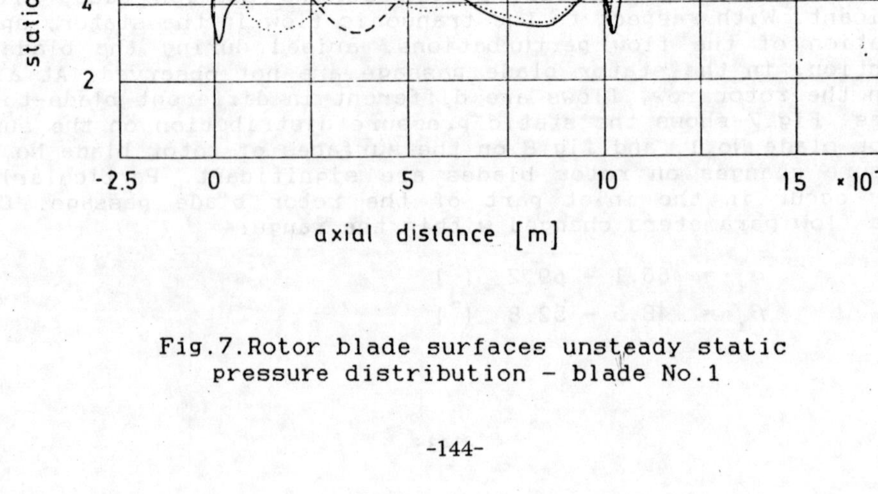

Fig.7.Rotor blade surfaces unsteady static pressure distribution - blade No.1

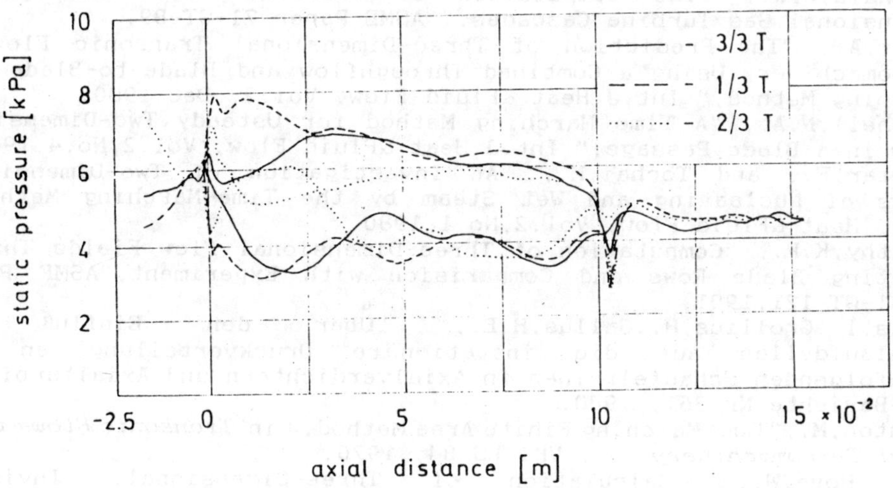

Fig.8.Rotor blade surfaces unsteady static pressure distribution - blade No.2

$$p_1 = 5295 - 5670 \quad [Pa]$$

$$\beta_2 = -51.9 \text{ -- } -52.5 \quad [^\circ]$$

$$p_{ov} = 8100 - 8117 \quad [Pa]$$

The obtained values approximately correspond to the values obtained from the steady flow calculations.

On the ground of the static pressure fluctuations on the stator and rotor blade surfaces, the dynamic load fluctuations can be determined [13].

CONCLUSIONS

The paper presents two methods to the model two-dimensional transonic flow through an axial turbine stage. These methods make possible a better knowledge of the two blade-row interaction. The steady-state can be obtained by using the first method. It lets determine the averaging in time parameters in the whole computational domain by the averaging procedure in the interrow spacing. The second method allows to consider the flow perturbation during the stator and rotor interaction. Modelling unstationary effects, occuring in the flow through a stage, is necessary because they considerably influence on the parameters distribution. The qualifications of the obtained results should be done, as all other numerical calculations, on the basis of measurements.

REFERENCES

1. Godunov,S.K., "Differential method for the numerical calculation of hydrodynamic shocks," Math. Coll., Vol.47,No.3.,1959 (in Russian).
2. McDonald,P.W., "The Computation of Transonic Flow Through Two Dimensional Gas Turbine Cascades," ASME Paper 71-GT-89.
3. Spurr,A., "The Prediction of Three-Dimensional Transonic Flow in Turbomachinery Using a Combined Throughflow and Blade-to-Blade Time Marching Method," Int.J.Heat &Fluid Flow, Vol.2, Dec.1980.
4. Mitchell,N.A., "A Time Marching Method for Usteady Two-Dimensional Flow in a Blade Passage," Int.J.Heat &Fluid Flow, Vol.2,No.4,1980.
5. Bakhtar,F., and Tochai,M., "An Investigation of Two-Dimensional Flows of Nucleating and Wet Steam by the Time-Marching Method," Int.J.Heat &Fluid Flow, Vol.2,No.1,1980.
6. Sarathy,K.P., "Computation of Three-Dimensional Flow Fields Through Rotating Blade Rows and Comparision with Experiment,"ASME Paper No.81-GT-121,1981.
7. Ispas,I.,Grollius,H.,Gallus,H.E., "Über den Einfluß von Nachlaufdellen auf die instationäre Druckverteilung an den nachfolgenden Schaufelreihen in Axialverdichtern und Axialturbinen," VDI-Berichte Nr.361, 1980.
8. Couston,M.,"Time Marching Finite Area Method," in *Transonic Flows in Axial Turbomachinery*, VKI LS 84, 1976.
9. Van Hove,W., "Calculation of Three-Dimensional, Inviscid, Rotational, Flow in Axial Turbine Blade Rows," ASME Journal of Engineering for Gas Turbines and Power, Vol.106,April 1984.
10. Arts,T.,"Cascade Flow Calculations Using a Finite Volume Method," in *Numerical Methods for Flows in Turbomachinery Bladings*, VKI LS 1982-05.
11. Hodson, H.P.,"An Inviscid Blade-to Blade Prediction of Wake-Generated Unsteady Flow," ASME Journal of Engineering for Gas Turbines and Power, Vol.107,April 1985,337.
12. Sokolowsky, G.A.,Gniesin, W.I., "Unsteady Transonic and Viscous Turbomachinery Flows," Naukowa dumka, Kiev, 1986 (in Russian).
13. Wrøblewski,W.,"Rozwiązanie zadania transonicznego przepływu przez układy łopatkowe stopnia," Praca doktorska, Gliwice, 1988.

NUMERICAL SIMULATION OF THREE-DIMENSIONAL NONSTATIONARY FLOWS AND VARIABLE AERODYNAMIC FORCES IN TURBOMACHINE STAGES

V. I. GNESIN
(Ukr. SSR Academy of Sciences, Institute for Problems in Machinery, Kharkov, USSR)

ABSTRACT

A numerical method for the solution of a direct problem on the space transonic flow of the ideal gas through moving one relative to another turbomachine bladings is presented in the nonstationary statement. A full system of the Euler equations, given as laws of conservation is integrated using the principle of transition to periodically nonstationary solution and the explicit monotonous finite difference scheme of Godunov-Kolgan. The peculiarity of an algorithm is to calculate turbomachine stage with an arbitrary ratio of numbers of stator and rotor blades without great increase of the required computer memory due to the application of the property of space-time (generalized) periodicity of the flow in each vane channel. The method proposed allows to receive time variable fields of gas-dynamic parameters of the flow and the nonstationary loads at the stator and rotor blades. The numerical investigation of the influence of the ratio of a number of blades of the nozzle cascade and the rotor wheel on the variable aerodynamic forces affecting the blades in the transonic gas flow has been carried out. An analysis of the space flows, nonstationary aerodynamic performances of the turbine stages at its design and off-design operating conditions is presented. The comparison with theoretical and experimental results of the other authors is given.

INTRODUCTION

Great interest to problems of nonstationary flow in aerodynamic cascades is evoked by the substantial effect of the nonstationarity on optimal constructions and the efficiency of turbomachine operation. Vibration of axial turbomachine blades inevitably arising at relative displacements of bladings is one of the most important and complicated problems that engineers have to be faced with at design and creation of steam and gas turbine power plants. Fatigue failure of blades due to great repeated stresses, whose levels, unfortunately, remain unknown until special tests are performed, may lead to great trouble. So there exist pressing demands for the development of extensive and reliable theoretical methods for the determination of variable aerodynamic forces and aerodynamically excited vibrations of turbomachine blades.

Non-steady flows in turbomachines may be divided into two large classes: flows, being caused by circumferential irregularity of mean (main)flow and flow disturbance, being caused by vibration of blades by variable aerodynamic forces. Disturbance of the first type results in excitation of forced (resonance) vibrations while disturbance of the second type is the basis of a physical mechanism of aerodamping or self-exciting nonstationarity, i.e. flutter.

Despite the fact that the phenomena considered are observed only jointly in real flows, the separate theoretical investigation of effects in particular, the study of aerodynamic and elastic performances of the system is negligible, is of keen interest. The interaction of irregular arranging one after the other and moving one relative to another nozzle cascade and rotor wheel, being passed by irregular periodically nonstationary flow of compressible fluid is classified to such phenomena. With the experimental research of nonstationary flows the basic difficulty is to measure parameters in rotating rotor wheel.At the same time the numerical investigations allow, if the calculation method has been developed, to predict the effect of geometric and operating conditions on nonstationary performances sufficiently quickly and cheap. Nowadays the calculations of viscous nonstationary flows are difficult because of the required high speed and computer memory volume. As a model of ideal gas is successfully applied in calculations of stationary flows in a turbine, it is also natural to use it for a case of nonstationary flows.

A problem of interference of aerodynamic cascade of profiles in two-dimensional transonic flow in the nonstationary statement was first considered in the works [1,2] using the MacCormack finite difference scheme and in the works [3,4] using the Godunov scheme. In the works [5,6] calculations of two-dimensional nonstationary flows including interaction of a moving cascade and blade wakes behind exit edges of the previous cascade have been carried out.

Calculations of three-dimensional periodically nonstationary transonic flows were first published in the works [7,8], in which the Euler equations have been solved by the Godunov-Kolgan and finite volumes methods. The modified algorithm for the calculation of periodical three-dimensional flow, using the property of generalized periodicity of flow in neighbouring channels as phase shift along space and time coordinate is presented in the paper. To integrate the Euler nonstationary equations the monotonous difference scheme 9, which is the generalization of the Godunov numerical method has been applied. The main peculiarities of the difference scheme are the application of the solution for the problem on break-down of discontinuity, linear space-time approximation of parameters within each mesh, the principle of the minimal derivative, practical insensivity of a break of lines of the calculated grid.

The algorithm proposed is thoroughly considered in the paper. Solutions for high pressure turbine stages and low pressure turbine stages are obtained at design and off-design conditions. Aerodynamic performances of blades with the description of nonstationary flow field are given.

DESCRIPTION OF THE ALGORITHM

Consider three-dimensional transonic flow of inviscous non-heat-conductive gas through axial turbine stage in the physical domain (Fig. I), including the fixed nozzle cascade and the rotor wheel rotating with constant angular velocity ω . In the general case both bladings have inequal number of blades of arbitrary configuration. Let the nozzle cascade be of z_1 blades, the rotor wheel be of z_2 blades. It is evident that the flow through the stage will not have periodicity from the blade to the blade (at cascade step), except for the case of equal number of blades in both bladings, and it usually is not applied in practice of turbines. Angular steps of bladings are determined by relationships:

$$t_1 = 2\pi/z_1 \quad ; \quad t_2 = 2\pi/z_2 \quad .$$

Refer the least general step of two bladings as the general step of the stage

$$T = k_1 t_1 = k_2 t_2 \quad ,$$

where k_1, k_2 are coprime numbers. If z_1 and z_2 are coprime numbers, then $k_1 = z_1$, $k_2 = z_2$, and a period equals $T = 2\pi$. So at each time moment the flow in stage has periodicity equal the step T of the stage.

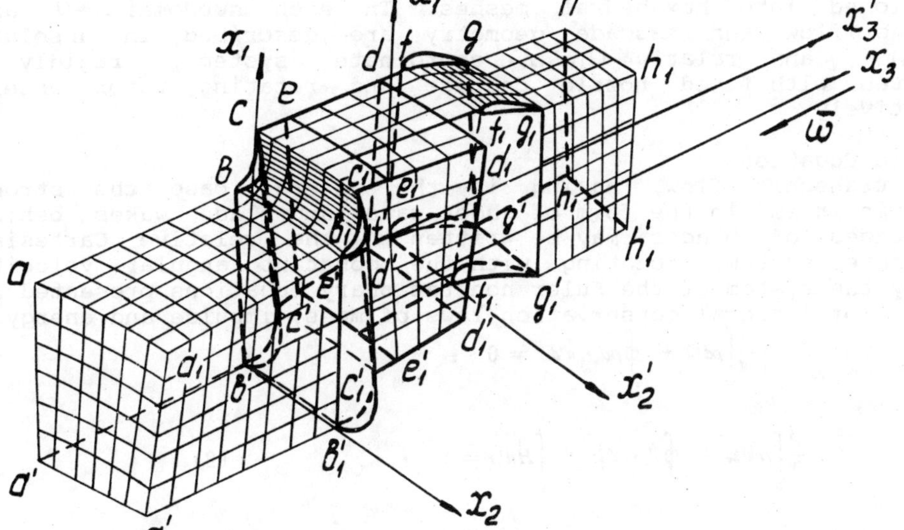

Fig. 1 Calculated domain.

Due to the relative displacement of bladings the flow structure in each vane channel varies in time with periodicity $\tau_1 = t_2/\omega$ for nozzle cascade and $\tau_2 = t_1/\omega$ for rotor wheel. Periodicity of flow along time and circumferential coordinate may conventionally be characterized by flow structure movement with respect to each blading with velocity ω_i , its value is determined by the ratio of a number of blades and the angular velocity of rotor wheel rotation. In the general case of arbitrary ratio of a number of blades it is easy to show that flow structure displacement in each blading in the direction opposite to the rotation of the rotor wheel occurs with the velocity equal respecitively

$$\omega_1 = \frac{\omega}{\frac{z_1}{z_2}\left[1 + E\left(\frac{z_2}{z_1}\right)\right] - 1} ; \qquad (1)$$

$$\omega_2 = \frac{\omega}{1 - \frac{z_2}{z_1}E\left(\frac{z_1}{z_2}\right)} ;$$

where $E(x)$ denotes the whole part of x .

Taking account of space and time periodicity of flow it is convenient to choose the calculated domain Ω , including one nozzle cascade vane channel and one rotor wheel vane channel, entry domain, blade clearance and exit (Fig. 1). The calculated domain is divided into two subdomains $\Omega_1(aa_1a'a'_1dd_1d'd'_1)$ and $\Omega_2(ee_1e'e'_1hh_1h'h'_1)$,having an extent in the circumferential direction t_1 and t_2 respectively and including the general part $\Omega_1 \cap \Omega_2$. The whole calculated domain is partitioned into hexahedral meshes. In each subdomains Ω_1 and Ω_2 gas flow and cascade geometry are described in absolute (x_1, x_2, x_3) and relative (x'_1, x'_2, x'_3) coordinate systems, rigidly connected with fixed nozzle cascade and rotating rotor wheel, respectively.

I. Basic Equations

Space transonic flow, having in the general case the strong discontinuities in the form of shock waves and blade wakes behind exit edges of blades may be written in the relative Cartesian coordinate system, rotating with the constant angular velocity $\vec{\omega}$ by the system of the Euler nonstationary equations presented in the form of integral conservation laws of mass, impulse and energy.

$$\frac{\partial}{\partial t}\int_{\Omega} \rho d\Omega + \oint_{\sigma} \rho\vec{v} \cdot \vec{n}^{\circ} d\sigma = 0 \quad ;$$

$$\frac{\partial}{\partial t}\int_{\Omega} \rho\vec{v} d\Omega + \oint_{\sigma} \hat{T} \cdot \vec{n}^{\circ} d\sigma + \int_{\Omega} \vec{H} d\Omega = 0 \quad ; \qquad (2)$$

$$\frac{\partial}{\partial t}\int_{\Omega} E d\Omega + \oint_{\sigma} (E + p)\vec{v} \cdot \vec{n}^{\circ} d\sigma = 0 \quad .$$

Here $\vec{n}^{\circ} = \{\alpha, \beta, \gamma\}$ is a vector of a unit normal to surface element $d\sigma$; $\hat{T} = \{T^{ik}\} = \{\rho v^i v^k + g^{ik} p\}$ is a tensor of impulse flow density (g^{ik} is a metric tensor); $H = \{\rho a^i_e + 2\rho e^{ijk}\omega_j v_k\}$ (e^{ijk} is the Levy-Chivita tensor); E is total energy of volume unit. According to the accepted coordinate system and the direction of the vector of the angular velocity $\vec{\omega}$, vector values entering into the equations (2) may be presented as the following components:

$$\hat{T}\cdot\vec{n}^{o}=\begin{Bmatrix}(\rho v_1^2+p)\alpha+\rho v_1 v_2\beta+\rho v_1 v_3\gamma\\ \rho v_2 v_1\alpha+(\rho v_2^2+p)\beta+\rho v_2 v_3\gamma\\ \rho v_3 v_1\alpha+\rho v_3 v_2\beta+(\rho v_3^2+p)\gamma\end{Bmatrix} \quad ;$$

$$\vec{H}=\begin{Bmatrix}-\rho\omega^2 r\cos(\vec{r},\vec{x}_1)+2\rho\omega v_2\\ -\rho\omega^2 r\cos(\vec{r},\vec{x}_2)+2\rho\omega v_1\end{Bmatrix} \quad ;$$

$$\vec{v}=\{v_1,v_2,v_3\} \quad ;\vec{\omega}=\{o,o,-\omega\} \quad .$$

The system of equations (2) is added with the equation of state

$$\varepsilon=\frac{1}{k-1}\frac{p}{\rho} \quad ,$$

where ε is an internal energy of mass unit; k is adiabat inde

2. Difference Scheme

Three half-integer numbers $\{m_i+1/2\}, i=1,2,3;\ 1\leqslant m_1\leqslant N;\ 1\leqslant m_2\leqslant M$ $1\leqslant m_3\leqslant L$ are assigned to each elementary mesh of the difference grid; one integer and two half-integer numbers are assigned to each side. the integer defines the axis, oriented along the normal to the given side. One may obtain difference analogues of conservation laws for mass, impulse, energy, presented in the symbolic vector form by integrating the equation (2) with respect to the volume of the elementary mesh $\Omega_{m_i+1/2}$ with the surface σ and averaging gas-dynamic parameters at lateral sides.

$$f^{t+\Delta t}_{m_i+1/2}=f^{t}_{m_i+1/2}-\frac{\Delta t}{\Omega_{m_i+1/2}}\Big[\sum_{i=1}^{3}(\vec{F}_i\cdot\vec{n}^{o}\sigma)_{m_i+1}-\sum_{i=1}^{3}(\vec{F}_i\cdot\vec{n}^{o}\sigma)_{m_i}\Big]-\Delta t H^{t}_{m_i+1/2} \quad , \tag{3}$$

where $f=\{\rho,\rho v_1,\rho v_2,\rho v_3,E\}$,

$$\vec{F}=\{\rho\vec{v},\hat{T},(E+p)\vec{v}\}$$

Flows of gas-dynamic values through lateral sides of meshes are defined from the solution of a problem on the break-down of arbitrary discontinuity at the boundary between neighbouring meshes. The piece-wise-constant approximation of parameters in meshes of the difference grid is supposed by S.K. Godunov [10] and used in the works [4,7].

The scheme modification [9] in which distributions of parameters in meshes is taken to be piecewise-linear, is used in this work. Parameters at central points of sides are defined at intermediate time layer using space-time extrapolation of parameters $a=\{p,\rho,v_1,v_2,v_3\}$ within neighouring meshes

$$a_{m_i}^{\mp} = a_{m_i \mp 1/2} + \left(\frac{\partial a}{\partial x_1}\right)_{m_i \mp 1/2} (x_{1m_i} - x_{1m_i \mp 1/2}) + \left(\frac{\partial a}{\partial x_2}\right)_{m_i \mp 1/2} (x_{2m_i} - x_{2m_i \mp 1/2})$$

$$+ \left(\frac{\partial a}{\partial x_3}\right)_{m_i \mp 1/2} (x_{3m_i} - x_{3m_i \mp 1/2}) + \left(\frac{\partial a}{\partial t}\right)_{m_i \mp 1/2} \frac{\Delta t}{2} \qquad (4)$$

and the solution of the problem on the break-down of discontinuity for values $a_{m_i}^-$ and $a_{m_i}^+$.

Time derivatives in (4) are approximated by one-sided differences and determined on the basis of values of gas-dynamic parameters in the layer t (3). Space derivatives are chosen using the principle of minimal one-sided derivatives 11. For their determination the values of each gas-dynamic parameter in the center of the given mesh $am_{m_i+1/2}$ and in the central points of six neighbouring meshes are of the sets of derivatives $(\partial a/\partial x_i)^1_{mi+1/2}$ may be found from the solution of the system of three linear equations connecting increments of the appropriate parameter in meshes $(m_i + 1/2)$, $(m_1 - 1/2, m_2 + 1/2, m_3 + 1/2)$, $(m_1 + 1/2, m_2 - 1/2, m_3 + 1/2)$, $(m_1 + 1/2, m_2 + 1/2, m_3 - 1/2)$ and unknown values of derivatives:

$$a_{m_1-1/2,m_2+1/2,m_3+1/2} - a_{m_i+1/2} = \left(\frac{\partial a}{\partial x_1}\right)^I_{m_i+1/2} \Delta x_{11} + \left(\frac{\partial a}{\partial x_2}\right)^I_{m_i+1/2} \Delta x_{21} + \left(\frac{\partial a}{\partial x_3}\right)^I_{m_i+1/2} \Delta x_{31} \ ;$$

$$a_{m_1+1/2,m_2-1/2,m_3+1/2} - a_{m_i+1/2} = \left(\frac{\partial a}{\partial x_1}\right)^I_{m_i+1/2} \Delta x_{12} + \left(\frac{\partial a}{\partial x_2}\right)^I_{m_i+1/2} \Delta x_{22} + \left(\frac{\partial a}{\partial x_3}\right)^I_{m_i+1/2} \Delta x_{32} \ ; \qquad (5)$$

$$a_{m_1+1/2,m_2+1/2,m_3-1/2} - a_{m_i+1/2} = \left(\frac{\partial a}{\partial x_1}\right)^I_{m_i+1/2} \Delta x_{13} + \left(\frac{\partial a}{\partial x_2}\right)^I_{m_i+1/2} \Delta x_{23} + \left(\frac{\partial a}{\partial x_3}\right)^I_{m_i+1/2} \Delta x_{33} \ ;$$

The system of equations (5) always has the only solution because line segments connecting a centre of the mesh $m_i+1/2$ and centers of neighbouring meshes are not parallel.

The second set of derivatives $(\partial a/\partial x_i)^{\mathrm{I}}_{m_i+1/2}$ is determined similarly from linear relationships, connecting increments of the parameter in the center of the mesh $m_i+1/2$ and centers of the second triplet of the neighbouring meshes $(m_1 + 3/2, m_2 + 1/2, m_3 + 1/2), (m_1 + 1/2, m_2 + 3/2, m_3 + 1/2)$, $(m_1 + 1/2, m_2 + 1/2, m_3 + 3/2)$, If one-sided derivatives have the same sign, then the least in absolute magnitude value of the derivative is taken; in case of different signs the derivative is assumed to be zero:

$$\left(\frac{\partial a}{\partial x_i}\right)=\begin{cases}\operatorname{sign}\left(\frac{\partial a}{\partial x_i}\right)^{I}\min\left\{\operatorname{abs}\left(\frac{\partial a}{\partial x_i}\right)^{I},\operatorname{abs}\left(\frac{\partial a}{\partial x_i}\right)^{II}\right\}, & \\ \qquad\text{if } \left(\frac{\partial a}{\partial x_i}\right)^{I}\cdot\left(\frac{\partial a}{\partial x_i}\right)^{II}>0; & \\ 0, \qquad \text{if } \left(\frac{\partial a}{\partial x_i}\right)^{I}\cdot\left(\frac{\partial a}{\partial x_i}\right)^{II}<0. & \end{cases}$$

So, according to the equations (3) to determine gas-dynamic values $(\rho,\rho\vec{v},E)$ at time moment $t+\Delta t$ for mesh number $(m_i+1/2)$ it is necessary to know these values at time moment t in the same mesh and flows of these values through sides of the mesh at time moment $t+\Delta t/2$. A step of time integration Δt is defined from the condition of stability of the difference scheme for the linearized system of equations and finally, may be presented in the form

$$\Delta t=\frac{\tau_{x_1}\tau_{x_2}\tau_{x_3}}{\tau_{x_1}\tau_{x_2}+\tau_{x_1}\tau_{x_3}+\tau_{x_2}\tau_{x_3}};$$

$$\tau_{x_i}=\frac{h_{x_i\min}}{\max(|\vec{v}_i|+a,a-|\vec{v}_i|)},$$

$i=1,2,3$; a is sound velocity.

The difference scheme plotted, is montonous, has the first order of time approximation and the second order of space coordinates approximation. The given scheme allows to obtain an acceptable accuracy with gas-dynamic calculations at irregular grids.

3. Boundary Conditions

The condition of slipping $\vec{v}\cdot\vec{n}=0$ is used at the boundaries of the domain, coinciding with internal and external contours of blading and blade surfaces.

A quantity and a type of boundary conditions in cross-sections of an entry $aa_1a'a'_1$ and an exit $hh_1h'h'_1$ are determined from the analysis of disturbance propagation in the nonstationary gas flow. At subsonic axial component of velocity the flow parameters at entry at moment $t+\Delta t$ are connected with flow parameters inside the calculated domain at moment t by a relationship (along the characteristic direction), i.e. four from five unknown flow parameters (or relationships between them) should be given. In case $v_3>a$ (a is sound velocity) all the five parameters should be given at the entry. AT the exit in case $v_3<a$ one from five parameters is given, in case $v_3>a$ the boundary conditions are not laid down.

The axial component of the velocity usually is subsonic in turbine stages. In this case at the entry into the stage are given the enthalpy i_o , and pressure p_o of the stagnated flow and the angles of the flow α and γ in tangential and meridional cross sections, at the exit behind the rotor wheel the distribution of the static pressure p is given. Relationships along performances in cross-sections of the entry and the exit are added to these conditions.

A full system of equations to determine parameters including relationships on performances may be written as follows:
at the entry

$$i_o = i_o(x_1, x_2) \quad ; \quad p_o = p_o(x_1, x_2) \quad ;$$
$$\alpha = \alpha(x_1, x_2) \quad ; \quad \gamma = \gamma(x_1, x_2) \quad ; \tag{6}$$

$$d\left(v_3 - \frac{2a}{k-1}\right) = 0$$

and at the ext

$$p = p(x_1, x_2) \quad ; \quad dp - a^2 d\rho = 0 \quad ;$$
$$dv_1 - (\omega^2 r - 2\omega v_2)dt = 0 \quad ; \tag{7}$$
$$dv_2 + 2\omega v_1 dt = 0 \quad ; \quad d\left(v_3 + \frac{2a}{k-1}\right) = 0$$

The boundary conditions (6,7) at the penetrable boundaries of the entry and the exit correspond to the complete (or partial) reflection of disturbance from these boundaries.

To simulate the ideal gas flow in the assumption of the wave absence, coming outside the calculated domain, the so-called "reflectless" boundary conditions may be used. In this case at the entry boundary instead of i_o, p_o the entropy function and the Riemann "right" invariant are given, the Riemann "left" invariant is given at the exit instead of the static pressure p [10]. The requirement to adaptability of the numerical algorithm for peculiarities of the solution for the problem under discussion (periodical nonstationarity including flow history) is realized by periodical change of "internal" boundary conditions in the blade clearance and at surfaces of periodicity according to instantaneous relative positions of bladings. At each time step as the boundary conditions at the exit of the domain Ω_1 are taken parameters from the corresponding internal meshes of the domain Ω_2. Similarly as the boundary conditions at the entry of the domain Ω_2 are taken parameters from the corresponding internal meshes of the domain Ω_1.

The statement procedure of periodicity conditions and the "internal" conditions in the clearance is considered as an example of two-row cascade, presented as the chart in Fig. 2, for which $z_1/z_2 = 2/3$. In the chart under discussion a number of meshes in circumferential direction has been chosen as follows: $M_1 = 3$ for the nozzle cascade and $M_2 = 2$ for the rotor wheel. In this case all the different relative positions of cascades are determined by time moments t, t' and t''. The domain Ω_1 is fixed, the domain Ω_2 is moved together with the rotor wheel, as shown in Figure, remaining at the fixed state during the time equal $t_1/M_1\omega$.

Let the domain have two last columns of the domain $\Omega_1(L_1 - 3/2$ and $L_1 - 1/2)$ and two first columns of the domain $\Omega_2(1/2$ and $2/3)$.

Consider meshes of two types: a type A-mesh at the given time moment belongs only to one domain Ω_1 or Ω_2 (e.g., $1/2, L_1-1/2$): type B-mesh belongs to $\Omega_1 \cap \Omega_2$ at the given moment (e.g., meshes $(M_1-1/2, L_1-1/2)$ and $(M_2-1/2, 3/2)$. Then the "internal" boundary conditions at moment t for the domain Ω_1 are laid down as follows: the paramenters from the mesh($M_2-1/2, 3/2$)of the domain Ω_2 at moment t' are taken to the mesh ($1/2, L_1-1/2$) (A type), the parameters from the mesh($M_2-1/2, 3/2$) of the domain Ω_2 at moment are taken to the mesh ($M_1-1/2, L_1-1/2$) (B type). The boundary conditions are laid down similarly at the entnry cross-section of the domain Ω_2 .

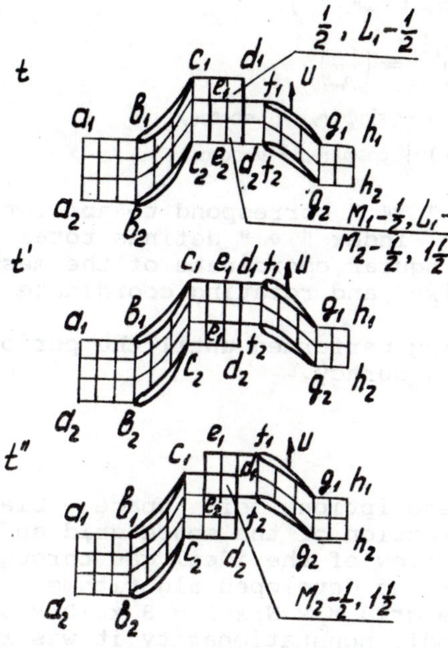

Fig. 2 Sequence of relative positions of two cascades

At the lines a_1b_1, a_2b_2, c_1d_1, c_2d_2, e_1f_1, e_2f_2, g_1h_1, g_2h_2 is used the generalized condition of periodicity, including the phase shift between flow structures in the neighbouring channels of the cascade under discussion due to different steps. In other words, the correspondence between the parameters is stated for values referring not to one but different time moments. E.g., flow parameters at boundaries a_1b_1, c_1d_1 of the domain Ω_1 at time moment t should equal the parameters of flow at the boundaries a_2b_2, c_2d_2 of the domain Ω_1 at moment t' and at boundaries a_2b_2, c_2d_2 at moment t equal the parameters at the boundaries a_1b_1, c_1d_1 at moment t' .Similarly, the parameters at the boundaries e_1f_1, g_1h_1 of the domain Ω_2 at time moment t should equal the parameters at the boundaries e_2f_2, g_2h_2 of the domain Ω_2 at moment t'' , and the parameters at the boundaries e_2f_2, g_2h_2 at moment t equal the corresponding parameters at the boundaries e_1f_1, g_2h_2 at moment t' .

The solution transition from time moment t to moment $t+\Delta t$ is calculated for each blading accounting their relative position and flow field in the previous layer.

When transitioning from the absolute movement (in Ω_1 domain) to the relative one (in Ω_2 domain) and vice versa, the static parameters do not vary and the total values of flow parameters and kinematic performances are defined by relationships

$$i_{oc} - v_{2c}\omega r = i_{ow} - \frac{(\omega r)^2}{2} = \text{const};$$

$$\frac{T_{ow}}{T_{oc}} = 1 - \frac{(wr)^2}{2i_{oc}}\left(2\frac{v_{2c}}{\omega r} - 1\right);$$

$$\frac{P_{ow}}{P_{oc}} = \left(\frac{T_{ow}}{T_{oc}}\right)^{\frac{k}{k-1}} = \left(\frac{\rho_{ow}}{\rho_{oc}}\right)^k;$$

$$v_{1w} = \cos\varphi(v_{1c} - \omega r\sin\varphi_c) - \sin\varphi(v_{2c} + \omega r\cos\varphi_c);$$

$$v_{2w} = \sin\varphi(v_{1c} - \omega r\sin\varphi_c) + \cos\varphi(v_{2c} + \omega r\cos\varphi_c);$$

Here indexes " c " and " w " correspond to absolute and relative coordinate systems, an index " o " defines total values of flow parameters; φ_c is the angular coordinate of the mesh center; φ is the angle between the fixed and rotating coordinate systems.

The calculation is being performed until the periodic solution is stated within the given accuracy.

RESULTS AND DISCUSSIONS

The programs of descriptions of space blading geometric performances, the construction of the space grid and the calculation of the threedimensional flow of the ideal gas through the stage are composed on the basis of the developed algorithms. The calculations were made at the space grid M x N x L = 8 x 10 x 40 for each vane channel. To state periodic nonstationarity it was required 2500 - 3000 iterations. Under the transition to the steady state conditions the maximum relative deviations of parameters in time intervals multiple with respect to the period did not exceed 0.1%, the maximum relative error of the turbine flow in the whole domain except for leadinig and exit edges did not exceed 1.5 - 2 % , in the vicinity of edges was 5-6%/ .

The results of the numerical investigation of the space flow in the stage of high pressure are presented and the stator and rotor of which are shown in Fig. 3 (digits correspond to the values of the nondimensional velocity). The main performances of the stage: D/l =8.3; ω =377 I/S. the stator blades are made with the inverse twist α_{1m} =13° 30; $\Delta\alpha_1 = \alpha_{1t} - \alpha_{1r} = -15°$; the rotor blades are made with variable in radius profile and constant angles of the entry β_1 = 26°30 and the exit β_2 =25° 8 . The space flow is shown by a family of lines of equal velocities (λ' = const) on blade surfaces and in vane channels of the nozzle cascade (Fig. 3a) and the rotor wheel (Fig. 3b). The velocity values are time averaged.

Fig. 4 presents the results of comparison of distributions along the radius of the static pressure in the blade clearance (Fig. 4 a) and the angles α_1 (Fig. 4b) and β_2 (Fig. 4c) obtained by the calculation (solid lines) and experiment (circles). The static pressure gradient p and the change of angles α_1 and β_2 at calculation and experiment are seen to be rather close. The reduced gradient of pressure is obtained at the given stage due to the use of the inverse twist of the stator blades $\alpha_r > \alpha_p$

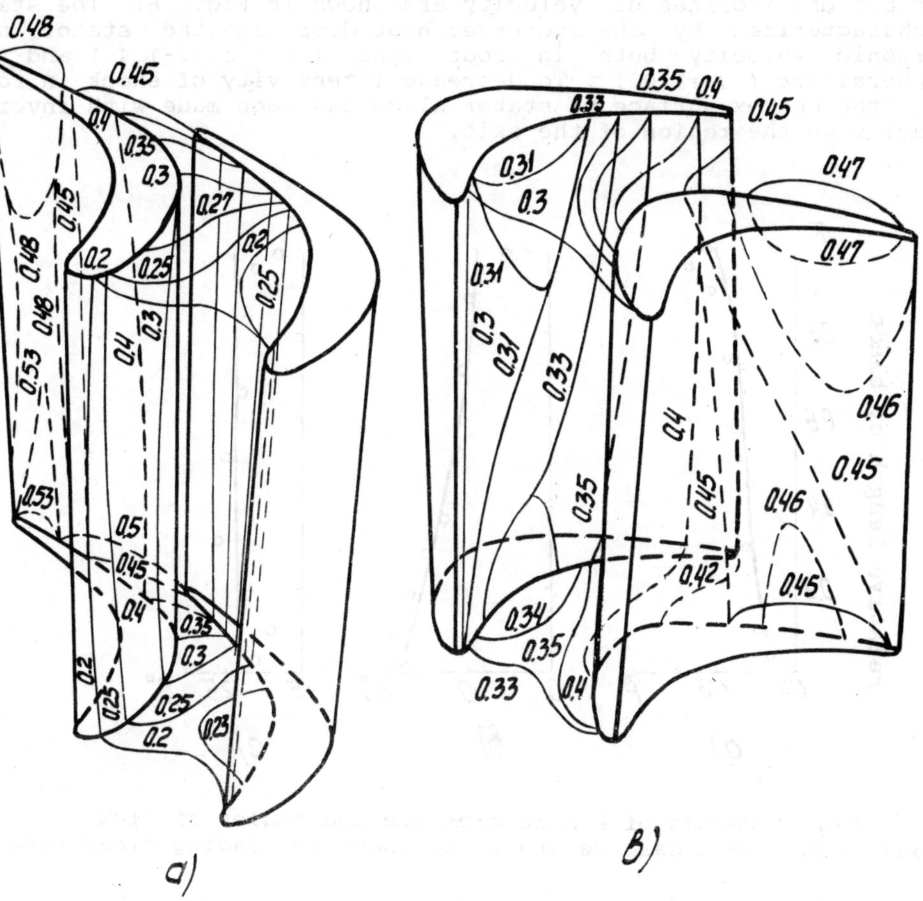

Fig. 3 Distribution of lines λ = const on blading surfaces a) nozzle cascade, b) rotor wheel.

Figure 5 gives static pressure distributions at the stator blade surface in root (Fig. 5a) , middle (Fig.5b) and peripheral cross-sections (Fig. 5c). The data of experimental measurements of the static pressure at the fixed points of the blade are plotted. The agreement of the calculated and experimental data is satisfactory, discrepancy of the results does not exceed 5-10%.

The method under discussion allows to calculate the space periodic changes of the flow parameters in the turbine stage and define amplitudes and phases of the main harmonics of the aerodynamic disturbance with the help of the harmonic analysis. The results of the numerical analysis of the transonic flow of gas through the last turbine stage at D/l =2.9 are given below. Instantaneous distributions isolines of velocity are shown in Fig. 6. The stage is characterized by the increased heat drop in the stator and supersonic velocity both in root zone (λ = 1.35-1.4) and the peripheral one (λ =1.05) . To decrease intensivity of shock in root domain the convex surface of stator blade has been made with inverse concavity in the region of the exit.

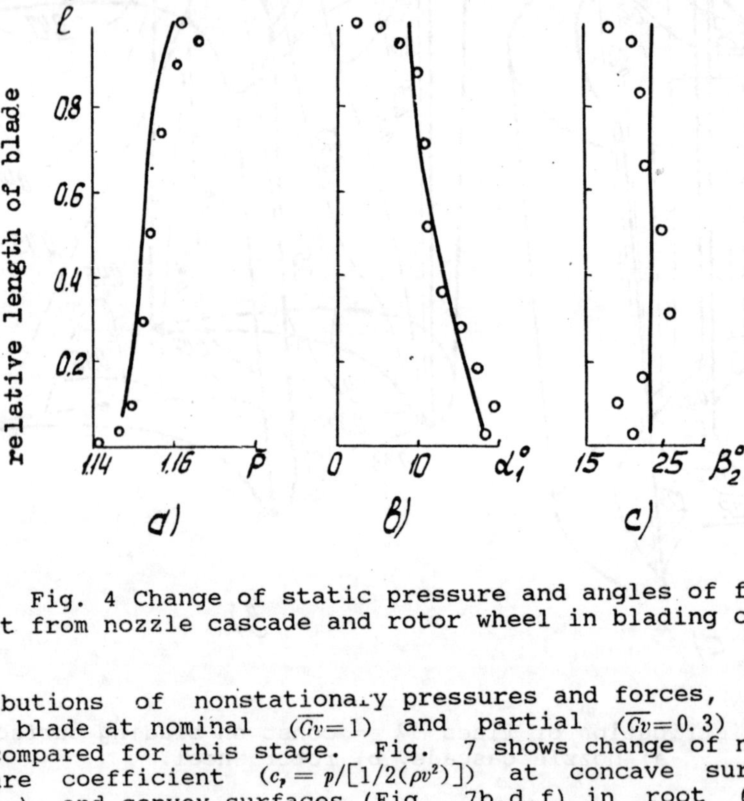

Fig. 4 Change of static pressure and angles of flow exit from nozzle cascade and rotor wheel in blading clearance.

Distributions of nonstationary pressures and forces, acting on a fixed blade at nominal $(\overline{Gv}=1)$ and partial $(\overline{Gv}=0.3)$ conditions are compared for this stage. Fig. 7 shows change of nonstationary pressure coefficient $(c_p = p/[1/2(\rho v^2)])$ at concave surfaces (Fig. 7a,c,e,) and convex surfaces (Fig. 7b,d,f) in root (0.I),middle (0.5 l), peripheral (0.9 l) blade sections, in the points stated at 0.05; 0.2; 0.5 ; 0.8 and 0.9 of blade width at nominal conditions. The curves 1,2 are the first and the second basic harmonics with frequencies of ωz_1 and $2\omega z_1$ (ω is circular frequency, z_1 is a number of stator blades). The maximum value of pressure pulsation amplitude is in root sections in the vicinitiy of the leading edge. The amplitude of the nonstationary

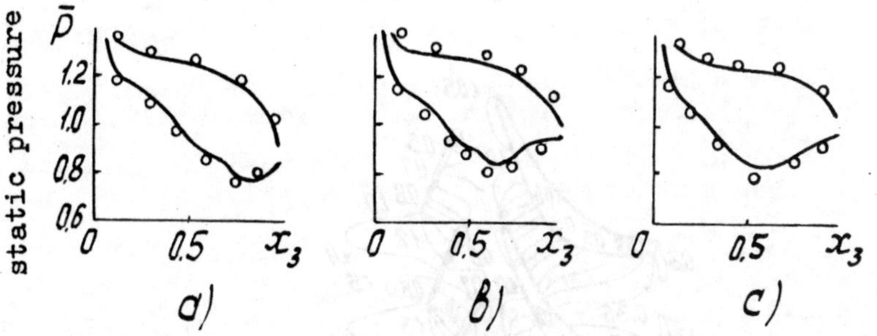

Fig.5 Distribution of static pressure along nozzle cascade blade contour - a,b,c - root middle, peripheral cross-sections.

pressure decreases chordwise and along radius from root to periphery. The levels of mean values and pulsation components of the nonstationary forces acting on the moving blade at nominal conditions are shown in Fig. 8. (here c_r , c_y and c_z are radial, circumferential and axial components of force, respectively). Figures 9, 10 give distributions of nonstationary pressures and forces at the moving blade at stage operation under partial conditions($\overline{G}v=0.3$) . The conditions are distinguished by the increased level of pressure pulsations, in particular, at the side of pressure. In contrast to the nominal conditions the amplitude of pulsations takes the maximum values for root and peripheral sections, that may be explained by infavourable conditions of the flow rate of the working medium (the appearance of root breakaway and transition of the peripheral domain to the enrgy input conditions). The analysis made witnesses that the variable aerodynamic loads on blades mainly depend on both turbine stage construction and operating conditions.

The ratio of blade numbers both of the nozzle cascade and the rotor wheel is one of the factors, effecting the level of the nonstationary forces. The numerical investigation of effect of the ratio of numbers of blades in the range of $0.3 \leqslant z_1 \leqslant z_2 \leqslant 1.2$ (Fig. 11) has shown that with approach z_1/z_2 to unit some growth of time averaged values of the coefficients of circulation (solid circles), forces (triangles the absolute value of the force, check marks - axial components, light circles-a lifting force) and moments (blocks) (Fig. 11 a) is accompanied by a sharp decrease of amplitudes of the first harmonics of the nonstationary pulsations of these parameters (Fig. 11 b). Accounting, that, on the one hand, the amplitudes of forces and moment at moving blades are decreased with increasing z_1/z_2 , and, on the other hand, at equal steps the forces acting on all the blades of the moving blade are cophasing and it is undesirable, one may recommend $z_1/z_2 = 0.7-0.8$ as the optimal ratio of numbers of blades. The results obtained qualitatively agree with the experimental and calculated data of the other authors [2,12,13].

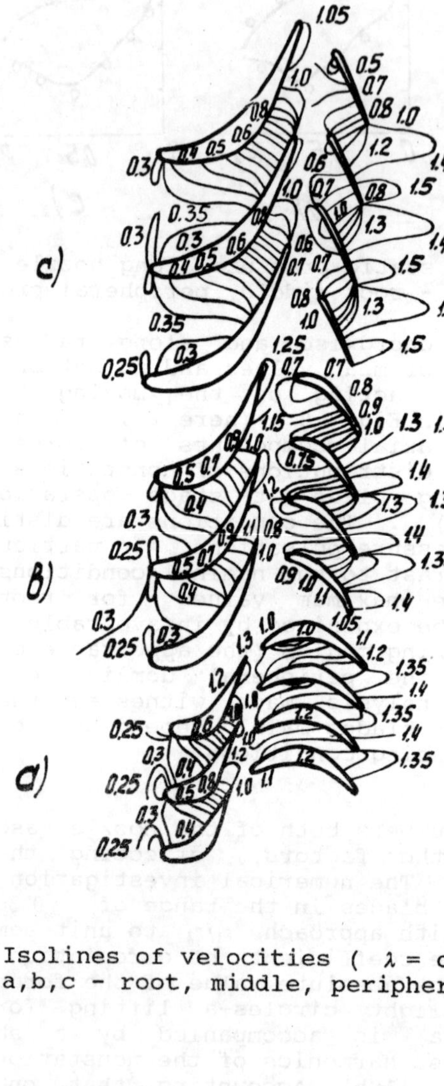

Fig. 6 Isolines of velocities (λ = const) in vane channels - a,b,c - root, middle, peripheral cross-sections.

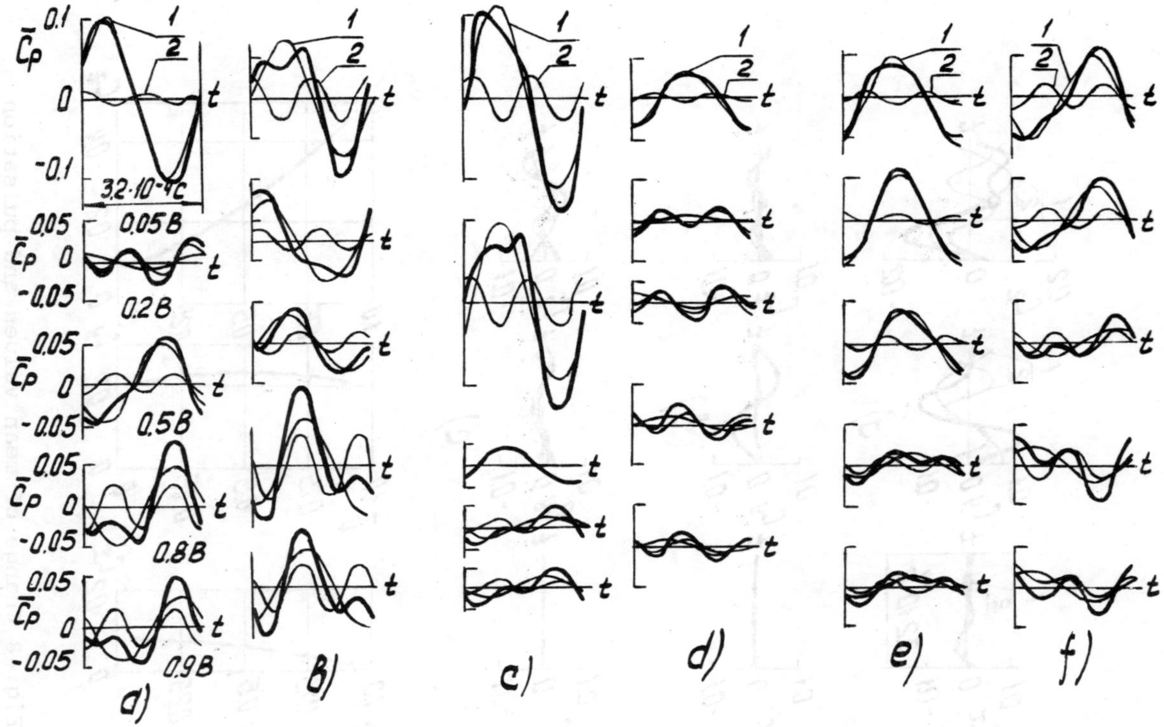

Fig. 7 Change of nonstationary pressure coefficient in root (a,b), middle (c,d) and peripheral (e,f) cross-sections of rotor wheel blade at design conditions - a,b,c - root, middle and peripheral crossections.

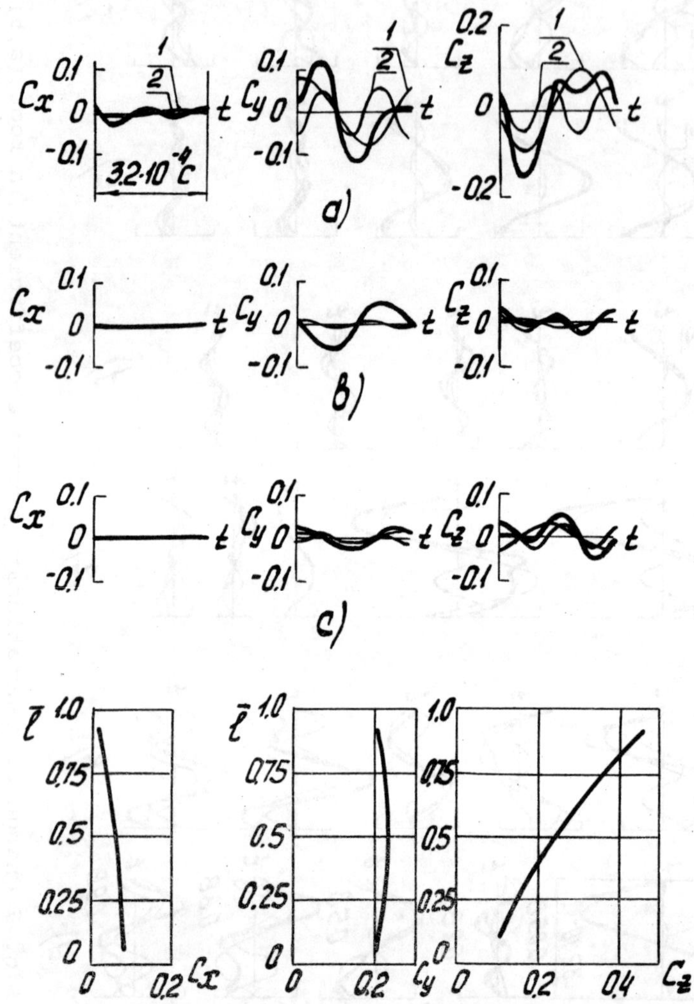

Fig. 8 Change of mean values and pulsation components of nonstationary forces at rotor wheel blade at design conditions - a,b,c - root, middle and peripheral cross-sections.

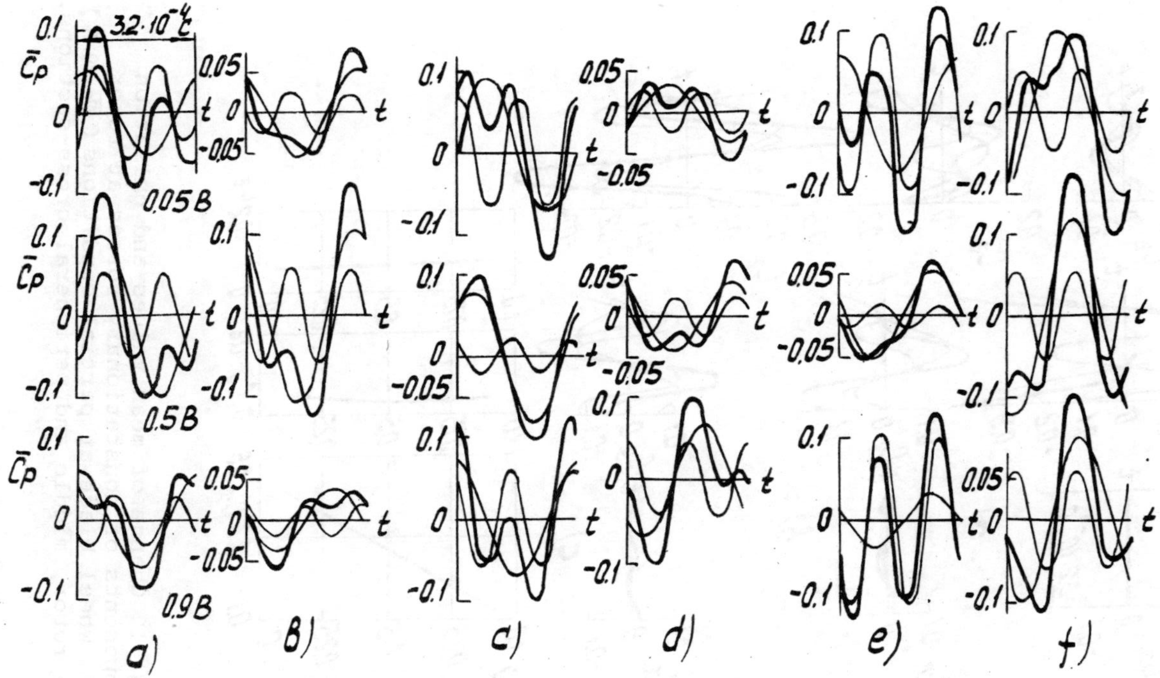

Fig. 9 Change of nonstationary pressure coefficient in root (a,b), middle (c,d) and peripheral (e,f) cross-sections of rotor wheel at partial conditions ($\overline{Gv}$ = 0.3).

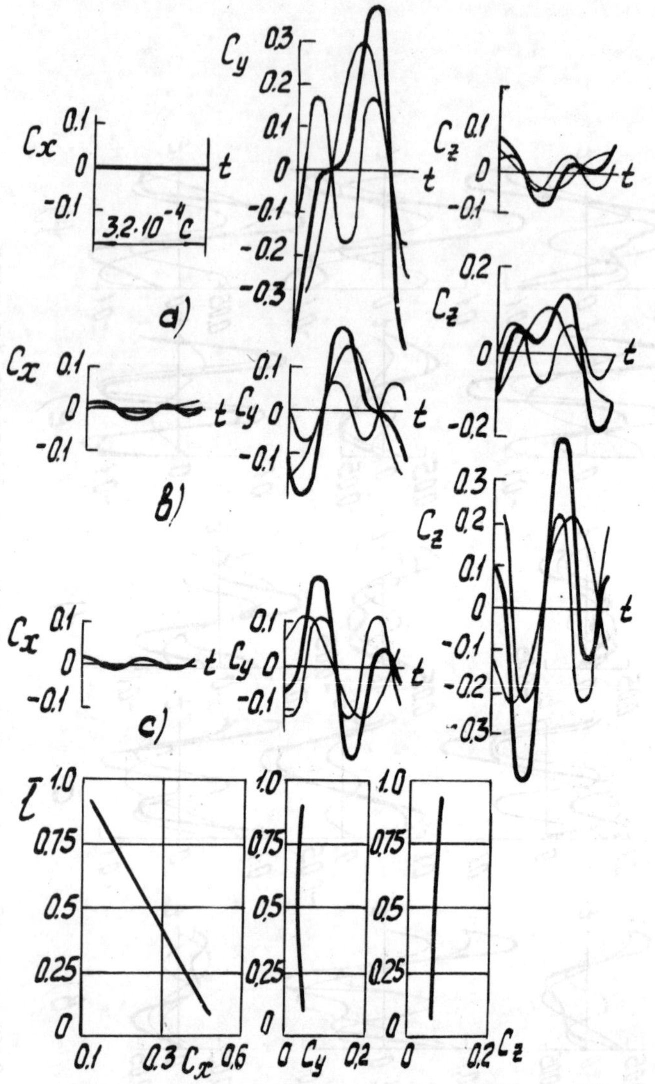

Fig. 10 Change of mean values and pulsation components of nonstationary forces at rotor wheel blade at partial conditions ($\overline{Gv}=0.3$)
a,b,c - rotor, middle and peripheral cross-sections.

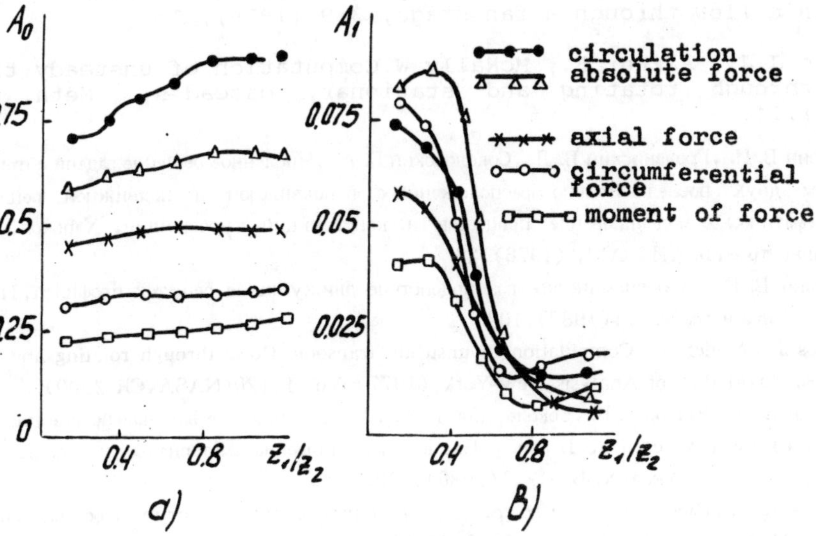

Fig. 11 Effect of ratio of blades on aerodynamic performances of the moving blade a) mean value, b) amplitude of the first harmonic.

CONCLUSIONS

The numerical method for the calculation of space transonic periodically nonstationary flow of ideal gas through the axial turbine stage has been considered.

Much attention has been paid to an algorithm of realization of the generalized condition of periodicity allowing to carry out calculations of turbine stages without the substantial increase of the required time and computer memory.

The calculation of three - dimensional flows in stages of high and low pressure under nominal and partial conditions with different geometric performances and gas-dynamic parameters have been carried out using the method developed. The comparative analysis of the nonstationary disturbing forces with different ratios of nozzle cascade and rotor wheel blades has been performed. Dynamic loads are shown to be decreased to some extent due to the choice of optimal ratio of blade numbers.

The author of this manuscript is a scientific worker.

REFERENCES

1.Erdos J.J., Alzner E., McNally W.Numerical solution of periodic transonic flow through a fan stage, 369 (1976), 16.

2.Erdos J.J., Alzner E., McNally W.Computation of unsteady transonic flow through rotating and stationary cascades, Meth. Anal., 1(1976),14.

3. Гнесин В. И. ,Гродзинский В. Л. ,Соколовский Г. А. ,Численное решение задачи обтекачия невязким газом двух последовательно расположенных неполвижной и подвижной решеток, В кн. : Математическое моделирование процессов и конструкций турбомашин, Харьков, Ин- т пробл. машиностроения АН УССР, (1976),20.

4. Гнесин В. И. , Аэродинамическое взаимодействие движушихся решеток профилей, Изв. АН СССР. Энергетика и трансп. ,1(1982),102.

5. Erdos J. , Alzner E. , Computation of unsteady transonic flows through rotating and stationary cascades. -In:Method of Analysis,New-York,(1977),Vol. 1,120(NASA,CR 2900).

6. Гнесин В. И. ,Ершов С. В. ,Соколовский Г. А. ,Аэродиамическое взаимодействие плоских решеток с учетом кромочых следов, В кн. : Научно- методические материалы по прикладным задачам азромеханики. Харьков:ХВВАИУ,1(1986),173.

7. Гнесик В. И. ,Расчет трехмѐрного трансзвукового потока газа через ступень осевой турбины,Изв. АН СССР. Механика жидкости и газа,6(1982),138.

8. Koya M. ,Kotake S. ,Numerical flow analysis of fully three-dimensional turbine cascades,ASME Pap. ,84(1984),16.

9. Гнесин В. И. ,Ершов С. В. ,Монотонная разностная схема второго порядка точности для двухмерных задач газовой динамики, В кн. : Моделирование в механике. Новосибирск: ИТПМ СО АН СССР, 2 (19),(1988),39.

10. Численное решение многомерных задач газовой динамики/А. В. Забродин,М. Я. Иванов и др. -М. : Наука,(1976),400.

11. Қолган В. П. , Применение принципа минималых значений производной и построение конечно-разностных схем для расчета разрывных решений газовой динамики, Учен. зап. Центр. азрогидродинам. ин-та,3(6) (1972),68.

12. Ласкин А. С. Исследование аэродинамического возбждения колебаний лопаточного аппарата и потерь энергии при нестационарных процессах в турбинах: Автореф. дис. ... д- ра техн. наук.- л. , (1979),40.

13. Соколовский Г. А. , Гнесин В. И. , Нестационарные трансзвуковые и вязкие течения в турбомашинах,Қиев:Наук. думка,(1986),260.

The Interaction between Distortion of Inlet Flow and Blade Stall Flutter in Axial–Flow Compressor

X.D.Yang Y.C.Feng

(Beijing University of Aeronautics and Astronautics)

ABSTRACT:

As we know, the distortion of inlet flow has a big effect on the performance of axial–flow compressor or fan. But the effect on the stall flutter is scarcely studied. This paper studied this effect by the testing. The testing rotor encounters the stall flutter when the inlet flow is uniform. The test shows that stall flutter disappears when inlet flow is in distortion. In order to study this phenomenon the numerical actuator disk method, which can predict three dimensional unsteady flow in axial flow compressors, is developed by authors. The strip theory and numerical actuator disk method are used to predict stall flutter in the inlet flow distortion. The computation indicates that numerical actuator disk method is effective for the stall flutter in the flow distortion. The works of the paper prove that the suitable inlet flow distortion can dispel the stall flutter in turbomachinery.

SYMBOLS:

h	amplitude of blade
i	$\sqrt{-1}$
k	reduced frequency
M	Mach number
P	pressure
q	velocity vector of flow
t	time
u	velocity component of coordinate r
v	velocity component of coordinate φ
w	velocity component of coordinate z
α	incidence angle
β	flow angle
γ	ratio of specific heats
δ	deviation angle
θ	stagger angle
σ	solidity
x	total pressure loss coefficient
ρ	density
φ	oscillating phase angle or circumference coordinate

Subscript:

n	normal components relative to actuator disk
–	the leading edge of deforming disk
+	the trailing edge of deforming disk

$-\infty$	upstream
$+\infty$	downstream
Superscript:	
$-$	steady component
$\sim$	unsteady component
$\simeq$	amplitude of unsteady component
rel	relative to deforming disk

INTRODUCTION:

The stall flutter of axial–flow compressor in aeroengine is a breakdown which can result in costly (both in time and money) overruns in aeroengine development programs. To solve this problem, engineers and scientists have paid numerous research. At present, a lot of predicting methods have been developed which can predict blade flutter at design of new aeroengines. But most of these methods can only treat the flutter under the uniform inlet flow.

In the axial–flow compressors or fans, the distortion of inlet flow have big effect on the performance of them. A lot of research works have been done and many useful results have been achieved. But the effect of distortion on the blade stall flutter is scarcely studied. The purpose of this paper is studing the interaction between inlet flow distortion and blade stall flutter by testing and computation.

The test was made in the single stage compressor facility. Fig.4 is the sketch map of the testing rotor. Fig.5 is measured stall flutter boundary without the distortion. So with the undistored inlet flow, the rotor encounters stall flutter at some rotating speed. In order to induce the inlet flow distortion, the distortion net which can cause the abrupt variation of velocity at radial or rotating direction is used (Fig.4). The types of distortion are radial and tangential distortion, the radial distortion is divided into the blade tip distortion in which the tip flow velocity is reduced and blade hub distortion in which the hub flow velocity is reduced. The testing results show that all kinds of distortion can weaken the stall flutter, especially the radial distortion at blade tip. This result is very useful to pretect blades from stall flutter.

In order to verify the phenomenon, the trip theory and numerical actuator disk method for three dimensional flow are developed on the base of actuator disk method from Ref.1 and Ref.2 by authors, which can predict stall flutter with the radial distortion. In fig.6, region I is the radial distortion zone and region II is undistored flow zone. Line A–A is the abrupt jumping plane of flow velocity. The parameters of steady inlet flow with the distortion are determined by testing. And the uniform flow in distored region is given by the inlet flow. Fig.7 has the computational results by strip theory. It indicates that the computational flutter boundary by strip theory is similar to the measured boundary with undistored inlet flow. But the testing results are that there is no flutter with the radial distortion at the blade tip. So the strip theory is not good for prediction of stall flutter with the distortion. The reason may be that there is the three dimensional effect at the distortion zone.

The actuator disk method in Ref.1 & Ref.2 features very simple computation and very small quantity of the computations. But the actuator disk method at present can only treat two dimensional unsteady flow. Its computation shows that the method is very effective for prediction of blade stall flutter. The authors intend to develop this method to predict three dimensional unsteady flow around the oscillating blade row. If this method is successful, the stall flutter under the inlet flow distortion can be predicted by the method.

For the two dimensional actuator disk, the analytic solution to the unsteady flow field around the oscillating cascade can be yielded.But it is very difficult to get analytic solution to three dimensional actuator disk method. So the numerical method has to be adapted. The authors use finite difference to solve the unsteady flow field around the oscillating blade row.

The computational results for above–mentioned testing rotor by the method indicates that the predicted flutter boundary is inside the surge zone, so the stall flutter does not appear with the tip distortion (Fig.7). This result is the same as above mentioned testing result. So, suitable inlet flow distortion can dispel the stall flutter.

In order to study the distribution of the distortion, another type of distribution of radial distortion is given (Fig.8). The computational results indicates that such a distortion which means radial continuous variation of inlet flow velocity from hub to tip can't weaken the stall flutter.

THE TEST OF STALL FLUTTER WITH INLET FLOW DISTORTION:

Ref.4 studied the effect of distortion on the unstall flutter in axial flow fan. Its results showed that the reduced velocity of the fan would be increased when stall flutter appeared with the inlet flow distortion. The uniform inlet flow corresponds to lower critical reduced velocity of stall flutter than the distored inlet flow. So the inlet flow distortion may weaken unstall flutter. This result is very interesting. But for the stall flutter, the similar test has not been published in the literature. So the authors used the testing rotor to do such a test of stall flutter with the inlet flow distortion.

The test was made in single stage compressor facility in BUAA. The sketch map of testing rotor is Fig.4.The aerodynamic damping of the testing rotor is zero in designing situation. The rotor encounters stall flutter without the distortion of inlet flow. Fig.5 is the boundary of stall flutter with uniform inlet flow. From the figure, it can be seen that the rotating speed in stall flutter zone is from 100% to 74%. So the testing rotor can be used to study the effect of the distortion on aeroelasticity. The main designing parameters of the testing rotor are :

frequency in the first bending mode:	363 Hz
flow rate of the rotor:	12.93 kg / sec
rotating speed:	16500 rpm
March number at blade tip:	1.08
solidity at the tip:	1.035
stagger angle at the tip:	50.1 °

The distortion net is used to induce inlet flow distortion.The types of the distortion are respectively the radial distortion at the tip and hub and the circumference distortion. This paper emphasizes the test of radial distortion especially at the tip. The sketch map of the distortion net is refered to in Fig.4 and Fig.6.

The testing results show that when the every types of distortions are induced, the stall flutter zone is reduced. Especially, the stall flutter with the radial distortion at the blade tip disappears at all rotating speed.This result conforms the test in Ref.4. Maybe, such a result can be used to dispel stall flutter.

In order to verify the effect of the distortion, the predicting method is developed to compute the stall flutter with the distortion. This paper uses the radial distortion at the tip as the typical computational example. In this paper the introduction about three dimensional actuator disk method is emphasized.

MATHEMATICS–PHYSICS MODEL OF THREE DIMENSIONAL ACTUATOR DISK METHOD:

According to the introduction, two dimensional actuator disk method can be developed into three dimension. The basic assumptions are :

1) Suppose the thick of blade row is zero. The oscillating blade row becomes a deforming actuator disk. Three dimensional unsteady flow field in front of and behind the actuator disk are computed and the flow inside the blade row is not considered. Fig.1 is the coordinate of unsteady flow field. Fig.2 and Fig.3 are the sketch maps of actuator disk.

2) the flow is inviscid.

3) There are five abrupt jumping conditions across the disk. And the quasi–steady assumption is supposed.

4) The blades vibrates in first bending mode. The motion of actuator disk is determined by the vibration of blade row:

$$\begin{cases} \tilde{h}_{x,\,i,\,j} = \tilde{h}_{x,\,i,\,j} e^{i\varphi j + ikt} \\ \tilde{h}_{y,\,i,\,j} = \tilde{h}_{y,\,i,\,j} e^{i\varphi j + ikt} \end{cases} \tag{1}$$

5) The unsteady flow is considered the small perturbation to the steady one. So every aerodynamic parameters can be divided into steady and unsteady components:

$$\begin{cases} f = \bar{f} + \tilde{f} \\ |\tilde{f} / \bar{f}| << 1.0 \end{cases} \tag{2}$$

6) Unsteady components are harmonic with time:

$$\tilde{f}(r,\ \varphi,\ z,\ t) = \tilde{\tilde{f}}(x,\ \varphi,\ z)e^{ikt} \tag{3}$$

THE BASIC EQUATIONS & BOUNDARY CONDITIONS:

Because the flow is inviscid, the unsteady three dimensional Euler's equation can be used as the basic equations. By assumption 5) the equation can be linearized :

$$\frac{\partial\tilde{\rho}}{\partial t} + \bar{u}\frac{\partial\tilde{\rho}}{\partial r} + \bar{v}\frac{\partial\tilde{\rho}}{r\partial\varphi} + \bar{w}\frac{\partial\tilde{\rho}}{\partial z} + \bar{\rho}\left[\frac{\partial\tilde{u}}{\partial r} + \frac{\partial\tilde{v}}{r\partial\varphi} + \frac{\partial\tilde{w}}{\partial z}\right] + \left(\frac{\bar{\rho}}{r} + \frac{\partial\bar{\rho}}{\partial r}\right)\tilde{u} + \frac{\partial\bar{\rho}}{r\partial\varphi}\tilde{v} + \frac{\partial\bar{\rho}}{\partial z}\tilde{w} + \left[\frac{\partial(\bar{u}r)}{\partial r} + \frac{\partial\bar{v}}{r\partial\varphi} + \frac{\partial\bar{w}}{\partial z}\right]\tilde{\rho} = 0 \tag{4}$$

$$\frac{\partial\tilde{u}}{\partial t} + \bar{u}\frac{\partial\tilde{u}}{\partial r} + \bar{v}\frac{\partial\tilde{u}}{r\partial\varphi} + \bar{w}\frac{\partial\tilde{u}}{\partial z} + \frac{1}{\bar{\rho}}\frac{\partial\tilde{P}}{\partial r} + \frac{\partial\bar{u}}{\partial r}\tilde{u} + \left(\frac{\partial\bar{u}}{r\partial\varphi} - 2\frac{\bar{v}}{r}\right)\tilde{v} + \frac{\partial\bar{u}}{\partial z}\tilde{w} - \frac{1}{\bar{\rho}^2}\frac{\partial\bar{P}}{\partial r}\tilde{\rho} = 0 \tag{5}$$

$$\frac{\partial\tilde{v}}{\partial t} + \bar{u}\frac{\partial\tilde{v}}{\partial r} + \bar{v}\frac{\partial\tilde{v}}{r\partial\varphi} + \bar{w}\frac{\partial\tilde{v}}{\partial z} + \frac{1}{\bar{\rho}}\frac{\partial\tilde{P}}{r\partial\varphi} + \left(\frac{\partial\bar{v}}{\partial r} + \frac{\bar{v}}{r}\right)\tilde{u} + \left(\frac{\partial\bar{v}}{r\partial\varphi} + \frac{\bar{u}}{r}\right)\tilde{v} + \frac{\partial\tilde{v}}{\partial z}\tilde{w} - \frac{1}{\bar{\rho}^2}\frac{\partial\bar{P}}{r\partial\varphi}\tilde{\rho} = 0 \tag{6}$$

$$\frac{\partial\tilde{w}}{\partial t} + \bar{u}\frac{\partial\tilde{w}}{\partial r} + \bar{v}\frac{\partial\tilde{w}}{r\partial\varphi} + \bar{w}\frac{\partial\tilde{w}}{\partial z} + \frac{1}{\bar{\rho}}\frac{\partial\tilde{P}}{\partial z} + \frac{\partial\bar{w}}{\partial r}\tilde{u} + \frac{\partial\tilde{w}}{r\partial\varphi}\tilde{v} + \frac{\partial\bar{w}}{\partial z}\tilde{w} - \frac{1}{\bar{\rho}^2}\frac{\partial\bar{P}}{\partial r}\tilde{\rho} = 0 \tag{7}$$

$$\frac{\partial\tilde{P}}{\partial t} + \bar{u}\frac{\partial\tilde{P}}{\partial r} + \bar{v}\frac{\partial\tilde{P}}{r\partial\varphi} + \bar{w}\frac{\partial\tilde{P}}{\partial z} - \bar{a}^2\left[\frac{\partial\tilde{\rho}}{\partial t} + \bar{u}\frac{\partial\tilde{\rho}}{\partial r} + \bar{v}\frac{\partial\tilde{\rho}}{r\partial\varphi} + \bar{w}\frac{\partial\tilde{\rho}}{\partial z}\right] + \left(\frac{\partial\bar{P}}{\partial r} - \bar{a}^2\frac{\partial\bar{\rho}}{\partial r}\right)\tilde{u} + \left(\frac{\partial\bar{P}}{r\partial\varphi} - \bar{a}^2\frac{\partial\bar{\rho}}{r\partial\varphi}\right)\tilde{v}$$

$$+ \left(\frac{\partial\tilde{P}}{\partial z} - \bar{a}^2\frac{\partial\tilde{\rho}}{\partial z}\right)\tilde{w} + \frac{\gamma}{\bar{\rho}}\left(\bar{u}\frac{\partial\bar{\rho}}{\partial r} + \bar{v}\frac{\partial\bar{\rho}}{r\partial\varphi} + \bar{w}\frac{\partial\bar{\rho}}{\partial z}\right)\tilde{P} - \frac{\bar{a}^2}{\bar{\rho}}\left(\bar{u}\frac{\partial\bar{\rho}}{\partial r} + \bar{v}\frac{\partial\bar{\rho}}{r\partial\varphi} + \bar{w}\frac{\partial\bar{\rho}}{\partial z}\right)\tilde{\rho} = 0 \tag{8}$$

These equations have four independent variables, so the computational quantity to solve is very large. Assumption 6) can simplify the equations.But as we know from computational aerodynamics, by such simplified equations it is difficult to solve in transonic flow. So the assumption similar to Ref.5 is adapted, in which unsteady amplitude of aerodynamic parameters tend to steady values with the march of time.So the amplitudes can be represented:

$$\tilde{f}(r,\ \varphi,\ z,\ t) = \tilde{\tilde{f}}(r,\ \varphi,\ z,\ t)e^{ikt} \tag{9}$$

Then, the linearized Euler's equation is changed into the form of unsteady amplitude with four independent variables (r, φ , z, t):

$$\frac{\partial\tilde{\tilde{\rho}}}{\partial t}+\bar{u}\frac{\partial\tilde{\tilde{\rho}}}{\partial r}+\bar{v}\frac{\partial\tilde{\tilde{\rho}}}{r\partial\varphi}+\bar{w}\frac{\partial\tilde{\tilde{\rho}}}{\partial z}+\bar{\rho}\left[\frac{\partial\tilde{\tilde{u}}}{\partial r}+\frac{\partial\tilde{\tilde{v}}}{r\partial\varphi}+\frac{\partial\tilde{\tilde{w}}}{\partial z}\right]+\left(\frac{\bar{\rho}}{r}+\frac{\partial\bar{\rho}}{\partial r}\right)\tilde{\tilde{u}}+\frac{\partial\bar{\rho}}{r\partial\varphi}\tilde{\tilde{v}}+\frac{\partial\bar{\rho}}{\partial z}\tilde{\tilde{w}}+\left[\frac{\partial(\bar{u}r)}{\partial r}+\frac{\partial\bar{v}}{r\partial\varphi}+\frac{\partial\bar{w}}{\partial z}+\bar{i}k\right]\tilde{\tilde{\rho}}=0 \tag{10}$$

$$\frac{\partial\tilde{\tilde{u}}}{\partial t}+\bar{u}\frac{\partial\tilde{\tilde{u}}}{\partial r}+\bar{v}\frac{\partial\tilde{\tilde{u}}}{r\partial\varphi}+\bar{w}\frac{\partial\tilde{\tilde{u}}}{\partial z}+\frac{1}{\bar{\rho}}\frac{\partial\tilde{\tilde{P}}}{\partial r}+\left[\frac{\partial\bar{u}}{\partial r}+\bar{i}k\right]\tilde{\tilde{u}}+\left(\frac{\partial\bar{u}}{r\partial\varphi}-2\frac{\bar{v}}{r}\right)\tilde{\tilde{v}}+\frac{\partial\bar{u}}{\partial z}\tilde{\tilde{w}}-\frac{1}{\bar{\rho}^2}\frac{\partial\bar{P}}{\partial r}\tilde{\tilde{\rho}}=0 \tag{11}$$

$$\frac{\partial\tilde{\tilde{v}}}{\partial t}+\bar{u}\frac{\partial\tilde{\tilde{v}}}{\partial r}+\bar{v}\frac{\partial\tilde{\tilde{v}}}{r\partial\varphi}+\bar{w}\frac{\partial\tilde{\tilde{v}}}{\partial z}+\frac{1}{\bar{\rho}}\frac{\partial\tilde{\tilde{P}}}{r\partial\varphi}+\left(\frac{\partial\bar{v}}{\partial r}+\frac{\bar{v}}{r}\right)\tilde{\tilde{u}}+\left(\frac{\partial\bar{v}}{r\partial\varphi}+\frac{\bar{u}}{r}\right)\tilde{\tilde{v}}+\bar{i}k\tilde{\tilde{v}}+\frac{\partial\bar{v}}{\partial z}\tilde{\tilde{w}}-\frac{1}{\bar{\rho}^2}\frac{\partial\bar{P}}{r\partial\varphi}\tilde{\tilde{\rho}}=0 \tag{12}$$

$$\frac{\partial\tilde{\tilde{w}}}{\partial t}+\bar{u}\frac{\partial\tilde{\tilde{w}}}{\partial r}+\bar{v}\frac{\partial\tilde{\tilde{w}}}{r\partial\varphi}+\bar{w}\frac{\partial\tilde{\tilde{w}}}{\partial z}+\frac{1}{\bar{\rho}}\frac{\partial\tilde{\tilde{P}}}{\partial z}+\frac{\partial\bar{w}}{\partial r}\tilde{\tilde{u}}+\frac{\partial\tilde{w}}{r\partial\varphi}\tilde{\tilde{v}}+\left(\frac{\partial\bar{w}}{\partial z}+\bar{i}k\right)\tilde{\tilde{w}}-\frac{1}{\bar{\rho}^2}\frac{\partial\bar{P}}{\partial z}\tilde{\tilde{\rho}}=0 \tag{13}$$

$$\frac{\partial\tilde{\tilde{P}}}{\partial t}+\bar{u}\frac{\partial\tilde{\tilde{P}}}{\partial r}+\bar{v}\frac{\partial\tilde{\tilde{\rho}}}{r\partial\varphi}+\bar{w}\frac{\partial\tilde{P}}{\partial z}-\bar{a}^2\left[\frac{\partial\tilde{\tilde{\rho}}}{\partial t}+\bar{u}\frac{\partial\tilde{\tilde{\rho}}}{\partial r}+\bar{v}\frac{\partial\tilde{\tilde{\rho}}}{r\partial\varphi}+\bar{w}\frac{\partial\tilde{\tilde{\rho}}}{\partial z}\right]+\left(\frac{\partial\bar{P}}{\partial r}-\bar{a}^2\frac{\partial\bar{\rho}}{\partial r}\right)\tilde{\tilde{u}}+\left(\frac{\partial\bar{P}}{r\partial\varphi}-\bar{a}^2\frac{\partial\bar{\rho}}{r\partial\varphi}\right)\tilde{\tilde{v}}$$
$$+\left(\frac{\partial\bar{P}}{\partial z}-\bar{a}^2\frac{\partial\bar{\rho}}{\partial z}\right)\tilde{\tilde{w}}+\left[\frac{\gamma}{\bar{\rho}}\left(\bar{u}\frac{\partial\tilde{\rho}}{\partial r}+\bar{v}\frac{\partial\tilde{\rho}}{r\partial\varphi}+\bar{w}\frac{\partial\tilde{\rho}}{\partial z}\right)+\bar{i}k\right]\tilde{P}+\left[-\frac{\bar{a}^2}{\bar{\rho}}\left(\bar{u}\frac{\partial\bar{\rho}}{\partial r}+\bar{v}\frac{\partial\bar{\rho}}{r\partial\varphi}+\bar{w}\frac{\partial\bar{\rho}}{\partial z}\right)-\bar{i}k\bar{a}^2\right]\tilde{P}=0 \tag{14}$$

Equations (10)–(14) are the basic equations for the problem.

There are four kinds of boundary conditions with the equations:

1) Boundary of flow field inlet:

According to the aerodynamics, this boundary has a characteristic condition and four conditions must be given:

$$\tilde{P}^{*}_{-\infty}=\tilde{T}^{*}_{-\infty}=\tilde{U}_{-\infty}=0 \tag{15}$$

$$\tilde{V}_{-\infty}=\tilde{W}_{-\infty}COS\beta_{-\infty} \tag{16}$$

2) Boundary of flow field exit:

By aerodynamics this boundary has four characteristic conditions and a condition must be given:

$$\tilde{P}^{*}_{-\infty}=0 \tag{17}$$

3) Boundary of end solid wall at blade hub and tip:

The solid wall condition can be expressed :

$$\tilde{\tilde{u}}=\tilde{\tilde{w}}\cdot tan\dot{\beta}_m \tag{18}$$

β_m is the slope angle at the end wall.

4) Actuator disk:

There are five abrupt jumping conditions:

a) The continuity of flow rate across the disk:

$$(\rho_{-,i,j}+\tilde{\rho}_{-,i,j})q^{rel}_{-,i,j}=(\rho_{+,i,j}+\tilde{\rho}_{+,i,j})q^{rel}_{+,i,j} \tag{19}$$

b) The conservation of total enthalpy aross the disk:.

$$\frac{\gamma}{\gamma-1}\frac{P_{-,i,j}}{\rho_{-,i,j}}+\frac{(q^{rel}_{-,i,j})^2}{2}=\frac{\gamma}{\gamma-1}\frac{P_{+,i,j}}{\rho_{+,i,j}}+\frac{(q^{rel}_{+,i,j})^2}{2} \tag{20}$$

c) The abrupt–jump of flow angle at the trailing edge:

As same as Ref.5, the quasi–steady assumption is used to determine unsteady flow angle by steady empirical deviation angle relation.The form of empirical relation is:

$$\delta = g(\alpha_{-},\ M_{-},\ \sigma,\ \theta) \tag{21}$$

d) The abrupt jump of total pressure loss across the disk:

The quasi–steady assumption is also used to solve the unsteady flow loss across the disk by empirical relation of steady total pressure. The form of the relation is:

$$x = x(\alpha_{-},\ M_{-},\ \sigma,\ \theta) \tag{22}$$

e) Stream line angle of surface S (the slope angle in Fig.2):

The steady angle can be determined by curvature method of streamline.This paper adapted quasi–steady assumption to solve the unsteady angle by steady curvature method of stream line.The details can be refered to in Ref.3.

The initial flow field is supposed zero. The computation of this paper proves that such a initial flow does not increase the computational quantity greatly.

THE FINITE DIFFERENCE SCHEME:

MacCormack finite differnce sheme is adapted to the basic equations. The scheme is as this:

$$\begin{cases} L_{i,j,k}(f) = \dfrac{f_{i,j,k+1} - f_{i,j,k}}{\Delta\xi(k+1,\ k)} \dfrac{\Delta\xi(k+1,\ k)}{z_{k+1} - z_k} - \dfrac{f_{i+1,j,k} - f_{i,j,k}}{\Delta r} \dfrac{r_{i,j,k+1} - r_{i,j,k}}{z_{k+1} - z_k} \\ M_{i,j,k}(f) = \bar{u}_{i,j,k} \dfrac{f_{i+1,j,k} - f_{i,j,k}}{\Delta r} + \bar{v}_{i,j,k} \dfrac{f_{i,j+1,k} - f_{i,j,k}}{r_i \Delta\varphi} + L_{i,j,k}(f) \end{cases} \tag{23}$$

$$\tilde{\bar{\rho}}^{n+\frac{1}{2}}_{i,j,k} = \tilde{\bar{\rho}}^{n}_{i,j,k} - \Delta t \cdot \left\{ M_{i,j,k}(\tilde{\bar{\rho}}^{n}) + \bar{\rho}_{i,j,k} \left[\frac{\tilde{\bar{u}}^{n}_{i+1,j,k} - \tilde{\bar{u}}^{n}_{i,j,k}}{\Delta r} + \frac{\tilde{\bar{v}}^{n}_{i,j+1,k} - \tilde{\bar{v}}^{n}_{i,j,k}}{r_i \Delta\varphi} + L_{i,j,k}(\tilde{\bar{w}}^{n}) \right] + A^{n}_{i,j,k} \right\} \tag{24}$$

$$\tilde{\bar{u}}^{n+\frac{1}{2}}_{i,j,k} = \tilde{\bar{u}}^{n}_{i,j,k} - \Delta t \cdot \left\{ M_{i,j,k}(\tilde{\bar{u}}^{n}) + \frac{1}{\bar{\rho}_{i,j,k}} \frac{\widetilde{\bar{P}}^{n}_{i+1,j,k} - \widetilde{\bar{P}}^{n}_{i,j,k}}{\Delta r} + B^{n}_{i,j,k} \right\} \tag{25}$$

$$\tilde{\bar{v}}^{n+\frac{1}{2}}_{i,j,k} = \tilde{\bar{v}}^{n}_{i,j,k} - \Delta t \cdot \left\{ M_{i,j,k}(\tilde{\bar{v}}^{n}) + \frac{1}{\bar{\rho}_{i,j,k}} \frac{\widetilde{\bar{P}}^{n}_{i,j+1,k} - \widetilde{\bar{P}}^{n}_{i,j,k}}{r_i \Delta\varphi} + C^{n}_{i,j,k} \right\} \tag{26}$$

$$\tilde{\bar{w}}^{n+\frac{1}{2}}_{i,j,k} = \tilde{\bar{w}}^{n}_{i,j,k} - \Delta t \cdot \left\{ M_{i,j,k}(\tilde{\bar{w}}^{n}) + \frac{1}{\bar{\rho}_{i,j,k}} L_{i,j,k}(\widetilde{\bar{P}}^{n}) + D^{n}_{i,j,k} \right\} \tag{27}$$

$$\widetilde{\bar{P}}^{n+\frac{1}{2}}_{i,j,k} = \widetilde{\bar{P}}^{n}_{i,j,k} - \Delta t \cdot \left\{ M_{i,j,k}(\widetilde{\bar{P}}^{n}) - \bar{a}^2 \cdot \left[\frac{\tilde{\bar{\rho}}^{n+\frac{1}{2}}_{i,j,k} - \tilde{\bar{\rho}}^{n}_{i,j,k}}{\Delta t} + M_{i,j,k}(\tilde{\bar{\rho}}^{n}) \right] + E^{n}_{i,j,k} \right\} \tag{28}$$

$$N_{i,j,k}(f) = \bar{u}_{i,j,k} \frac{f_{i,j,k} - f_{i-1,j,k}}{\Delta r} + \bar{v}_{i,j,k} \frac{f_{i,j,k} - f_{i,j-1,k}}{r_i \Delta\varphi} + L_{i,j,k-1}(f) \tag{29}$$

$$\tilde{\bar{\rho}}^{\overline{n+1}}_{i,j,k} = \tilde{\bar{\rho}}^{n+\frac{1}{2}}_{i,j,k}$$

$$-\Delta t\cdot\left\{N_{i,j,k}(\tilde{\rho}^{n+\frac{1}{2}})+\bar{\rho}_{i,j,k}\left[\frac{\tilde{u}^{n+\frac{1}{2}}_{i,j,k}-\tilde{u}^{n+\frac{1}{2}}_{i-1,j,k}}{\Delta r}+\frac{\tilde{v}^{n+\frac{1}{2}}_{i,j,k}-\tilde{v}^{n+\frac{1}{2}}_{i,j-1,k}}{r_i\Delta\varphi}+L_{i,j,k-1}(\tilde{w}^{n+\frac{1}{2}})\right]+A^{n+\frac{1}{2}}_{i,j,k}\right\} \tag{30}$$

$$\overline{\tilde{u}^{n+1}_{i,j,k}}=\tilde{u}^{n+\frac{1}{2}}_{i,j,k}-\Delta t\cdot\left\{N_{i,j,k}(\tilde{u}^{n+\frac{1}{2}})+\frac{1}{\bar{\rho}_{i,j,k}}\frac{\bar{P}^{n+\frac{1}{2}}_{i,j,k}-\bar{P}^{n+\frac{1}{2}}_{i-1,j,k}}{\Delta r}+B^{n+\frac{1}{2}}_{i,j,k}\right\} \tag{31}$$

$$\overline{\tilde{v}^{n+1}_{i,j,k}}=\tilde{v}^{n+\frac{1}{2}}_{i,j,k}-\Delta t\cdot\left\{N_{i,j,k}(\tilde{v}^{n+\frac{1}{2}})+\frac{1}{\bar{\rho}_{i,j,k}}\frac{\bar{P}^{n+\frac{1}{2}}_{i,j,k}-\bar{P}^{n+\frac{1}{2}}_{i,j-1,k}}{r_i\Delta\varphi}+C^{n+\frac{1}{2}}_{i,j,k}\right\} \tag{32}$$

$$\overline{\tilde{w}^{n+1}_{i,j,k}}=\tilde{w}^{n+\frac{1}{2}}_{i,j,k}-\Delta t\cdot\left\{N_{i,j,k}(\tilde{w}^{n+\frac{1}{2}})+\frac{1}{\bar{\rho}_{i,j,k}}L_{i,j,k-1}(\bar{P}^{n+\frac{1}{2}})+D^{n+\frac{1}{2}}_{i,j,k}\right\} \tag{33}$$

$$\overline{\bar{P}^{n+1}_{i,j,k}}=\bar{P}^{n+\frac{1}{2}}_{i,j,k}-\Delta t\cdot\left\{N_{i,j,k}(\bar{P}^{n+\frac{1}{2}})-\bar{a}^{2}\cdot\left[\frac{\overline{\tilde{\rho}^{n+1}_{i,j,k}}-\tilde{\rho}^{n+\frac{1}{2}}_{i,j,k}}{\Delta t}+N_{i,j,k}(\tilde{\rho}^{n+\frac{1}{2}})\right]+E^{n+\frac{1}{2}}_{i,j,k}\right\} \tag{34}$$

$$\begin{cases}\tilde{\rho}^{n+1}_{i,j,k}=\left(\tilde{\rho}^{n+\frac{1}{2}}_{i,j,k}+\overline{\tilde{\rho}^{n+1}_{i,j,k}}\right)/2\\ \tilde{u}^{n+1}_{i,j,k}=\left(\tilde{u}^{n+\frac{1}{2}}_{i,j,k}+\overline{\tilde{u}^{n+1}_{i,j,k}}\right)/2\\ \tilde{v}^{n+1}_{i,j,k}=\left(\tilde{v}^{n+\frac{1}{2}}_{i,j,k}+\overline{\tilde{v}^{n+1}_{i,j,k}}\right)/2\\ \tilde{w}^{n+1}_{i,j,k}=\left(\tilde{w}^{n+\frac{1}{2}}_{i,j,k}+\overline{\tilde{w}^{n+1}_{i,j,k}}\right)/2\\ \bar{P}^{n+1}_{i,j,k}=\left(\bar{P}^{n+\frac{1}{2}}_{i,j,k}+\overline{\bar{P}^{n+1}_{i,j,k}}\right)/2\end{cases} \tag{35}$$

In the formulas, L, M, N are the operators. A–E are the known constants.

COMPUTATIONAL ANALYSIS:

In order to analyze the testing results, authors used respectively the strip theory and numerical three dimensional actuator disk method to predict stall flutter with the inlet flow distortion.

The strip theory features that the computational aerodynamic work by two dimensional actuator disk method in Ref.1 and Ref.2 are integrated from the hub to tip:

$$W_{aero}=\int_0^{l_A}w_i(l)dl+\int_{l_A}^{l(w)}w_i(l)dl \tag{36}$$

Waero is the aerodynamic work.

In Fig.6, region I is the zone of flow distortion and region II is non distortion zone. To simplify the computation, the flow of each region is assumed to be uniform. The steady parameters in the distortion zone is determined by the measured parameters at the test.

Only unsteady flow with the tip distortion is simulated because this type of distortion dispels stall flutter in the test.Predicted flutter boundary by the strip theory is in Fig.7. From the figure, It can be seen that at some rotating speed, the stall flutter disappears and in other rotating speed stall flutter is more violent. The computational results of strip theory is not satisfactory with the test. Maybe the reason is the strong three dimensional effect at line A−A of Fig.6. The strip theory can't treat three dimensional effect. So for the prediction of stall flutter with the distortion, this method is not effective.

In Fig.7 there is the computational flutter boundary with the distortion by three dimensional actuator disk theory.The wholl stall flutter boundary predicted by this method is inside the surge line. This proves that the distortion at the tip can dispel stall flutter and three dimensional actuator disk method can reflect unsteady flow under three dimensional effect. so this method is effective to predict stall flutter with the radial distortion.

In order to study the effect of the distortion ditribution on the stall flutter, a kind of the distribution is given in Fig.5.The flow velocity is increased from tip to hub gradually.The computation of three dimensional actuator disk method proves that such a distribution does not have obvious effect on the stall flutter (Fig.9). So if the distortion is adapted to dispel flutter, such a distribution should not be used.

CONCLUDING REMARKES:

(1) The inlet flow distortion at blade tip or hub can weaken stall flutter, especially the suitable distortion at the blade tip will dispel it.

(2) The distortion with continuous radial variation of inlet flow velocity from hub to tip does not have such effect, it can't dispel the stall flutter.

(3) By using strip theory, the computational results are not satisfactory. The reason may be three dimensional effect in the boundary of distortion zone and undistored flow.

(4) The three dimensional actuator disk method developed by authors can get the satisfactory results which conform to the testing results.

REFERENCE:

1. J.J.Adamczyk:
   Analysis of Supersonic Stall Flutter in Axial−flow Compressor by Actuator Disk Theory.
   NASA TP−1345, 1978
2. Yang Xiaodong; Tang Zhiming; Zhou Sheng:
   An Actuator Disk model for Predicting Stall Flutter of Steam Turbines.
   ASME Paper 85−IGT−99
3. Yang Xiaodong:
   The Research for the Basic Model of Aeroelasticity in Turbomachinery.
   Ph.D. Thesis in Beijing University of Aeronautics and Astronautics. November, 1988
4. D.G.Halliwell
   AIAA Paper 79−44
5. X.D.Yang, D.P. Tao and S. Zhou:
   A Method for Predicting Stall Flutter under Variable Interblade Phase Angle along Rotating Direction.
   Another paper presented to the Symposium

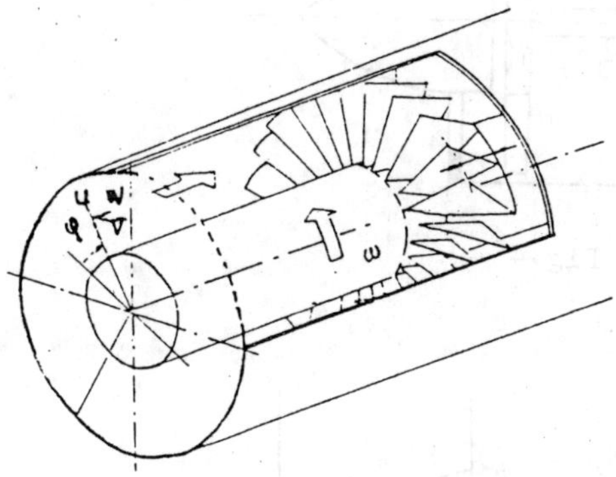

Fig.1

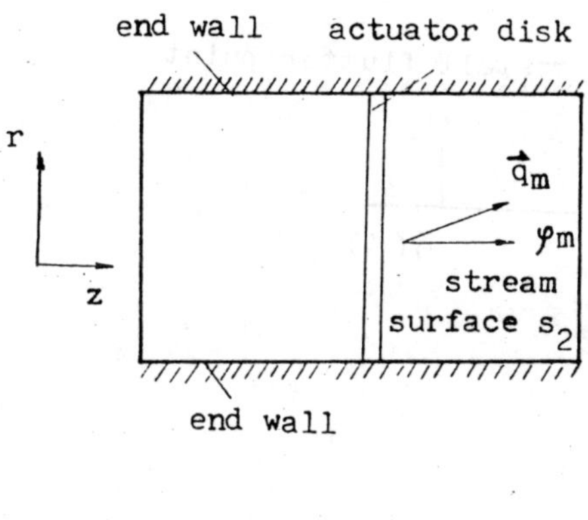

Fig.2

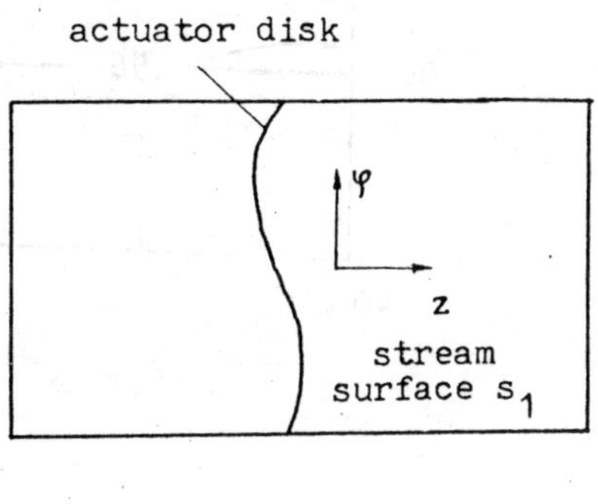

Fig.3

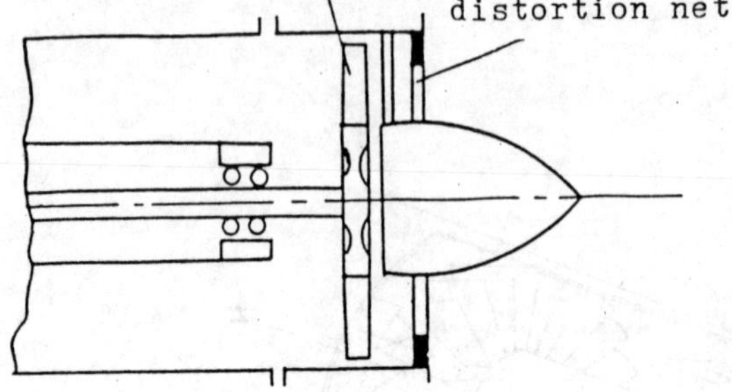

Fig.4

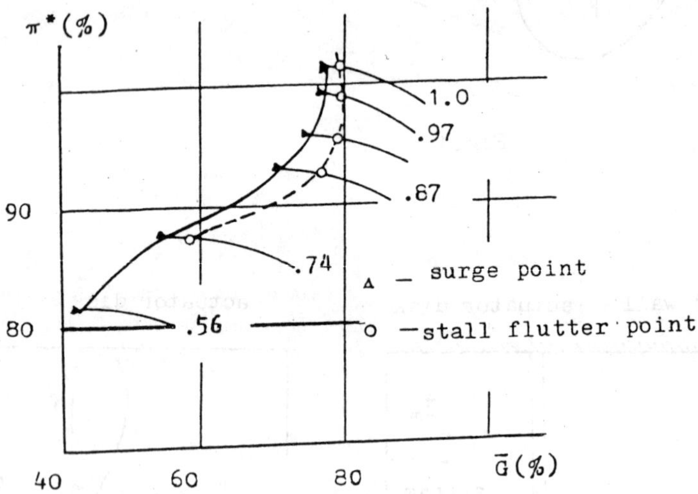

Fig.5

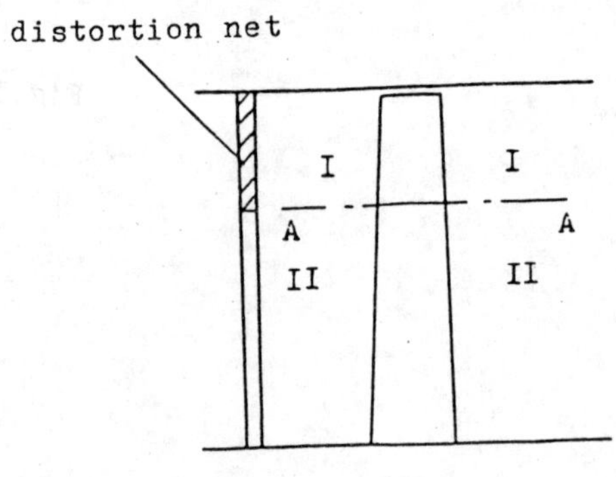

Fig.6

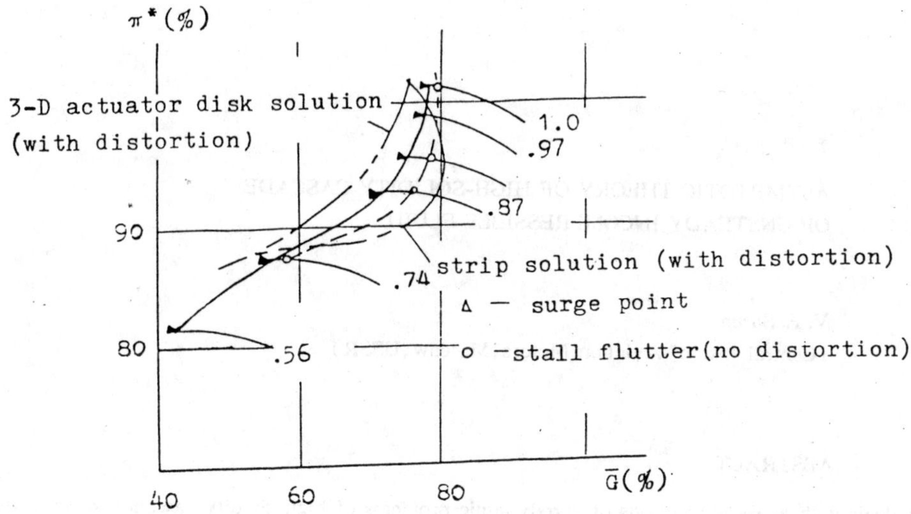

Fig.7

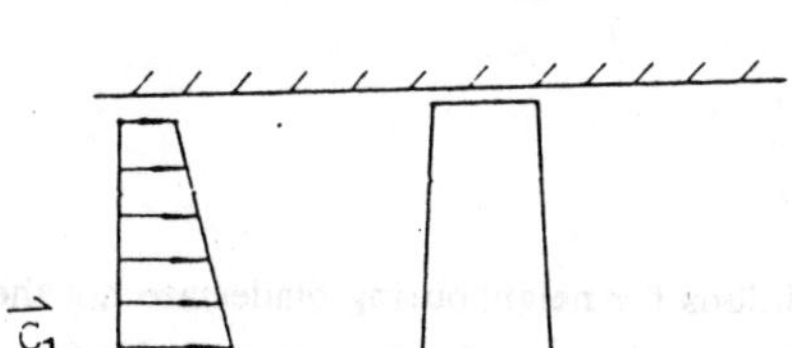

Fig.8

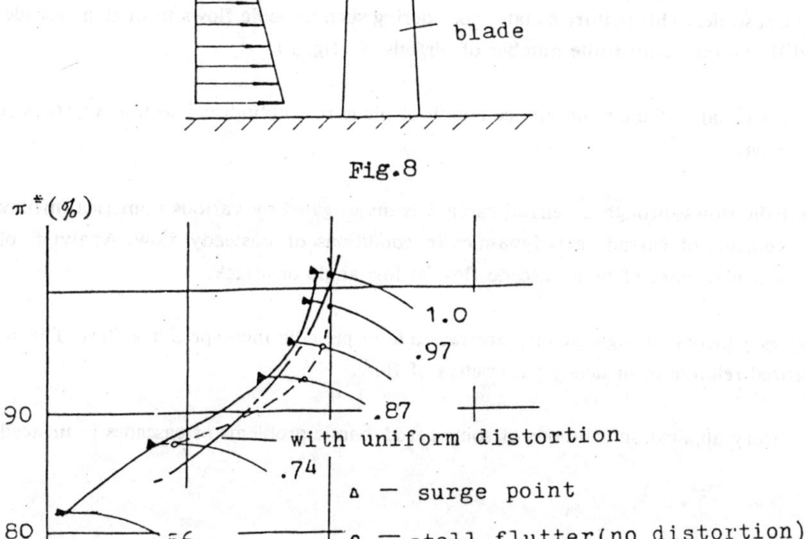

Fig.9

ASYMPTOTIC THEORY OF HIGH-SOLIDITY CASCADE OF UNSTEADY INCOMPRESSIBLE FLUID

V. E. Saren
(Central Inst. of Aviation Motors, Moscow, USSR)

ABSTRACT

This paper deals with analytic solutions of aerodynamic problems of high-solidity cascades in unsteady incompressible flow that arise from the turbomachinery blading aeroelasticity effect. Problems of a cascade of thin vibrating airfoils in unstalled and stalled flows, a cascade moving in nonuniform flow, two interacting cascades are studied.

INTRODUCTION

In the above-mentioned problems flow conditions for neighbouring blades are not the same due to the differences in blade-to-blade vibration, flow nonuniformity period versus cascade blade spacing and blade spacing of two interacting cascades. This feature dictates considering such periodic flows through a cascade where the flow base period H embraces any finite number of airfoils N (fig. 1).

In general case H is equal to the circumference length of blade row cylindrical section and N is equal to the cascade blade number.

Behaviour of periodic flows through an airfoil cascade is investigated by various numerical methods depending on specific problems of cascade aerodynamics in conditions of unsteady flow. Analytic solutions are known for few particular cases of plate cascade flow at low angle of attack.

This paper provides a theory of high-solidity are casecade in periodic incompressible flow. The aerodynamic problem is linearized relative to unsteady parameters of flow.

As examples of theory application solutions of some aerodynamic problems of cascades in unsteady flow are given.

RESULTS AND DISCUSSION

1. **High-sodility cascade of thin airfoils in potential incompressible flow.**

Let us consider an airfoil cascade in potential flow of incompressible fluid (fig. 1).

We shall admit that the complex conjugate fluid speed $V_{1\infty}=v_{1\infty}e^{-i\alpha_0}$ at infinity upstream of the cascade is given and the flow base period is defined as $H=Nh$, where h—airfoil spacing.

The complex conjugate fluid speed $V(z)$ in the vicinity of flow is an analytic function of the complex coordinate $z=x+iy$ and can be expressed by the Cauchy formula. If the base flow period covers airfoils having numbers $k=0,1,\cdots,N-1$, the the function $V(z)$ can by written as

$$V(z)=\frac{1}{2Hi}\sum_{k=0}^{N-1}\int_{l_0}V_k(\zeta)\mathrm{cth}\left[\pi\left(\frac{z-\zeta}{H}-i\frac{K}{N}\right)\right]d\zeta+V_\infty$$

$$V_\infty=\mathrm{const} \tag{1}$$

where V_k is equal to the complex conjugate speed of fluid on k-th airfoil.

Let us transform the right part of the equation (1). Since for any functions $\{V_k,k=0,1,\cdots,N-1\}$ the following formulas are valid

$$V_k=\sum_{l=0}^{N-1}V^l e^{j2\pi lk/N} \quad ; \quad V^l=\frac{1}{N}\sum_{k=0}^{N-1}V_k e^{-j2\pi lk/N}$$

$$j=\sqrt{-1} \quad ; \quad i\cdot j\neq -1$$

then the speed of periodic through the cascade can be presented as the sum

$$V(z)=\sum_{l=0}^{N-1}V^l(z)+V_{1\infty}$$

$$V^l(z)=\frac{1}{2Hi}\int_{l_0}V^l(\zeta)K_l(z,\zeta)d\zeta$$

where the functions $V^l(z)$ comply with the periodicity condition such as

$$V^l(z+ih)=V^l(z)e^{j2\pi l/N}$$

Asymptotic character of the functions beyond the cascade is due to the kernel K_l behaviour at high values of $æ=\pi\tau/z-\zeta$, where b—airfoil chord length and $\tau=b/l$—cascade solidity. At that the following equations are valid: in the vicinity before the cascade $(Reæ<0)$

$$\frac{1}{N}K_l=\begin{cases}-2e^{4æ}+O(e^{L_1æ}) & ; \quad l=0\\ -(1+ij)e^{2æ(1-\nu)}-(1-ij)e^{2æ\nu}+O(e^{2æ}) & ; \quad l=1,2,N-1 \quad ; \quad \nu=l/N\end{cases} \tag{2}$$

in the vicinity after the cascade $(Reæ>0)$

$$\frac{1}{N}K_l=\begin{cases}2(1+e^{-2æ})+O(e^{-1æ}) & ; \quad e=0\\ (1+ij)e^{-2æ\nu}+(1-ij)e^{-2æ(1-\nu)}+O(e^{-2æ}) & ; \quad l=1,2,\cdots,N-1\end{cases} \tag{3}$$

As seen from (2) and (3) the cascade-caused, flow speed disturbance level is largely dependent on the parameter $\gamma = l/N$. At that periodic flow (as in contrast to the case $l=0$) does not distort the stream at the infinity of the cascade ($\lim\limits_{æ\to\infty} V^l \big|_{l\neq 0} = 0$). Besides at fixed value of æ the speed limit for $\gamma \to 0$ does not tend to approach its value for the flow with the period $H=h$

$$\lim_{\nu\to +o} V^l \neq \lim_{\nu\to 1-o} V^l \neq V^o$$

We shall regard disturbances of the normal fluid speed at cascade airfoils as given

$$\varphi(s) = Jm_i[\Phi(s)\cdot e^{ia}] \quad ; \quad (\Phi(z+ih) = \Phi(z)e^{j2\pi\nu})$$

Then the unstalled airfoil flow condition leads to the integral equation

$$Jm_i\left\{\frac{e^{ia(s)}}{2i}\tau \int_{l_o} \gamma^l(\sigma)K_l(s,\sigma)d\sigma\right\} = \varphi(s) \tag{4}$$

where γ^l — intensity of the vortex layer simulating the cascade airfoil. The equation (4) is solved in accordance with the condition of pressure continuity at the profile trailing edge as applied to the functions that achieve their limit value at the end of the arc l_o. Let us study an extreme case when the cascade has a considerable solidity ($\tau \gg 1$). Asymptotic solution $\gamma^l(s)$ of the equation (4) for high values of τ takes the form

$$\gamma^l(s) = 2E^l \cdot \delta(s) + \gamma_1^l(s) \tag{5}$$

$\delta(s)$ — Dirac δ — function and $\gamma_1^l(s)$ — continuous function.

Asymptotic integral equations for different values $l=0,1,2,\cdots,N-1$ based on the limit relationships (2) and (3) are as follows:
at $l=0$

$$\int_o^s \gamma^o(\sigma)d\sigma = \frac{\varphi(s)}{\tau\cos\alpha(s)} \quad , \quad (o \leqslant s \leqslant s^o) \tag{6}$$

at $l=1,2,\cdots,N-1$

$$-e^{-ja(s)}\int_o^s \gamma^l(\sigma)e^{-2\pi\nu\tau\overline{(z-\zeta)}}d\sigma + e^{ja(s)}\int_s^{s^o} \gamma^l(\sigma)e^{2\pi\nu\tau(z-\zeta)}d\sigma - e^{-ja(s)}\int_o^s \gamma^l(\sigma)e^{-2\pi(1-\nu)\tau(z-\zeta)}d\sigma$$

$$+ e^{-ja(s)}\int_s^{s^o} \gamma^l(\sigma)e^{2\pi(1-\nu)\overline{(z-\zeta)}}d\sigma = \frac{2\varphi(s)}{\tau}$$

$$z(s) = x(s) + jy^{(s)} \quad ; \quad \zeta(\sigma) = \xi(\sigma) + j\eta^{(\sigma)} \tag{7}$$

Solutions of the equations (6) and (7) are determined for generalized functions as given by (5). At that when $l=0$ we get

$$\gamma^{o}(s) = 2\frac{\varphi(o)}{\tau \cos\alpha(o)} \cdot \delta(s) + \frac{\partial}{\partial s}\left[\frac{\varphi(s)}{\tau \cos\alpha(s)}\right] \tag{8}$$

An analytic solution of the equation (7) is obtained for a practically important particular case $\nu \ll 1$, when the base flow period covers a big number of cascade airfoils. In this case we have

$$E^{l} = -\frac{e^{j\alpha(o)}}{\cos\alpha(o)}\left[\mu\int_{s}^{s^{o}}\varphi d\sigma + \frac{1}{\tau}\varphi(s^{o})e^{j\alpha(s^{o})} + + \frac{1}{\tau}\varphi(o)e^{-j\alpha(o)}\right] \quad ; \quad \mu = 2\pi\frac{l}{N}$$

$$\gamma_1^{l}(s) = e^{\mu\tau z}\frac{\partial}{\partial s}\left\{\frac{e^{-\mu(z+\bar{z})}}{2\mu\tau\cos\alpha(s)}\frac{\partial}{\partial s}\left[e^{\mu\tau\bar{z}}\left(2\mu\int_{s}^{s^{o}}\varphi d\sigma + \frac{2}{\tau}\varphi(s^{o})e^{j\alpha}\right]\right\} \tag{9}$$

Solution of the equations (6) and (7) allows to express the complex conjugate speed of the generalized periodic flow through a high-solidity arc cascade as quadratures of the function φ. In particular for a cascade of straight line segments ($\alpha=\text{const}$) the fluid speed at a cascade airfoil is defined by the relationship

$$V^{l}_{(s)} = -\frac{\mu\tau j e^{-i\alpha(s)}}{\cos\alpha(s)}\int_{s}^{1}\varphi d\sigma - \frac{j e^{j\alpha(s)}}{\cos\alpha(s)}\varphi(1)e^{-i\alpha(s)} - \frac{\varphi(s)}{i\cos\alpha(s)}$$

2. Vortex wake iduced sped

If the right part of the equation (4) is time-dependent, then the fluid speed circulation is a variable and according to the Tompson theorem free vortices appear in the flow. We shall admit that unsteady disturbances of the fluid speed is much less than the incoming flow speed ($|V_1/V_{1\infty}| \ll 1$). Then we may consider that the flow speed within the accuracy of the terms that are linear relative to the value $|V_1/V_{1\infty}|$ can be expressed as follows

$$V = V_o(z) + V_1(z,t)$$

where V_o—speed of the main flow with a period equal to the airfoil spacing h. In such case vortex wakes are simulated by tangential speed jump lines located along stream lines of the main flow.

The pressure continuity condition at vortex wakes leads to the following equation of the length unit density of wake vorticies

$$\varepsilon^{l}(s,t) = -\frac{1}{v_o(s)}\frac{\partial\Gamma^{\iota}}{\partial t}\bigg|_{t=t_1} \quad ; \quad \Gamma^{\iota} = \int_{l_o}\gamma^{l}ds \quad ; \quad t_1 = t - \int_{s^{o}}^{s}\frac{d\sigma}{v_o(\sigma)} \qquad)(10)$$

where v_o—main flow speed. Integration is made along the contour of vortex wake downstream of a given airfoil. The flow speed disturbance in this case is expressed as

$$V_1 = V_{1np.} + V_{1cl.}$$

The speed $V_{1cl.}$ induced by wake vortices having the density as defined by the equations (10) can be calculated by using the above-mentioned asymptotic relationships for high values of τ.

In particular in case of harmonic variation of the fluid speed circulation at a cascade airfoil $(\Gamma^t = \overline{\Gamma}^t e^{j\omega t})$ at the point $z \in l_o$ we get

$$V_{1cl} = \begin{cases} 0 \quad ; \quad l = 0 \\ q\tau \dfrac{\cos\alpha(s^o)}{\cos\alpha_o} \overline{\Gamma}^t e^{j\omega t} \left[(1 - ij) \dfrac{e^{2\pi\nu\tau(z - z_o)}}{2\pi\nu\tau e^{j\alpha(s^o)} + jq} + (1 + ij) \dfrac{e^{2\pi(1-\nu)\tau(z - z_o)}}{2\pi(1 - \nu)\tau e^{-j\alpha(s^o)} + jq} \right]; \\ l > 0 \end{cases}$$

where $q = \omega b / v_{1\infty}$ — Stouhal number.

In solving the equation (7) the speed $V_{1cl.}$ should be included into the right part φ' for convenience. Then the unknown function can be written as $\gamma' = \gamma_{np.}' + \overline{\Gamma}^t \cdot \gamma_{cl.}'$ where the constant $\overline{\Gamma}^t$ is calculated from the following formula after having solved the equation (7)

$$\overline{\Gamma}^t = \frac{1}{1 - \int\limits_{l_o} \gamma_{cl.} ds} \cdot \int\limits_{l_o} \gamma_{np.} ds$$

3. Unsteady aerodyanmic reactions

Integrating the Cauchy-Lagrange equation in case of fluid speed harmonic oscillations at any smooth surface airfoil, within the accuracy to terms that are linear relative to unsteady disturbances we obtain

$$C_x - iC_y = 2i\left\{ \frac{1}{2} \int\limits_L V_o^2 dz + e^{j\omega t} \int\limits_L (V_{1np.} + V_{1cl.} + \Phi) dz - jqe^{j\omega t} \int\limits_L (V_{1np.} + V_{1cl.} + \Phi) \overline{(z - z_o)} dz \right\}$$

$$C_M = 2\, Re_i \left\{ - \frac{1}{2} \int\limits_L V_z^2 dz - e^{j\omega t} \int\limits_L V_o (V_{1np.} + V_{1cl.} + \Phi) z dz - jqe^{j\omega t} \int\limits_L (V_{1np.} + V_{1cl.} + \Phi) \int\limits_z^{z_o} z_1 dz_1 \cdot dz \right\} \tag{11}$$

where C_x, C_y, C_M denote dimensionless projections of aerodynamic force and moment with respect to the origin of coordinates.

The limit transition in the equations (11) to an infinitively thin arc l_o brings about the generation of tractive force concentrated on the airfoil leading edge. Taking this force into account in our case of unsteady flow through a thin airfoil loading the neglect of the tractive force leads to a big error in calculations of unsteady aerodynamic characteristics.

Relationships for factors C_x, C_y, C_M at the arc l_o take the shape

$$C_x - iC_y = 2i\int_{l_o} \gamma_o \cdot v_o e^{-i\alpha} ds + 2ie^{j\omega t}\int_{l_o} (\gamma_o v - v_o \gamma) e^{-i\alpha} ds$$

$$- 2ijqe^{j\omega t}\int_{l_o} \gamma \cdot (z - z_o) e^{j\alpha} ds - \frac{e^{i\alpha(o)}}{\tau} C_o^2 - \frac{e^{j\alpha(o)}}{\tau} C_o C e^{j\omega t}$$

$$C_M = -2\int_{l_o} (\gamma_o \cdot v_o) Re(z d\bar{z}) - 2e^{j\omega t}\int_{l_o} (\gamma_o v \cdot \gamma v_o) Re(z d\bar{z}) + ijqe^{j\omega t}\int_{l_o} \gamma(|z|^2 - |z_o|^2) ds \quad ;$$

$$C_o, C = \text{const} \tag{12}$$

where last terms in the first equation define a steady and an unsteady components of the tractive force respectively.

The equations (12) are suitable for the asymptotic theory under consideration. The constants C_0, C_1 are related to the values E_o, E_l by the equations

$$C_o = \tau \cdot E_o \quad , \quad C = \tau \cdot E^l$$

4. Flow through a cascade with airfoils vibrating in unstalled stream.

Let high-solidity cascade airfoils vibrate in a synchronous harmonic mode with the phase shift $\mu = 2\pi l/N$ $(l = 0, 1, 2, \cdots, N-1)$ between oscillations of adjacent airfoils in ideal fluid flow. The oscillation formula is considered present as follows

$$f(z) = Ae^{ig} + il(z - z_{k_T}) \quad , \quad 0 \leqslant g \leqslant \pi/2$$

where A, l small amplitudes of advancing and torsional oscillations respectively and z—complex coordinate of a point on an airfoil in its medium position.

In our case the right part of the equation (4) is given by

$$\varphi(s) = jq\varphi_1(s) + jq\tau\overline{\Gamma} \cdot \varphi_2(s) + \varphi_3(s) \tag{13}$$

where asymptotic values of the functions $\varphi_1, \varphi_2, \varphi_3$ are defined from the formulas below

$$\varphi_1 = Jm[\bar{f}e^{i\alpha(s)}] \quad ;$$

$$\varphi_2(s) = \frac{\cos\alpha(s^o)}{\cos\alpha_o}\left[\frac{e^{i\alpha(s)}e^{2\pi\nu\tau(z-z_o)}}{2\pi\nu\tau e^{i\alpha(s^o)} + jq} - \frac{e^{-i\alpha(s)}e^{2\pi(1-\nu)\tau(\overline{z-z_o})}}{2\pi(1-\nu)\tau e^{-i\alpha(s^o)} + jq}\right]$$

$$\varphi_3(s) = -l\frac{\cos\alpha_o}{\cos\alpha} - 2\pi\tau\frac{\sin(\alpha(o) - \alpha_o)}{\cos\alpha(o)}[\nu \cdot \bar{f}(o)e^{-j\alpha(s)}e^{-2\pi\nu\tau\bar{z}} - (1-\nu)f(o)e^{j\alpha(s)}e^{-2\pi(1-\nu)\tau z}]$$

$$- 2\pi\tau\cos\alpha_o\int_0^s [j\nu\bar{f}(\sigma)e^{-j\alpha(\sigma)}e^{-2\pi\nu\tau(\overline{z-\zeta})} - j(1-\nu)f(\sigma)e^{j\alpha(\sigma)}e^{-2\pi(1-\nu)\tau(z-\zeta)}]\frac{\alpha^1}{\cos^2\alpha(\sigma)}d\sigma$$

$$- 2\pi\tau\cos\alpha_o \int_{s}^{s^o} [j(1-\nu)\overline{f}(\sigma)e^{-j\alpha(\sigma)}e^{2\pi(1-\nu)\tau(\overline{z-\zeta})} - j\nu f(\sigma)e^{j\alpha(\sigma)}e^{2\pi\nu\tau(z-\zeta)}] \frac{\alpha'}{\cos^2\alpha(\sigma)};$$

$$f = Ref + jJmf \quad ; \quad z = x + jy$$

The first and second terms in the right part of the equation (13) are related to the airfoil oscillation rate and vortex wakes downstream of the vibrating cascade. The third term is due to a mutual shift of vibrating airfoils. Its addition can be explained by a steady disturbance of the flow owing to the angle of attack or to the curvature of the arc l_o. In the specific case of advancing synphase oscillation $(f=\text{const}, l=o)$ this term turns to zero.

Solution of the equation (7) and following computation of unsteady aerodynamic response at cascade airfoils lead to relationships of the kind as given below

$$C_{x,y,m} = \frac{\mu}{\mu\tau e^{j\alpha(s^o)} + jqF_o}[jF_{02} + q(F_{11} + jF_{12}) + q^2(F_{21} + jF_{22}) + q^3(F_{31} + F_{32})] \quad (14)$$

where real values $F_o, \cdots, F_{32}$ depend on the parameters $\mu\tau, \alpha_o$, the arc shape l_o and the mode of oscillation f only.

As an example of use of the equation (14) let us consider a flutter problem of a cascade of rigidly fixed high-aspect ratio plates vibrating in bending mode perpendicular to the chord under lateral aerodynamic force action $(f=Ae^{i(\pi/2+\alpha)})$.

For pure bending deformation from the plate vibration equation stems out the flutter equation i. e.

$$q^2 + \frac{1}{M}C_o(q) - q_o^2 = 0 \quad ; \quad q_o = \omega_o b/v_{1\infty}$$

where $M=2\rho_M S/\rho_\Gamma b^2$ mass criterion, ρ_M, ρ_Γ—plate material density and fluid density respectively, S—plate cross-section area, ω_o—plate natural frequency in vacuum, C_ν—dimensionless aerodynamic reaction in the direction of plate displacement due to harmonic oscillation. The value C_ν is defined by the formula such as (14).

Self-oscillation of the plate system under study occurs if there is a real positive number that ensures validity of the following two equations

$$\left.\begin{aligned} Jm_jC_o(q) &= 0 \\ q^2 + \frac{1}{M}Re_jC_o(q) - q^2o &= 0 \end{aligned}\right\} \quad (15)$$

Let us introduce the designation

$$\lambda = M\left[1 - \left(\frac{\omega_o}{\omega}\right)^2\right]$$

where ω—flutter frequency. Then the relationship $\lambda = -Re\left(\frac{1}{q^*}C_v(q^*)\right)$ where q^* is a real positive solution of the first equation of the set (15), defines self-oscillation domain boundary in the parameter space. Examples of the relationship $\lambda=\lambda(a_o-a)$ for different domains of the parameter $\mu\tau$ variation are depicted at fig. 2.

It is characteristic that if there is no aerodynamic loading $(a_o-a=0)$ one degree of freedom flutter is impossible. Within a domain of real mass criterion values $(M\sim10^1\div10^3)$ and in conditions of diffuser type flow through a cascade the flutter is possible beginning from a cirtain value of the attack angle a_o-a. It is noteworthy that there is always $q^*>q_o$, i. e. the flutter Strouhal number exceeds the Strouhal number based on natural frequency in vacuum.

5. Cascade stalled flow

Let us consider the flow through a cascade of arcs l_o when the stall zone boundary downstream of the cascade forms a free streamline extending from point s^* on the airfoil suction side to the infinity after the cascade (fig. 3). The other stall zone boundary begins at the airfoil trailing edge. The flow speed at the boundaries is given as $v_2=v_o(s^*)>v_{o\infty}$, where v_o denotes the cascade unstalled flow speed. The concept of analysis of such flow is that a free streamline position is defined as a result of minor displacement of a basic streamline due to small variation of flow speed on it.

Suppose the streamline L is obtained by virtue of displacement by $\varepsilon\cdot\Delta(0<\varepsilon\ll1)$ of the streamline L_o in normal direction $\vec{\nu}_o$.

The radius-vector $\vec{r}(s)$ of a point on the line L is equal

$$\vec{r}(s) = \vec{r}_o(s) + \Delta(s)\cdot\vec{\nu}_o(s) \quad , \quad \vec{r}_o \in L_o$$

Let us describe the speed vector of the flow with the streamline L as

$$\vec{V} = \vec{V}_o + \varepsilon\vec{V}_1$$

where $\vec{V}_o$—speed of the flow with the streamline L_o. Then the following relationships are valid

$$\vec{V}_o = v_o(s)\cdot\vec{s}_o \quad ; \quad \left(\vec{s}_o = \frac{d\vec{r}_o}{ds}\right)$$

$$\vec{V}_1 = [v_1(s) - \text{æ} v_o\Delta(s)]\cdot\vec{s}_o + \left[\frac{\partial(v_o\cdot\Delta)}{\partial s} - \Delta\cdot div\vec{V}_o\right]\cdot\vec{\nu}_o + O(\varepsilon) \qquad (16)$$

where æ(s)—curvature of the line L_o.

If the L_o line displacement value $\varepsilon\Delta$ is known, then from the equations (16) we get the boundary condition

$$\vec{V}_1 \cdot \vec{v}_o|_{L_o} = \frac{\partial(v_o \cdot \Delta)}{\partial s} - \Delta \cdot div\vec{V}_o$$

for the speed change $(\varepsilon \cdot \vec{V})$ due to L_o streamline displacement. On the contrary if the speed variation $(\varepsilon \cdot v_1(s))$ is known then the displacement may be determined by solving in succession the boundary problems

$$\vec{V}_{1n} \cdot \vec{s}_o|_{L_o} = v_1(s) - æ(s)v_o(s)\Delta_{n-1}(s) \quad ; \quad n = 1,2,\cdots \quad ; \quad \Delta_o \equiv 0$$

$$\frac{\partial(v_o \cdot \Delta)}{\partial s} - \Delta_n(s) \cdot div\vec{V}_o = \vec{V}_{1n} \cdot \vec{v}_o \qquad (17)$$

It has been shown that at adequately low values $|æ|$ the iteration of (17) converges. For an incompressible fluid $(div\vec{V}_o = 0)$ the boundary conditions (16) take the form

$$V_o^{\pm} = \pm\, v_o^{\pm}(s)e^{-ia(s)}$$

$$V_{1n}^{\pm} = \pm\left[V_{1n}^{\pm}(s) - æ^{\pm} v_o^{\pm}\Delta_{n-1}^{\pm} \mp i\frac{\partial(v_o^{\pm}\Delta_n^{\pm})}{\partial s}\right]e^{-ja(s)}$$

$$n = 1,2,\cdots \quad ; \Delta_o = 0 \qquad (18)$$

where "+" and "−" symbols denote respectively pressure and suction sides of the arc l_o tracing from the leading edge. At that

for $s \in (0, s^*)$

$$\Delta^+ = \Delta^- = 0 \quad v_1^+ + v_1^- = \gamma_1(s) \quad ; \quad \frac{v_1^+ - v_1^-}{2} = v_{o1} \qquad \text{— unknown values}$$

for $s \in (s^*, s^o)$

$$\Delta^+ = 0 \quad ; \quad v_1^- = -\frac{v_2 - v_o^-(s)}{\varepsilon} \quad ; \quad \Delta^+(s) \quad ; \quad \Delta^-(s) \qquad \text{— unknown values}$$

for $s \in (s^o, \infty)$

$$v_1^+ = -v_1^- = \frac{v_2 - v_o(s)}{\varepsilon} \quad ; \quad \Delta^+(s) \quad ; \quad \Delta^-(s) \qquad \text{— unknown values} \qquad (19)$$

Here the arc coordinate s in the domain (s^o, ∞) is defined starting from the trailing edge and further a long the unstalled flow critical streamline downstream of the cascade airfoil.

If cascade airfoils execute harmonic oscillation according to the law

$$\varepsilon \cdot \Delta(s) = A \cdot f(s)e^{j\omega t}e^{jk\mu} \quad ; \quad k = 0, \pm 1, \pm 2, \cdots$$

then in general case $v_o(s)$ in the equation (18) should be regarded as the steady stalled flow speed determined at an airfoil and stall zone boundaries. Taking into consideration the linear terms relative to the oscillation amplitude A let us express unsteady disturbances of the flow speed the stall zone boundaries at k-th airfoil as

$$\Delta^{\pm}(s,t) = f^{\pm}(s)e^{j\omega t}e^{jk\mu}$$
$$v_1^{\pm}(s,t) = v_1^{\pm}(s)e^{j\omega t}e^{jk\mu}$$

An unsteady stalled flow features that at the stall zone boundaries f, v_1 are unknown. Let us denote by

s_{kp} the coordinate of a critical point on the arc $l_o(v_o(s_{kp})=0)$. Then in linear approximation the condition of pressure continuity at the stall zone boundaries produces the following equations

$$v_1^+ = -j\frac{\omega}{v_2}\Gamma^+ e^{j(\omega/v_2)(s-s^o)} \quad ; \quad (s^o < s < \infty)$$

$$v_1^- = j\frac{\omega}{v_2}\Gamma^- e^{j(\omega/v_2)(s-s^*)} \quad ; \quad (s^* < s < \infty)$$

where

$$\Gamma^+ = \int_{s_{kp}}^{s^o} v_1^+ ds + \begin{cases} 0 & ; \quad v^+(s_{kp}) = 0 \\ \int_0^{s_{kp}} (v_1^+ + v_1^-) ds & ; \quad v^-(s_{kp}) = 0 \end{cases}$$

$$\Gamma^- = \int_{s_{kp}}^{s^*} v_1^- ds + \begin{cases} 0 & ; \quad v^-(s_{kp}) = 0 \\ \int_0^{s_{kp}} (v_1^+ + v_1^-) ds & ; \quad v^+(s_{kp}) = 0 \end{cases}$$

Thus for unsteady stalled flow at different parts of an airfoil we have

for $s \in (o, s^*,)$

$$f^+ = -f^- = f \quad v_1^+ + v_1^- = \gamma_1(s) \quad ; \quad \frac{v_1^+ - v_1^-}{2} = v_1^o \qquad \text{—unknown values}$$

for $s \in (s^*, s^o)$

$$f^+ = f \quad ; \quad v_1^-(s) = j\frac{\omega}{v_2}\Gamma^- e^{-j(\omega/\tau_2)(s-s^*)} \qquad \text{—unknown values}$$

$$f^-(s) \quad ; \quad v_1^+(s)$$

for $s \in (s^o, \infty)$

$$v^+(s) = -j\frac{\omega}{v_2}\Gamma^+ e^{-j(\omega/\tau_2)(s-s^o)} \quad ; \quad v^-(s) = j\frac{\omega}{v_2}\Gamma^+ e^{-j(\omega/\tau_2)(s-s^*)} \qquad \text{—unknown values} \tag{20}$$

$$f^+(s) \quad ; \quad f^-(s)$$

Within the admitted limits of asymptotics integral equations that are equivalent to the stalled flow problem take the shape

at $l=0$

$$\frac{V_{1n}^+ + V_{1n}^-}{2} e^{i\alpha(s)} = \frac{\tau e^{i\alpha(s)}}{i} \int_0^s (V_{1n}^+ + V_{1n}^-) e^{i\alpha(\sigma)} d\sigma \tag{21}$$

at $l = 1, 2, \cdots, N-1$

$$\frac{V_{1n}^+ + V_{1n}^-}{2} e^{i\alpha(s)} = \frac{\tau e^{i\alpha(s)}}{i} \left\{ \int_0^s (V_{1n}^+ + V_{1n}^-) \left[(1+ij) e^{-2\pi\tau(l/N)\overline{(z(s)-\zeta(\sigma))}} + (1-ij) e^{-2\pi\tau(1-l/N)(z(s)-\zeta(\sigma))} \right] e^{i\alpha(\sigma)} d\sigma \right.$$

$$
-\int_{s}^{\infty}(V_{1n}^{+}+V_{1n}^{-})\left[(1-ij)e^{-2\pi\tau(l/N)(z(s)-\zeta(\sigma))}+(1+ij)e^{-2\pi\tau(1-l/N)\overline{(z(s)-\zeta(\sigma))}}\right]e^{i\alpha(\sigma)}d\sigma\Big\}
\tag{22}
$$

where $V_{1n}^{\pm}$ are obtained from the relationships (18~20).

The equations (21) and (22) can be univalently solved in case of generalized functions $\gamma_1(s)=v_1^{+}+v_1^{-}$ of the type (5), where the constant E' is defined by the condition of continuous downstream stall zone change. After having solved the equation (21) and (22) the functions $\Delta^{\pm}(s), f^{\pm}(s)$ are determined as solutions of the equations as follows

$$
\frac{\partial(v_0^{-}\Delta^{-})}{\partial s}=-Jm(V_1^{-}e^{i\alpha}) \quad ; \quad \frac{\partial(v_0^{-}f^{-})}{\partial s}-j\omega f^{-}=-Jm(V_1^{-}e^{i\alpha})
$$

$$
s\in(s^{*},\infty) \quad ; \quad \Delta^{-}(s^{*})=0 \quad ; \quad f^{-}(s^{*})=-f(s^{*})
$$

$$
\frac{\partial(v_0^{+}\Delta^{+})}{\partial s}=-Jm(V_i^{-}e^{i\alpha}) \quad ; \quad \frac{\partial(v_0^{+}f^{+})}{\partial s}-j\omega f^{+}=-Jm(V_i^{+}e^{i\alpha})
$$

$$
s\in(s^{0},\infty) \quad ; \quad \Delta^{+}(s^{0})=0 \quad ; \quad f^{+}(s^{0})=f(s^{0})
$$

To illustrate stall zone shape variation due to cascade airfoil oscillation given below is a solution of the equation (21) and (22) for cascade of straight plates $(\alpha\equiv 0)$ that oscillate perpendicular to the chord $(f=1)$ with a small phase shift $(l/N\ll 1)$. Separation point is located at the leading edge $(s^{*}=0)$.

The desired solution for the base airfoil of the cascade is obtained as follows

$$
\gamma(s,t)=-2\left(\frac{\sin\alpha_0}{\tau}-\varepsilon\frac{1}{\tau}\right)\delta(s)-2\varepsilon e^{2\tau s}+Ajq(F_1+\mu F_2)e^{-\omega\tau s}e^{j\omega t} \quad ; \quad s\in(0,1)
$$

$$
\Delta^{+}(s,t)=\varepsilon\frac{e^{-2\pi}}{\cos\alpha_0}(s-1)+A\left[e^{-jq(s-1)}+F_3(1-e^{-jq(s-1)})+F_4(s-1)e^{-jq(s-1)}\right]e^{j\omega t};s\in(1,\infty)
$$

$$
\Delta^{+}(s,t)=\begin{cases}-\varepsilon\dfrac{1-e^{2\tau s}}{\cos\alpha_0}+A\left[-j\dfrac{q}{2\tau}F_1e^{2\tau s}+F_5e^{-jqs}+F_6\right]\cdot e^{j\omega t} & ; \quad s\in(0,1)\\ -\varepsilon\dfrac{e^{-2\tau}\tau(s-1)+1}{\tau\cos\alpha_0}+A\left[F_5e^{-jqs}+F_6e^{-jk(s-1)}+F_7(1-e^{-jq(s-1)})\right. & \\ \qquad\left.+F_8(s-1)e^{-jq(s-1)}\right]e^{j\omega t} & ; \quad s\in(1,\infty)\end{cases}
$$

$$
F_1=1-j+j\frac{q^3}{\tau} \quad ; \quad F_2=\frac{1+j}{2}+q^2 \quad ; \quad F_3=\frac{j(\mu\tau+q)-4\tau}{4\tau}F_1-\frac{2\tau-jq}{2\tau}(e^{\mu\tau}-1) \quad ;
$$

$$
F_4=jq\frac{2\tau-j\mu}{4\tau}F_1 \quad ; \quad F_5=j\frac{qF_1+2\mu\tau}{2\tau} \quad ; \quad F_6=-(j\mu+1) \quad ;
$$

$$
F_7=j\frac{\mu\tau-q}{4\tau}F_1-\frac{2\tau+jq}{2\tau}-(e^{\mu\tau}-1) \quad ; \quad F_8=jq\frac{2\tau+j\mu}{4\tau}F_1
\tag{23}
$$

Here $0<\varepsilon=(v_2-\cos\alpha_0)\ll 1$; $q=\omega b/v_{1\infty}/\cos\alpha_0$; $\mu=2\pi l/N$; $(l/N\ll 1; l=1,2,\cdots)$ — constant values and terms of an order of values $e^{-2\pi}, \varepsilon A$ are deleted. For comparison it should be noted that with an unstalled flow the function $\gamma(s,t)$ takes the form

$$\gamma(s,t) = -2\left(\frac{\sin\alpha_o}{\tau} - A\frac{q}{\tau}E_1e^{j\omega t}\right)\cdot\delta(s) + A[jq\mu^2\tau s + q\mu(E_1 - j)]e^{j\omega t}$$

$$E_1 = \frac{1}{2}q[4(1+\mu\tau) + \mu^2\tau^2] : [jq(2+\mu\tau) + \mu\tau] - j(2+\mu\tau)$$

As seen from the equations (23) the stall zone boundary makes a finite angle $(\neq 0)$ with the airfoil chord at the separation point. The stall zone thickness downstream of the cascade does not remain constant and the stall zone boundaries oscillate in a mode of wave propogating downstream.

6. Airfoil cascade moving in nonuniform flow

When a blade cascade rotates in circumferentially nonuniform flow the stream is unsteady since each blade flow pattern varies as it moves. The fundamental circular frequency of this variation for a given cylindrical section is defined by the value $\omega = 2\pi u/H$ where u—linear speed of section movement, H—nonuniformity period alone the cascade front. The reduced frequency of parameter variation (Strouhal number) in coordinates related to a moving blade is defined as

$$q = 2\pi\frac{ub}{Hv_{1\infty}}$$

Another feature of cascade flow pattern stems from the fact that at a given time t the flow conditions for adjacent blades are different and this difference corresponds to the time shift by $h/2\pi u = 1/N\omega$ A relative level of flow nonuniformity is a significant parameter either. This level of flow may be characterized vy the value $|\tilde{v}/V_{1\infty}|$, where the vector $\tilde{v}$ is equal to the maximum cascade flow entry speed variation relative to its mean value within the period. For practical problems the parameter $|\tilde{v}/V_{1\infty}|$ may be often considered to be low and it allows to suppose that the dependence of blade unsteady aerodynamic reactions versus this parameter is linear.

The complex conjugate speed of fluid at k-th cascade airfoil obtained by means of cylindrical section of the blade row is given as

$$V_\Sigma = v_{1\infty}e^{-j\alpha_o} + v(z_k,t) + iu + V(z_k,t)$$

where entry flow distortion is expressed as

$$v(z_k,t) = \sum_{l=-\infty}^{\infty} v_e(z)e^{j2\pi\frac{l}{N}\left(y+\frac{u}{h}t+k\right)} \quad ; \quad v_o = 0; v_{-n} = v'_n - jv''_n$$

Unknown speed V is supposed to be an analytic function of variable z and this makes it possible to write the following relationship

$$v(z_k,t) = \sum_{l=-\infty}^{\infty} v_e(z)e^{jl\omega t}\cdot e^{2j\pi\frac{l}{N}k} \tag{24}$$

Definition of each component of the expansion (24) comes to analysis of an unsteady periodic flow through a cascade with harmonic disturbance of fluid speed at airfoils. Accrding to the asymptotic theory of high-solidity cascades the right part of the equation (1) for first harmonics of the expansion (21) $(l \ll N)$ takes the form

$$\varphi(s) = - Jm_i[ve^{j\alpha(s)}]e^{j\mu\tau y} + \overline{\Gamma}\tau æ^{j\alpha(s)}e^{\mu\tau(z-z(s^\circ))}$$

$$\mu = \frac{2\pi l}{N}; \quad æ = \frac{jql}{\mu\tau e^{j\alpha(s^\circ)} + jql} \qquad (25)$$

The second term in the right part of (25) relates to the presence of vortex wakes downstream of the cascade. It should be noted that for the given flow taking into account the formula for q the factor æ can be easily transformed as follows

$$æ = \frac{\cos\alpha_o}{\cos\alpha_o - je^{j\alpha(s^\circ)}C_a}$$

where $C_a = v_{1\infty}\cos\alpha_o/u$ — reduced cascade flow rate.

Thus for the problem under consideration the reduced flow rate becomes the main parameter defining the unsteady flow on aerodynamic characteristics of a cascade moving in a nonuniform stream. This in particular leads to the fact that the limit transition at $\mu \to 0$ when the quasiuniform flow model is a priori supposed applicable, does not make it possible to use quasisteady description of the cascade flow pattern.

An unsteady component of dimensionless aerodynamic force projections (ratio of the force and the ram in relative motion) in case of first harmonics of the expansion (21) $(l \ll N)$ is defined as

$$C^l_{x,y} = Re_j\left\{\frac{|v|}{\tau}\cdot\frac{C_a^2}{1+C_a^2}\cdot\frac{e^{j\omega t}}{F_o - je^{j\alpha(s^\circ)}C_a}\left[\frac{F_{-11}+jF_{-12}}{C_a} + (F_{01}+jF_{02}) + (F_{11}+F_{12})C_a\right]\right\}$$

where the real values $F_0, \cdots, F_{12}$ are the functions of parameters $\mu\tau, \alpha_o$, airfoil shape and parameters of incoming flow nonuniformity $(\arg v)$.

Fig. 1 shows the dependence of λ on the reduced flow rate C_a where λ is defined as

$$\lambda = \tau\left|\frac{Jm_i[(C^l_x - iC^l_y)e^{i\alpha}]}{v}\right|$$

for a plate cascade witn the airfoil angle $\alpha = -15°(u>0)$ and axial pattern of the incoming flow $(\alpha_o = 0, \arg v = 0)$.

The λ—value appears to be an averaged over the cascade spacing coefficient of proportionality between the amplitude of force normal to the airfoil chord and the nonuniformity amplitude with sine-type nonuniformity of cascade entry flow. It is interesting to note that at some values of C_a nonuniformity harmonic at low $\mu\tau$ has a weak interaction with the cascade $(\lambda \ll 1)$ because of nonmonotony of the function $\lambda(C_a)$

This phenomenon was observed during experiments and the test results presented here allow to explain it.

Fig. 5 gives the results of comparison of test data and asymptotic theory analysis data. Unsteady loads were measured at the medium rotor radius of an axial compressor stage model with present circumferential nonuniformity of the entry flow. The experiment was carried out in conditions $N=10; a_o = argv = 0$. Flow nonuniformity distribution at cascade entry was close to rectangular. The given results relate to the first harmonic for which $\mu\tau = 0.6283$.

7. Hydrodynamic interaction of cascades

Problem of hydrodynamic interaction between two moving cascades located in tandem with a spacing was investigated by many authors using various numerical methods. The reference 1 contains some analysis data obtained by methods of the proposed asymptotic theory. In that case the solution is found on the assumption that one cascade is a high-solution arc lattice and the other is a low-solidity lattice of straight line, segments, circles or ellipses.

We shall take one little studied inverse problem construction of a nonuniform cascade deleting flow distortion at the entry of the second cascade located downstream of the first one. This problem airses in connection with the task to reduce resonatnt stresses in axial compressor rotors placded upstream of struts, pylons or other relatively rugged structural elements choking the flowpath.

In such case two fixed cascades can be seen a flowpath cylindrical section. One made of strust is a low-solidity oval cascade (number 2) and the other made of straightening vanes is a high-solidity thin airfoil cascade (number 1) located upstream. At that $h_1/h_2 \ll 1$ and the flow features the common period $H = N_1 h_1 = N_2 h_2$. Let medium lines of the cascade 1 be deformed according to a cirtain present shape f and $l_k (k = 0,1,2,\cdots,N_1-1)$ — value of corresponding non-dimentional generalized coordinate at k-th airfoil. Then the disturbance of complex conjugate speed of unstalled incompressible flow can be expressed as

$$V_\Sigma - V_{1\infty} = V_{10} + V_{20} + V_{12} + V[f, l_k] \qquad (26)$$

where $V_{\nu 0}(\nu = 1,2), V_{12}$ and V denote respectively disturbances caused by ν-th cascade at $\Delta = \infty$, hydrodynamic iteraction of the cascades 1 and 2, cascade 1 airfoil deformation. On the assumption of relatively small needed deformation of values $l_k (k = 0,1,2,\cdots,N_1-1)$.

Let us expand the right part of the equation (26) in a series by powers of the value $\exp(2\pi \frac{z}{H})$ in the vicinity of flow before the cascade 1 $(Rez < 0)$ and let us require that the first $[N_1/2]$ terms of the expansion become zero. Then we get the set of linear (relative to l_k) equations as follows

$$V_{20}^l + V_{12}^l + V^l(l_k) = 0 \quad ; \quad l = 1,2,\cdots,\left[\frac{N}{2}\right] \qquad (27)$$

Solution of the equation set (27) allows to define desired cascade 1 airfoil deformations which ensure the absence of a large-scale (with the period $H, H/2, \cdots H/\left[\frac{N_1}{2}\right]$) flow nonuniformity before the cascade 1.

Thus the task of construction of a necessary nonuniform cascade amounts to a successive solution of the following problems:

A. analysis of flow through the cascade 2 (definition of V_{20}),

B. analysis of hydrodynamic iteraction between the high-solidity cascade 1 and low-solidity cascade 2,

C. cascade 1 flow analysis with minor airfoil deformation according to mode f,

D. determination of gēneralized coordinates l_k as a solution of the equation set (27).

It the value V_{20} is known than within the scope of the proposed asymptotic theory the solution of problems B and C, i. e. definition of the set matrix amounts to quadratures. At that for the problem B the right part of the equation (4) takes the shape

$$\varphi(s) = - Jm_i\left\{\frac{e^{i\alpha(s)}}{2Hi}\frac{1}{2}\sum_{n=0}^{N-1} e^{2\pi j\frac{n}{N}l}[(1-ij)B_n + (1+ij)B_{N-n}]\right\}$$

$$B_n = \sum_{k=1}^{N_2}\int_{L_k} V_{02}(\zeta)e^{2\pi n\frac{z-\zeta}{H}}[\operatorname{cth}\frac{\pi N_1}{H}(z_o - \zeta) - 1]d\zeta \quad ; \quad B_{N-n} = \overline{B}_n$$

wher $L_k(k=1,2,\cdots,N_2)$ cascade 2 airfoils embraced within the flow base period H.

In solving the problem C the right part of the equation (4) is given as

$$\varphi(s) = - l^l v_{01}(s) Jm_i\left\{\frac{df}{dz}\right\} + l^l \cdot \varphi_3(s) \quad ; \quad l^l = \sum_{k=0}^{N_1-1} l_k e^{j2\pi\frac{K}{N_1}l}$$

where the $\varphi_2(s)$ relates to the mutual airfoil displacement under the action of deformation and it is defined by the equation (13) As an example given below is a solution of the set (27) for the case when the cacscade 1 is a high-solidity cascade of circles with radius R.

It is assumed that cascade that cadcade 1 deformation is obtained by rotating of airfoils about the leading edge. Then the generalized coordinate of l_k represent the rotation angle (positive, counterclockwise) of k-th airfoil of the cadcade.

Mutual positionninockwise) of k-th airfoil of the cadcade.

Mutual positionning of the cascade 1 and the basic circle is depicted at fig. 6.

Unknown rotation angles are approximately equal

$$a_k = 2\sum_{l=1}^{3} \alpha^l \cdot \sin 2\pi^l / N_1 k \quad ; \quad k = 0,1,2,\cdots,N_1 - 1$$

$$\alpha^l = 2\pi^2\left(\frac{R}{h_2}\right)^2 l \cdot e^{-2\pi\frac{l}{N_1}\tau\frac{R+\Delta}{b_1}} \cdot \frac{2 - e^{-2\pi\frac{l}{N_1}\tau}}{2\pi\frac{l}{N_1}\tau + 1} \quad ; \quad l = 1,2,3 \qquad (28)$$

The table 1 below contains values of (in degrees) calculated by use of the equation (28) for $R=6mm; l_1 = 24mm; h_1 = 12mm; h_2 = 144mm; \Delta = 6mm$ For comparison there are also given the results obtained by application of the exact theory.

Table 1.

Airfoil number	1	2	3	4	5	6	7	8	9	10	11
Asymptotic theory	+0.5	+1.0	+1.75	+2.25	+3.50	0	−3.5	−2.25	−1.75	−1.0	−0.5
Exact theory	+0.3	+0.7	+1.4	+2.2	+3.9	0	−3.9	−2.2	−1.4	−0.7	−0.3

Fig. 6 depicts measured values of

$$\lambda = \frac{v - v_{1\infty}}{v_{1\infty}} \cdot 100\%$$

Measurements were made at a distance of 3,5 mm form the forward front of the above-described cadcade. Points interconnected with solid line relate to an uniform cascade and points interconnected with dash line relate to a nonuniform cascade having the rotation angles α_k as given in the first line of the table 1. As it can be seen a considerable (upto 13%) flow speed distortion with a uniform cascade disappears with a nonuniform cascade disappears with a nonuniform cascade having preserved a step-like (with period h_1) speed disturbance.

CONLUSION

As it can be seen form the results presented above the proposed asymptotic theory of high-solidity airfoil cascade permits carring out a qualitative analysis of a number of problems of cascade unsteady aerodynamics and aeroelasticity as well as getting in some cases the relationships which are suitable for engineering applications.

REFERENCES

1. Saren V. E. Isvestija Akademii nauk SSSR, MJG, 4 (1971), 75

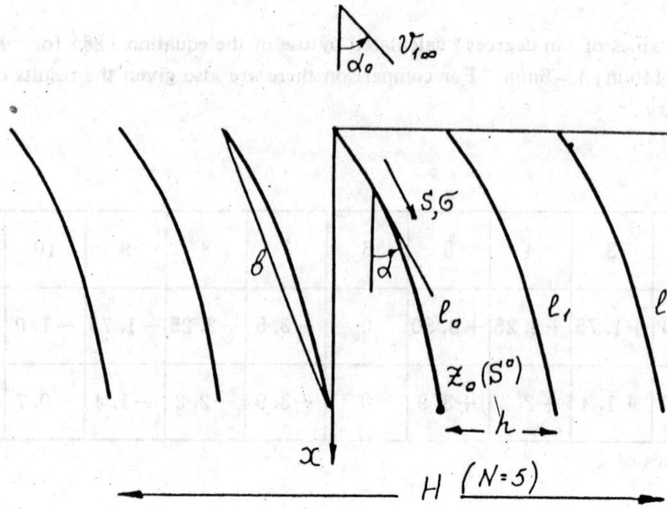

Fig.1 High-solidity arc cascade

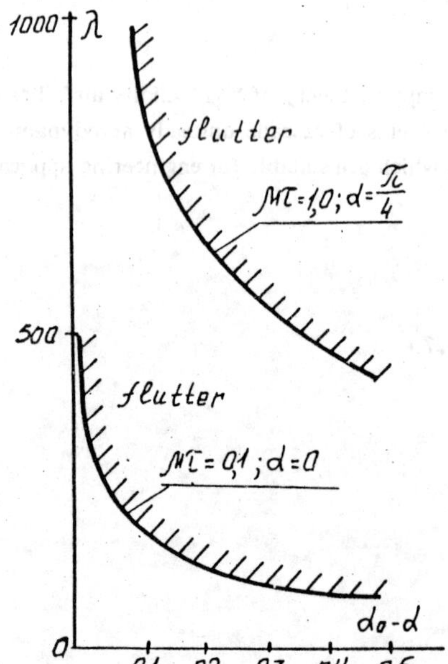

Fig.2 Dependence of flutter boundary on aerodynamic loading.

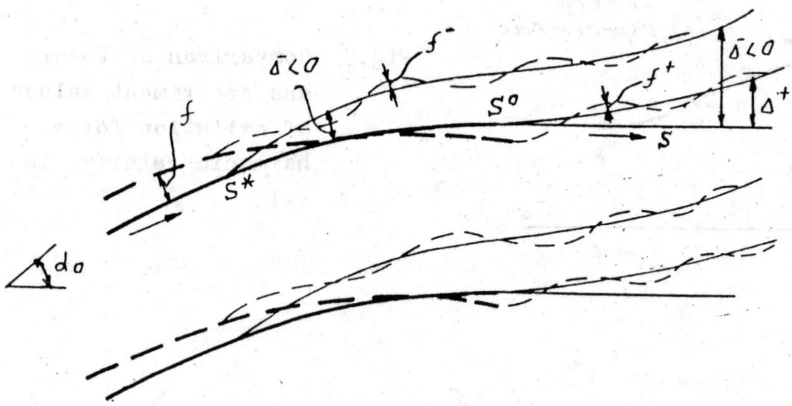

Fig.3 To a model of unsteady stalled flow around vibrating arc cascade.

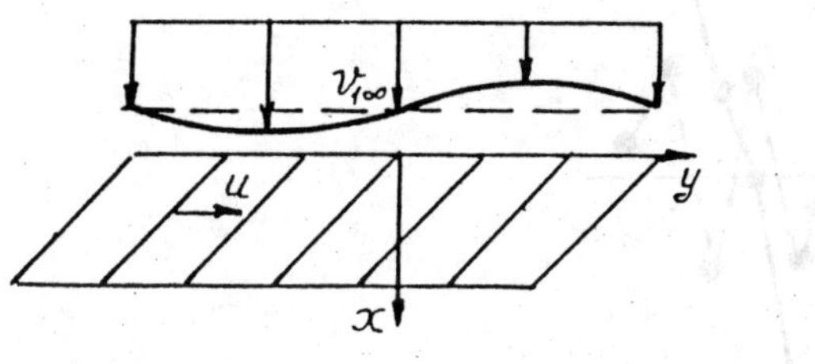

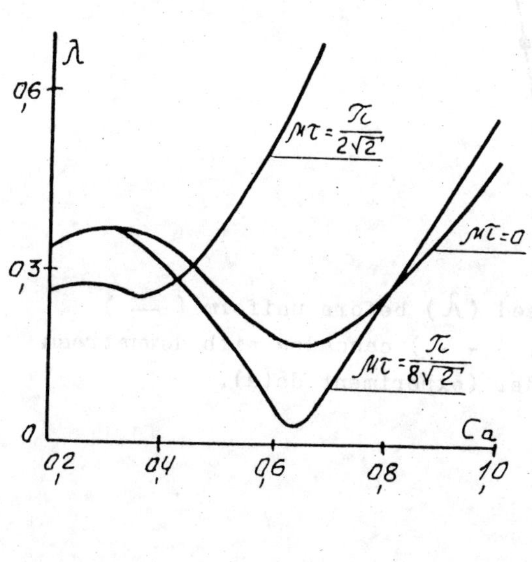

Fig.4 Relative level of exitation force harmonic (λ) versus flow rate factor (C_a) for various scales of incoming flow sine nonuniformity ($\mu\tau$).

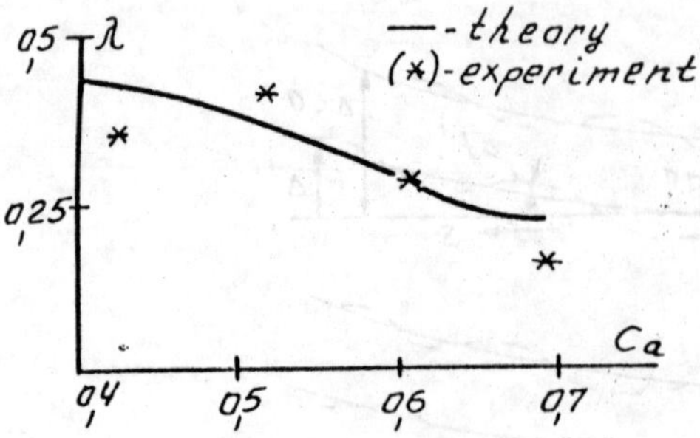

Fig.5 Comparison of theory and experiment values of exitation force harmonic relative level.

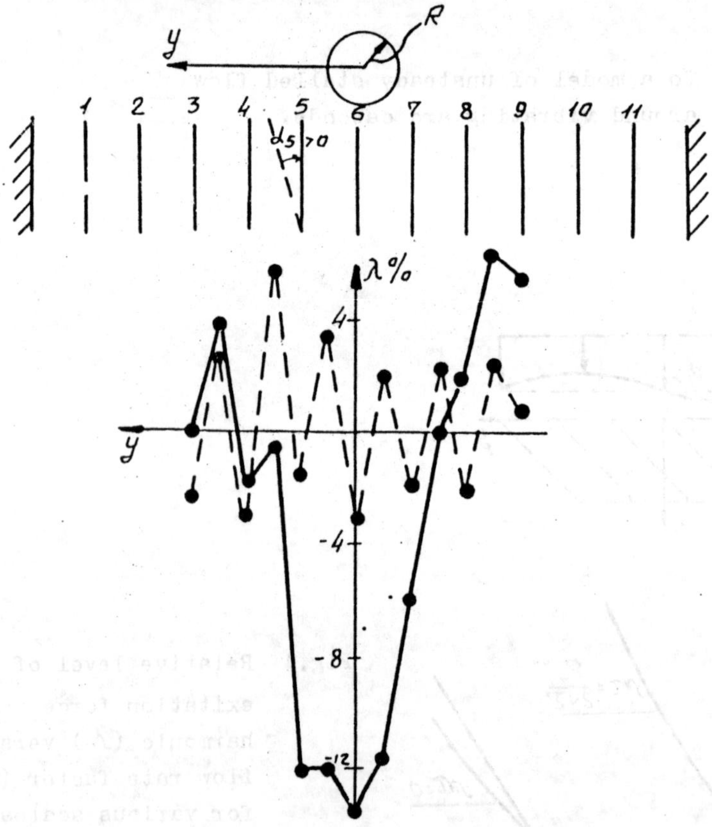

Fig.6 Relative flow speed (λ) before uniform (—) and nonuniform (- - -) cascades with downstream positioned cylinder (experiment data).

THE SOLVING OF ONE DIMENSIONAL UNSTEADY FLOW

WITH DIFFERENCE METHODS

Shao Guangri, Shang Yu * and Wang Rongsheng
(Shenyang Institute of Aeronautical Engineering, Shenyang, China)

ABSTRACT

In this paper, one dimensional unsteady gas flow equations are formulated with Riemann variables as the functions of time and space to be solved. By using the 'de Haller' test and a simple pipe flow with a temperature discontinuity, the equations are solved by difference methods of first–order accuracy and second order accuracy for non–homentropic cases. As compared with the commonly used characteristic method, the results show the above mentioned formulation and difference methods have the advantages of consuming less computer time and being easier to reach higher order accuracy. The application to turbocharged internal combustion engines is also discussed in this paper.

NOTATIONS

a acoustice velocity

a_{ref} reference acoustic velocity

a_A acoustic velocity after isentropic change of pressure to reference pressure

A non–dimensional acoustic velocity $A = a / a_{ref}$

A_A non–dimensional value of a_A $A_A = A_A / a_{ref}$

D pipe diameter

f friction factor(f= 0.005)

F cross–sectional area of flow

g friction force per unit mass $g = f\frac{u|u|}{2}\frac{4}{D}$

G non–dimensional friction force $G = \frac{g x_{ref}}{a_{ref}^2}$

p pressure

p_{ref} reference pressure

q rate of heat transfer per unit mass per unit time

t time

u gas velocity

U non–dimensional gas velocity $U = \frac{u}{a_{ref}}$

x distance

U non–dimensional distance $X = \frac{x}{X_{ref}}$

x_{ref} reference length

z non–dimensional time $Z = a_{ref} t / X_{ref}$

k specific heat ratio $k = C_p / C_v$

Δt time step

Δx mesh length

Δz non–dimensional time step

λ,β Riemann variables

ρ density

INTRODUCTION

The performance prediction of a turbocharged internai–combustion engine is concerned in the analysis of unsteady pipe flows, and the governing equations of a one dimensional unsteady compressible flow in a pipe with variable cross–section area, friction and heat transfer are known as

$$\frac{\partial}{\partial t}\begin{bmatrix}\rho \\ u \\ p\end{bmatrix} + \begin{bmatrix} u & \rho & o \\ o & u & \rho^{-1} \\ o & \rho a^2 & u\end{bmatrix}\frac{\partial}{\partial x}\begin{bmatrix}\rho \\ u \\ p\end{bmatrix} = \begin{bmatrix} -\frac{\rho u}{F}\frac{dF}{dx} \\ -G \\ (k-1)\rho(q+uG) - \frac{a^2 u}{F}\frac{dF}{dx}\end{bmatrix} \tag{1}$$

The right–hand side vector if caused by variable crosss–section area, friction and heat transfer. For a flow in uniform cross–sectioin area pipe where friction and heat transfer may be neglected, the right–hand side vector vanishs and the system reduces to a homogeneous one. The most popular technique used for the solving of the system is recognized as the method of characteristics (mch), which was fully discussed by Rowland S. Benson in his book. In this technique, the λ,β and path line characteristics were introdiced and along each characteristic, the respective compatibiliity conditions in non–dimensional forms were shown as

$$d\lambda = -\frac{(k-1)}{2}\frac{(\lambda+\beta)(\lambda-\beta)}{2(k-1)}\frac{1}{F}\left[\frac{dF}{dx}\right]dZ + \frac{(\lambda+\beta)}{2A_A}dA_A + \frac{(k-1)^2}{2}\frac{qx_{ref}}{a^3_{ref}}\frac{2}{(\lambda+\beta)}dZ$$
$$-\frac{(k-1)}{2}\frac{2fx_{ref}}{D}\left[\frac{\lambda-\beta}{k-1}\right]^2\frac{(\lambda-\beta)}{|\lambda-\beta|}\left[1-(k-1)\frac{2(\lambda-\beta)}{(k-1)(\lambda+\beta)}\right]dZ \tag{2}$$

$$d\beta = -\frac{(k-1)}{2}\frac{(\beta+\lambda)(\beta-\lambda)}{2(k-1)}\frac{1}{F}\left[-\frac{dF}{dx}\right]dZ + \frac{(\beta+\lambda)}{2A_A}dA_A + \frac{(k-1)^2}{2}\frac{qx_{ref}}{a^3_{r\cdot f}}\frac{2}{(\beta+\lambda)}dZ$$
$$-\frac{(k-1)}{2}\frac{2fx_{ref}}{D}\left[\frac{\beta-\lambda}{k-1}\right]^2\frac{(\beta-\lambda)}{|\beta-\lambda|}\left[1-(k-1)\frac{2(\beta-\lambda)}{(k-1)(\lambda+\beta)}\right]dZ \tag{3}$$

$$dA_A = \frac{(k-1)}{2}\frac{A_A}{A^2}\left[\frac{qx_{ref}}{a^3_{ref}} + \frac{2fx_{ref}}{D}\left|\frac{\lambda-\beta}{K-1}\right|^3\right]dz \tag{4}$$

The values λ,β,A_A at the grids for time Z+ΔZ can be caluculatedfrom those at the grids for time Z. Of course, in the implementations, the C.F.L stability criterion should be observed, that is

$$\frac{\partial Z}{\partial X} \leqslant \frac{1}{A+|U|} \tag{5}$$

Bu using this method and the models of quasi–steady flow for various boundary linkings, the performannce predictions of turbocharged internal–combustion engines have been successively achieved.

As in the last two decades, the solving of th first–order hyperbolic system with finite difference methods was studied by many investigators and a number of papers were presented, the difference methods have been intrfducing into the area concerned in this paper. In the existing literatures, the governing equations of one dimensional unsteady flow are formulated with ρ, u, p as functions to be solved. The co–called conservation–form system is still formulated with variables which are combinations of ρ, u and p. In the implementation of difference methods to the governing equations formulated with ρ,u and p, the values ρ, u and p at grids close to pipe ends at every time step should be transformed to Riemann variables λ,β and A_A so as to meet the needs of boundary calculations.

REFORMULATION OF GOVERNING EQUATION AND DIFFERENCE SCHEMES APPLIED

The system of equation(1) can be transformed to a form with Riemann variables λ,β and A_A as functions to be solved, the system after transformation is as follows:

$$\frac{\partial}{\partial Z}\begin{bmatrix}\lambda\\ \beta\\ A_A\end{bmatrix} + \begin{bmatrix} c\lambda - b\beta & o & o\\ o & b\lambda - c\beta & o\\ o & o & \frac{1}{2}(c+b)(\lambda-\beta)\end{bmatrix}\frac{\partial}{\partial X}\begin{bmatrix}\lambda\\ \beta\\ A_A\end{bmatrix} = \begin{bmatrix}\varphi_\lambda\\ \varphi_\beta\\ \varphi_{A_A}\end{bmatrix}$$

where b = (3−k) / 2(k−1), c = (k+1) / 2(k−1).

$$\varphi_\lambda = -\frac{(k-1)}{2}\frac{(\lambda+\beta)(\lambda-\beta)}{2(k-1)}\frac{1}{F}\left[\frac{dF}{dx}\right] + \frac{(\lambda+\beta)}{2A}\frac{dA_A}{dZ} + \frac{(k-1)^2}{2}\frac{qx_{ref}}{a^3_{ref}}\frac{2}{(\lambda+\beta)}$$

$$-\frac{(k-1)}{2}\frac{2fx_{ref}}{D}\left[\frac{\lambda-\beta}{k-1}\right]^2\frac{(\lambda-\beta)}{|\lambda-\beta|}\left[1-(k-1)\frac{2(\lambda-\beta)}{(k-1)(\lambda+\beta)}\right]$$

$$\varphi_\beta = -\frac{(k-1)}{2}\frac{(\beta+\lambda)(\beta-\lambda)}{2(k-1)}\frac{1}{F}\left[\frac{-dF}{dx}\right] + \frac{(\beta+\lambda)}{2A_A}\frac{dA_A}{dZ} + \frac{(k-1)^2}{2}\frac{qx_{ref}}{a^3_{ref}}\frac{2}{(\beta+\lambda)}$$

$$-\frac{(k-1)}{2}\frac{2fx_{ref}}{D}\left[\frac{\beta-\lambda}{k-1}\right]^2\frac{(\beta-\lambda)}{|\beta-\lambda|}\left[1-(k-1)\frac{2(\lambda-\beta)}{(k-1)(\lambda+\beta)}\right]$$

$$\varphi_{AA} = \frac{(k-1)}{2}\frac{A_A}{A^2}\left[\frac{qx_{ref}}{a^3_{ref}} + \frac{2fx_{ref}}{D}\left|\frac{\beta-\lambda}{K-1}\right|^3\right] \tag{6}$$

The system (6) is really congruent with equations (2), (3) and (4), for it is easy to see that

$$\frac{d\lambda}{dz} = \frac{\partial\lambda}{\partial z} + (U+A)\frac{\partial\lambda}{\partial x} = \partial\lambda/\partial z \quad + (c\lambda - b\beta)\frac{\partial\lambda}{\partial x}$$

$$\frac{d\beta}{dz} = \frac{\partial\beta}{\partial z} + (U-A)\frac{\partial\beta}{\partial x} = \partial\beta/\partial z \quad + (b\lambda - c\beta)\frac{\partial\beta}{\partial x}$$

The $\frac{dA_A}{dz}$ terms in φ_λ and φ_β are different with each other, the expanded form of the former is as

$$\frac{dA_A}{dZ} = \partial A_A/\partial Z \quad + (U+A)\frac{\partial A_A}{\partial X}$$

while for the latter is as

$$\frac{dA_A}{dZ} = \partial A_A/\partial Z \quad + (U-A)\frac{\partial A_A}{\partial X}$$

Both the difference schemes of first–order accuracy and second–order accuracy (SDM) have been applied to system(6). The scheme of first–order accuracy used is the upwind scheme (UDM). for the single equation $\partial V/\partial t + m(\partial V/\partial X) = 0$, it has the form

$$V^{n+1}_j = V^n_j - (\tau/h)\left(\frac{1}{2}(m+|m|)(V^n_j - V^n_{j-1}) + \frac{1}{2}(m-|m|)(V^n_{j+1} - V^n_j)\right) \tag{7}$$

its stability condition was proved to be $|m|(\tau/h) < 1$, which is just the same as the C.F.L. condition. The notations τ and h stand for the grid sizes along t and x axis respectively. In the implementation of solving system(6), if the values of λ, β, A_A are set to be frozen at the just solved time step, the system is decoupled into three separate equations. All of them can then be solved separately by a scheme somewhat like (7) with taking account of the terms $\varphi_\lambda, \varphi_\beta, \varphi_{AA}$. From the structure of the system one is easy to recognize the third equation should be solved first. The second–order accu.acy scheme (SDM) used is a mixed type of Lax–Wendroff and Maccormack schemes. In the process, the third equation is first solved by the Laxwendroff scheme, which for a single equation with constaat coefficient m, $\partial V/\partial t + m(\partial V/\partial X) = 0$ is as

$$V^{n+1}_j = V^n_j - \frac{1}{2}(\tau/h)m(V^n_{j+1} - V^n_{j-1}) + \frac{1}{2}(\tau/h)^2 m^2)(V^n_{j+1} - 2V^n_j + V^n_{j-1}) \tag{8}$$

and this is proved conditionally stable. By the same principle with which (8) is established and with the values of the variables set to be frozen at the just solved time step, the scheme for the equation with variable coefficient M and nonhomogeneous term N, $\partial V/\partial t + M(\partial V/\partial X) = N$ can be obtained as

$$V^{n+1}_j = V^n_j - \frac{1}{2}\left[\frac{\tau}{h}\right]M^n_j(V^n_{j+1} - V^n_{j-1}) + \frac{\tau^2}{2}\left[M^n_j\frac{(M^n_{j+1} - M^n_{j-1})}{2h}\frac{(V^n_{j+1} - V^n_{j-1})}{2h}\right] + \frac{1}{2}\left[\frac{\tau}{h}\right]^2(M^n_j)^2(V^n_{j-1}$$

$$-2V^n_j + V^n_{j+1}) - \frac{\tau^2}{2}M^n_j\frac{(N^n_{j+1} - N^n_{j-1})}{2h} + \tau N^n_j \tag{9}$$

After the third equation is solved, the first and second equations are then solved by a scheme similar to

MacCormack two–step schems. The MacCormack scheme was originally for the conservation–form system, $\partial w / \partial t + \partial H / \partial x = 0$, and the two steps are

$$\overline{W_j^{n+1}} = W_j^n - (\tau / h)(H_{j+1}^n - H_j^n) \tag{10}$$

$$W_j^{n+1} = \frac{1}{2}(W_j^n + \overline{(w_j^{n+1}})) - \frac{1}{2}(\tau / h)(\overline{H_{j+1}^n} - \overline{H_{j-1}^{n+1}}) \tag{11}$$

where $\overline{H_j^{n+1}} = H(\overline{W_j^{n+1}})$.For the first and second equations of system(6),

$$\frac{\partial}{\partial Z}\begin{bmatrix} \lambda \\ \beta \end{bmatrix} + \begin{bmatrix} c\lambda - b\beta & o \\ o & b\lambda - c\beta \end{bmatrix} \frac{\partial}{\partial x}\begin{bmatrix} \lambda \\ \beta \end{bmatrix} = \begin{bmatrix} \varphi_\lambda \\ \varphi_\beta \end{bmatrix}$$

Let

$$W = \begin{bmatrix} \lambda \\ \beta \end{bmatrix}$$

$$B = \begin{bmatrix} c\lambda - b\beta & o \\ o & b\lambda - c\beta \end{bmatrix}, D = \begin{bmatrix} \varphi_\lambda \\ \varphi_\beta \end{bmatrix}$$

and follwoing the similar steps of the formulation of MacCormack scheme, one can obtain the scheme for the problem concerned as follows:

$$\overline{W_j^{n+1}} = W_j^n - (\tau / h)B_j^n(W_{j+1}^n - W_j^n) + \tau D_j^n \tag{12}$$

$$W_j^{n+1} = \frac{1}{2}(W_j^n + \overline{W_j^{n+1}}) - \frac{1}{2}(\tau / h)\overline{B_j^{n+1}}(\overline{W_j^{n+1}} - \overline{w_{j-1}^{n+1}}) + (\tau / 2)\overline{D_j^{n+1}} \tag{13}$$

where $\overline{B_j^{n+1}} = B(\overline{W_j^{n+1}}), \overline{D_j^{n+1}} = \overline{W_j^{n+1}})$.

NUMERICAL EXPERIMENT OF FIRST–ORDER ACCURACY SCHEME

The reformulated governing equations and the difference schemes cited are tested by unsteady flows in simple pipes. The so–called 'de Haller' test is shown in Fig. 1(a). The initial state in the pipe is the same as that in the vessel. At the beginning, the right end of the pipe is closed and an unsteady flow in the pipe is induced by a sudden openning of the right closed end. Let the initial pressure and temperature by denoted by P_0, T_0 the ambient pressure by P_{ref}, the pipe wall temperature by T_w, the length of pipe by 1, the diameter of pipe by D, and their values are assigned as follows:

$P_{ref} = 0.1MPa$

$P_0 = 0.15MPa$

$T_0 = 300K$

$T_w = 400K$

$L = 1m$

$D = 0.01m$

Fig. 1(b) and (c) show the comparisons between the results obtained by MCH and UDM, the results obtained by the two methods at the two different points in the pipe are about the same. The comparison of the computer time of the two methods shows the UDM scheme is more effective, it consumes about 15 less computer time in this particular example.

Another test of an unsteady flow in a simple pipe is shown in Fig. 2. The initial state in the vessel and that in the pipe are different with each other. At the beginning, the vessel is closed to the pipe, and a sudden openning of the vessel causes an unsteady flow in the pipe. Let the initial pressure and temperature in the pipe be denoted by P_0, T_0, that of vessel by P_1, T_1 the pipe wall temperature by T_w, and their values are assigned as

$$P_0 = 0.1MPa$$
$$P_0 = 0.15MPa$$
$$T_0 = 300K$$
$$T_1 = 500K$$
$$T_w = 500K$$
$$L = 1m$$
$$D = 0.01m$$

This example might be thought as a simulation of the unsteady flow in the exhaust pipe which is adjacent to the exhaust valve where discontinuity of temperature always exists. The results of the two methods show they are almost coincident with each other.

NUMERICAL EXPERIMENT OF SECOND–ORDER ACCURACY SCHEME

The comparison of the results of MCH and SDM in 'de Haller' test is shown in Fig. 3, the curves shown in Fig. 3 make one hard to compare the accuracy of the two methods. But as one makes tests with tapered pipes, the fact that SDM is superior in accuracy is obvious. Let d1,d2 be the diameters of the left and right ends of a pipe respectively, $\dot{m}_{in}$ and $\dot{m}_{out}$ be the mass flow per unit time at left and right ends of pipe at the time the pipe flow reaches a steady state. The values of mass flow ratio $(\dot{m}_{in} - \dot{m}_{out}) / \dot{m}_{in}$ of different d1 / d2 for the two methods are shown in Talbe 1.

Table 1 Mass flow ratio of different d1 / d2 at steady state for MCH and SDM

d1 / d2	1.00	1.25	1.5	1.75	2.00
MCH	–0.36%	0.40%	1.20%	2.00%	2.80%
SDM	0.54%	–0.01%	–0.32%	–0.41%	–0.38%

From Fig. 3 one can find the drawback of SDM, the Method Produces overshoot at somewhere sudden teperature or (and) pressure change takes place. If the scheme is applied to the test shown in Fig. 2(a), there will be an oscillation occuring, this is clearly shown in fig. 4. But this drawback can be effectively diminished by a mixed scheme of first–order and second–order accuracy (MDM).

APPLICABILITY TO INTERNAL–COMBUSTION ENGINES

The up–wind scheme is used to analyse the unsteady flows in exhaust pipes of a three–cylinder turbocharged diesel engine in a software, which originally used characteristic method to analyse the unsteady flows in pipes. In the replacement all theboundary calculations are kept unchanged. Fig. 5 shows the results obtained by the UDM method are almost congruent with that obtained by MCH.

CONCLUSIONS

1. The governing equation system of unsteady pipe flow formulated with λ, β and A_A as functions to be solved is applicable in turbocharged internal–combustion engine analysis.
2. The particular first–order accuracy difference scheme, the up–wind scheme can replace the commonly lused characteristic method in analysing turbocharged internal combustion egines with same accuracy and less computer–time consuming.
3. The second–order accuracy scheme reaches higher order accuracy, but needs to be mixed with first order accuracy scheme wherever gas temperature discontinuity takes place.

REFERENCES

1. R.S.Benson, the Thermodynamics and Gas Dynamics of Internal combustion Engines, Vol.1, Clarendon Press, Oxford (1982).
2. A.H.Shapiro, The Dynamics and Thermodynamics of Compressible Fluid Flow, Vol.2, The Ronald Press Company, New York (1954).
3. M.Poloni et al., Int. J.Mech. Sci., 28(4) (1987), 367–378.

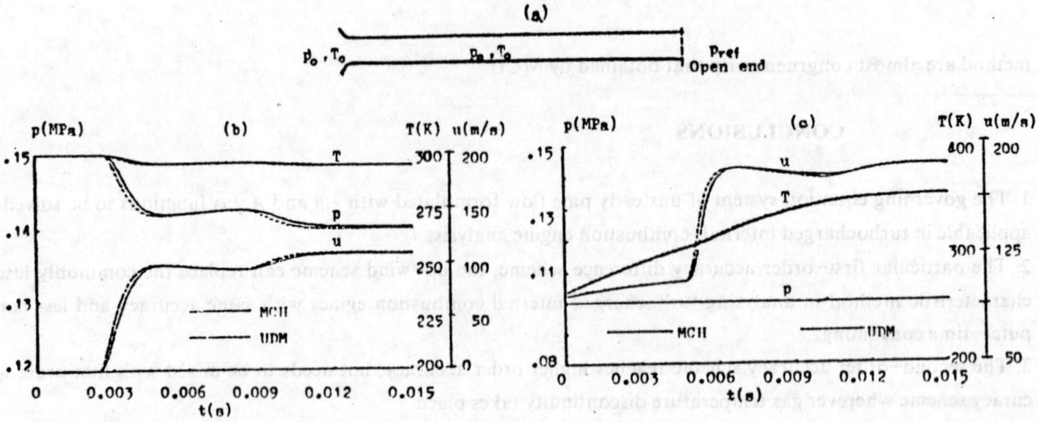

Fig. 1

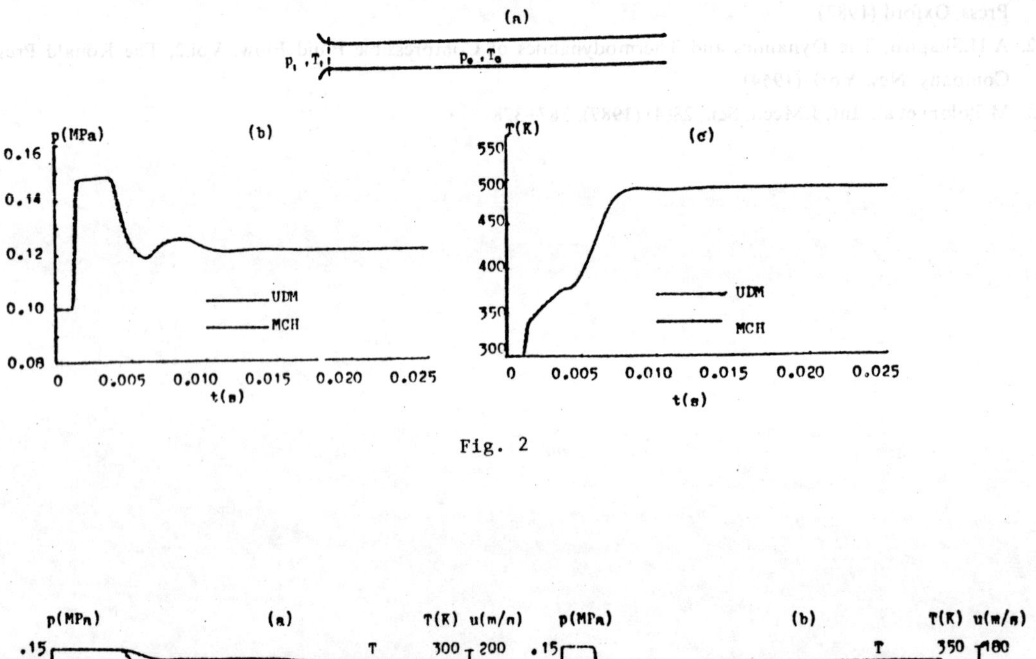

Fig. 2

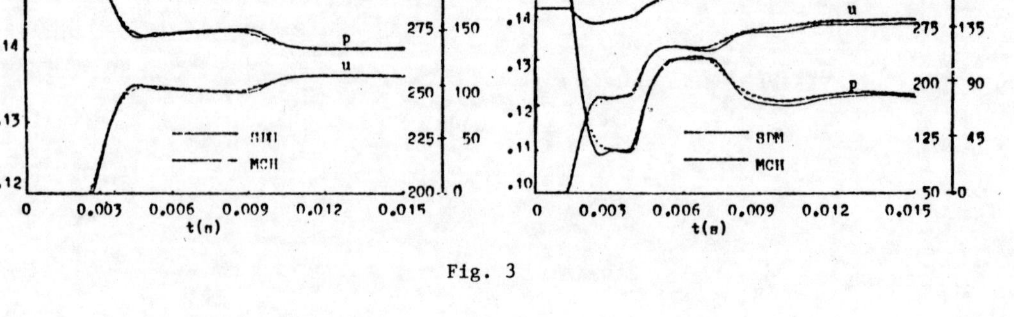

Fig. 3

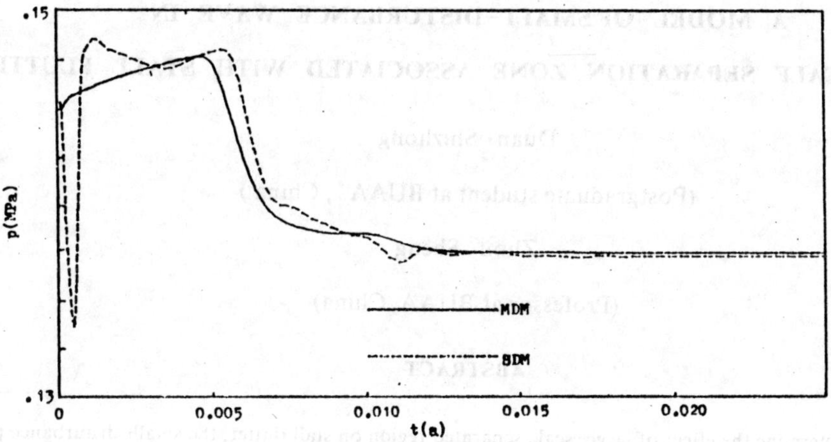
.15
.13
p(MPa)
MDM
SDM
0
0.005
0.010
0.015
0.020
t(s)

Fig. 4

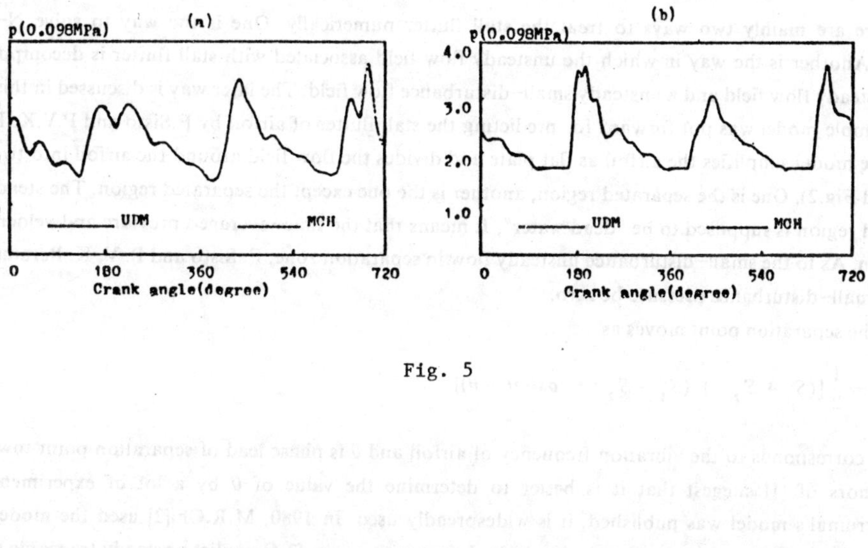
(a)
p(0.098MPa)
4.0
3.0
2.0
1.0
UDM
MCH
0
180
360
540
720
Crank angle(degree)
(b)
p(0.098MPa)
4.0
3.0
2.0
1.0
UDM
MCH
0
180
360
540
720
Crank angle(degree)

Fig. 5

A MODEL OF SMALL–DISTURBANCE WAVE IN LARGE–SCALE SEPARATION ZONE ASSOCIATED WITH STALL FLUTTER

Duan Shizhong

(Postgraduate student at BUAA *, China)

Zhou Sheng

(Professor at BUAA, China)

ABSTRACT

In order to determine the effect of large–scale separated region on stall flutter, the small–disturbance pressure in the region should be obtained by some way. In this paper, a model is proposed to express the characteristics of small–disturbance wave in the region. The method depends on the time–filtered equations of Navier–Stokes equations.

INTRODUCTION

There are mainly two ways to treat the stall flutter numerically. One is the way to solve N–S equations directly. Another is the way in which the unsteady flow field associated with stall flutter is decomposed into two parts–a steady flow field and a unsteady small–disturbance flow field. The later way is discussed in this paper only.

A simple model was put forward for predicting the stall flutter of airfoil by F.Sisto and P.V.K. Perumal[1] in 1974. The model simplifies the airfoil as flat plate and divides the flow field around the airfoil into two region (See Fig.1 and Fig.2). One is the separated region, another is the one except the separated region. The steady flow in the separated region is supposed to be "dead water", It means that the time--averaged pressure and velocity are zero in the region. As to the small–disturbance unsteady flow in separation zone, F. Sisto and P. V. K. Perumal assumed,

1. Small–disturbance pressure be zero.
2. The separation point moves as

$$S(t)=\frac{1}{2}[(S_1+S_2)+(S_1-S_2)\cdot cos(\omega t-\theta)] \tag{1}$$

where ω corresponds to the vibration frequency of airfoil and θ is phase lead of separation point towards motion. The authors of [1] suggest that it is better to determine the value of θ by a lot of experiments. Since the Sisto–Perumal's model was published, it is widespreadly used. In 1980, M.R.Chi[2] used the model to treat the unsteady subsonic flow around vibrating cascade. Later some papers[3,4] predict unsteady transonic stall flutter of cascade by using the same model. But Sisto–Perumal's model could be improved, because the small–disturbance pressure in separated region may not be zero. So a new model is advanced in this paper to solve the small– disturbance wave in separation region during stall flutter.

It is found that the wake flow of vibrating cascade could be composed of two parts during stall flutter. One is unsteady harmonic main flow corresponding to vibrating cascade, another is the high–frequency pulsation corresponding to turbulent flow in the separation zone. If we make a time filter of wake flow, and use some method like small vortices viscous simulation to describe the effect of high–frequency pulsation on harmonic main flow, then, we can get the time–filtered equations as the governing equations of the small–disturbance flow in the separation region. The new method is different from Large Eddy Simulation method, but there are some similarities between the two methods, which could be mentioned in detail later.

PHYSICAL PROBLEM

Let us to see the physical picture (Fig. 3) of flow around cascade during stall flutter. A physical parameter Ψ could be decomposed as following.

$$\Psi = \Psi_0 + \tilde{\Psi} + \Psi' \tag{2}$$

where Ψ_0 corresponds to steady rotational flow in the separation region, $\tilde{\Psi}$corresponds to the unsteady harmonic flow caused by small–amplitude vibration of cascade, $\tilde{\Psi}$ is much less than Ψ_0,Ψ'corresponds to the high–frequency pulsation. This phenomenon could be observed at a static point in separation zone with time flying. If we look into the same flow in separated region by a different way, that is, watch the whole flow field at a time. Then, we can find large eddies and many small vortices which correspond to turbulent flow. So, the high–frequency pulsation is similar to small vortices, and some method similar to small vortices viscous simulation could be used to express the high–frequency pulsation. The method will be mentioned in next chapters. It is the second term $\tilde{\Psi}$in Eq. (2) which does work to aeroelastic work coefficient and effects the onset of stall flutter directly. So the problem is how to get the second term $\tilde{\Psi}$in Eq. (2).

GOVERNING EQUATION

In this chapter we will get the governing equation for $\tilde{\Psi}$in Eq (2).

1. Filter Function

we choose filter function $G(\tau-t)$ as

$$G(\tau - t) = \begin{cases} 0 & \left(-\infty, -\frac{\Delta T}{2}\right) \\ \frac{1}{\Delta T} & \left[-\frac{\Delta T}{2}, \frac{\Delta T}{2}\right] \\ 0 & \left(\frac{\Delta T}{2}, +\infty\right) \end{cases} \tag{3}$$

where ΔT is the time–gap which is much less than the cycle of harmonic flow $\tilde{\Psi}$, but much greater than the cycle of high–frequency pulsation Ψ'. Then, we have

$$\overline{\Psi}(\tau) = \int_t G(\tau - t) \cdot \Psi(t) \cdot dt \doteq \Psi_0(\tau) + \tilde{\Psi}(\tau) \tag{4}$$

where $\Psi(t)$ and$\overline{\Psi}(\tau)$are expressed in Fig. 4.

2. Governing equations

The N–S equations are

$$\begin{cases} \frac{\partial U_i}{\partial t} + \frac{\partial}{\partial x_j}(U_i U_j) = -\frac{1}{\rho} \cdot \frac{\partial P}{\partial x_i} + \nu \cdot \nabla^2 U_i \\ \frac{\partial U_i}{\partial x_i} = 0 \end{cases} \tag{5}$$

we assume that the flow in separated zone is incompressible. After making a time filter of Eq. (5), we have

$$\begin{cases} \frac{\partial \overline{U_i}}{\partial \tau} + \frac{\partial}{\partial x_j}(\overline{U_i U_j}) = -\frac{1}{\rho}\frac{\partial \overline{P}}{\partial x_i} + \nu \cdot \nabla^2 \overline{U_i} \\ \frac{\partial \overline{U_i}}{\partial x_i} = 0 \end{cases} \tag{6}$$

According to Eq. (2) and Eq. (4), we have

$$U_i = \overline{U}_i(t) + U'_i$$

Then

$$\overline{U_i U_j}(\tau) = \overline{\overline{U}_i \overline{U}_j} + \overline{\overline{U}_i U'_j} + \overline{U'_i \overline{U}_j} + \overline{U'_i U'_j} = \overline{\overline{U}_i \overline{U}_j} - \overline{\tau}_{ij} \tag{7}$$

where

$$\overline{\tau}_{ij} = -\left(\overline{\overline{U}_i U'_j} + \overline{U'_i \overline{U}_j} + \overline{U'_i U'_j}\right) \tag{8}$$

$$\overline{\overline{U}_i \overline{U}_j}(\tau) = \int_t G(\tau - t) \cdot \overline{U}_i(t) \cdot \overline{U}_j(t) \cdot \alpha t = \overline{U}_i(\tau) \cdot \overline{U}_j(\tau) + \frac{\Delta T^2}{24} \cdot \frac{\partial^2 (\overline{U}_i \cdot \overline{U}_j)}{\partial \tau^2} + 0(| t - \tau |^3) \tag{9}$$

This is because of

$$\overline{U}_k(t) = \overline{U}_k(\tau) + (t-\tau) \cdot \frac{\partial}{\partial \tau}\overline{U}_k(\tau) + \frac{(t-\tau)^2}{2} \cdot \frac{\partial_2}{\partial \tau^2}\overline{U}_k(\tau) + 0(| t - \tau |^3) \tag{10}$$

After accumulating Eq. (6) to Eq. (9), we have

$$\begin{cases} \frac{\partial \overline{U}_i}{\partial \tau} + \frac{\partial(\overline{U}_i \cdot \overline{U}_j)}{\partial x_j} = -\frac{1}{\rho}\frac{\partial \overline{P}}{\partial x_i} - \frac{\Delta T^2}{24} \cdot \frac{\partial^3}{\partial \tau^2 \partial x_j}(\overline{U}_i \cdot \overline{U}_j) + \frac{\partial}{\partial x_j}\overline{\tau}_{ij} + \nu \cdot \nabla^2 \overline{U}_i \\ \frac{\partial \overline{U}_i}{\partial x_i} = 0 \end{cases} \tag{11}$$

where

$$\overline{\tau}_{ij} = -\left(\overline{\overline{U}_i U'_j} + \overline{U'_i \overline{U}_j} + \overline{U'_i U'_j}\right)$$

we notice that The high–frequency pulsation stress term $\overline{\tau}_{ij}$ has some similar characteristics with Reynold's Stress in Large Eddy Simulation method, which has been mentioned before. The term $\frac{\partial}{\partial x_j}\overline{\tau}_{ij}$ express physically the effect of pulsation flow on unsteady harmonic flow. The high–frequency pulsation flow gets energy from main flow and dissipate it. So, we can assume,

$$\bar{\tau}_{ij} = -K \cdot \left(\frac{\partial \bar{U}_i}{\partial x_j} + \frac{\partial \bar{U}_j}{\partial x_i}\right) \tag{12}$$

As we do in Large Eddy Simulation method, there will be many models to dertermine the value of K in Eq. (12), Here we choose the simplist model to make calculation easier. It is the constant vortices viscous coefficient model.

$$K = \frac{1}{2}C \tag{13}$$

where, C is constant.

Now, we can get.

$$\begin{cases} \dfrac{\partial \bar{U}_i}{\partial \tau} + \dfrac{\partial(\bar{U}_i \cdot \bar{U}_j)}{\partial x_j} = -\dfrac{1}{\rho}\dfrac{\partial \bar{P}}{\partial x_i} - \dfrac{\Delta T^2}{24} \cdot \dfrac{\partial^3}{\partial \tau^2 \partial x_j}(\bar{U}_i \cdot \bar{U}_j) + (\frac{1}{2}C + \gamma) \cdot \nabla^2 \bar{U}_i \\ \dfrac{\partial \bar{U}_i}{\partial x_i} = 0 \end{cases} \tag{14}$$

If we assume

$$\tilde{\psi} = \varphi \cdot {}^{i\omega t}$$

Then, we have

$$\tilde{\psi} = \psi_0 + \varphi \cdot {}^{i\omega t} \tag{15}$$

we can get the governing equation for $\tilde{\Psi}$, i. e. φ from Eq. (14) and Eq. (15)

$$\begin{cases} i\omega u_i + (1 - \omega^2 \cdot \dfrac{\Delta T^2}{24}) \cdot \dfrac{\partial}{\partial x_j}(U_{io} u_j + U_{jo} u_i) = -\dfrac{1}{\rho} \cdot \dfrac{\partial p}{\partial x_i} + (\frac{1}{2}C + \gamma)\dfrac{\partial^2 u_i}{\partial x_j^2} \\ \dfrac{\partial u_j}{\partial x_i} = 0 \end{cases} \tag{16}$$

THE NEW MODEL

Within the zone of ABCDTSGHA (Fig. 3), the goverñing equation could be expressed as following [4]

$$\begin{cases} (1 - M_{ox}^2) \cdot \varphi_{xx} + \left(1 - M_{0y}^2\right) \cdot \varphi_{yy} - 2M_{0x} M_{0y} \varphi_{xy} + C_4 \varphi_x + C_5 \cdot \varphi_2 + C_6 = 0 \\ a^2 = a_0^2 - (k-1) \cdot (U_0 \cdot \varphi_x + V_0 \cdot \varphi_y + i\omega\varphi) \cdot e^{i\xi t} \end{cases} \tag{17}$$

where

$$\begin{cases} C_4 = -\dfrac{1}{a_0^2}\left[(k-1) \cdot U_0 \dfrac{\partial V_0}{\partial y} + (k+1) \cdot U_0 \dfrac{\partial U_0}{\partial x} + 2V_0 \dfrac{\partial U_0}{\partial y} + 2i\omega U_0\right] \\ C_5 = -\dfrac{1}{a_0^2}\left[(k-1) \cdot V_0 \dfrac{\partial U_0}{\partial x} + (k+1) \cdot V_0 \dfrac{\partial V_0}{\partial y} + 2U_0 \dfrac{\partial U_0}{\partial y} + 2i\omega V_0\right] \\ C_6 = \dfrac{1}{a_0^2}\left[\omega^2 - i\omega \cdot (k-1)\left(\dfrac{\partial V_0}{\partial y} + \dfrac{\partial U_0}{\partial x}\right)\right] \end{cases}$$

In the separation zone of STEFS (Fig. 3), the governing equations are Eq. (16).The boundany conditions are

1. On the dividing streamline ST or FE,

$$\begin{cases} \dfrac{\partial \varphi}{\partial x} = u \\ \dfrac{\partial \varphi}{\partial y} = v \\ \dfrac{kP_0}{a_0^2} \cdot (U_0\varphi_x + V_0\varphi_y + i\omega\varphi) = p \end{cases} \quad (18)$$

2. On blade surface SF, the flow attached condition is assumed, i.e.

$$\left(\frac{\partial}{\partial t} + \vec{V} \cdot \nabla\right) F = 0 \quad (19)$$

3. Far field TE, perturpation pressure is zero.

$$p = 0 \quad (20)$$

CONCLUDING REMARKS

1. A new model is proposed in this paper to predict the stall flutter of cascade instead of Sisto–Perumal's model. a program with this new model has been finished, numerical work is just doing.
2. As we know, The unsteady Kutta condition is not correct in some case. It is possible to give out a new unsteady wake condition from this new model. A lot of work on the model should be done.

NOMENCLATURE

ψ:	Physical parameter .
$\overline{\psi}$:	time–filtered physical parameter .
ψ_0:	Steady physical parameter .
$\tilde{\psi}$:	unsteady harmonic physical parameter.
ψ':	high–frequency pulsation parameter.
φ:	unsteady perturbation velocity potential.
ω:	vibration frequency.
u:	flow velocity component in the direction of X axis.
v:	flow velocity component in the direction of Y axis.
p:	Unsteady perturbation pressure amplitude
$\vec{V}$:	Flow velocity
a:	Sound speed.
M_o:	Steady Mach Number.
γ:	viscid coefficient
$\overline{\tau}_{ij}$:	high–frequency pulsation flow stress.
k:	Specific heat ratio
K:	Vortices viscous coefficient
i:	$\sqrt{-1}$
F:	F(x,y,t), the moving equation of oscillating blade

REFERENCE

1. F. Sisto and P.V.K. Perumal " Lift and Moment Prediction for an Oscillating Airfoil with a Moving Separation Point" ASME 74–GT–28, 1974.
2. M.·R. Chi, ASME Paper 80–C2 / Aero–1, 1980
3. He Li and Zhou Sheng "Research on the Mechanism of Stall Flutter of blade", Aeronautic Journal, Vol. 6, No5, 1985 .
4. Duan S. Z., Wei X. L. and Zhou S "A Numerical Method Solving 2–D Unsteady Flour Field Around Cascade of Oscillating Airfolfs with Arbitrary Camber and Thickness" Proceedings of the Fourth International Symposium of Unsteady Aerodynamics and Aeroelasticity of Turbomachines and Propellers. P. 271–292

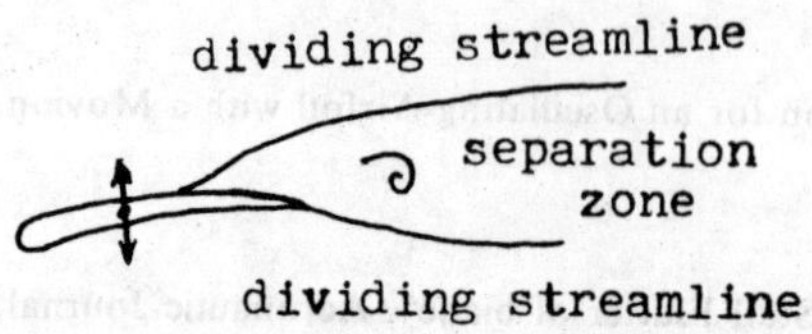

Fig.1 Flow with a largescale separation zone around airfoil

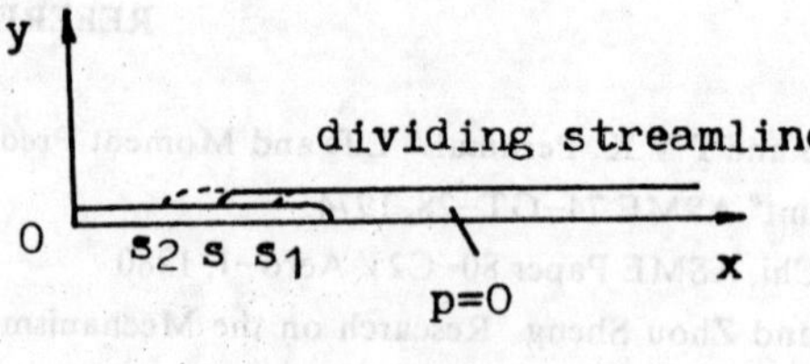

Fig.2 Sisto-Perumal's analytical model

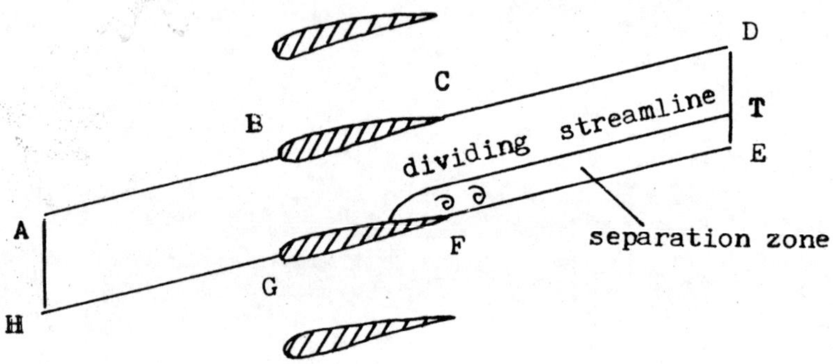

Fig.3 Flow around cascade during stall flutter

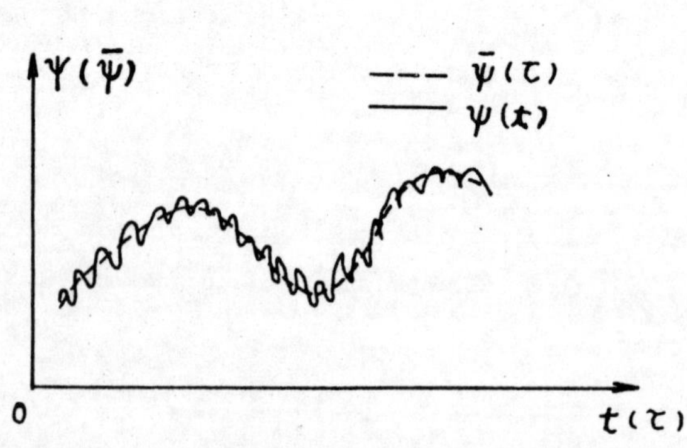

Fig.4 Physical parameter before and after time-filtered

Some Explorations on the Mechanism of Blade Flutter Suppression by Porous Wall Casing

by Chen Rong–guo, Wei Xing–lu, Zhou Sheng

Beijing University of Aeronautics and Astronautics

Abstract

The technique of blade flutter suppression by porous wall casing has been put forward for several years.However, a lot of works regarding the mechanism and the technical feasibility are needed to continue. On the basis of preceding references, a model of flow around 2D plane cascade, in which the spanwise unsteady disturbance velocity component is considered, and some corresponding numerical results have been presented. The numerical results have shown that the porous wall casing could produce a great influence on aeroelastic stability of blade row. However, the relations between geometric parameters of porous wall casing and the aeroelastic stability of blade row are rather complicated. Sometimes using porous wall casing should deteriorate the aeroelastic stability of blade row, if the acoustic impedance values of porous wall casing are not correctly selected.

§ 1 Introduction

Some research works on the mechanism of blade flutte; suppression by porous wall call casing have been published for several years [1,2,3]. Yamasaki and Namba [2] have established an analytical model by using the method of unsteady lifting surface, and calculated the effects of the characteristics of sound attenuance of porous wall on the unsteady aerodynamic forces acting on blade surfaces. The numerical results showed that the effect of acoustic admittance of porous wall on spanwise local lift distribution is important, but the effect of acoustic admittance on total aerodynamic work doing to blade during an oscillation cycle is minor. Watanabe and Kaji [3] analyzed aeroelastic stability of 2D oscillating cascade by using 3D semi–actuator disk model, and the numerical results showed that there are some effective values of acoustic admittance of porous wall which could suppress onset of blade flutter. However, there are some other values of admittance too, Which will deteriorate the aeroelastic stability of blade row.

The conclusions from Ref [2] and Ref [3] are consistent qualitatively in some sides but the quantitative differences between them are rather larger. of course, the differences could be understood, because the method of unsteady lifting surface and the method of semi–actuator disk are different after all, both in models and in calculating processes, especially in processing acoustic impedance of porous wall. Therefore, perhaps it is beneficial doing some work to explore the complicated mechanism from another point–view.

Wei and Zhou analyzed the mechanism of blade flutter suppression by using porous wall casing too [4]. They came from 3D Euler equations and then considered the different effects of porous side wall and solid side wall on spanwise unsteady disturbance velocity component while the 3D flow field was simplified into 2D plane flow field. Afterwards, the effects of porous side wall on aeroelastic stability of oscillating cascade could be analyzed. However, the values of spanwise unsteady disturbance velocity component are specified in advance. Obviously, it is a shortcoming of Ref. [4].

On the basis of Ref. [4], We will relate the spanwise unsteady disturbance velocity component with acoustic characteristics of side wall, then the influence of porous wall can appear in the governing unsteady potential equation as some terms of spanwise disturbance. So, the acoustic impedance of side wall will appear in the main potential equation , even though the potential equation is two–dimensional.

§ 2 Governing Equation

For unsteady flow around an oscillating plane cascade with porous side, the governing equation is [4]

$$\left\{a_0^2\frac{\partial^2\Phi}{\partial r^2}+\frac{1}{r_t}(a_0^2+V_{0\theta}^2)\frac{\partial\Phi}{\partial r}\right\}+(a_0^2-u_0^2)\frac{\partial^2\Phi}{\partial x^2}+(a_0^2-u_0^2)\frac{\partial^2\Phi}{\partial y^2}-2u_0v_0\frac{\partial^2\Phi}{\partial x\partial y}+c_1\frac{\partial\Phi}{\partial x}+c_2\frac{\partial\Phi}{\partial y}+c_3\Phi=0 \quad (1)$$

where the coefficients

$$c_1=-\left\{(k-1)u_0\frac{\partial v_0}{\partial y}+(k+1)\frac{\partial u_0}{\partial y}+2v_0\frac{\partial u_0}{\partial y}+2i\omega u_0\right\}$$

$$c_2=-\left\{(k+1)u_0\frac{\partial v_0}{\partial y}+(k-1)\frac{\partial u_0}{\partial x}+2v_0\frac{\partial u_0}{\partial y}+2i\omega v_0\right\}$$

$$c_3=\left\{\omega^2-i\omega(k-1)\left(\frac{\partial u_0}{\partial x}+\partial v_0/\partial y\right)\right\}$$

The unknown quantity is $\Phi=\Phi(x,y)$, the amplitude of unsteady disturbance potential. After solving the $\Phi(x,y)$, the unsteady disturbance pressure p can be solved as follows:

$$P=-\frac{kP_0}{a_0^2}\left(u_0\frac{\partial\Phi}{\partial x}+v_0\frac{\partial\Phi}{\partial y}+i\omega\Phi\right)e^{i\omega t} \quad (2)$$

Then, the unsteady force and moment acting on blade can be obtained, and we can judge the onset of flutter according to the value of unsteady aerodynamic work during an oscillating period.

§ 3 The Correction of Spanwise Unsteady Flow with Acoustic Characteristics of Porous Wall

For solid wall, the boundary condition at side wall is

$$\frac{\partial\Phi}{\partial r}=0, \qquad (r=r_t)$$

For porous wall, the boundary condition at side wall is

$$\frac{\partial\Phi}{\partial r}\neq 0, \qquad (r=r_t)$$

It means that the term

$$\left\{a_0^2\frac{\partial^2\Phi}{\partial r^2}+\frac{1}{r_t}(a_0^2+V_{\infty}^2)\frac{\partial\Phi}{\partial r}\right\}$$

has been added in Eq. (1) for the situation of porous side wall by comparison with the corresponding equation for the situation of solid side wall. That term reflects the effect of porous side wall on unsteady flow in cascade. of course that term should relate with acoustic characteristics of porous side wall.

Under action of unsteady pressure $\tilde{p}$, the normal unsteady velocity $\tilde{v}_n$ to porous wall are not equal to zero. In order to describe acoustic characteristics of porous side wall, let's introduce specific acoustic impedence.

$$z=\frac{\tilde{P}}{\rho_0 a_0 \tilde{v}_n} \tag{3}$$

where $\tilde{p}$ the acoustic pressure, and the normal velocity $\tilde{v}_n$ can be expressed as

$$\tilde{v}_n=\frac{\partial\tilde{s}_n}{\partial t}$$

which means the time partial derivative of normal displacement of fluid particles. The normal displacement of fluid particles is equal to the radial displacement of fluid particles.

$$\tilde{s}=\tilde{s}_n$$

the radial unsteady velocity

$$\tilde{s}_r=\frac{D\tilde{s}}{Dt}=\tilde{v}_n+u_0\frac{\partial\tilde{s}}{\partial x}+v_0\frac{\partial\tilde{s}}{\partial y}$$

According to the assumption of harmonic oscillation, the amplitudes of above–mentioned quantities are

$$z=\frac{P}{i\omega\rho_0 a_0 s_n} \tag{4}$$

$$s=s_n \tag{5}$$

$$v_r=\frac{\partial\Phi}{\partial y}=i\omega s+u_0\frac{\partial s}{\partial x}+v_0\frac{\partial s}{\partial y} \tag{6}$$

Eq. (4) (5) (6) represent the boundaries at porous side wall. The terms $\partial\Phi/\partial r$ and $\partial^2\Phi/\partial r^2$ in Eq. (1) can be expressed from Eq. (4) (5) (6), therefore, Eq (1) can be written as

$$c_0(\vec{W}_0\cdot\nabla)^2\Phi+(1-M_x^2)\frac{\partial^2\Phi}{\partial x^2}+(1-M_y^2)\frac{\partial^2\Phi}{\partial y^2}-2M_xM_y\frac{\partial^2\Phi}{\partial x\partial y}+c_4\frac{\partial\Phi}{\partial x}+c_5\frac{\partial\Phi}{\partial x}+c_6\Phi=0 \tag{7}$$

where

$$c_0=-\frac{1}{za_0 i\omega}[1+(1+M_y^2)(1-\bar{d})]/H$$

$$c_4=\frac{c_1}{a_0^2}+c_0[2i\omega u_0+\frac{k+1}{2}u_0\frac{\partial u_0}{\partial x}+\frac{k-1}{2}u_0\frac{\partial v_0}{\partial y}+v_0\frac{\partial v_0}{\partial x}]$$

$$c_5 = \frac{c_2}{a_0^2} + c_0[2i\omega v_0 + \frac{k+1}{2} v_0 \frac{\partial v_0}{\partial y} + \frac{k-1}{2} u_0 \frac{\partial v_0}{\partial x} + u_0 \frac{\partial u_0}{\partial y}]$$

$$c_6 = \frac{c_3}{a_0^2} + c_0[i\omega + \frac{k-1}{2}(\frac{\partial u_0}{\partial x} + \frac{\partial v_0}{\partial y})]i\omega$$

$$M_x = \frac{u_0}{a_0}$$

$$M_y = \frac{v_0}{a_0}$$

A code solving unsteaey flow field around an oscillating cascade of arbitrary camber and thicness profiles which oscillate harmonically has been picked up [6].

§ 4 Correlations of Acoustic Impedance of porous Wall with structural parameter

As shown Fig. 1, in general, the porous wall casing consists of porous wall plane and some chambers. The main structural parameters of porous wall casing are: thickness of porous wall b; diameter of holes of porous wall d; hole surface ratio σ, depth of chambers D. The specific acoustic impedance of porous wall casing can be expressed as

$$z = \frac{\tilde{P}}{\rho_0 a_0 \tilde{v}_m} = R^* + iX^* = |z| e^{i\arg z}$$

where

$$R^* = \frac{A}{\sigma} + \frac{1-\sigma^2}{\sigma}(\frac{\beta}{\sigma} + c) \qquad (8)$$

$$X^* = \frac{\omega b}{\sigma a_0}(1 + \frac{1}{\sqrt{3^2 + \frac{\beta^2}{2}}} + 0.85\frac{d}{D}) - cot(\frac{\omega D}{a_0}) \qquad (9)$$

and

$$A = \frac{32 \nu b}{a_0^2 d^2}(\sqrt{1 + \frac{\beta^2}{32}} + \frac{\sqrt{2}}{8}\beta\frac{d}{B})$$

$$\beta = (\frac{\omega}{\nu})^{\frac{1}{2}}(\frac{d}{2})$$

$$B = \frac{2|P|}{\rho_0 a_0^2[(R^* + \frac{1}{cos\varphi})^2 + X^{*2}]^{\frac{1}{2}}}$$

$$c = 0.3M$$

and ν the kinetic viscous coefficient of fluid; M the tangential Mach number along porous wall; φ the angle of incidence; |p| the absolute value of incident sound pressure. When a porous wall casing is designed, Eq. (8) and Eq. (9) can be used to determine two among four main structural parameters (b, d, σ, D), another two parameters

should be choiced in advance, while the acoustic environment (frequency ω, sound pressure amplitued |p| etc.,) and flow environment (tangential mach number M, kinetic viscous coefficient ν etc.,) and specific acoustic impedance (|z|, Argz, $\dot{R}$, $\dot{x}$, which are determined from unsteady aerodynamic calculation described at § 3) are the known numbers.

§ 5 Some Numerical Results

Some numerical results which showed the correlations of unsteady aerodynamics around an oscillating cascade with porous side wall have been presented at Fig. 2~4. From those figures, we can see that the different values of acoustic impedance of porous side wall can change the unsteady pressure distribution along blade surface, and the unsteady aerodynamic work doing on the blade during an oscillating period.

In order to study the effect of acoustic impedance on aerodynamic work, the acoustic impedance might be decomposed into its absolute value and its phase angle. The phase angle of acoustic impedance Argz determines the increase or decrease of aerodynamic work, and the absolute value of acoustic impedance |z| determines the value of aerodynamic work. Fig.5 corresponds to the first bending mode of blade vibration, and the interblade phase angle equals −45 ° (that is the case of the least aeroelastic stability of oscillating cascade). If we choise the phase angle of acoustic impedance Argz at −8 ° ~90 ° , then the use of porous side wall will decreasse the unsteady aerodynamic work, that means the improvement of aeroelastic stability. Fig. 6 corresponds to the first torsion mode with the interblade phase angle $\sigma=90°$. The aeroelastic stability of cascade will be improved if the Argz=−45 ° ~90 ° .

From Fig. 7 and Fig. 8 we can see that the absolute value of acoustic impedance is smaller, the effect of using porous wall is larger. However, the model of our method will be not correct while the absolute value of acoustic impedance decreases to certain limit.Fig.7 and Fig.8 correspond to the first torsion mode with $\sigma=90°$ and $\sigma=0°$ respectively.

In addition, the effect of acoustic impedance of porous side wall on aerdynamic work of oscillating cascade depends on the value of aerodynamic work of the same oscillating cascade but with solid side wall. The value of aerodynamic work of corresponding oscillating cascade is greater, the effect of acoustic impedance is greater, the effect of acoustic impedance of porous wall on aerodynamic work is larger.

§ 6 Symbols

a_0 − sound speed of steady mean flow.
Δcp(I) − imaging part of unsteady pressure coefficient at blade surface.
Δcp(R) − real part of unsteady pressure coefficient at blade surface.
$\bar{d}$− hub / tip diameter ratio.
H−height of blade
u_0 − x component of steady mean flow velocity.
v_0 − y component of steady mean flow velocity.
w_a − unteady aerodynamic work doing at blade during an oscillating period.
$\vec{w}_0$ − relative velocity vector of steady meanflow.
σ−specific hole surface; interblade phase angle.
ω− circular frequency.

§ 7 References

[1] Kurosaka, M., Journal of the Gas Turbime Society of Japan, Vol.8, No. 32, P.31, 1984 (in Japanese)

[2] Yamasaki, N. and Namba, M., Effect of Sound absorbing Wall on aerodynamic forces of a subsonic oscillating cascade, Transactions of Japan Socity of Mechanical Engineers, Ser.B, Vol.50, No–449, pp. 19–27. 1984 (in Japanese)

[3] Watanate, T. and Kaji, S., Possibility of cascade flutter suppresion by use of non–rigid duct walls, Proceedings of the Third International Symposium on Aeroelasticity in Turbomachines, pp. 261–276. 1984.

[4] Wei, X. L., and Zhou, S., The effect of 3D unsteady flow on aeroelastic stability of cascade of oscillating profiles, Paper presented at second Chinese National Computational Fluid Mechanics Conference, 1984.

[5] Duan, S. Z., A Numerical Method Solving 2D Unsteady Flow Fiefd around cascade of Oscillating Airfoils with Arbitrary Camber and Thickness, M. Sc. Thesis Beijing Institute of Aero. Astron. 1986.

[6] Chen, R.G., A Reseavch for the Blade Flutter Suppression Effect of an Acoustical Casing, M. Sc. Thesis, Beijing University of Aero Astron 1988.

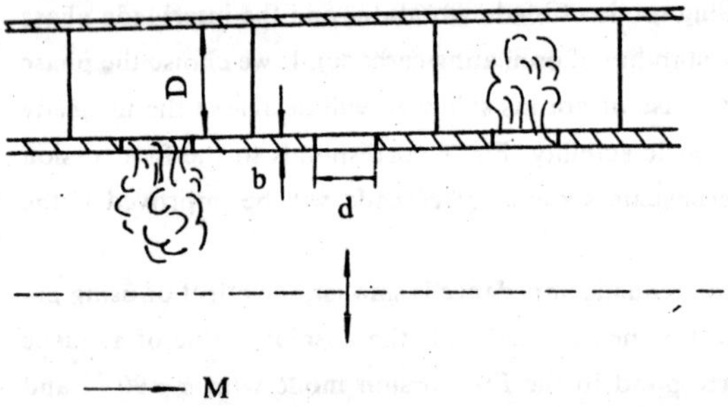

Fig. 1

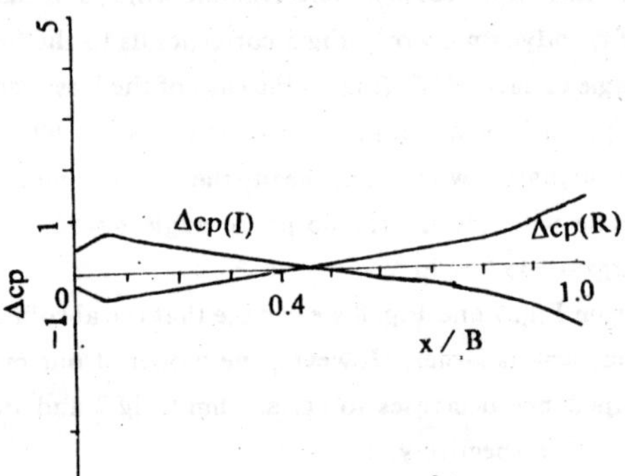

z = 0.04 (90 °)

Fig. 2

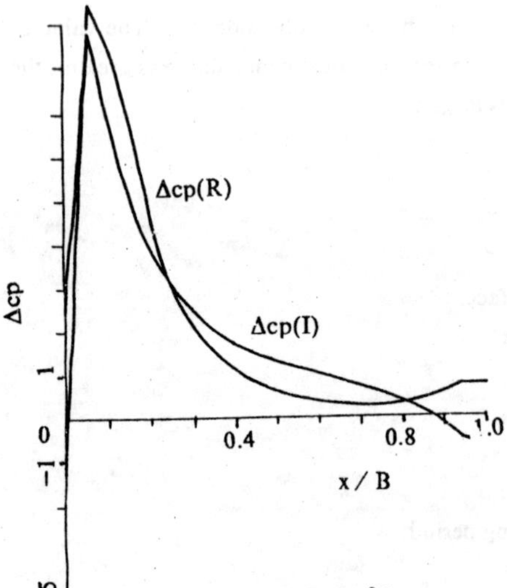

z = 1.0 (90 °)

Fig. 3

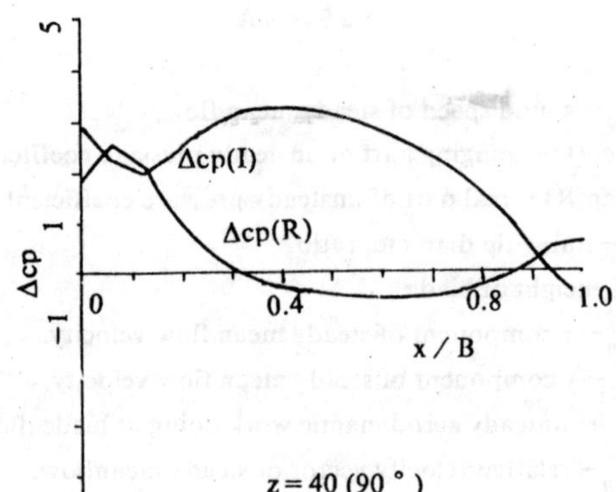

z = 40 (90 °)

Fig. 4

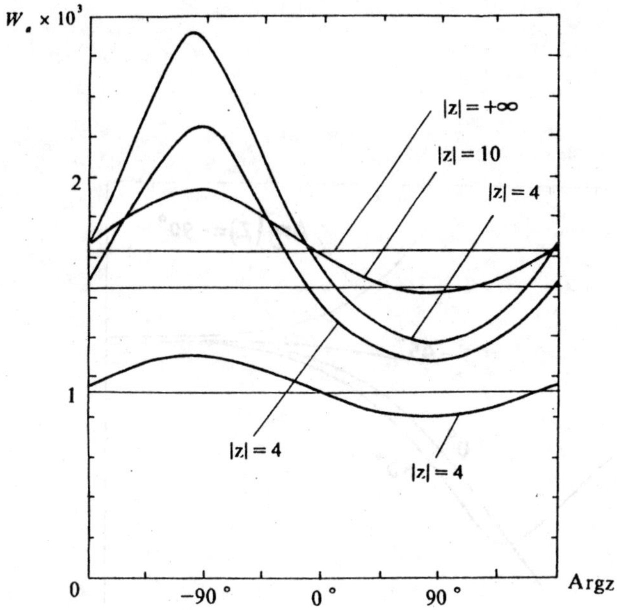

$W_a \times 10^3$
|z| = +∞
|z| = 10
|z| = 4
|z| = 4
|z| = 4
2
1
0
-90 °
0 °
90 °
Argz

Fig. 5

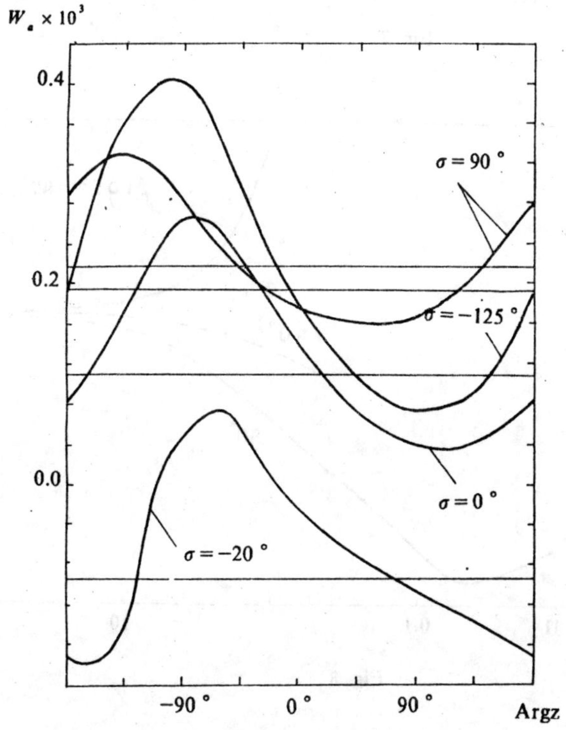

$W_a \times 10^3$
0.4
0.2
0.0
σ = 90 °
σ = −125 °
σ = 0 °
σ = −20 °
-90 °
0 °
90 °
Argz

Fig. 6

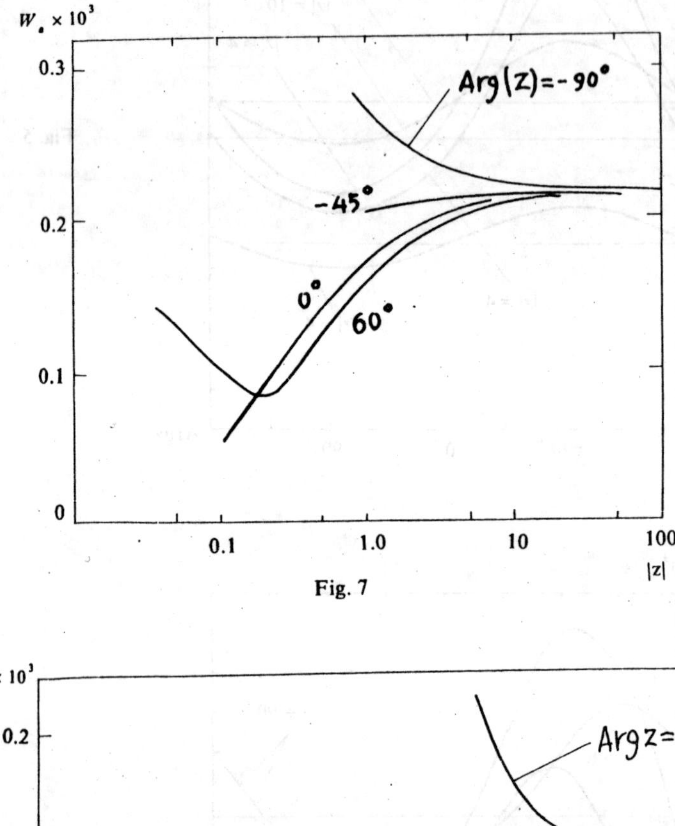

$W_e \times 10^3$
0.3
0.2
0.1
0
0.1
1.0
10
100
|z|
Arg(Z)=-90°
-45°
0°
60°

Fig. 7

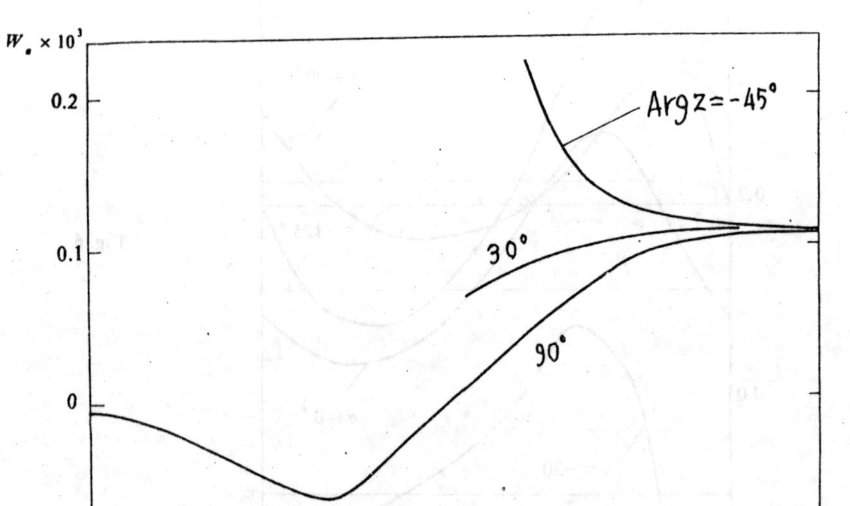

$W_e \times 10^3$
0.2
0.1
0
0.01
0.1
10
|z|
Arg z=-45°
30°
90°

Fig. 8

Part Two

Unsteady Aerodynamics: Experiment

EXPERIMENTAL INVESTIGATION OF OSCILLATING CASCADE AERODYNAMICS

Daniel H. Buffum
(National Aeronautics and Space Administration
Lewis Research Center, Cleveland, Ohio 44135)

Sanford Fleeter
(Thermal Sciences and Propulsion Center
School of Mechanical Engineering, Purdue University
West Lafayette, Indiana 47907)

ABSTRACT

A series of experiments are performed in the NASA Lewis Transonic Oscillating Cascade Facility to provide fundamental data quantifying the high subsonic and transonic steady and oscillating aerodynamics of a biconvex airfoil cascade at realistic reduced frequency values for all interblade phase angles. This is accomplished by developing and utilizing an unsteady aerodynamic influence coefficient technique in which only one cascaded airfoil is oscillated at a time. The vector summation of the resulting airfoil surface unsteady pressures measured on a dynamically instrumented airfoil enables the unsteady aerodynamics of an equivalent cascade with all airfoils oscillating at any specified interblade phase angle to be determined.

NOMENCLATURE

C	airfoil chord
$\underline{C}_p$	unsteady pressure coefficient, $p_1/(\frac{1}{2}\rho V^2 \alpha_1)$
$\bar{C}_p$	steady pressure coefficient, $(p_{in}-p_o)/(\frac{1}{2}\rho V^2)$
C_p^n	unsteady pressure influence coefficient for n^{th} airfoil
k	reduced frequency, $\omega C/2V$
M	inlet Mach number
p_{in}	inlet static pressure
p_j	j^{th} harmonic of airfoil surface static pressure
S	airfoil spacing
V	inlet velocity
x	chordwise coordinate
y	coordinate normal to chordwise direction
α_o	mean flow incidence angle
α_1	torsional oscillation amplitude
β	interblade phase angle (positive when airfoil n leads airfoil n-1)
γ	stagger angle
ΔC_p	unsteady pressure difference coefficient
ρ	inlet density
ω	airfoil oscillation frequency

INTRODUCTION

The ability to predict flutter has not kept pace with recent advances in the design of advanced turbomachines and propellers. As a result, the development of advanced unsteady aerodynamic cascade models is of current research interest. However, to both direct the development of these models as well as to evaluate these and existing models, fundamental oscillating cascade data are required. Unfortunately, relatively few experimental results are available, particularly at realistic high values of the reduced frequency in the high subsonic and transonic flow regimes. This is due to the inherent complexity of the experiments. Namely, measurements must be obtained not only for each steady flow condition and reduced frequency value, but also over a range of interblade phase angles. Because these experiments are so time consuming, results are obtained only for a very limited number of specified interblade phase angle values.

Oscillating cascade data may be obtained for all interblade phase angle values through simpler experiments. In particular, when the unsteady disturbances are small, as in flutter stability problems, an unsteady influence coefficient technique can be utilized. In this technique, only one airfoil in the cascade is oscillated, with the resulting airfoil surface unsteady pressure distributions measured on the oscillating airfoil and its stationary neighbors. The unsteady aerodynamics of an equivalent cascade with all airfoils oscillating at any specified interblade phase angle value is then determined through a vector summation of these influence coefficient data.

Several investigations have been directed at validation of this technique through correlation of unsteady aerodynamic influence coefficient results with corresponding oscillating cascade data acquired with all airfoils oscillating at specified interblade phase angles. Hanamura, Tanaka and Yamaguchi obtained good results for such experiments in a water channel [1]. Davies and Whitehead [2] performed experiments in an annular cascade at high subsonic inlet conditions and reduced frequencies up to 0.1, but the measurements were limited to unsteady aerodynamic moment coefficients. The effect of oscillating both an isolated airfoil and three cascaded airfoils at low speed was studied by Tanaka, Yamamoto and Fujimoto [3]. Although a summation of the influence coefficients was not presented, the technique appears promising in regions of attached flow. For a subsonic compressible flow field, Buffum and Fleeter [4,5] found good correlation of influence coefficient airfoil surface unsteady pressure distributions and corresponding data acquired with all airfoils oscillating at specific interblade phase angle values. In supersonic inlet Mach number experiments at ONERA [6], the summation of influence coefficients has been compared to data for a linear cascade with two airfoils oscillating.

In this paper, the steady and unsteady aerodynamics of a cascade of biconvex airfoils executing torsion mode oscillations are investigated for two incidence angles at high subsonic and transonic Mach numbers at realistic reduced frequency values. This is accomplished in the NASA Lewis Transonic Oscillating Cascade Facility by developing and utilizing an unsteady aerodynamic influence experimental technique. Detailed steady airfoil surface pressure distributions quantify the mean flow field. Unsteady airfoil surface pressure influence coefficients are measured for one oscillating airfoil in the cascade. These influence coefficients are then summed vectorally for correlation with data obtained with all airfoils oscillating at specified interblade phase angle values. The associated unsteady pressure difference data are also correlated with the predictions of a linearized subsonic oscillating cascade analysis.

INFLUENCE COEFFICIENT TECHNIQUE

Figure 1 depicts the two-dimensional finite thin airfoil cascade representation of a rotor blade row. For a given mean flow field and reduced frequency, the assumption of small unsteady disturbances allows the cascade unsteady aerodynamics to be expressed as linearly combined influence coefficients which can be determined both experimentally and analytically.

Consider a finite airfoil cascade with 2N+1 airfoils executing constant amplitude harmonic oscillations with a constant interblade phase angle β. The unsteady airfoil surface pressure, expressed as a pressure coefficient $C_p(x)$ acting at a point on the reference airfoil (Airfoil 0 in Figure 1), can be expressed as a Fourier series, Equation 1. Thus these influence coefficients define the unsteady pressure coefficient developed on the reference airfoil due to the motion of airfoil n with all of the other airfoils stationary.

$$C_p(x, \beta) = \sum_{n=-N}^{N} C_p^n(x)e^{in\beta} \quad (1)$$

where C_p^n are the complex unsteady aerodynamic pressure influence coefficients.

OSCILLATING CASCADE FACILITY

The NASA Lewis Transonic Oscillating Cascade Facility, Figure 2, combines a linear cascade wind tunnel capable of test section inlet Mach numbers approaching unity with a high speed mechanical drive system which imparts controlled torsional oscillations to any or all of the airfoils. Air drawn from the atmosphere passes through a smoothly contracting inlet section, into a constant area 9.78 by 19.21 cm test section and then through a diffuser and exhaust header. The flow rate is controlled by two valves located in the header. Upstream of the test section, a partitioned bleed system applies suction to the end wall boundary layers, thus reducing end wall viscous effects. Suction is also applied through tailboard slots to reduce effects due to the upper and lower cascade walls.

To obtain realistic values of the reduced frequency, the mechanical drive system must provide high frequency controlled oscillations of the airfoils. Nine barrel cams, each with a six cycle sinusoidal groove machined into its periphery, are mounted on a common rotating shaft driven by a 74.6 kW electric motor. Connecting arms, joined to one end of each airfoil by trunnions, have buttons on the opposite end to follow the camshaft. The amplitude of the airfoil motion is 1.2 degrees, dictated by the cam and follower geometry. When all of the airfoils are oscillated, the various specified constant interblade phase angle airfoil oscillations are achieved by rotating the cams to new relative positions. In this investigation, the maximum airfoil oscillatory frequency is 350 Hz, corresponding to a reduced frequency of 0.39 with a 0.65 cascade inlet Mach number and 0.323 at Mach 0.80.

AIRFOILS AND INSTRUMENTATION

The cascade is comprised of nine uncambered biconvex airfoils with a chord of 7.62 cm, a thickness-to-chord ratio of 0.076, a solidity of 1.3 and a 53 degree stagger angle. The radius of curvature of both airfoil surfaces is 27.4 cm, with the leading and trailing edges rounded with a 0.025 cm radius of curvature. The airfoils are supported by two midchord trunnions, resulting in a midchord elastic axis location.

Steady airfoil surface pressure distributions are measured with conventional static pressure taps. There ae sixteen chordwise measurement locations, with four additional taps to assure the spanwise uniformity of the pressure distribution. Rows of sidewall static pressure taps located upstream and downstream of the cascaded airfoils, Figure 2, are used to determine the mean inlet and exit static pressures.

The primary dynamic data quantify the complex unsteady surface pressures on the oscillating cascaded airfoils. These data are obtained via six Kulite dynamic pressure transducers flush mounted symmetrically about the midchord of one surface of the airfoil, Table 1. These transducers, having an active sensor diameter of 0.097 cm (1.3% of the airfoil chord), are placed in milled slots and potted in RTV for isolation from airfoil strain. A thin coating of RTV protects the sensor surface and fairs it into the surface contour of the airfoil.

The time-variant position of the reference oscillating airfoil is determined by a capacitance-type proximity sensor which produces a voltage proportional to the air gap between it and an adjacent object. This sensor is positioned to face a six cycle sinusoidally-shaped cam mounted on the airfoil drive camshaft so as to be in phase with the reference airfoil motion.

DATA ACQUISITION AND ANALYSIS

Conventional instrumentation is used to quantify the steady flow field. An average of the upstream sidewall static pressures along with the atmospheric pressure are used to calculate the inlet Mach number. Steady flow airfoil surface static pressures are calculated from an average of approximately 100 samples.

Unsteady signals are a.c. coupled and recorded on magnetic tape for post-experiment processing. During tape playback, the signals are simultaneously digitized at rates sufficient to capture at least three harmonics of the oscillation frequency, with 32,768 samples taken per channel. Each data channel is divided into blocks, typically with 4096 samples, and then

Fourier decomposed and referenced to the airfoil motion by subtracting the motion phase from the unsteady pressure phase. With all of the transducer signal blocks decomposed, the results are averaged and, for the influence coefficient data, the transducer acceleration calibration responses subtracted vectorally. To minimize errors due to spectral leakage, an interpolation scheme is applied to the decomposed results in conjunction with a Hanning window [7].

Final unsteady pressure data are defined by the complex dynamic pressure and pressure difference coefficients, C_p and ΔC_p, Equation 2. Dynamic pressure coefficients for any interblade phase angle are determined by summing influence coefficients per Equation 1.

$$C_p(x) = \frac{p_1(x)}{1/2\rho V^2 \alpha_1} \; ; \; \Delta C_p = C_{pl} - C_{pu} \qquad (2)$$

where p_1 is the first harmonic of the unsteady static pressure, α_1 is the torsional oscillation amplitude and C_{pl} and C_{pu} are the lower and upper airfoil surface unsteady pressure coefficients.

RESULTS

The oscillating cascade aerodynamics are experimentally investigated at flow conditions defined by incidence angles of 0 and 7 degrees with an inlet Mach number of 0.65 and 7 degrees incidence with an inlet Mach number of 0.80. With the instrumented reference airfoil in Position 0 as defined in Figure 1, influence coefficient data are acquired on this airfoil with the airfoils in Positions -2, -1, 0, +1 and +2 individually oscillating at reduced frequencies of 0.390 for M = 0.65 and 0.323 for M = 0.8.

The steady airfoil surface pressure distributions are presented first, then the unsteady pressure influence coefficients on the individual surfaces of the reference instrumented airfoil are considered. These influence coefficient data are then summed to predict the unsteady aerodynamics of an equivalent cascade with all airfoils oscillating at a fixed interblade phase angle value. For several different interblade phase angles, the resulting unsteady pressure distributions are correlated with: (1) baseline data obtained in experiments in which all of the airfoils were simultaneously oscillating at interblade phase angles of -90, 0 and +90 degrees; (2) predictions from the classical unsteady, small perturbation, subsonic, zero mean incidence flat plate analyses of references 8 and 9. Additional interblade phase angle results are obtained utilizing the influence coefficient technique and correlated with the flat plate cascade predictions. In these experiments, the airfoil motion is defined by the change in the incidence angle with time:

$$\alpha(t) = \alpha_o + \alpha_1 \mathrm{Re}\{e^{(i\omega t)}\} \qquad (3)$$

where α_o is the mean steady incidence angle, α_1 is the oscillatory amplitude of 1.2 degrees and ω is the frequency.

Steady State Aerodynamics

To demonstrate periodicity at the steady state conditions, airfoil surface pressure distributions were obtained for multiple passages in the cascade. For example, Figure 3 presents data at M=0.8 for the four cascade passages surrounding the center airfoil. The Passage 1 data are the pressure distributions on the upper surface of Airfoil 0 (the center airfoil) and the lower surface of Airfoil 1; Passage -1 data are for the lower surface of Airfoil 0 and the upper surface of Airfoil -1, etc. The good cascade periodicity is readily apparent. Analogous results were obtained at M=0.65.

Figures 4 and 5 present the steady flow airfoil surface pressure coefficient distributions on the center airfoil with the cascade mean flow incidence angle set at 0 and 7 degrees. The zero pressure difference conditions at the trailing edge is satisfied. Also, the pressure coefficient distributions are nearly identical for the two Mach numbers, with loading only on the forward half. At Mach 0.8, there is a small region of supersonic flow on the upper surface near the leading edge which is terminated by a shock, as determined from schlieren images.

UNSTEADY PRESSURE INFLUENCE COEFFICIENTS

Figures 6, 7 and 8 present the chordwise distribution of the unsteady pressure influence coefficients on the individual surfaces of the Position 0 reference airfoil, with the oscillating airfoil in the three relative positions defined by -1 through +1 at incidence and an inlet Mach number of 0.80.

The self-induced oscillating airfoil unsteady pressure response is shown in Figure 6. Namely, this figure presents the unsteady pressures on the surfaces of the reference airfoil, with the reference airfoil itself oscillating. The magnitude of the unsteady pressure on each surface attains a maximum at the leading edge, tending towards zero at the trailing edge, with the magnitude generally larger on the upper surfae. Also, the upper surface phase data are out-of-phase with the airfoil motion.

Figures 7 and 8 show the unsteady pressure effect on the surfaces of the stationary instrumented reference airfoil due to individually oscillating its nearest neighbors, i.e., individually oscillating the airfoils in Positions +1 and -1, respectively. As seen, oscillating the neighboring airfoil has a relatively large effect on the magnitude of the unsteady pressure on the reference airfoil surface nearest to the oscillating airfoil. In particular, Figure 6 shows that oscillating the airfoil located immediately above the reference airfoil results in relatively large unsteady pressure fluctuations over the aft half of the reference airfoil upper surface, with the lower surface pressure coefficient magnitude nearly constant and a nearly linear variation in the lower surface phase angle. With the oscillating airfoil positioned immediately below the reference airfoil, Figure 8 shows that there are relatively large pressure fluctuations over the leading quarter of the reference airfoil lower surface, while the upper surface has only a small response in the leading and trailing edge regions, with the lower surface oscillations out-of-phase with the motion.

The effect of oscillating the airfoils which are in the +2 and -2 positions, i.e., the oscillating airfoils are further distanced from the reference airfoil, is to further reduce the magnitude of the unsteady pressures on the reference airfoil as compared to the previous results where the nearest neighbors were oscillated.

Unsteady Pressure Differences

Summation of the unsteady pressure influence coefficients to determine the unsteady pressure difference coefficient is demonstrated in Figure 9. The 0.65 inlet Mach number, 7° incidence, 0.39 reduced frequency data are presented as a dynamic pressure difference coefficient for an interblade phase angle of 0 degrees, with N specifying the limits of the sum per Equation 1. Thus N=0 corresponds to the self-induced unsteady aerodynamic response. The influence coefficient series is rapidly convergent, with only the reference airfoil and its two immediate neighbors having a significant effect on the resulting dynamic pressure differences.

The influence coefficient determined unsteady airfoil surface pressure difference data are correlated with: (1) corresponding data obtained in experiments in which all of the cascaded airfoils are oscillating simultaneously with constant interblade phase angle values of -90, 0 and +90 degrees, and (2) the classical thin airfoil cascade predictions. The results for an inlet Mach number of 0.65 and a reduced frequency of 0.39 are presented in Figueres 10, 11, and 12 for 0 degrees incidence and in Figures 13, 14 and 15 for 7 degrees of incidence. For an inlet Mach number of 0.80 and a reduced frequency of 0.323, the unsteady pressure difference results are presented in Figures 16, 17, and 18.

With regard to the unsteady pressure difference magnitude data, at zero incidence and an inlet Mach number of 0.65, the complex unsteady aerodynamic influence coefficient magitude data generally exhibit good correlation with both the oscillating cascade data and the linearized theory. In contrast, at 7 degrees of incidence with both 0.65 and 0.80 inlet Mach number values, the linearized theory tends to overpredict the magnitude data. However, for these two cases, there is generally very good agreement between the two sets of experimental magnitude data, with the exception of the leading edge region where the magnitudes of the influence coefficient data are significantly larger than the others. At the lower subsonic Mach number, this is due to a small oscillating separation bubble near the leading edge of the upper surface during the influence coefficient experiments. This separation was first indicated by significant higher harmonic content in the leading upper surface pressure transducer signal, with subsequent steady state flow visualization with Airfoil 0 set at 8.2 degrees of incidence indicating a small separated flow region near the leading edge. At Mach 0.80, the oscillating leading edge shock is the cause of the leading edge data variations.

With regard to the unsteady pressure difference phase data, at 0 incidence with an inlet Mach number of 0.65, the phase data are in good agreement with the prediction at an interblade phase angle of -90 degrees. However, for in-phase motions, the phase data are decreased relative to the prediction, with the influence coefficient data centered between the prediction and the oscillating cascade data at 40% and 60% of chord. At an interblade phase angle of +90 degrees, the influence coefficient phase data are in good agreement with the prediction, with the oscillating cascade data increased in value.

At an incidence angle of 7 degrees at inlet Mach numbers of 0.65 and 0.80, the -90 degree interblade phase angle influence coefficient data are in good agreement with the all airfoil oscillating data and also the predictions. For in-phase motions, the correlation of these 7 degree incidence phase angle data is good at a Mach number of 0.65, with the Mach 0.80 data exhibiting scatter. For a +90 degree interblade phase angle of 0.65 Mach number, the phase data fall within a fairly narrow band, with the influence coefficient data typically between the prediction and the all airfoil oscillating data. This trend is also evident at a Mach number of 0.80, although there is now more offset among the various results.

SUMMARY AND CONCLUSIONS

Fundamental experiments have been performed in the NASA Lewis Transonic Oscillating Cascade Facility to investigate the steady and torsion mode oscillating aerodynamics of a biconvex airfoil cascade. For subsonic and transonic mean flow conditions and realistic high values of reduced frequency, an unsteady aerodynamic influence coefficient technique was developed and utilized in which only one airfoil in the cascade is oscillated at a time and the resulting unsteady pressures measured on the oscillating airfoil and its stationary neighbors. The unsteady aerodynamics of an equivalent cascade with all airfoils oscillating for arbitrary interblade phase angles is then determined by means of a vector summation of these data.

These unsteady aerodynamic influence coefficient experiments were performed at an inlet Mach number of 0.65 with a reduced frequency of 0.39 at both 0 and 7 degrees of incidence as well as at an inlet Mach number of 0.80 with a reduced frequency of 0.323 at 7 degrees of incidence. The analysis of these unique data and correlation with both the predictions from the unsteady, small perturbation, subsonic flat plate cascade analyses and baseline data obtained in experiments where all of the airfoils are oscillating simultaneously revealed the following.

* The unsteady aerodynamic influence coefficient series is rapidly convergent, with only the reference airfoil and its two neighbors having a significant effect on the resulting unsteady pressure difference.

* The unsteady pressure difference magnitude data determined from the influence coefficient experiments generally exhibit good correlation with both the oscillating cascade data and the linearized theory, the primary exception being in the vicinity of the leading edge at 7 degrees of incidence at both 0.65 and 0.8 Mach numbers.

* The correlation of the unsteady pressure difference phase data determined from the influence coefficient experiments with the oscillating cascade data and the predictions were generally good, but were a function of the interblade phase angle as well as the mean flow conditions.

REFERENCES

1. Hanamura, Y., Tanaka, H. and Yamaguchi, K., "A Simplified Method to Measure Unsteady Forces Acting on the Vibrating Blades in Casade," *Bulletin of the JSME*, Vol. 23, No. 180, June 1980, pp. 880-887.

2. Davies, M.R.D. and Whitehead, D.C., "Unsteady Aerodynamic Measurements in a Transonic Annular Cascade," *Unsteady Aerodynamics of Turbomachines and Propellers*, Cambridge University, Cambridge, England, September, 1984, pp. 487-502.

3. Tanaka, H., Yamamoto, K. and Fujimoto, I., "Unsteady Aerodynamic Response of Cascade Blades in Pitching Oscillation with Flow Separation," *Unsteady Aerodynamics of Turbomachines and Propellers,* Cambridge University, Cambridge, England, September, 1984, pp. 397-410.

4. Buffum, D.H., and Fleeter, S., "Investigation of Oscillating Cascade Aerodynamics by an Experimental Influence Coefficient Technique," *NASA Technical Memorandum 101313, AIAA Paper 88-2815*, July 1988.

5. Buffum, D.H., and Fleeter, S., "Experimental Investigation of Transonic Oscillating Cascade Aerodynamics", *NASA Technical Memorandum 101993, AIAA Paper 89-0321*, January 1989.

6. Szechenyi, E., "Fan Blade Flutter-Single Blade Instability or Blade to Blade Coupling?", *ASME Paper 85-GT-216*, March 1985.

7. Burgess, J.C., "On Digital Spectrum Analysis of Periodic Signals," *Journal of the Acoustical Society of America*, Vol. 58, No. 3, September 1975, pp. 556-567.

8. Smith, S.N., "Discrete Frequency Sound Generation in Axial Flow Turbomachines," *Cambridge University Report CUED/A-Turbo/TR29*, 1971.

9. Fleeter, S., "Fluctuating Lift and Moment Coefficients for Cascaded Airfoils in a Nonuniform Compressible Flow," *AIAA Journal of Aircraft*, Vol. 10, No. 2, February 1973, pp. 93-98.

AIRFOIL	
Type	biconvex, no camber
Surface radius of curvature	27.4 cm
Leading and trailing edge radii of curvature	0.025 cm
Maximum airfoil thickness	0.58 cm
Chord, C	7.62 cm
Elastic axis	midchord
Dynamic pressure transducer locations, % chord	12, 25, 40, 60, 75, 88
CASCADE	
Number of airfoils	9
Solidity, C/S	1.3
Spacing, S	5.86 cm
Stagger angle, γ	53 degrees
Amplitude of motion	1.2 degrees

Table 1. Airfoil and cascade geometry

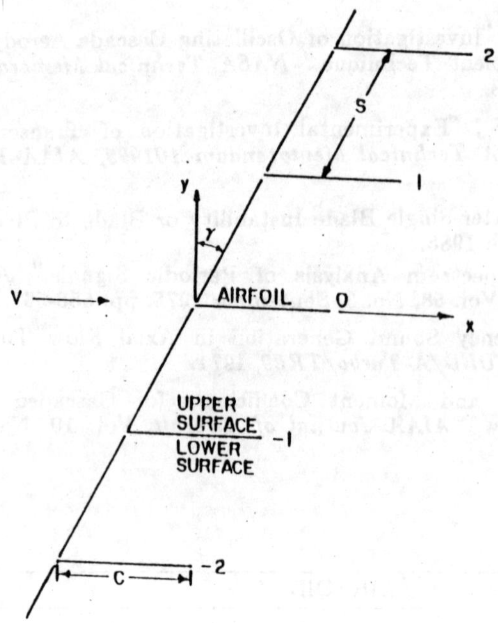

Figure 1. Cascade geometry.

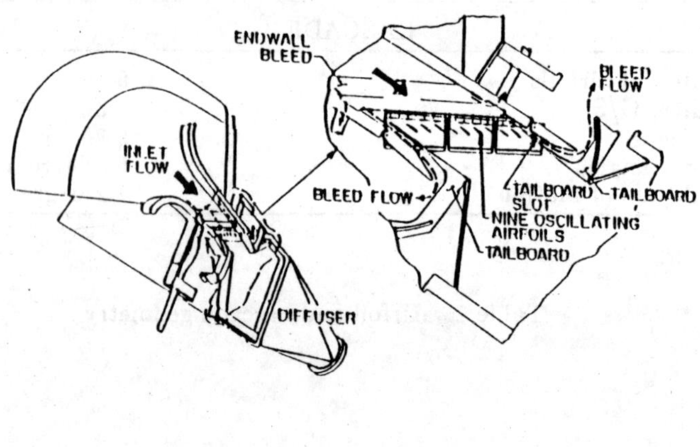

Figure 2. NASA Lewis Transonic Oscillating Cascade Facility.

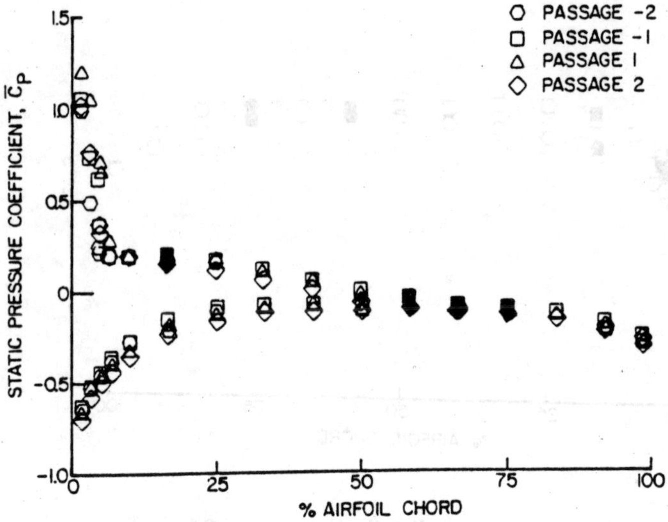

Figure 3. Periodicity of steady flow airfoil surface pressure distribution, M=0.8, $\alpha_o = 7°$.

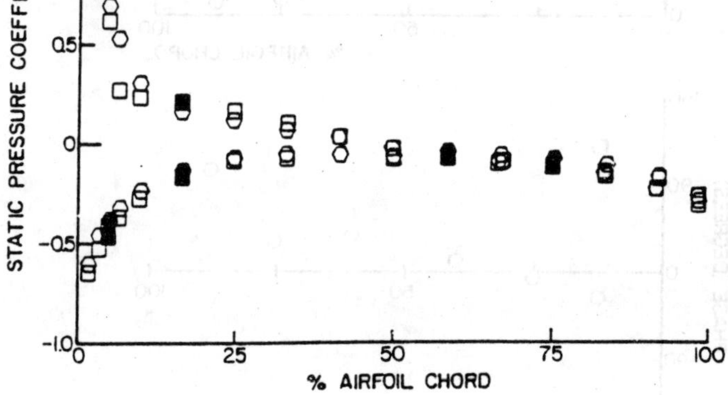

Figure 4. Steady flow airfoil surface pressure distributions, $\alpha_o = 7°$.

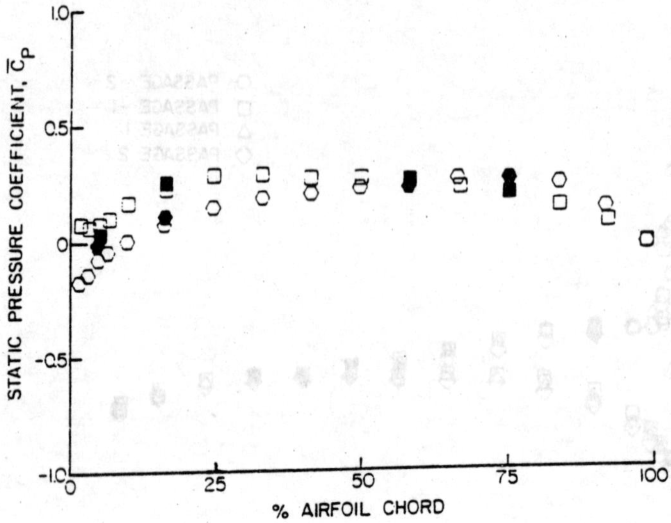

Figure 5. **Steady pressure distribution, $\alpha_o = 0°$.**

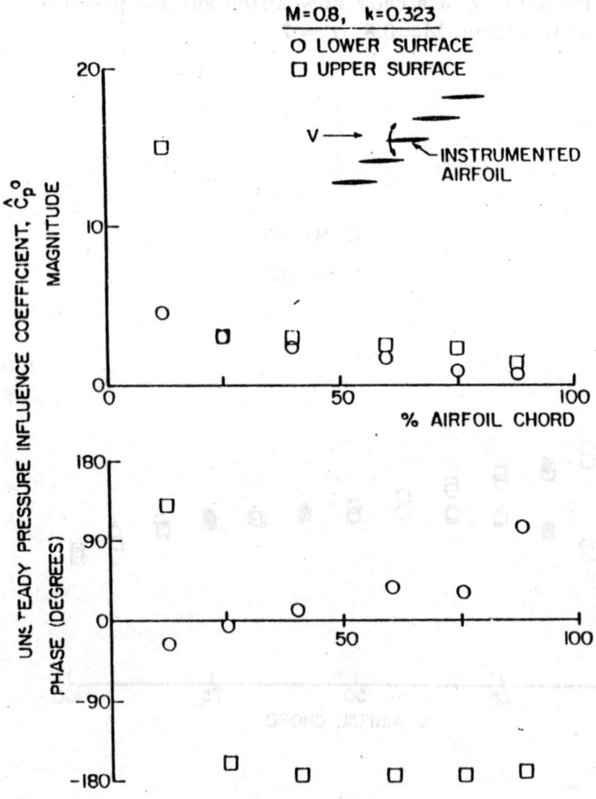

Figure 6. **Unsteady pressure influence coefficient, airfoil 0 oscillating.**

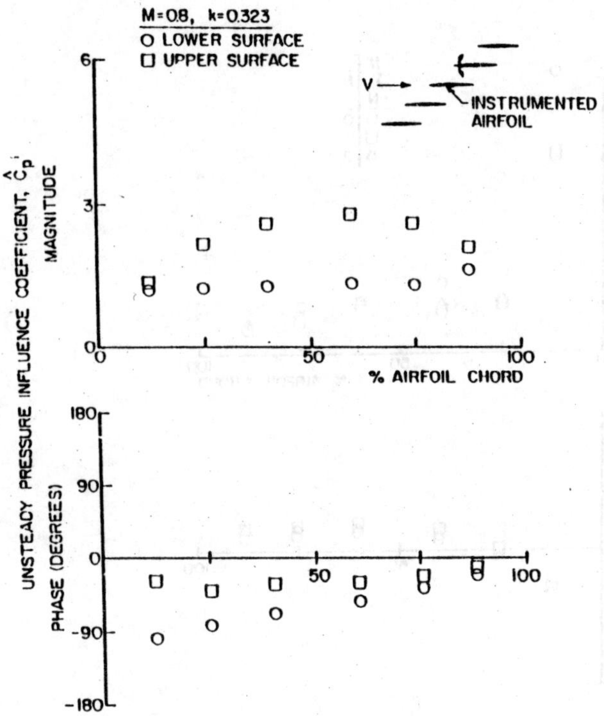

Figure 7. **Unsteady pressure influence coefficient, Airfoil 1 oscillating.**

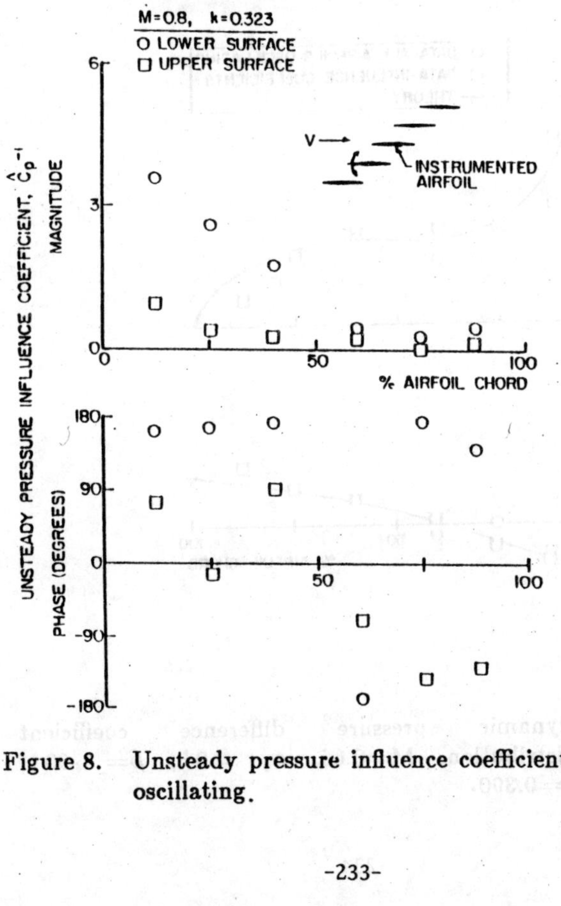

Figure 8. **Unsteady pressure influence coefficient, Airfoil -1 oscillating.**

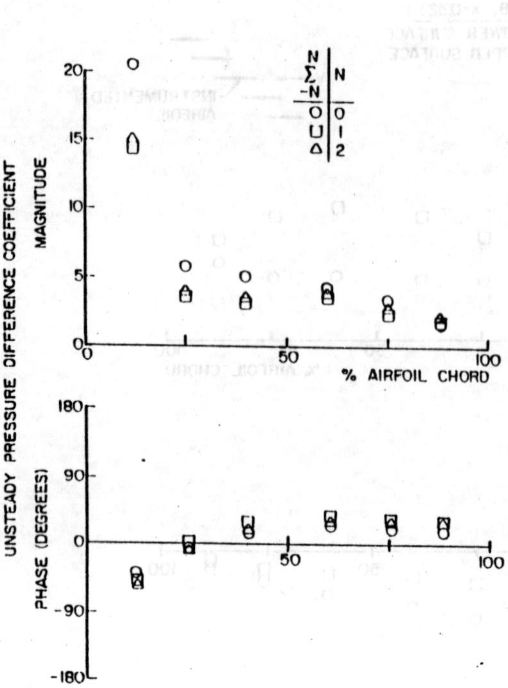

Figure 9. Summation of unsteady pressure influence coefficients, M=0.65, α_o =7°, k=0.39, β=0°.

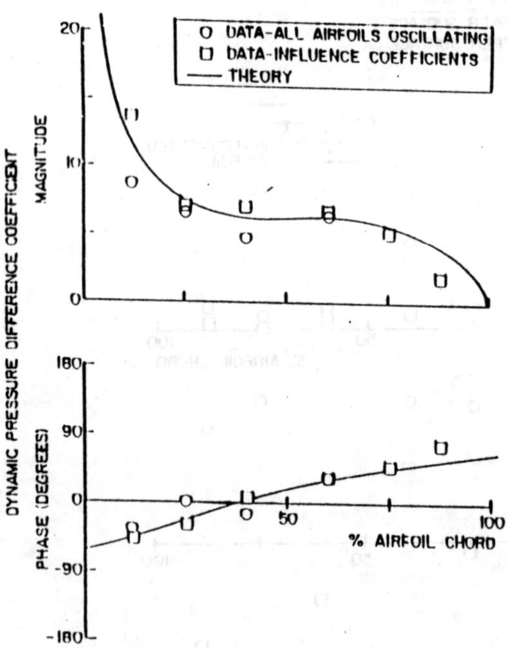

Figure 10. Dynamic pressure difference coefficient distribution, M=0.65, α_o =0°, β= -90°, k=0.390.

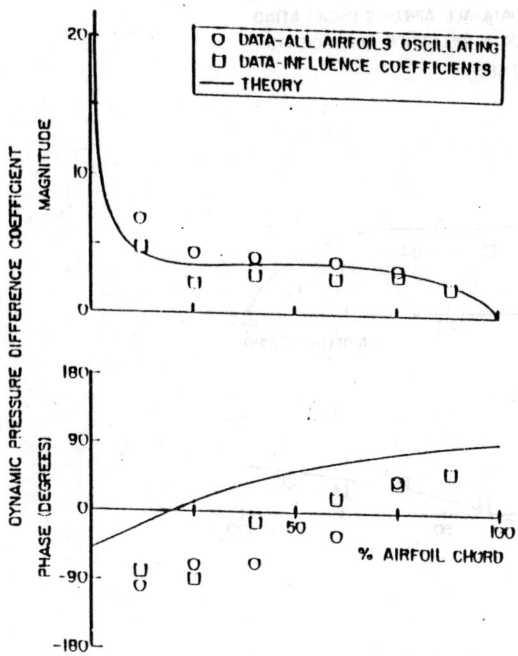

Figure 11. Dynamic pressure difference coefficient distribution, M=0.65, $\alpha_o=0^\circ$, $\beta=0^\circ$, k=0.390.

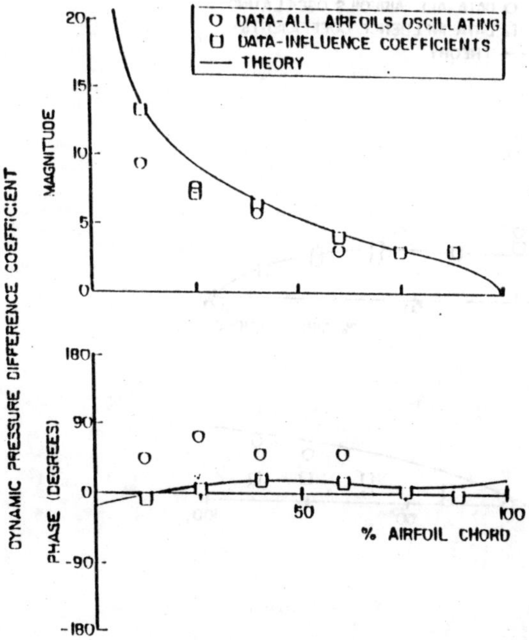

Figure 12. Dynamic pressure difference coefficient distribution, M=0.65, $\alpha_o=0^\circ$, $\beta=90^\circ$, k=0.390.

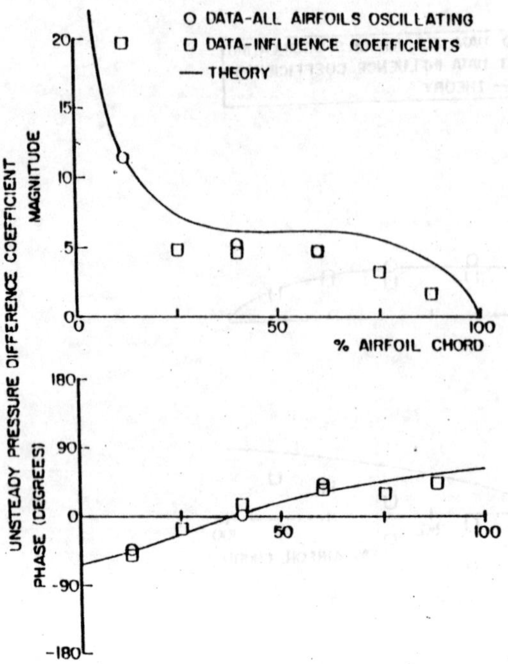

Figure 13. Unsteady pressure difference coefficient distribution, M=0.65, $\alpha_o=7°$, k=0.39, $\beta=-90°$.

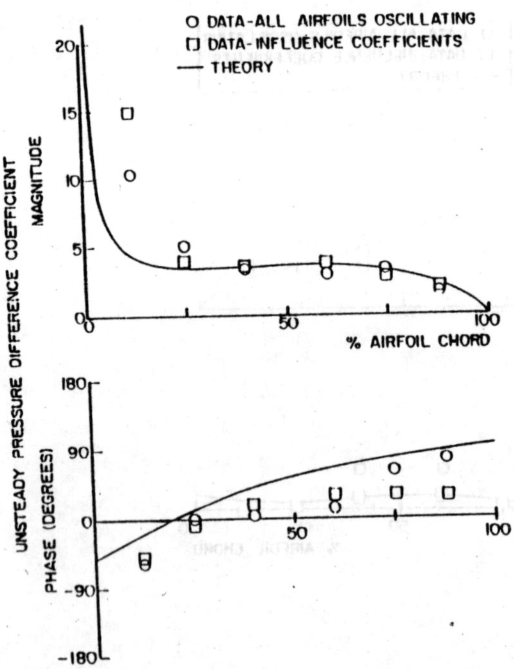

Figure 14. Unsteady pressure difference coefficient distribution, M=0.65, $\alpha_o=7°$, k=0.39, $\beta=0°$.

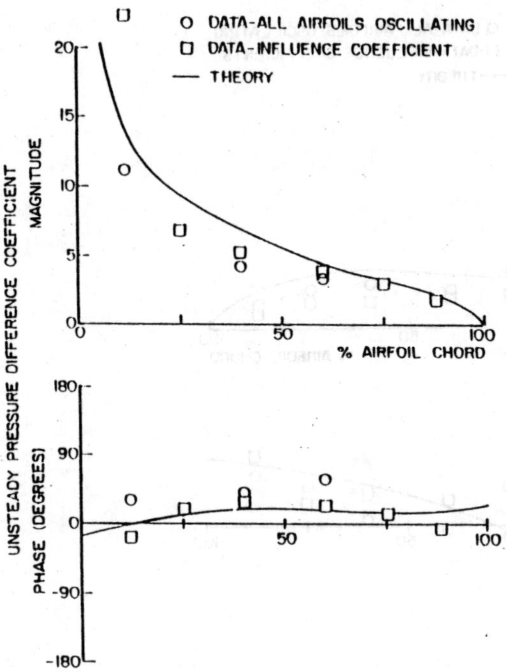

Figure 15. **Unsteady pressure difference coefficient distribution, M=0.65, α_o=7°, k=0.39, β=90°.**

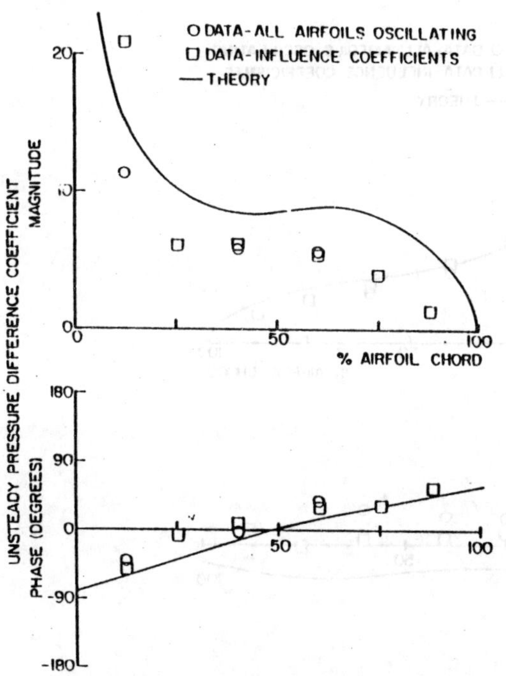

Figure 16. **Unsteady pressure difference coefficient distribution, M=0.80, α_o=7°, k=0.323, β= 90°.**

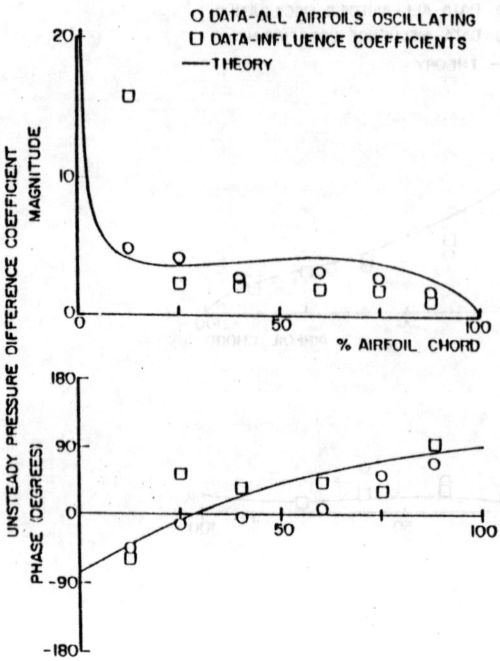

Figure 17. Unsteady pressure difference coefficient distribution, M=0.80, α_o=7°, k=0.323, β=0°.

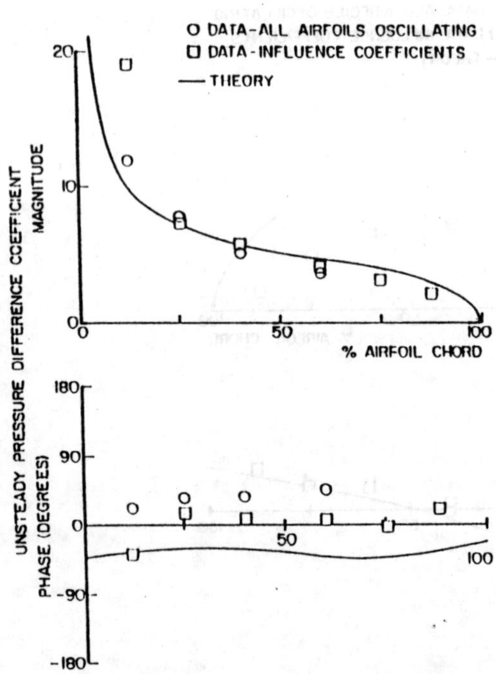

Figure 18. Unsteady pressure difference coefficient distribution, M=0.80, α_o=7°, k=0.323, β=90°.

UNSTEADY WAKE-BOUNDARY LAYER INTERACTION IN A COMPRESSOR CASCADE

J.S.Addison and Y.Dong*.
(Whittle Laboratory, Cambridge University, England)

SUMMARY

The effect of wakes on the development of compressor blade boundary layers is investigated in a low Reynolds number linear cascade with a rotating bar mechanism. Hot wire traverses are used to show how the boundary layer undergoes unsteady transition, laminar separation and turbulent reattachment.

Laminar separation exists between the wake passings but is intermittently and rapidly suppressed by the wake turbulence. Recovery to the steady state occurs more slowly with a timescale of the same order as the wake passing period and is not quite complete before the arrival of the next wake. Boundary layer transition in the separated shear layer occurs through the formation of Tollmein-Schlichting waves whereas the transition initiated by wake turbulence bypasses this stage.

INTRODUCTION

Conventionally, turbomachinery blading design has been based on cascade tests. Implicit in this is the assumption that the flow into the bladerows is steady. However, in a real application, the proximity of other bladerows means that this is clearly not the case and the effect of the unsteadiness needs to be ascertained. Leaving aside the effects of shock waves in supersonic machines and secondary flows in machines of low aspect ratio, the principal forms of unsteadiness are the potential and wake interactions. The former acts both upstream and downstream of a bladerow and comprises a periodic pressure fluctuation which decays exponentially away from the blades with a length scale roughly equal to the pitch or chord. On the other hand, the wakes shed by a bladerow act only on a downstream bladerow and decay much more slowly. With the inter-row spacings of most practical applications (typically between 20% and 50% of an axial chord in aeroengines), the wake effect will dominate in downstream rows, and it is this effect which is the subject of the work reported here.

The wake from a blade comprises a velocity defect due to momentum loss, and turbulence originating in the blade boundary layers. When this is translated from one frame of reference to another - e.g. between the rotor and stator, it results in a travelling velocity fluctuation and

---

* Currently at Rustons Gas Turbines, Lincoln, England.

pulse of turbulence in the freestream above a developing boundary layer[1]. Several studies[2,3,4] have shown that the principal effect is the initiation of an early start to boundary layer transition due to the turbulence originating in the wake. This increases profile loss and, through interaction of the boundary layer with the inviscid velocity field, generates unsteady lift forces. The latter effect will be most pronounced in the presence of boundary layers either close to or undergoing separation - a situation common in the diffusing boundary layers of compressor blades.

The question is made much more pertinent by the increasing use of controlled diffusion (or supercritical) compressor blades with prescribed velocity distributions. The aim is to reduce the profile loss by maintaining a laminar boundary layer for as long as possible. This results in a free-stream velocity distribution resembling a "ski-jump" such as in Fig.2, and a boundary layer which is close to separation over a large area of the blade. If the prediction of transition or separation behaviour is in error due to the presence of unsteady inlet flow in a real machine, the resulting change in pressure distribution might lead to early separation of the boundary layer with resultant catastrophic increase in profile loss and loss of lift. Thus, for the successful application of new design techniques, a more detailed understanding of the effects of wakes on boundary layers is needed.

The work to be reported here uses a linear cascade of prescribed velocity distribution (PVD) compressor blades and a rotating bar mechanism to simulate the wakes from upstream bladerows. It follows earlier investigations of the suction surface boundary layer on these blades at their design Reynolds number (3.0×10^5) with various combinations of clean inflow, grid generated turbulence and incident wakes[5,6]. These showed that clean inlet flow produced boundary layer behaviour in agreement with the predictions of an integral boundary layer code which included separation bubble and transition correlations. Laminar separation occurred on the suction surface and was followed by transition in the separated shear layer and rapid reattachment. Inflow with high levels of grid generated (stationary) turbulence ($Tu \approx 6\%$ at cascade inlet) led to transition ahead of the separation and the boundary layer remained attached. The passage of the wakes (instationary turbulence) caused intermittent suppression of the separation with little change to the profile loss. It is this latter configuration that will be considered in more detail here. The work reported in this paper uses the same cascade but running at half the Reynolds number (1.5×10^5). Since the length of a separation bubble scales roughly as $Re^{-1/2}$ the bubble will be longer, and the effects of the wakes should more pronounced.

EXPERIMENTAL ARRANGEMENTS

1. Cascade

A schematic diagram of the linear cascade is shown in Fig.1. Hollow aluminium bars are mounted on two toothed belts running on pulley wheels around the working section. The blades are of 450mm chord which has the advantage that the boundary layers are large enough for the behaviour to be very easy to investigate, but the disadvantage that only two blades can be fitted in the working section. An arrangement of splitter plates is used to give periodicity; particular attention being paid to the leading edge. The boundary layers at inlet to the working section are removed, and a grid at the cascade exit gives a back pressure which is used to bleed off the endwall boundary layers through slots next to the blade suction surfaces. A more detailed description of this facility can be found in the earlier publications[5,6].

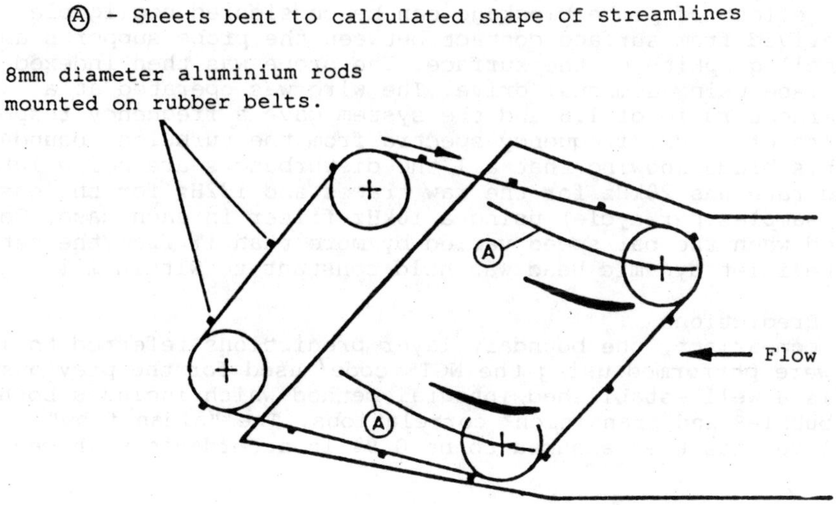

Fig.1 Schematic diagram of the test facility.

The blades under investigation are prescribed velocity distribution (PVD) blades with about 20° turning and a solidity of 0.8 designed specially for tests at low Mach numbers. In order to emphasise the effect of the wakes on the separation bubble they are tested here at half the design Reynolds number. Blading details are summarised in figure 2.

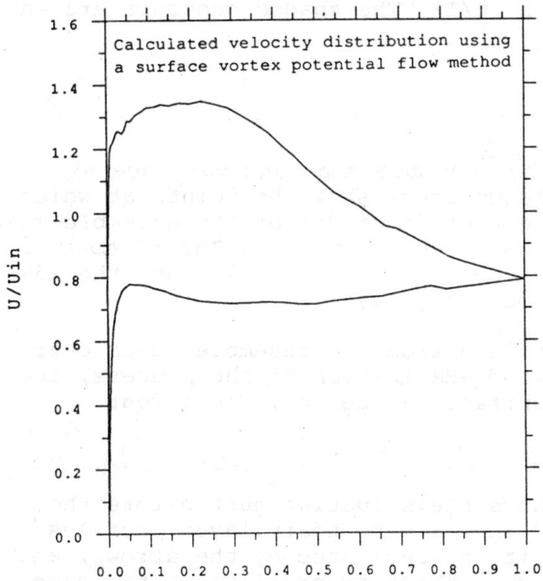

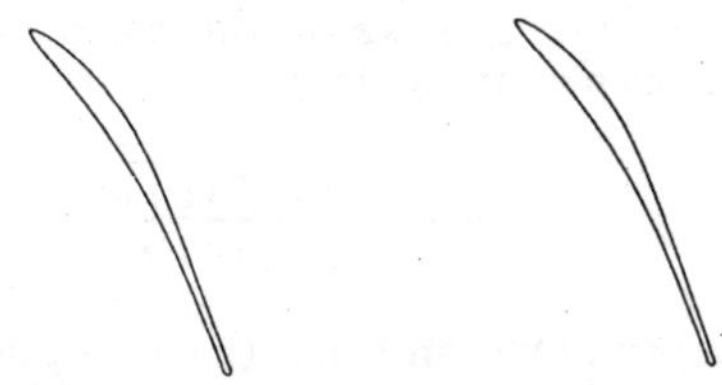

Design Conditions

Reynolds Number	3.1×10^5
Flow Coefficient	0.62
Stagger Angle	29.4°
Blade Inlet Angle	43.5°
Blade Exit Angle	20.0°
Chord	450 mm
Pitch/Chord Ratio	1.25

Test Conditions

Reynolds Number	1.5×10^5
Inlet Flow Velocity	4.88 m/s
Bar Passing Frequency	8.2 Hz

Fig.2 Blading Details.

2. Instrumentation and Logging
A 5μm tungsten wire in a DANTEC P15 probe was used in a traverse gear protruding through the blade surface. Due to the large scale of the cascade, the effect of probe blockage can be considered negligible. A datum was derived from surface contact between the probe supports and a layer of pencil graphite on the surface. The probe was then indexed away from the surface using a manual drive. The wire was operated at a constant overheat ratio of 1.8 and the system gave a frequency response well in excess of 25kHz; frequency spectra from the turbulent boundary layers of this blade showing that all the disturbances are below 10kHz. The sampling rate was 20kHz for the raw traces and 132Hz for the ensemble signals (16 samples per cycle) using a 10kHz filter in each case. Data were rejected when the bar speed varied by more than 1% from the set value and the inlet dynamic head was held constant to within ±1%.

3. Computer Predictions
For ease of comparison, the boundary layer predictions referred to in this paper were performed using the NGTE code[7] used for the previous work. This is a well established integral method which includes both separation bubble[8] and transition[9] correlations. The "clean flow" turbulence level has been assumed to be 0.9% in accordance with measured values.

RESULTS

The interaction of the wake with the suction surface boundary layer will be illustrated with instantaneous cross-sections of the boundary layer derived from the ensemble averaged results - summarised in Fig.3. In these plots the ordinate is the surface distance and the abcissa is the normal-to-surface distance both of which are normalised with respect to the total surface distance i.e. $s^*=s/s_{max}$ and $y^*=y/s_{max}$. For clarity, the scale normal to the surface is enlarged. Time is non-dimensionalised with respect to the wake passing period ($t^*=t/T$). The shaded contours are the local rms intensity

$$Tu = \frac{\sqrt{\langle\langle u^2(t)\rangle\rangle}}{\langle u(t)\rangle} \qquad (1)$$

where $\langle u(t)\rangle$ and $\langle\langle u^2(t)\rangle\rangle$ represent the ensemble mean and variance as defined in the nomenclature. The line contours show the points at which the local ensemble mean velocity is either 5% or 95% of the ensemble mean freestream velocity. These are defined as $\langle\delta_5\rangle$ and $\langle\delta_{95}\rangle$. The 5% contour gives an indication of the extent of the separation bubble, and the 95% contour indicates the "edge" of the boundary layer.

Fig.3 summarises the information obtained from the ensembled data over one complete cycle. For a more detailed examination of the process, the passage of the wake over the blade surface can be split into four (arbitrary) stages:

1. Start of the Story
Fig.4, corresponding to $t^*=0.063$, shows the situation just before the wake enters the picture, stage-left. The five boundary layer profiles plotted above are taken from the positions indicated by the arrows, and show the ensemble mean velocity and rms intensity profiles at the same instant.

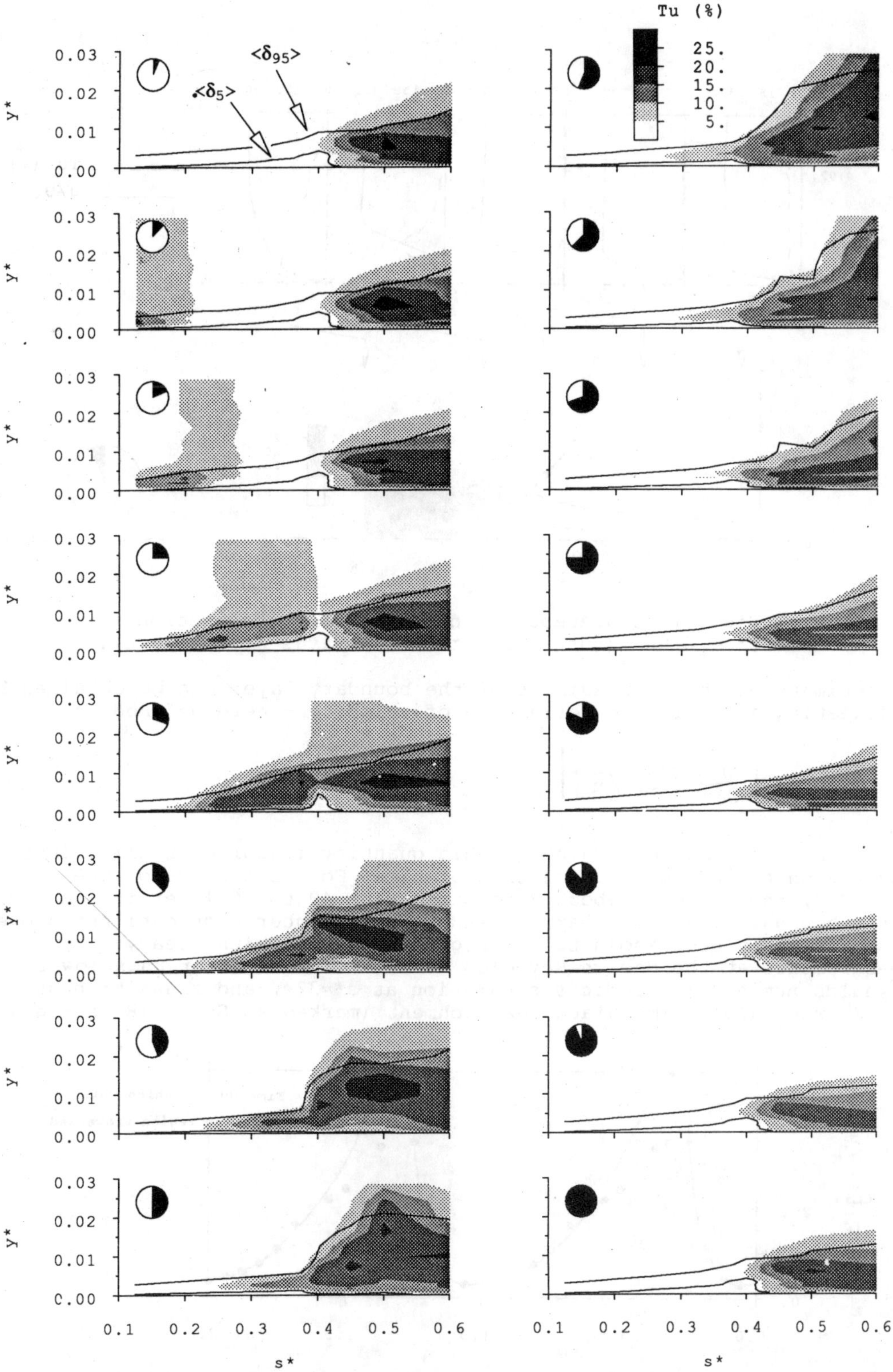

Fig.3 Instantaneous boundary layer cross sections over one complete wake passing cycle.

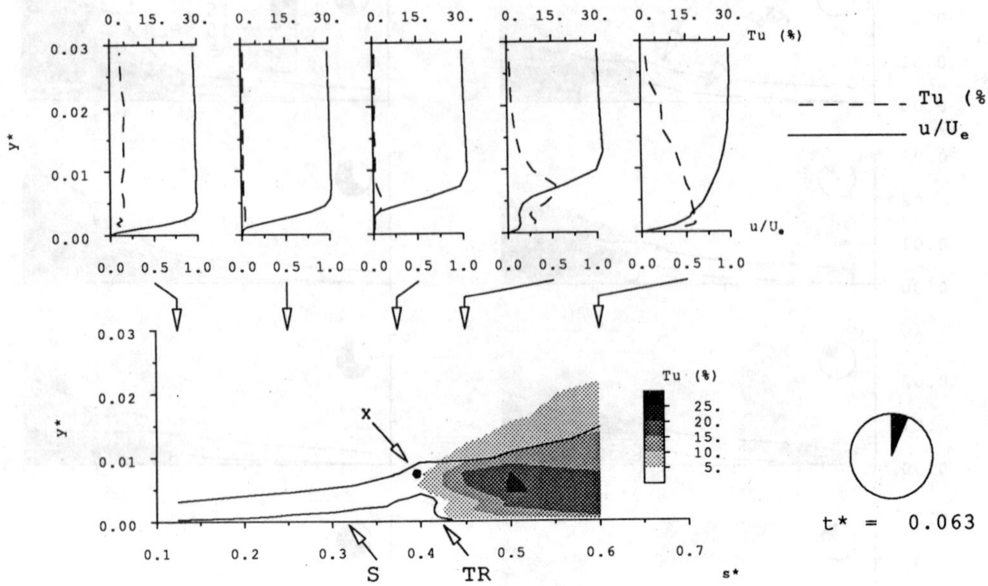

Fig.4 Instantaneous boundary layer cross section for t* = 0.063 (just prior to the arrival of the wake).

An estimate of the unsteadiness of the boundary layer can be obtained by calculating a local time derivative of velocity - here defined as

$$\left| \frac{\partial}{\partial t^*} \left[\langle u \rangle / \bar{u} \right] \right| \tag{2}$$

Fig.5 shows the maximum value of this quantity measured at all points in the region traversed as a function of time. For this stage of the cycle, t*=0.063, the value is about 2 compared with 40 to 50 later in the cycle - the boundary layer is nearly steady. The computer code outlined above has previously been shown to be in good agreement with steady measurements at the design Reynolds number[5]. When run for this lower Reynolds number, it predicts separation at s*≈32%, and transition at s*≈42% with almost immediate reattachment (marked as S and TR in Fig.4).

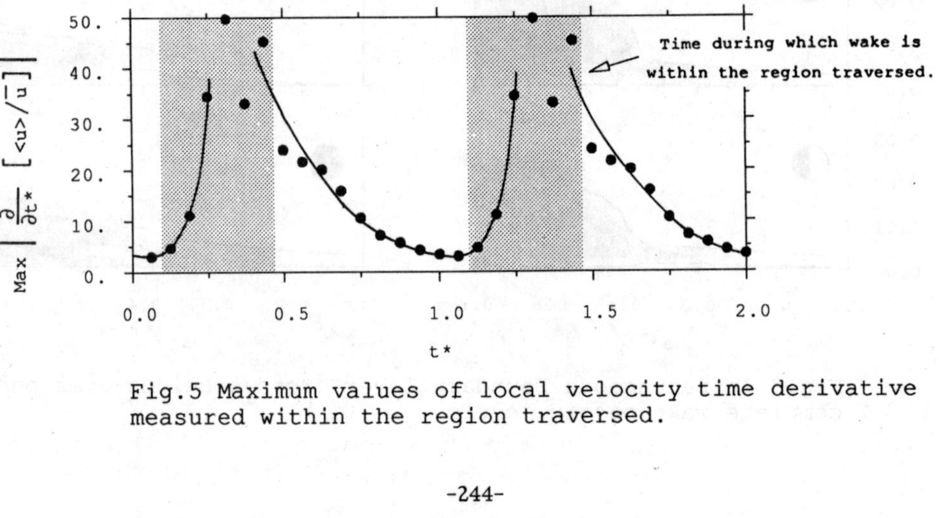

Fig.5 Maximum values of local velocity time derivative measured within the region traversed.

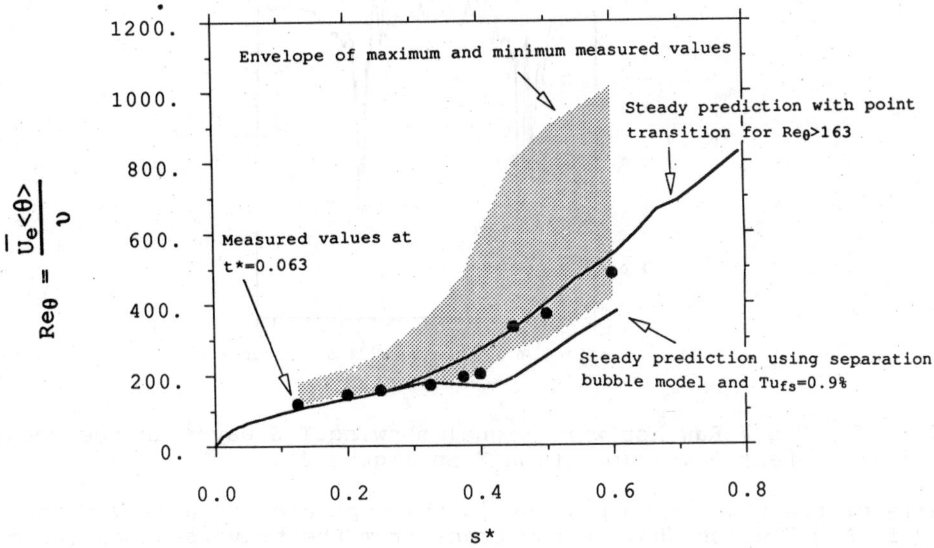

Fig.6 Comparison of predicted and measured values of boundary layer Reynolds number Re_θ.

Fig.6 compares the measured boundary layer Reynolds number Re_θ, here defined as

$$Re_\theta \equiv \frac{\overline{U}_e\langle\theta\rangle}{\upsilon} \qquad \text{where } \langle\theta\rangle \equiv \int_0^\infty \frac{\langle u\rangle}{\langle U_e\rangle}\left\{1 - \frac{\langle u\rangle}{\langle U_e\rangle}\right\} dy \qquad (3)$$

with the values predicted using the computational scheme in two ways:

1) "turbulent" prediction - a steady prediction specifying point transition for Re_θ=163 (a lower limit found in flat plate tests[9]), and

2) "laminar" prediction - a steady prediction using the measured "clean flow" freestream turbulence level (Tu_{fs}=0.9%) and the separation bubble and transition correlations in the code.

The measured values are presented as the envelope of the maximum and minimum values of Re_θ recorded, and the values measured for t*=0.063 (corresponding to Fig.4). Whilst the trend for the minima is closest to the "laminar" prediction, these results suggest that the boundary layer over the latter part of the blade has almost but not quite recovered to the laminar value. It is thus in agreement with Fig.5 which shows that there is still some small degree of unsteady behaviour within the region traversed. The results of Fig.6 suggest that this time level (t*=0.063) represents the laminar boundary layer over the first part of the blade, but a relaxing turbulent boundary layer from s*≈45%. The maxima of Fig.6 show that the wakes do not have a significant impact on the boundary layer until s*≈30% following which they result in values well in excess of the "turbulent" prediction. This change corresponds to the beginning of the wake-induced transition to be discussed in §2 below.

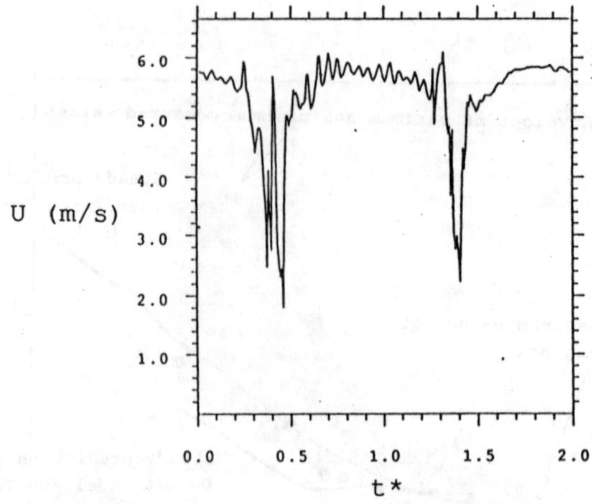

Fig.7 Raw hot wire signal showing T-S waves in the separated shear layer (position X in figure 4).

Details of the transition process in the separated shear layer can be found in Fig.7 which shows a raw trace from the traverse location marked with an X in Fig.4. Before the arrival of the wake transition results from the breakdown of sinusoidal oscillations - Tollmein-Schlichting (T-S) waves. Evidence for this interpretation can be adduced from comparison with the predicted frequency of T-S waves in a free-shear layer[10]

$$f_{TS} = \frac{1}{4\pi}\frac{U_e}{\Delta} \qquad (4)$$

where Δ is the thickness of the shear layer. From the boundary layer traverses it is estimated that $\Delta \approx 3$mm and $U_e \approx 5.8$m/s, giving a predicted value of 153Hz compared to the experimentally measured 147Hz.

Referring once again to Fig.4, the shaded area indicates the normalised ensemble rms intensity. Transition can be seen to start in the free-shear layer and results in almost immediate reattachment of the boundary layer. Care must be taken in interpreting values of the rms intensity since it represents the sum total of all forms of non-phase locked unsteadiness whether arising from turbulence or from other events[4]. In this case, it includes the unsteadiness due to the stochastic formation of turbulent spots during the transition process and this explains the peak at $s^* \approx 50\%$. The subsequent fall in rms intensity to a level appropriate to a turbulent boundary layer indicates the merging of the turbulent spots formed in the free shear layer. However, transition is not fully completed before the ensemble averaging indicates reattachment.

2. Arrival of the Wake

Fig.8, corresponding to $t^*=0.25$, shows the arrival of the wake over the laminar boundary layer. Transition is still occurring downstream in the separated shear layer, unaware of the approach of the wake, and the wake turbulence is within the boundary layer above the separated region although the separation bubble is not yet affected.

It is apparent from the leading edge of the wake footprint that the diffusion of wake turbulence into the boundary layer is very rapid. (The trailing edge lags to a greater extent due to the transition process to be discussed below). This is not entirely unexpected. It has already been shown[4] that the unsteady frequency parameter ϖ can be written in terms of

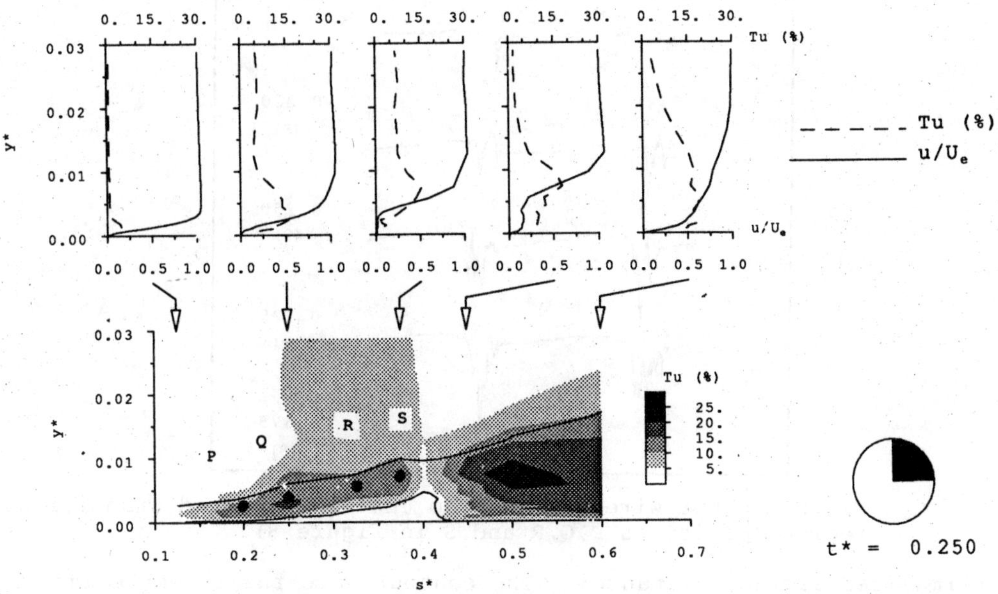

Fig.8 Instantaneous boundary layer cross section for t* = 0.250

the ratio of diffusion time (δ^2/υ) to the unsteady period ($1/\omega$) for a laminar boundary layer

$$\varpi = \frac{\omega x}{U_e} \approx \frac{1}{25}\left[\frac{\omega\delta^2}{\upsilon}\right] \qquad (5)$$

and that the effects of wake turbulence on a laminar boundary layer were better represented by replacing the viscosity with an eddy viscosity υ_t so that

$$\varpi' = \varpi\left[\frac{\upsilon}{\upsilon_t}\right] \qquad (6)$$

For this case, $\varpi \sim 3$ suggesting viscous diffusion timescales of the same order as the wake passing period. However, the modified form $\varpi' \sim 0.03$ indicates that the timescales associated with the turbulent fluctuations are very much shorter than the wake passing period and these dominate the diffusion of disturbances from the wake.

Fig.9 shows raw signals taken at the locations marked with P,Q,R and S in Fig.8. Over the first part of the boundary layer surface (s*<20%), the passage of the wake is associated with low frequency disturbances. However, further downstream the traces show the appearance of high frequency bursts. These signify the beginning of transition induced by the wakes.

The effect can be seen more clearly in Fig.10a - a surface-distance time (s-t) plot where the abcissa is the normalised time and the ordinate is

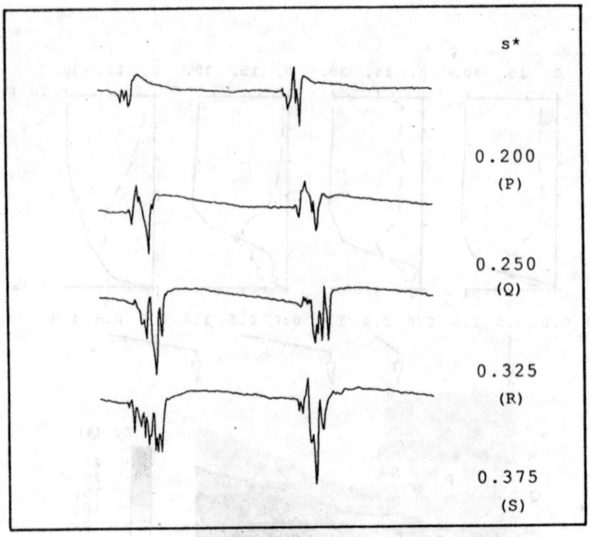

Fig.9 Raw hot wire traces from the wake-induced transitional region (positions P,Q,R and S in figure 8).

the normalised surface distance‡ . The contours are the ensemble rms intensities measured at traverse points 50% of the way through the time mean boundary layer (δ_{50}). Also shown is the predicted trajectory of a feature moving at 100% and 50% of U_e. There is a distinct change of behaviour at about s*≈25% corresponding to transition where a wedge indicating streamwise spreading of the wake footprint begins. That this spreading is not a result of wake dispersion in the freestream can be seen from Fig.10b which shows a similar s-t plot derived from the values of rms intensity in the freestream. Here the traces can be seen to be parallel tracks at the mean freestream velocity with little (if any) streamwise spreading. This effect has also been shown in other work[4].

The motion of line A-A ahead of the convective velocity should be noted. This does not signify the motion of turbulent spot leading edges ahead of the wake but is a consequence of the ensemble averaging process and is associated with progressively earlier formation of spots due to increasing susceptibility of the boundary layer as it grows. The feature is not significant here, and has been discussed in more detail in other publications[4]. The trailing edge of the wedge has an angle consistent with the rear of turbulent spots found in other experiments[11,12] at about 50% U_e.

---

‡ Strictly speaking, a better choice of ordinate would be

$$\int_0^s \frac{ds}{U_e(s)}$$

being the time taken for a particle to convect in the freestream from the leading edge to s. Then the convective velocity (and any constant fractions of it) would appear as straight lines on the plot. In practice, the additional complication is unwarranted.

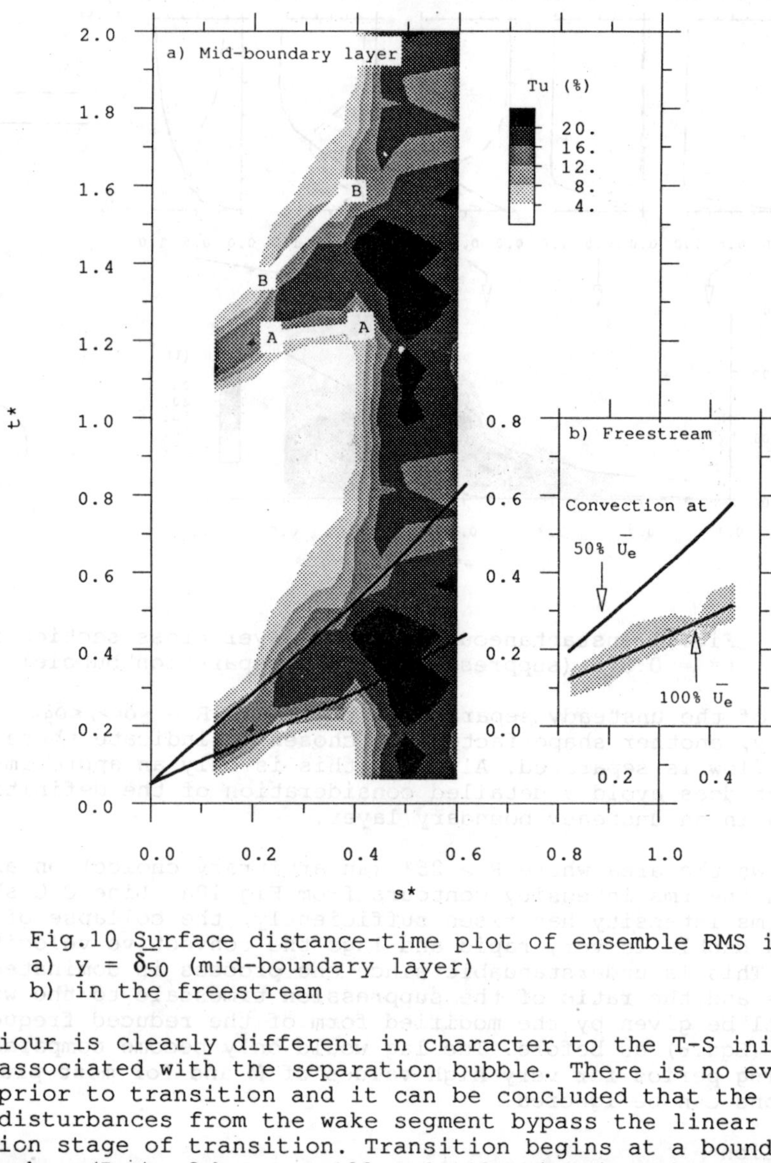

Fig.10 Surface distance-time plot of ensemble RMS intensity
a) $y = \delta_{50}$ (mid-boundary layer)
b) in the freestream

This behaviour is clearly different in character to the T-S initiated breakdown associated with the separation bubble. There is no evidence of T-S waves prior to transition and it can be concluded that the high levels of disturbances from the wake segment bypass the linear amplification stage of transition. Transition begins at a boundary layer Reynolds number (Re_θ) of between 130 and 140 which is slightly below the lower limit for transition of $Re_\theta \geq 163$ often accepted[9]. In view of the highly unstable nature of near-separating flow and the high levels of disturbances in the boundary layer resulting from the wake, this is not entirely surprising.

3. Suppression of the Bubble

In Fig.11, corresponding to t*=0.375, the increased level of momentum transfer in the transitional boundary layer at the rear of the wake has suppressed the separation. As the shaded area of Fig.5 shows, during the passage of the wake the normalised time derivatives defined by Eqn.2 are between 40 and 50 indicating that the flow is truly unsteady. Consequently, the points of zero wall shear stress can no longer be regarded as separation and reattachment points[13]. In order to look at the

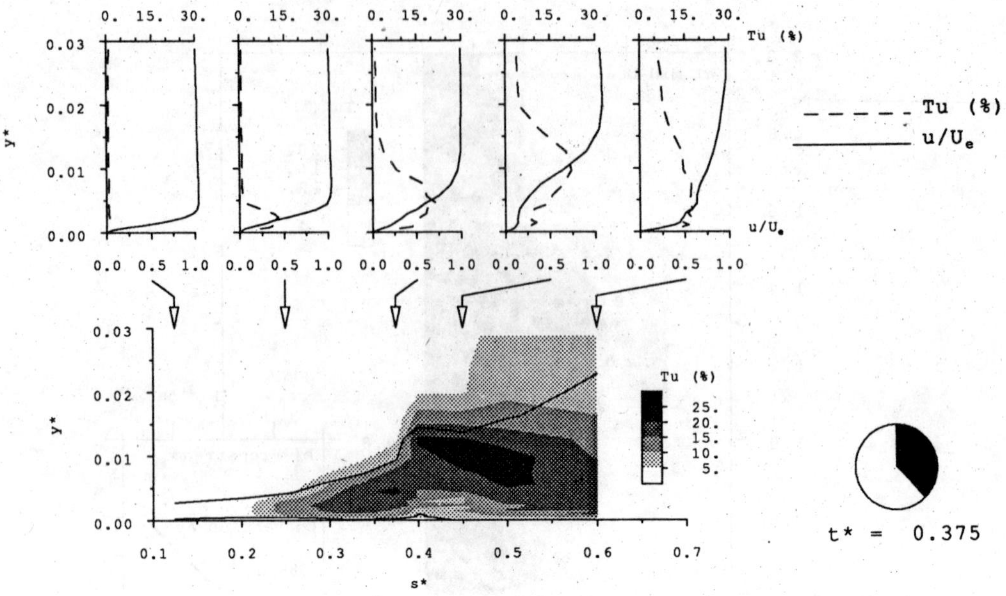

Fig.11 Instantaneous boundary layer cross section for $t^* = 0.375$ (suppression of the separation bubble)

behaviour of the unsteady separation, the ratio $R = \langle\delta_5\rangle/\langle\delta_{95}\rangle$ (which is, effectively, another shape factor) is chosen to indicate those regions where the flow is separated. Although this is only an approximate measure, it does avoid a detailed consideration of the definition of separation in an unsteady boundary layer.

Fig.12 shows the area where R > 25% (an arbitrary choice) on an s-t plot along with the rms intensity contours from Fig.10a. Line C-C shows that, once the rms intensity has risen sufficiently, the collapse of the separation bubble is very rapid and lags the convective velocity $\bar{U}_e$ only slightly. This is understandable since the process is dominated by turbulence and the ratio of the suppression timescale to the wake passing period will be given by the modified form of the reduced frequency parameter (Eqn.6) as before. The lag would only become comparable to the wake passing period for very high values of ϖ and for most practical applications can be ignored.

4. Relaxation

Fig.13 shows the situation a short time after the passage of the first wake but before the arrival of the next (t^*=0.750). The laminar boundary layer is beginning to recover from the trauma and is losing its attachment to the surface. As this happens, the point of laminar separation moves up the blade towards its initial position.

The effect is shown in Fig.12 where the start of the separated region is seen to move up the blade. However the slope of lines on the s-t plot parallel to D-D indicate that this movement is at speeds very much less than the convective speed. This situation is concerned with (laminar) viscous diffusion, and thus the ratio of the boundary layer diffusion time to the wake passing time will be given by the conventional form of the reduced frequency parameter - Eqn.5. Since, for this cascade, $\varpi \sim 3$, the recovery timescale for the separation bubble will be of the order as

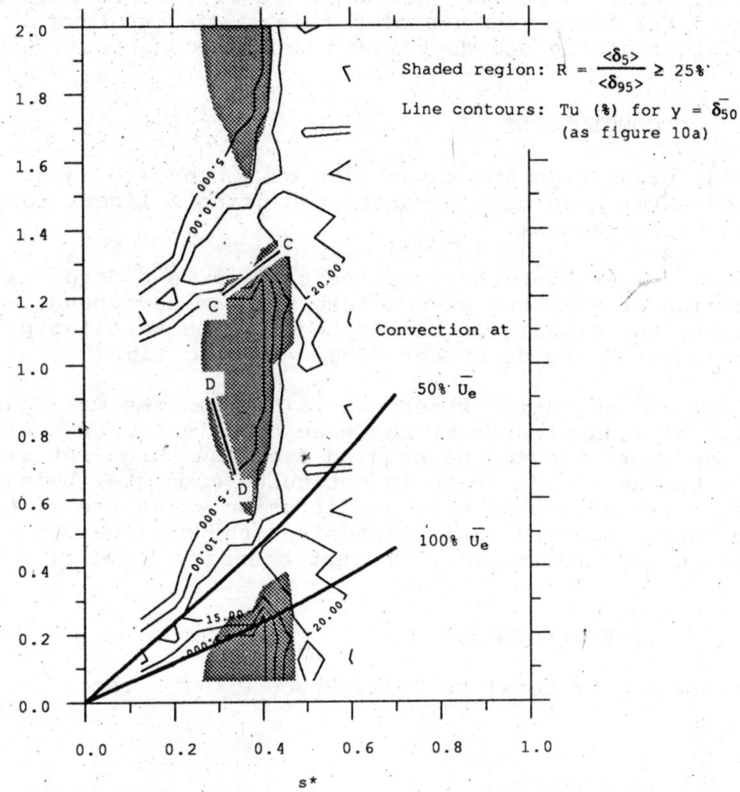

Fig.12 Unsteady behaviour of the separation bubble.

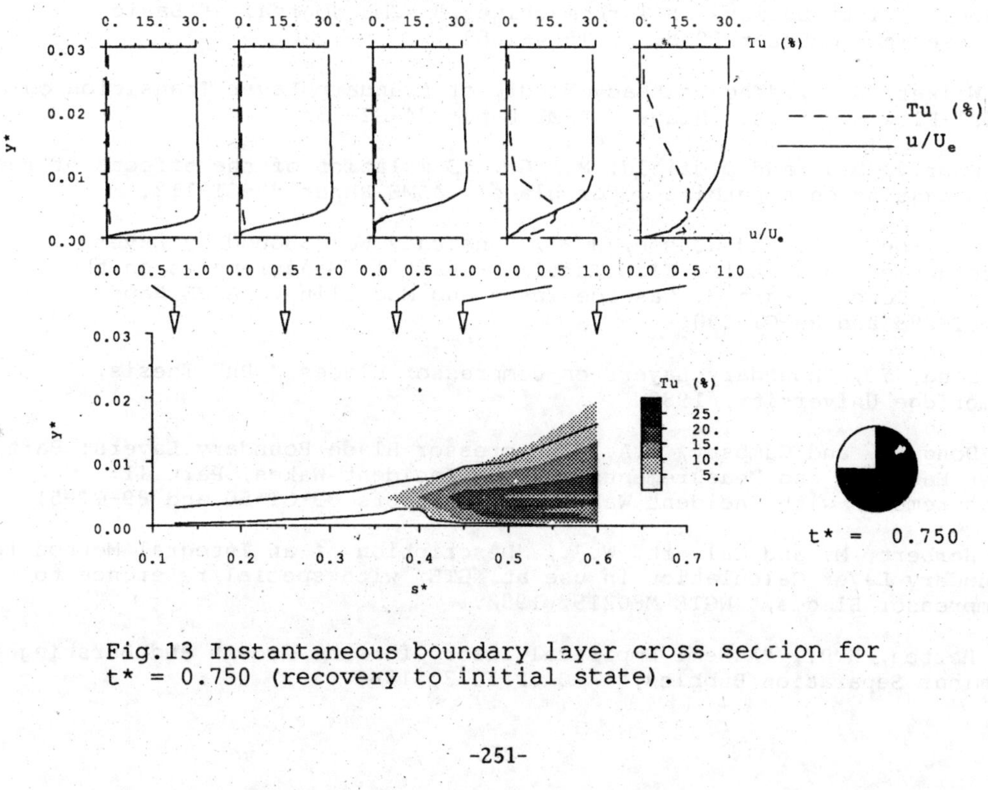

Fig.13 Instantaneous boundary layer cross section for t* = 0.750 (recovery to initial state).

the periodic disturbance time. Figs.5 and 6 show that the boundary layer has almost returned to its "quiescent" state before the arrival of the next wake. For higher values of ϖ this state would not be reached, and the behaviour of the boundary layer would become truly unsteady.

CONCLUSIONS

- Unsteady transition and separation of a boundary layer subjected to simulated wakes have been demonstrated using a linear cascade and a rotating bar mechanism.

- Parallel traces over the first part of the s-t diagrams show that there is diffusion of wake turbulence into the laminar boundary layer without transition, but transition begins beneath the wake footprint at a boundary layer Reynolds number (Re_θ) of about 135.

- The laminar separation which exists between the wake passings is intermittently suppressed by the wake turbulence; this effect being rapid due to the short timescales arising from the turbulent fluctuations. Recovery to the steady state is not quite complete, being dominated by the longer timescales of viscous diffusion which are of the same order as the bar passing period. This emphasises the dual nature of the reduced frequency ϖ in laminar and turbulent boundary layer phenomena.

ACKNOWLEDGMENTS

The authors are grateful to Rolls Royce PLC for their support of this research.

REFERENCES

1. Meyer, R.X., "The Effect of Wakes on the Transient Pressure and Velocity Distributions in Turbomachines," ASME Journal of Basic Engineering, October 1958, pp 1544-1552.

2. Walker, G.J., "The Unsteady Nature of Boundary Layer Transition on an Axial-Flow Compressor Blade," ASME Paper 74-GT-135.

3. Doorly, D.J. and Oldfield, M.L.G., "Simulation of the effects of shock wave passing on a turbine rotor blade", ASME Paper 85-GT-112.

4. Addison, J.S. and Hodson, H.P., "Unsteady Wake-Boundary Layer Interaction in an Axial Flow Turbine - Part 1: Measurements on the Turbine Rotor - Part 2: Cascade Tests and Modelling," ASME Papers 89-GT-289 and 89-GT-290.

5. Dong, Y., "Boundary Layers on Compressor Blades," PhD Thesis, Cambridge University, 1988.

6. Dong, Y. and Cumpsty, N.A., "Compressor Blade Boundary Layers: Part I; Test Facility and Measurements with No Incident Wakes; Part II; Measurements with Incident Wakes," ASME Papers 89-GT-50 and 89-GT-51.

7. Herbert, M. and Calvert, W.J., "Description of an Integral Method for Boundary Layer Calculation in use at NGTE, with special reference to Compressor Blades," NGTE M80219, 1982.

8. Horton, H.P., "A Semi-Empirical Theory for the Growth and Bursting of Laminar Separation Bubbles," ARC CP 1073, 1969.

9. Abu-Ghannan, B.J. and Shaw, R., "Natural transition of boundary layers - the effects of turbulence, pressure gradient and flow history," Journal of Mechanical Engineering Science, Vol. 22, No. 5, 1980, pp 213-228.

10. Gaster, M.,"The Role of Spatially Growing Waves in the Theory of Hydrodynamic Stability," Progress in Aeronautical Science, Vol 6, 1965, pp 251-270.

11. Obremski, H.J. and Fejer, A.A., "Transition in Oscillating Boundary Layer Flows," Journal of Fluid Mechanics, Vol 29, 1967, pp 93-111.

12. Schubauer, G.B., and Klebanoff, P.S., "Contributions on the Mechanics of Boundary Layer Transition," NACA TN 3489 (1955) and NACA Rep. 1289 (1956).

13. Telionis, D.P., "Introduction to Unsteady Boundary Layers", in AGARD Report No. 679 "Special Course on Unsteady Aerodynamics", June 1980.

NOMENCLATURE

f	Frequency (in Hz)
H	Shape factor = δ/θ
R	Shape factor = $\langle\delta_5\rangle/\langle\delta_{95}\rangle$
Re_θ	Boundary layer Reynolds number
s^*	Normalised surface distance = s/s_{max}
T	Wake passing period
t^*	Normalised time = t/T
Tu	RMS intensity = $\sqrt{\langle\langle u^2(t)\rangle\rangle}/\langle u(t)\rangle$
u	Streamwise velocity (within boundary layer)
U_e	Freestream velocity
y^*	Surface normal distance = y/s_{max}
δ	Boundary layer thickness
δ_n	Point at which local ensemble velocity is n% of freestream ensemble velocity
Δ	Thickness of separated shear layer
θ	Momentum thickness
υ	Kinematic viscosity
υ_t	Eddy viscosity
ω	Wake passing frequency
ϖ	Reduced frequency parameter = $\omega x/U_e$
ϖ'	Modified reduced frequency parameter = $\varpi(\upsilon/\upsilon_t)$
$\sim$	.....is of the order of.....

Time Mean

$$\overline{x(t)} \equiv \frac{1}{T}\int_0^T \left[\frac{1}{N}\sum_{n=1}^{N} x(t,n)\right] dt$$

Ensemble Mean

$$\langle x(t)\rangle \equiv \frac{1}{N}\sum_{n=1}^{N} x(t,n)$$

Ensemble Variance

$$\langle\langle x(t)^2\rangle\rangle \equiv \frac{1}{N}\sum_{n=1}^{N} (x(t,n) - \langle x(t)\rangle)^2$$

INTERACTION BETWEEN VIBRATING CASCADE BLADES AND SHEAR FLOW

Arkady A. Kaminer, Alexander G. Chervonenko
and Boris G. Reznik
(Institute for Problems of Strength, Academy of Sciences,
Ukrainian SSR, Kiev, USSR)

ABSTRACT

The paper presents the results of experimental investigation of the effect of nonuniform profile of flow velocity along blades on unsteady aerodynamic characteristics of the cascade. The tests were carried out on a straight cascade of vibrating blades flown about by the shear flow. The results of measurements of unsteady aerodynamic forces acting on the blades in the shear flow were compared with the corresponding data obtained on plane models. Disagreements with the two-dimensional section hypothesis are revealed and analyzed.

INTRODUCTION

The three-dimensional analysis of the flow action on vibrating blades is one of the most urgent problems of turbomachine aeroelasticity. Since its solution is rather complex, a simplified approach based on the use of the two-dimensioal section hypothesis is used in most cases. At the same time the applicability of such an approach and its error call for an experimental verification. The model factorial experiment is the most convenient form of such a verification. The performance of the model experimental investigation aimed at studying the effect of certain factors causing the three-dimensional character of interaction between the vibrating blades and flow as well as a verification of the two-dimensional section hypothesis is a significant contribution to the solution of the general problem.

EXPERIMENTAL

The effect of nonuniform profile of the flow velocity on the pattern of interaction with it of vibrating cascade blades may be studied on a certain experimental model. The model representing a straight cascade of untwisted blades flown about with the velocity varying along their length may be considered as a simple and convenient one for the analysis. Such a model being a straight cascade of blades vibrating in the uneven or "shear"[1] flow was the object to study (Fig. 1). The paper presented deals with the subsonic flow condition, linear law of changes in velocity profile of the shear flow at the inlet to the cascade as well as with the plane-parallel model of vibrations of blades towards the axis being normal to their chords.

The technique suggested for creating the shear flow is based on the establishing in the accelerating nozzle of the wind tunnel of the sloping gauze with the specially selected drag coefficient (Fig. 2). The shear flow in the working part is reached due to redistribution of the flow in front of the gauze and nonuniform acceleration of its layers in the nozzle site behind the gauze. The following parameters are taken as characteristics for such a flow: the middle layer velocity V_m and the flow shear coefficient q_v which characterizes the intensity of velocity variation along blades

$$q_v = \frac{V_e/V_m - 1}{l} \qquad (1)$$

where V_e — the velocity of the layer being at the distance l from the middle layer. The measurements performed under different flow rate have shown that the flow shear coefficient is a constant value for the fixed position of the gauze.

Unsteady aerodynamic forces were the parameters characterizing the vibrating blade interaction with the flow and adjacent blades[2]. The schematic diagram of their determination is given in Fig. 3. Since the aerodynamic influence of the adjacent blades vibrating in the cascade predominates, three middle blades were designed to vibrate, the central blade ("0") being the measuring one. the vibrations were excited by elastic systems and electrodynamic vibrators. Modulus F and phase angle φ of the exciting force of the vibrator maintaining oscillations of the measuring blade with the constant frequency and amplitude were determined in the experiment. The vector diagram of exciting forces corresponding to different conditions of the measuring blade oscillations was plotted in the phase plane: $\overline{F}_0$ — the blade "0" vibrates with the absence of flow; $\overline{F}_1$ — the blade "0" vibrates in the flow; $\overline{F}_2$ — blades "0" and "1" vibrate in the flow; $\overline{F}_3$ — blades "0" and "+1" vibrate in the flow. Vectors of unsteady aerodynamic forces were determined in terms of dynamic equilibrium of the vibrating system: $\overline{R}_0$ is the force acting on the measuring blade with its individual vibrations in the flow; $\overline{R}_{-1}$ is the force with which the vibrating blade "-1" acts on the measuring blade through the flow; $\overline{R}_{+1}$ is the force acting from the blade "+1" on the measuring blade through the flow. The phase plane is simultaneously a coordinate one for finding imaginary Im and real Re parts of unsteady aerodynamic forces. The method based on the use of the complex blade was developed to evaluate distribution of unsteady aerodynamic forces along the blade length. The central blade was made of two parts separated by a rather small gap (Fig. 3). The both parts have performed synchronous vibrations immitating the integral blade movement. The lower part was the measuring one. Displacing the small pitch gap along the blades and making measurements for each fixed position, it was possible to plot the distribution of unsteady aerodynamic forces $\overline{R}_s$ per unit length along the central blade length using the difference of the next and previous measurements.

The study involved several stages. First unsteady aerodynamic forces and their distribution along the blade length were determined for the cascade flown about by the shear flow. Then the forces on the two-dimensional cascade were measured for veloities corresponding to those of individual layers of the shear flow. After that the results were compared.

Nozzles creating the shear flow were changeable. When making measurements on the two-dimensional cascade, they were replaced by the nozzle without gauze.

The blades had a compressor profile. The elongation of blades $\bar{h}=h/b=2.2$; the blade vibration frequency was 42,95 Hz, the oscillation amplitudes $\overline{A}_n=A_n/b=0.022$. The following parameters were varied in the experiment: the flow shearing coefficient $q_v=0.12; 0.008; 0.004$ I/cm; the Strouhal number $sh=p \cdot b/V_m=0.065 \cdots 0.2$; the attack angle $i=0° \cdots 15°$; the relative blade spacing $\bar{t}=t/b=0.75; 1.0; 1.25$; the angle of cascade staggering $\beta=30°; 45°$. The Strouhal number was changed by varying the middle layer flow velocity.

RESULTS AND DISCUSSION

Experimental data from measurements in the shear flow are obtained in the form of vectors of uns. .dy aerodynamic forces $Im\bar{R}^s_o$, $\bar{R}^s_{-1}$, $\bar{R}^s_{+1}$. The analysis was made of aerodynamic force $Im\bar{R}^s_o$ acting on the measuring blade during its individual vibrations as well as of force moduli $/\bar{R}^s_{-1}/$ and $/\bar{R}^s_{+1}/$ acting on it due to the adjacent blade vibration (see Fig. 3).

Fig. 4 presents distribution of the unsteady aerodynamic forces per unit length for the cascade flown about by the shear flow with the maximal shearing intensity for one of the intermediate values of the attack angle (the regions of the end effect are excluded from the consideration). Here there are data obtained on the two-dimensional cascade and plotted according to the two-dimensional section hypothesis (TDH). The comparison has shown a decrease in the unsteady aerodynamic forces in the shear flow as against the two-dimensional section hypothesis and the greatest disagreements being observed in the blade sections flown about with the maximal velocity. In order to certify this effect for quantity, we have introduced coefficients K_*, they are a ratio of the unsteady aerodynamic force acting on the blade in the shear flow R^s_* to the R^Σ_* value obtained by summation along the blade of data based on the use of the two-dimensional section hypothesis

$$K_o = \frac{ImR^s_o}{ImR^\Sigma_o} \quad ; \quad K_{-1} = \frac{/R^s_{-1}/}{/R_{-1}/^\Sigma} \quad ; \quad K_{-1} = \frac{/R^s_{+1}/}{/R_{+1}/^\Sigma} \qquad (2)$$

The studies have shown that the attack angle, the Strouhal number and intensity of the flow shear are basic parameters determining the mentioned effect.

Dependences of coefficients K_* on the attack angle have the characteristic regions (Fig. 5). The convergence of unsteady aerodynamic forces in the shear flow with the data of the two-dimensional section hypothesis is observed at small values of the attack angle. The growth of the attack angle causes the origin and increase of origin and increase of disagreements, in this case the two-dimensional section hypothesis sets results too high. The further growth of the attack angle again decreases the disagreements. Visual control of the flow about the blade has shown from blades and full stalled flow from the leading edges, respectively. Thus, the effect of a decrease of unsteady aerodynamic forces in the shear flow manifests itself to a maximal extent under partial stalled flow from blades. It is marked that partial stalled flow of the shear flow is accompanied by the intiation of vertical flows on the blade surface as well as by instability of the position of the flow stalled point in certain sections along the blade length.

The influence of the conditions of the flow about blade on the effect of a decrease in the unsteady aerodynamic forces in the shear flow has been confirmed in studies of cascades with different geometry. The existence of maximal disagreements with the two-dimensional section hypothesis under the partial stalling of the shear flow from the blades has been observed for all values of the relative blade spacing and the angle of staggering. In this case in cascade with a smaller relative blade spacing these disagreements are shifted towards the region of large attack angles. At the same time the visual control has shown that these cascades are characterized by a rise of the attack angles which corresponds to the appearance of the stalled flow of blades.

A comparison of the results for cascades with different angles of separation has shown that the qualitatively analogous shift of the minimal values of coefficients K_* towards the region of large attack angles is observed with a rise of the angle of cascade staggering. The investigations revealed that the Strouhal number along with the attack angle is another most important which determines a decrease in unsteady aerodynamic forces in the flow shear as against the data of the two-dimensional section hypothesis (Fig. 6). The observed effect manifests itself to the greatest extent in the range of low values of the Strouhal number. If there is partial stalling of the shear flow, a decrease in the Strouhal number leads to a rise of disagreements both for the aerodynamic damping of the blade with its individual vibrations and for the aerodynamic action on it from the adjacent blades.

A comparison of the results obtained in flows with different shearing intensity has shown that an enhancement of shearing intensities results in an increase of disagreements with the two-dimensional section hypothesis (Fig. 7). The ranges of attack angles corresponding to the maximal and minimal disagreements remain unchanged.

The characteristic feature revealed by the investigations is that in the staggering cascade flown about by the shear flow there are the most considerable disagreements for the unsteady aerodynamic force $/R^s_{-1}/$ acting from blade situated upstream on the measuring one (Figs 5~7). Those disagreements are smaller for the aerodynamic damping force ImR^s_o. At last the unsteady aerodynamic force $/R^s_{+1}/$ evoked by vibrations of the downstream blade has minimal differences.

The studies also involved the experimental verification of the linearity of unsteady aerodynamic forces for the vibrating cascade flown about by the shear flow. Measurement conducted at different amplitudes of blade vibrations $\overline{A}_o=\overline{A}_{-1}=\overline{A}_{+1}=0.01\cdots 0.04$ have shown that variations in the amplitudes do not lead to changes in the phase angles of vectors of unsteady aerodynamic forces $\overline{R}^s_o, \overline{R}^s_{-1}, \overline{R}^s_{+1}$, but here the proportional change in their moduli occurs.

The analysis of the data obtained under different combinations of amplitudes of blade vibrations has shown that unsteady aerodynamic forces $\overline{R}^s_{-1}$ and $\overline{R}^s_{+1}$ acting from the adjacent blades on the measuring ones do not practically depend on the measuring blade oscillations.

The test of superposition of unsteady aerodyanmic forces was performed as follows. At first, according to the procedure described we have determined unsteady aerodynamic forces $R^s_o, \overline{R}^s_{-1}, \overline{R}^s_{+1}$, the latter—at the phase angles of blade vibrations $\mu_{-1}=\mu(\text{“}-1\text{”}-\text{“}0\text{”})=0°, \mu_{+1}=\mu(\text{“}-1\text{”}-\text{“}0\text{”})=0°$. Basing on these data positions of the total vector $\overline{R}^s_\Sigma=\overline{R}^s_o+\overline{R}^s_{-1}+\overline{R}^s_{+1}$ for the arbitrary equal phase angles of blade vibrations, in particular 0°, 90°, 180°, 270° were plotted geometrically in the phase plane (Fig. 8). Then oscillations were excited simultaneously to three blades and the total unsteady aerodynamic force acting on the measuring blade at the corresponding phase angles were measured. The results of geometrical summation and direct measurements were compared. Practically complete agreement of the result was observed in the whole range of parameters, that evidences for superposition of unsteady aerodynamic forces for vibrating cascade flown about by the shear flow.

Here we may draw the following conclusions.

A. The vibrating cascade shear flow may cause a decrease in unsteady aerodynamic forces relative to the data obtained with the two-dimensional section hypothesis. This effect is manifested to the best extent during partial stalling of the shear flow and is enhanced with an increase in the shear intensity and with a reduction in the Strouhal number. In the case of unstalled flow and full shear flow stalling disagreements with the two-dimensional section hypothesis were observed to the lesser extent.

B. The vibrating cascade blade aerodynamic interaction in the shear flow is characterized by the proportionality of the unsteady aerodynamic forces to small amplitudes of blade oscillations and their superpositions.

The authors of the paper are research workers.

REFERENCES

1. Kaji S. ,Takata H. and Niwa N. ,Trans. Jap. Soc. Mech. Eng. ,B47 (419) (1981),1177.
2. Kaminer A. A. ,Chervonenko A. G. and Tsimbalyuk V. A. ,A Technique for Investigating Unsteady Aerodynamic Characteristics of Cascade Blades Vibrating in a Three-Dimensional Flow,preprint Instituta Problem Prochnosti AN Ukr. SSR (1988),47.

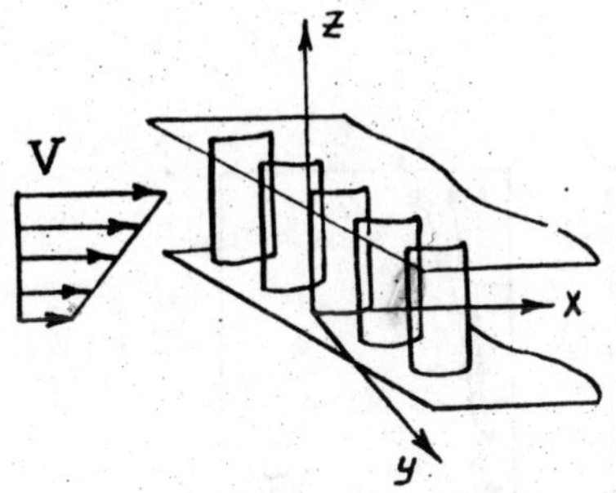

Fig.1. Test model.

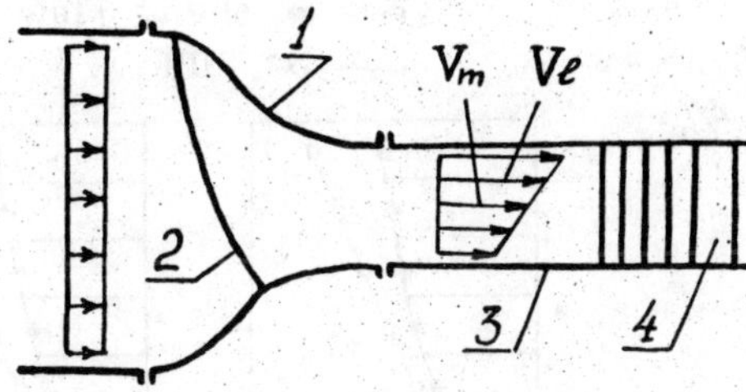

Fig. 2. Mode for obtaining a shear flow. 1 - nozzle; 2 - gauze; 3 - working part; 4 - cascade.

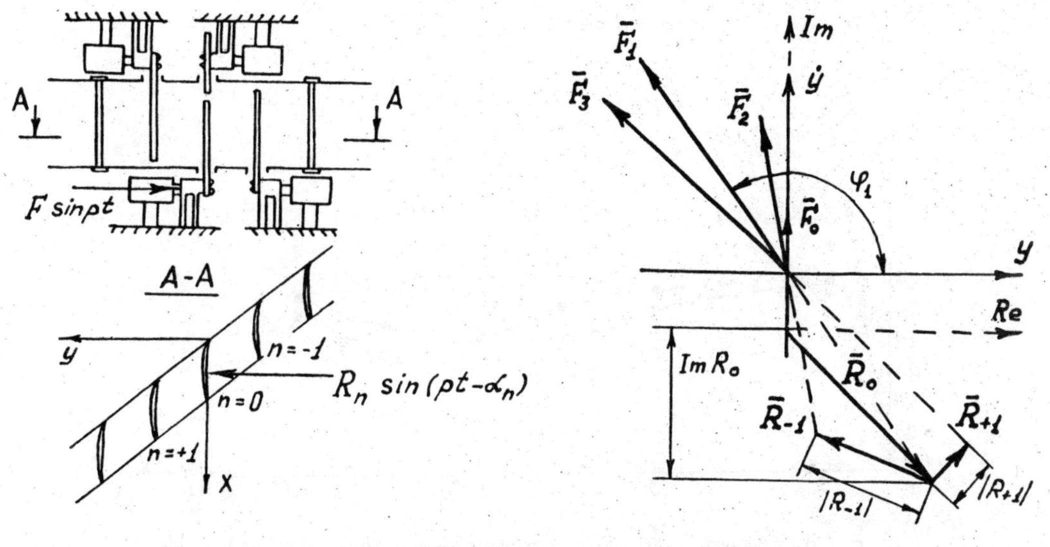

Fig. 3. Schematic diagram for determination of unsteady aerodynamic forces.

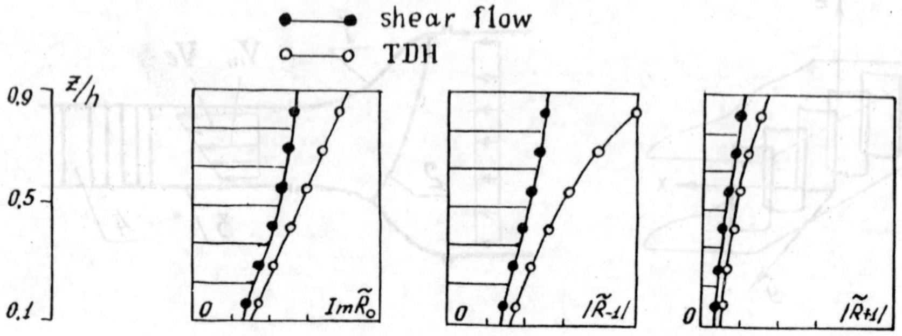

Fig. 4. Distribution of unsteady aerodynamic forces along the blade length (q_V = 0.12; i = 10°; Sh = 0.07; $\bar{t}$ = 1; β = 30°).

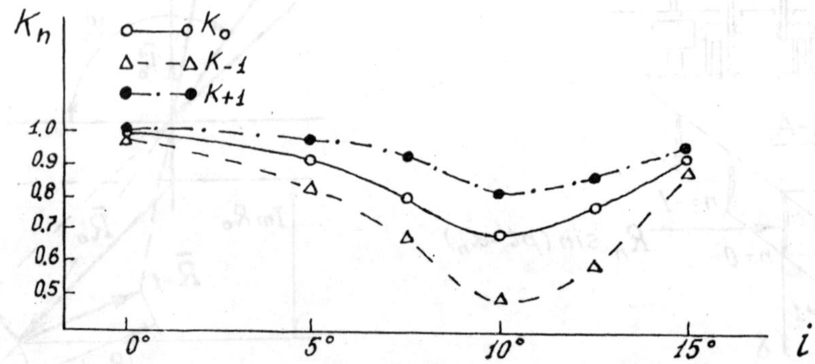

Fig. 5. Dependences of coefficients K_n on the attack angle (q_V = 0.12; Sh = 0.07; $\bar{t}$ = 1, β = 30°).

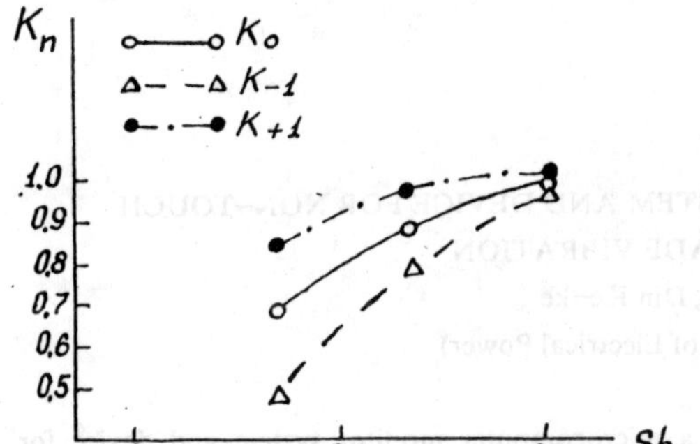

Fig. 6. Dependence of coefficients K_n on the Strouhal number (q_v = 0.12; i = 10°; $\bar{t}$ = 1; β = 30°).

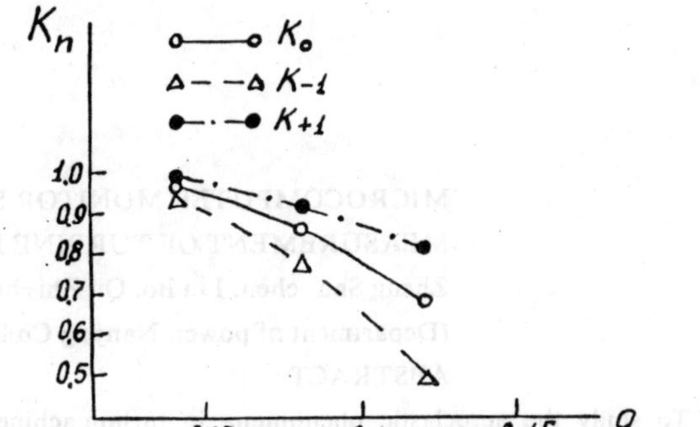

Fig. 7. Dependence of coefficients K_n on the flow shearing intensity (i = 10°; Sh = 0.07; $\bar{t}$ = 1; β = 30°).

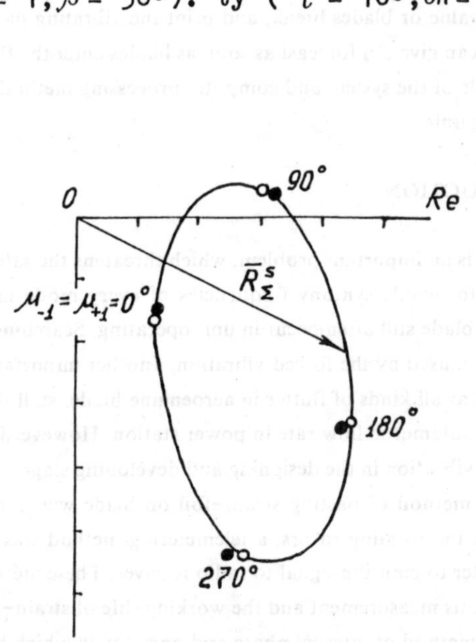

Fig. 8. Summation of unsteady aerodynamic forces:
—o— - geometrical summation;
● - measurements for three vibrating blades.

MICROCOMPUTER MONITOR SYSTEM AND DEVICE FOR NON–TOUCH MEASUREMENT OF TURBINE BLADE VIBRATION

Zheng Shu–chen, Liu Bo, Qu Zhi–huan, Din Ke–ke
(Department of power, Nanjing College of Electrical Power)

ABSTRACT

To study the aeroelastic phenomena in turbomachinery, a microcomputer monitor system and device for non–touch measurement of turbine blade vibration is developed by authors. The system can, continuously and with non–touch way, measure blade amplitude of vibration, phase angle and torsional angle at one stage of turbines when the machinery blades encounter vibration. In the case of turbine operation, it can display and print the vibrating parameters measured by the system, automatically give out the warning when blade amplitude of vibration is biger than safety value or blades break, and print the vibrating parameters in a span of time before the break occur. Especially, it can give out forecast as soon as blades enter the flutter boundary.
The paper describs the principle of the system and computer processing method with the measuring reports from test–table and full scale testing unit.

INTRODUCTION

Vibration of the turbine blade is an important problem, which threatens the safety operation of the unit. Though engineers regulated carefully to avoide syntony frequencies of every mode in design and research of turbine blades, the failuers of broken–blade still often occur in unit operating. Searching for its reason, besides design errors in avoiding syntong zone caused by the forced vibration, another important reason is blade self–excited vibration in turbomachine, such as all kinds of flutter in aeroengine blade, stall flutter of turbine blades under the high–back pressure and small volumetric flow rate in power station. However,it's difficult to predict these blade failures caused by self–excited vibration in the designing and developing stage.

Formerly, a touch measuring method of pasting strain–foil on blade was adopted to measure blade dynamic stress. To get the signals from the rotating rotors, a telemetering method was used, in which strain gauge was connected with radio transmitter to emit the signal to radio receiver. These methods have a shortcoming that the number of blades in simultaneous measurement and the working–life of strain–gauge and transceiver set are limited. The measuring vibration method of interval phase and non–touch which began to develop in 1970s [1], has opened up a wide prospeet for measurement of blade vibration. But as known from documents, there has still been the electronic beam device displaying blade amplitude value with bright–line length on the screen. The shortcoming for this device is that the pictures of the bright–line on srreen have to be taken for the measurement, which would not be easy . Such a device can't carry out continuous measurement, giving warning and printing the messeges after the failure. On the basis of developing electronic beam device (EBD) for measuring blade vibration amplitude value [2], we have independently developed a computer system and device for non–touch monitoring and measuring vibration of all blades at a rotor stage under operating.

FUNCTIONS OF THE SYSTEM

The functions achieved are as follows:

(1) Monitoring tangential vibration amplitude of all blades in a stage.

(2) Automatically giving warning when amplitude of vibration overpasses safety value.

(3) When blade breaks, automatically giving warning. and printing blade vibrating parameters before the failure occurs.

(4) Under the operating, monitoring the torsional angle of all blades in one stage and analyzing elastic torsional angle component and torsional vibration component of blade under the centrifugal force.

The four functions above–mentioned are effective on constant or non–constant rotational speed of unit. Therefore, in process of changing rotational speed from starting to stopping in unit, the system can successfully monitor blade vibration.

PRINCIPLE OF THE SYSTEM

A diagram of monitoring system for blade tangential vibration amplitude value is as following:

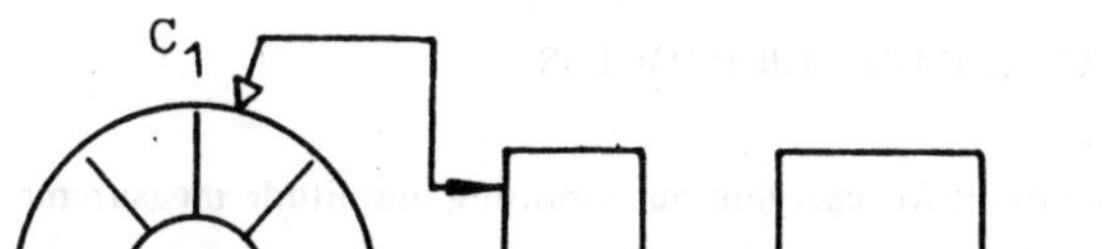

Diagram 1. Computer Monitoring System for measuring tangential vibration amplitude value of all blades in a stage under operating state.

A blade sensor C_1 is laid at the case oppsite to a blade tip measured. And when the blade passes through C_1, the blade pulse signal will be gained. When the rotor rotates for one circle, several pulse signals will be got (Number Z of the pulse signals is equal to the number of blades in a stage). If blades do not vibrate, blade will pass the sensor with the same phase every time. If blades vibrate, they pass the sensor with different phases. These phases are transmitted into computer 'with a form of time interval, then we can get $Z \times n$ time interval within n circles.

If the circle number is given previously, such as $n = N$, then within N circles we may find out the maximum Δt_{inmax} and the minimum Δt_{inmin} of time interval of blade No.i ($i = 1,2,...Z$). Through following simple calculation, amplitude of vibration Ain in blade No.i within circle may be found:

$$A_{IN} = \frac{L_{inmax} - L_{inmin}}{2} \qquad (i = 1,2,\cdots,z)$$

$$L_{inmax} = MAX\left(\frac{S\Delta t_{ij}}{T_j}\right) \qquad (j=1,2,\cdots,N)$$

$$L_{inmin} = MIN\left(\frac{S\Delta t_{ij}}{T_j}\right) \qquad (j=1,2,\cdots,N)$$

In the formula,
S—circle Length of rotor blade tip
T—period of blade wheel rotating for per circle (measured by sensor C_2) So, vibration amplitude value of all blades in a stage within N circles may be gained.
When vibrating amplitued of a certain blade or some blades overpass safety value, or have blade failure, judgement may be made by computer, and warning automatically can be given.
When the rotor rotates, some longer blades, besides blade tangential vibration, may have the elastic torsion under the action of centrifugal force and torsional vibration. The former plays an important role in determining the blade outlet angle under the operating state, while the latter is concerning vibration type measurement of blade torsional vibration. Through pulse sensors fitted in case, we may measure out the torsional angle of blade under operating state, and may separate the elastic torsional angle and torsional vibration angle.

PRINCIPLE OF COMPUTER PROCESS

A Principle of computer interface circuit for carrying out vibrating amplitude measurement, giving warning of broken blade and overpassing limitation is given as following:

1. Shaping circuit:
The signals emitted by all sensors are a series of irregular pulse signal. Shaping circuit receives these signals and converts them to some pulse signals with certain width and amplitude, which will emit interrupting requesting signal and provide control signal for later concerning circuit.

2. Counter:
It is used to produce timing signal. Measuring accuracy is concerned in the frequency of clock pulse signal. The higher frequency is, the higher distinguishability will be got. To keep distinguishability unchanged while rotating speed is raised, the clock pulse signal frequency also must be correspondingly raised.

3. Locked buffer circuit:
There are two functions for this part circuit.

(1) Segregation action: it's used to segregate counter output end from computer data bus because it presents high–resistance state to computer data bus.

(2) Lock action: Every sensor emits pulse signals, which are turned into a certain width and amplitude signal through shaping circuit. Wiith control of the signal, counter value is stored in the storage.

4. Translation circuit:
It is used to produce translation control signal, so as to read data in lock buffer. The diagram of circuit principle as following:

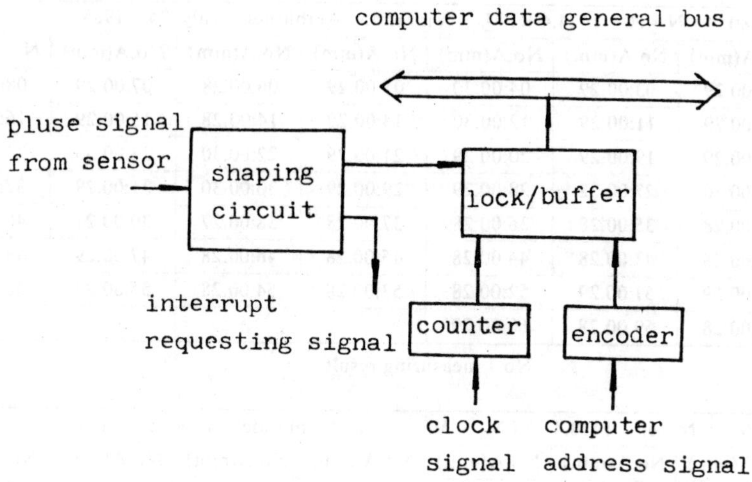

Diagram (2) The diagram of circuit principle for computer interface

With the support of hardware parts above mentioned, computer may then collect the needed data, based on which vibration amplitude of every blade would be derived.

The program deriving blade vibration amplitude includes following pasts: main program, interrupting service routine of blade signal and interrupting service routine of synchronous signal.

The work which main program completes ineludes: asking for user to input concerning parameters. These parameters are blade wheel radius. blade numbers and warning value for overpassing limitation. The works also includes initialization of every variable and state.

These interrupting service routines are main parts of the measuering system program. The interrupting service routine of blade signal collect the counter value dumped by these signals into computer. Besides collecting data, interrupting service routine of synchronous signal have to finish the following works: 1). judging whether the setting number of rotating circle has been reached. If it is achieved, deriving the vibration amplitude of every blade according to the collected data. Then some variable and state marks are set again. 2). making the computer repeat the process above mentioned.

The work of caculating vibration amplitude is carried on in such way: According to blade wheel rotating speed and its radius, the line speed of blade movement is derived. Every blade and synchronous signal detector respectively arrive at the correspond place where sensors are fitted in a certain time. According to the time difference and rotating line speed, the relative position of blade tip is determined. In N cricles, the maximum variation of this relative position for every biade can be found out and this value is divided by two. By this way, blade vibration amplitude is got.

After the data collected within a certain number of rotating circle N is processed, vibrating amplitude value of every blade will be gained. Then these values are put in a appointed memory block.When needed, these values can be displayed and printed.

REPORTS OF MEASUREMENT

The measuring and testing results are as follows:

R=80 N=60 1424.4 RPM Vibration Amblitude July 24 1989							
No.A(mm)	No.A(mm)	No.A(mm)	No.A(mm)	No.A(mm)	No.A(mm)	No.A(mm)	No.A(mm)
01:00.29	02:00.29	03:00.29	04:00.30	05:00.29	06:00.28	07:00.29	08:00.29
09:00.36	10:00.29	11:00.29	12:00.30	13:00.29	14:00.28	15:00.29	16:00.28
17:00.29	18:00.29	19:00.29	20:00.29	21:00.29	22:00.30	23:00.30	24:00.30
25:00.30	26:00.30	27:00.32	28:00.29	29:00.29	30:00.30	31:00.29	32:00.30
33:00.28	34:00.28	35:00.28	36:00.28	37:00.28	38:00.27	39:00.27	40:00.28
41:00.28	42:00.28	43:00.28	44:00.28	45:00.28	46:00.28	47:00.29	48:00.29
49:00.29	50:00.29	51:00.29	52:00.28	53:00.28	54:00.28	55:00.28	56:00.28
57:00.27	58:00.28	59:00.28	60:00.29				

No.1 measuring result

R=80 N=60 1428.4 RPM Vibration Amblitude July 24 1989							
No.A(mm)	No.A(mm)	No.A(mm)	No.A(mm)	No.A(mm)	No.A(mm)	No.A(mm)	No.A(mm)
01:00.14	02:00.14	03:00.13	04:00.14	05:00.13	06:00.13	07:00.13	08:00.13
09:00.13	10:00.13	11:00.13	12:00.12	13:00.14	14:00.14	15:00.14	16:00.14
17:00.14	18:00.15	19:00.15	20:00.14	21:00.14	22:00.14	23:00.14	24:00.13
25:60.14	26:00.13	27:00.13	28:00.13	29:00.13	30:00.14	31:00.15	32:00.13
33:00.13	34:00.13	35:00.13	36:00.13	37:00.13	38:00.13	39:00.13	40:00.13
41:00.14	42:00.14	43:00.14	44:00.13	45:00.13	46:00.14	47:00.14	48:00.13
49:00.13	50:00.13	51:00.14	52:00.13	53:00.14	54:00.18	55:00.13	56:00.15
57:00.14	58:00.15	59:00.15	60:00.15				

No.2 measuring results

CONCLUDING REMARKS

The system and device of the measurement which is developed by aut.. rs are successful in the respect of measuring blade vibration of turbomachinery. It features non touch, interval phase sample and data process of computer through interface circuit. Compared with traditional touching measurement, the system has the advantages as follows:

1. The working life of the system is very long, and it is convenient to fix and manipulate.
2. The system can measure and monitor vibration parameters of all blades ina stage.
3. The system can give warning when blade vibration overpasses the limitation, and give out the vibration parameters befor the failure. This function can be used to analyze the reason of failures. So this system will take a important role in research of aeroelasticity in turbomachinery andkeeping the unit safty.

REFERENCE

1. Zabrolzkey,I.I, Korostzerov,Yu.A and Shipov,L.A: Non–Constact Measurements of Vibratng Blades in Turbomachinery "Mechanical Manufacture of Moscow" 1977
2. Wu Shixiang: The priciple and usage of the electronic monitor of blade vibration on turbines. (in chinese) Chinese society of aeronautics: The symposium of technology in engine measurement, Dong Fang Steam Turbine Factory, Oct. 1982

2-E FORCING FUNCTION EFFECTS ON TURBOMACHINE GUST UNSTEADY AERODYNAMICS

Steven R. Manwaring[+] and Sanford Fleeter [*]
(Thermal Sciences and Propulsion Center
School of Mechanical Engineering
Purdue University, West Lafayette, Indiana 47907)

ABSTRACT

To investigate the forcing function effects on the fundamental 2-E gust unsteady aerodynamic response of a multistage turbomachine blade row, a series of fundamental experiments are performed in an axial flow research compressor which is extensively instrumented. The 2-E unsteady aerodynamic forcing function excitations are generated by both a circumferential inlet flow distortion and the wakes from airfoil-type obstructions which are representative of an airfoil, strut, or probe excitation.

NOMENCLATURE

b	Rotor blade semichord
$\overline{C}_l$	Rotor blade steady loading $= \int_0^c (\overline{C}_{p,\text{pressure}} - \overline{C}_{p,\text{suction}})dx$
$\overline{C}_p$	Rotor blade steady pressure coefficient
C_{pi}	Rotor blade harmonic unsteady pressure coefficient
$C_{\Delta pi}$	Rotor blade harmonic unsteady pressure difference coefficient
$\overline{i}$	Rotor blade mean incidence angle
k	Reduced frequency $= \omega b/\overline{V}_x$
p	Rotor blade surface unsteady pressure
$\overline{P}_{exit}$	Rotor drum steady pressure
$\overline{P}_s$	Rotor blade surface steady pressure
$\hat{p}_i$	Harmonic unsteady pressure
U_t	Rotor blade tip speed
u^+	Streamwise unsteady velocity component
$\hat{u}_i^+$	Streamwise gust harmonic component
v^+	Transverse unsteady velocity component
$\hat{v}_i^+$	Transverse gust harmonic component
$\overline{V}_x$	Mean axial velocity
ω	Two-per-rev forcing function frequency, radians

+ Engineer - Aerodynamic Design Methods with General Electric, Evandale
* Professor, School of Mechanical Engineering and Director, Thermal Sciences and Propulsion Center

INTRODUCTION

Nonuniform inlet flow to a rotor represents the most common aerodynamic excitation source for aerodynamically forced vibrations. These nonuniformities can be broadly categorized as wake excitations or inlet distortions. These nonuniformities result in detrimental aeromechanical effects, with the distortion and wakes both representing unsteady aerodynamic forcing functions to downstream turbomachine blade rows.

Distortions and wakes are both high energy aerodynamic forcing functions which are characterized by low to moderate values of the reduced frequency. On a first principle basis, they are both analyzed by first defining the unsteady aerodynamic forcing function in terms of harmonics. The periodic response of an airfoil row to each harmonic is then assumed to be comprised of two components. One is due to the harmonic components of the unsteady aerodynamic forcing function being swept past the nonresponding airfoil row, termed the streamwise and transverse gust responses. The second, the self-induced unsteady aerodynamics, arises when a vibrational response of the airfoil row is generated.

Current state-of-the-art unsteady aerodynamic models for the prediction of forced response do not consider differences in the various aerodynamic forcing functions. Also, the gust and motion-induced unsteady aerodynamic models involve many physical and numerical assumptions. Therefore, experimental modeling of the fundamental distortion and wake generated blade row periodic unsteady aerodynamic response, including both the forcing function and the resulting blade row unsteady aerodynamics, is needed for validation and enhancement of theoretical and numerical models.

In this paper, the fundamental 2-E gust unsteady aerodynamic response of a multistage turbomachine blade row is experimentally investigated for the first time. This is accomplished by means of a series of experiments performed in an extensively instrumented axial flow research compressor. The 2-E unsteady aerodynamic forcing function excitations are generated by both a circumferential inlet flow distortion and the wakes from airfoil-type obstructions which are representative of an airfoil, strut, or probe excitation.

As schematically depicted in Figure 1, two different two-per-rev aerodynamic forcing functions are considered: (1) the velocity deficit from two 90 degree circumferential inlet flow distortions and (2) the wakes from two upstream airfoil-type obstructions. These aerodynamic forcing functions to the first stage rotor blade row are measured with a rotating cross hot wire probe, with these data then analyzed to determine the streamwise and transverse velocity components, u^+ and v^+ shown in Figure 1. The resulting unsteady aerodynamic gust generated rotor blade surface unsteady pressure chordwise distributions are measured with embedded ultra-miniature high response dynamic pressure transducers.

RESEARCH COMPRESSOR

The Purdue Axial Flow Research Compressor experimentally models the fundamental turbomachinery unsteady aerodynamic multistage interaction phenomena including the incidence angle, the velocity and pressure variations, the aerodynamic forcing function, the reduced frequency, and the unsteady blade row interactions. The compressor is driven by a 15 HP DC electric motor and is operated at a speed of 2,250 RPM. Each identical stage of the baseline compressor contains 43 rotor blades and 31 stator vanes having a British C4 profile, with the first stage rotor inlet flow field established by a row of 36 variable setting inlet guide vanes. The overall airfoil and compressor characteristics are presented in Table 1. For these experiments, the first stage rotor blade row was extensively instrumented. Due to the large gap between the IGV row and the first stage rotor row, approximately 75% vane chord, the potential flow effects are negligible compared to the viscous wakes. To eliminate any potential flow effects from the downstream stator row on the instrumented first stage rotor blades, the first stage stators and second stage rotors were removed, as schematically depicted in Figure 2.

INSTRUMENTATION

Both steady and unsteady first stage rotor blade row data are required. The steady data quantify the rotor mean inlet flow field and midspan steady loading distribution. The unsteady data define the periodic aerodynamic forcing function and the resulting midspan blade surface periodic pressure distributions.

The inlet flow field, both steady and unsteady, is measured with a rotating cross hot-wire probe mounted 30% of blade chord upstream of the rotor row. The cross hot-wire probe is calibrated and linearized for velocities from 18.3 m/sec to 53.4 m/sec and +/-35 degrees angular variation, with the accuracy of the velocity magnitude and flow angle determined to be 4% and +/-1.0 degree, respectively. Centrifugal loading effects on the rotating hot-wire sensor resistances and, thus, responses were found to be negligible.

The detailed steady aerodynamic loading on the rotor blade surfaces is measured with a chordwise distribution of 20 midspan static pressure taps, 10 on each surface. The static pressure at the rotor exit plane, measured with a rotor drum static tap, is used as the blade surface static pressure reference. These static pressure measurements are made using a rotor based 48 port constant speed drive Scanivalve system located in the rotor drum.

The measurement of the midspan rotor blade surface unsteady pressures is accomplished with 20 ultra-miniature, high response transducers embedded in the rotor blades at the same chordwise locations as the static pressure taps. To minimize the possibility of flow disturbances associated with the inability of the transducer diaphragm to exactly maintain the surface curvature of the blade, a reverse mounting technique is utilized. The pressure surface of one blade and the suction surface of the adjacent blade are instrumented, with transducers embedded in the nonmeasurement surface and connected to the measurement surface by a static tap. The embedded dynamic transducers were both statically and dynamically calibrated. The static calibrations showed good linearity and no discernible hysteresis. The dynamic calibrations demonstrated that the frequency response, in terms of gain attenuation and phase shift, were not affected by the reverse mounting technique. The accuracy of the unsteady pressure measurements, determined from the calibrations, is +/-4%

The rotor based static pressure Scanivalve transducer, rotating hot-wire probe and 20 blade surface dynamic pressure transducers are interfaced to the stationary frame-of-reference through a 40 channel slip ring assembly. On-board signal conditioning of the transducer output signals is performed to maintain a good signal-to-noise ratio through the slip ring. The remaining 17 channels of the slip-ring assembly are used to provide excitation to the transducers and on/off switching excitations to the Scanivalve DC Motor.

DATA ACQUISITION AND ANALYSIS

STEADY DATA

The rotor blade surface static pressure data, measured with the rotor-based Scanivalve system, are defined by a root-mean-square error analysis of 20 samples with a 95% confidence interval. The reference for these midspan blade pressure measurements is the static pressure at the exit of the rotor measured on the rotor drum. Thus, the blade surface and the reference static pressures are measured at different radii. Hence, a correction for the resulting difference in the radial acceleration is applied in calculating the blade surface static pressure coefficient defined in Equation 1.

$$\overline{C}_p = \frac{\overline{P}_s - \overline{P}_{exit}}{\frac{1}{2}\rho U_t^2} \tag{1}$$

where U_t is the rotor blade tip speed.

PERIODIC DATA

The periodic data of interest are the harmonic components of the aerodynamic forcing function to the first stage rotor blade row together with the resulting rotor blade surface unsteady pressures and unsteady pressure differences. These are determined by defining a digitized ensemble averaged unsteady aerodynamic data set consisting of the rotating cross hot-wire probe and blade surface dynamic pressure transducer signals at each steady operating point. In particular, these time-variant signals are digitized with a high speed A-D system at a rate of 20 kHz and then ensemble averaged. The key to this averaging technique is the ability to sample data at a preset time, accomplished by an optical encoder mounted on the rotor shaft. The microsecond range step voltage signal from the encoder is the data initiation time reference and triggers the high speed A-D multiplexer system. To significantly reduce the random fluctuations superimposed on the periodic signals of interest, 200 averages are used. A Fast Fourier Transform (FFT) algorithm is then applied to these ensemble averaged signals to determine the harmonic components of the unsteady aerodynamic forcing function and the resulting rotor blade surface harmonic unsteady pressures and pressure differences.

The unsteady inlet flow field to the rotor row, measured with the rotating cross hot-wire probe, is quantified by the inlet relative velocity and flow angle. The velocity triangle relations depicted in Figure 1 are then used to determine the detailed unsteady velocity into the rotor row, in particular, the streamwise and transverse velocity components, u^+ and v^+ respectively. These are then Fourier decomposed to determine their harmonic components, $\hat{u}_i^+$ and $\hat{v}_i^+$.

The various unsteady aerodynamic gust mathematical models reference the gust generated airfoil aerodynamic response to a transverse gust at the leading edge of the airfoil. However, in the experiments described herein, the time-variant data are referenced to the initiation of the data acquisition shaft trigger pulse. Thus, for consistency with the models, the periodic data are further analyzed and referenced to a transverse gust at the leading edge of the first stage rotor blade. This is accomplished by assuming that: (1) the aerodynamic forcing function remains fixed in the stationary reference frame; and (2) the forcing function does not change from the rotating hot-wire probe axial location to the rotor row leading edge plane.

The rotor blade surface unsteady pressure data, measured with the embedded high response pressure transducers, are analyzed to determine the harmonics of the chordwise distribution of the unsteady pressure coefficient, C_{pi}, and the unsteady pressure difference coefficient, $C_{\Delta pi}$. These are defined in Equation 2 and are specified from the Fourier coefficients of the digitized unsteady pressure transducer signals. Thus, the final form of the gust generated rotor blade row unsteady aerodynamics is the chordwise distribution of the harmonic complex unsteady pressure and pressure difference coefficients.

$$C_{pi} = \frac{\hat{p}_i}{\rho \overline{V}_x^2 \left(\frac{\hat{v}_i^+}{\overline{V}_x}\right) \overline{\beta}} \tag{2a}$$

$$C_{\Delta pi} = C_{pi,\text{pressure}} - C_{pi,\text{suction}} \tag{2b}$$

where $\hat{v}_i^+$ is the harmonic transverse gust component, $\overline{V}_x$ is the mean axial velocity, and $\overline{\beta}$ is the relative mean flow angle.

RESULTS

A series of experiments are performed to investigate and quantify the effects of different aerodynamic forcing functions on periodic 2-E gust unsteady aerodynamics of a first stage rotor blade row. The 2-E unsteady aerodynamic forcing function excitations are generated by a circumferential inlet flow distortion and the wakes from airfoil-type obstructions. In these experiments, the two two-per-rev forcing functions are equivalent in terms of aerodynamically forced response as each has approximately the same nominal value for the first harmonic streamwise-to-transverse gust amplitude ratio.

PERIODIC AERODYNAMIC FORCING FUNCTIONS

The Fourier decomposition of the inlet distortion and wake aerodynamic forcing functions to the first stage rotor in terms of the gust components are presented in Figures 3 and 4. The inlet distortion results in a dominant two-per-rev excitation with smaller higher harmonics. With the wake forcing function, a strong two-per rev signal is again present. However, the higher harmonic amplitudes are approximately the same magnitude as the generating wake two-per-rev fundamental for the streamwise gust components and are larger for the transverse gust component.

BLADE SURFACE STATIC PRESSURES

The effect of steady aerodynamic loading, characterized by the mean incidence angle, on the rotor blade surface static pressure coefficient for both the inlet distortion and the obstruction wake forcing function are shown in Figure 5. The level of steady loading only affects the static pressure distribution on the pressure surface over the front 40% of the chord. On the suction surface, the steady loading variation has a large effect on the static pressure distribution over the entire suction surface. Also, these data exhibit no indication of flow separation. Of particular interest to the unsteady experiments, these data clearly show that these different aerodynamic forcing functions have no effect on the steady aerodynamic performance of the rotor, i.e., the airfoil steady surface static pressures are independent of the unsteady aerodynamic forcing function.

ROTOR ROW PERIODIC AERODYNAMIC RESPONSE

The periodic aerodynamic first harmonic response of the first stage rotor blade row to the inlet distortion and wake forcing functions, including the effect of steady aerodynamic loading, are presented in Figures 6, 7 and 8. In particular, these figures present the chordwise distributions of the complex unsteady pressure coefficient on the individual rotor blade surfaces as well as the corresponding complex unsteady pressure difference coefficients generated by the inlet distortion first harmonic two-per-rev aerodynamic forcing function.

Pressure Surface Unsteady Pressures

The effect of steady aerodynamic loading level on the first harmonic of the complex unsteady pressure distribution on the rotor blade pressure surface for the inlet distortion is shown in Figure 6. The unsteady pressure magnitude data decrease monotonically with increasing chord for all steady loading levels. Also, this form of the dimensionless unsteady pressure coefficient, Equation 2a, results in the compression of these magnitude data for all mean flow incidence angle values. Thus, the magnitude of the unsteady pressure response on the blade pressure surface, i.e., the low camber airfoil surface, is primarily due to the level of steady loading as characterized by the mean flow incidence angle. With regard to the unsteady pressure phase data on this airfoil surface, they increase over the front half of the chord and are independent of the level of steady loading.

Suction Surface Unsteady Pressures

The effect of steady aerodynamic loading on the first harmonic complex unsteady pressure on the rotor blade suction surface is shown in Figure 7 for the inlet distortion. In contrast to the pressure surface, the form of the dimensionless unsteady pressure coefficient, Equation 2a, does not compress the suction surface magnitude data with mean flow incidence angle, the exception being between 13% and 20% of the chord for the inlet distortion. The unsteady pressure magnitude data decrease to a minimum and then increase as the chord is traversed. Also, these magnitude data are dependent on the steady loading, increasing with increased steady loading. The corresponding phase data decrease over the chord and, in the leading edge and mid-chord regions, are dependent on the steady loading. Near the leading edge, increased loading results in an increase in the phase whereas in the mid-chord region the phase decrease in value.

Unsteady Pressure Differences

The effect of steady loading on the first harmonic complex unsteady pressure difference across the rotor blade is shown for the inlet distortion forcing function in Figure 8. Also presented as a reference are the flat plate cascade predictions. The magnitude data generally decrease with increasing chord, in trendwise agreement with the predictions. Also, these magnitude data increase with increasing steady loading. The corresponding unsteady pressure difference phase data are nearly constant and independent of the level of steady loading in the front chord region. There is a sharp increase in the phase data in the mid to aft chord regions of the airfoil, with the onset location of this sharp increase moving forward towards the leading edge with increasing mean flow incidence angle.

SUMMARY AND CONCLUSIONS

The fundamental 2-E gust unsteady aerodynamic response of a multistage turbomachine blade row has been experimentally investigated for the first time. This was accomplished by means of a series of experiments performed in an extensively instrumented axial flow research compressor. The 2-E unsteady aerodynamic forcing function excitations were generated by a circumferential inlet flow distortion and the wakes from airfoil-type obstructions. General experimental results are noted in the following.

* The inlet distortion forcing function shows a dominant two-per-rev, with much smaller higher harmonic content. In contrast, the wake forcing function shows a strong two-per-rev with higher harmonics of larger amplitudes

* The rotor steady aerodynamic performance is independent of the aerodynamic forcing function.

* On the pressure surface, the inlet distortion generated unsteady pressure magnitude data decrease monotonically with chord and are compressed with respect to the mean flow incidence angle. The phase data increase over the front half of the chord and are independent of the level of steady loading.

* On the suction surface, the inlet distortion generated unsteady pressure magnitude data decrease to a minimum and then increase with chord. Also, they are dependent on the steady loading, increasing with increased steady loading. The phase data decrease over the chord and are dependent on the steady loading.

* With regard to the inlet distortion generated unsteady pressure difference data, the magnitude data decrease with increasing chord and increase with increased incidence angle. The phase data are nearly constant and independent of the loading in the front chord region. There is a sharp increase in phase in the midchord region, with the onset location moving forward with increasing incidence angle.

ACKNOWLEDGEMENTS

Research sponsored by the Air Force Office of Scientific Research (AFSC) under Contract F49620-88-C-0022. The United States Government is authorized to reproduce and distribute reprints for governmental purposes notwithstanding any copyright notation hereon.

REFERENCES

1. Capece, V.R., Manwaring, S.R., and Fleeter, S., "Unsteady Blade Row Interactions in a Multi-Stage Compressor" *AIAA Journal of Propulsion and Power*, Vol. 2, No. 2, March-April 1986, pp. 168-174.
2. Capece, V.R. and Fleeter, S., "Unsteady Aerodynamic Interactions in a Multi-Stage Compressor," *ASME Journal of Turbomachinery*, Vol. 109, No. 3, July 1987, pp. 420-428.
3. Datko Jr., F.T. and O'Hara, J.A., "The Aeromechanical Response of an Advanced Transonic Compressor to Inlet Distortion," *ASME Paper 87-GT-189*. 1987.

4. Fleeter, S., Jay, R.L., and Bennett, W.A., "Rotor Wake Generated Unsteady Aerodynamic Response of a Compressor Stator," *ASME Journal of Engineering for Power*, Vol. 100, No 4, October, 1978, pp. 664-675.

5. Fleeter, S., Bennett, W.A., and Jay, R.L., "The Time Variant Aerodynamic Response of a Stator Row Including the Effects of Airfoil Camber," *ASME Journal of Engineering for Power*, Vol. 102, No. 2, April 1980, pp. 334-343.

6. Gallus, H.E., Lambertz, J., and Wallman, Th., "Blade-Row Interaction in an Axial-Flow Subsonic Compressor Stage," *ASME Journal of Engineering for Power*, Vol. 102, No. 1, January 1980, pp. 169-177.

7. Smith, S.N., "Discrete Frequency Sound Generation in Axial Flow Turbomachines," *ARC R&M 3709*, 1971.

	ROTOR	STATOR	IGV
Airfoil Type	C4	C4	C4
Number of Airfoils	43	31	36
Chord, C (mm)	30	30	30
Solidity, C/S	1.14	1.09	0.96
Camber, θ	27.95	27.70	36.9
Stagger Angle, λ	36	-36.1	18.5
Aspect Ratio	2.0	2.0	2.0
Thickness/Chord (%)	10	10	10
Design Flow Rate (kg/s)		2.03	
Design Axial Velocity (m/s)		24.4	
Rotational Speed (RPM)		2250	
Number of Stages		2	
Stage Pressure Ratio		1.003	
Inlet Tip Diameter (mm)		420	
Hub/Tip Radius Ratio		0.714	
Stage Efficiency (%)		85	

Table 1. Overall airfoil and compressor characteristics

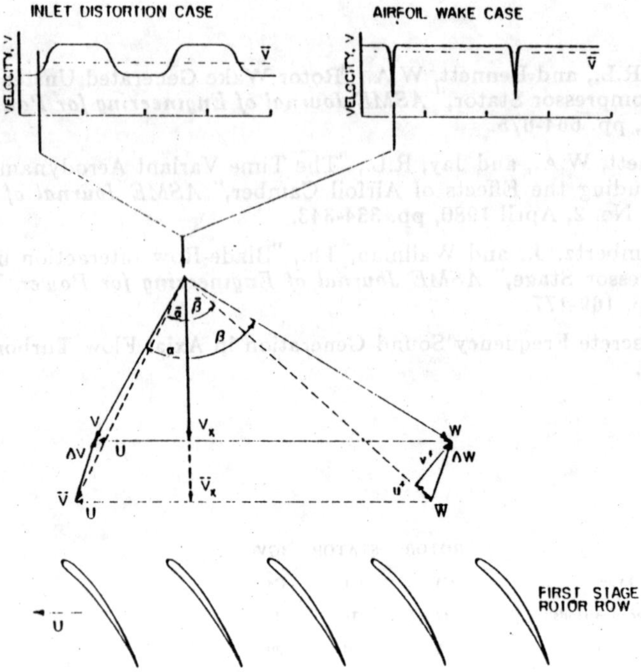

Figure 1 Aerodynamic forcing function decomposition

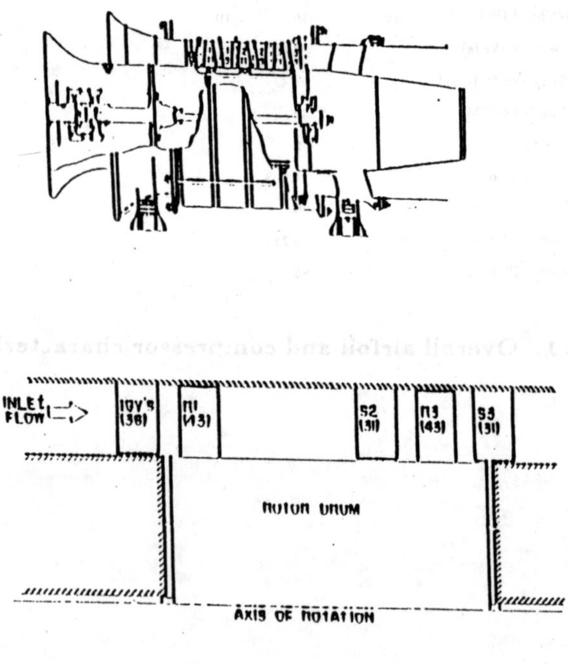

Figure 2. Research compressor configuration

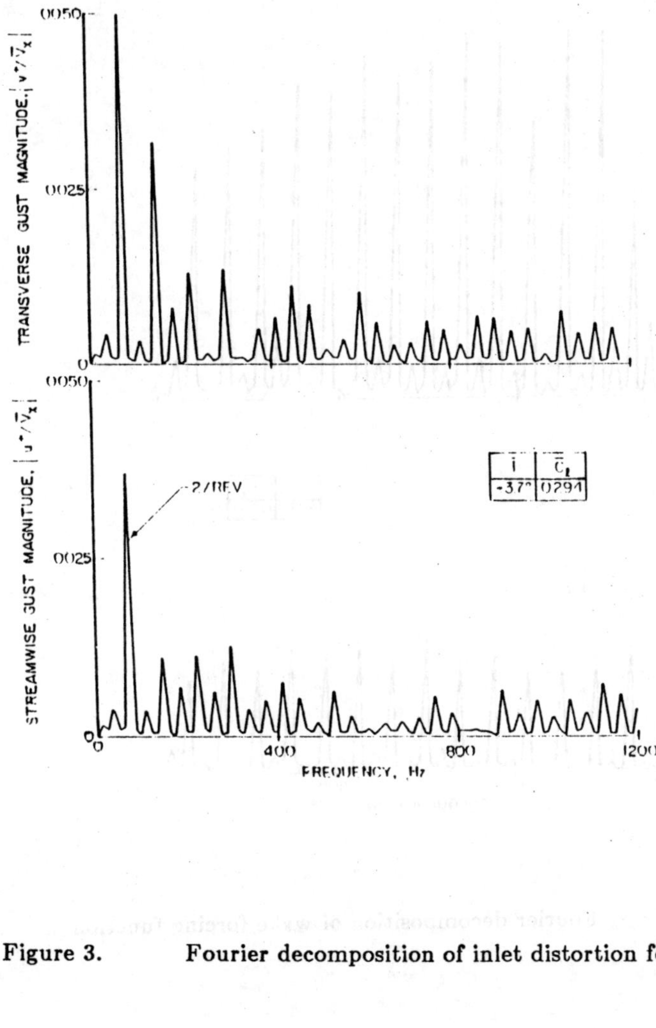

i	$\bar{C}_l$
-37°	0294

Figure 3. **Fourier decomposition of inlet distortion forcing function**

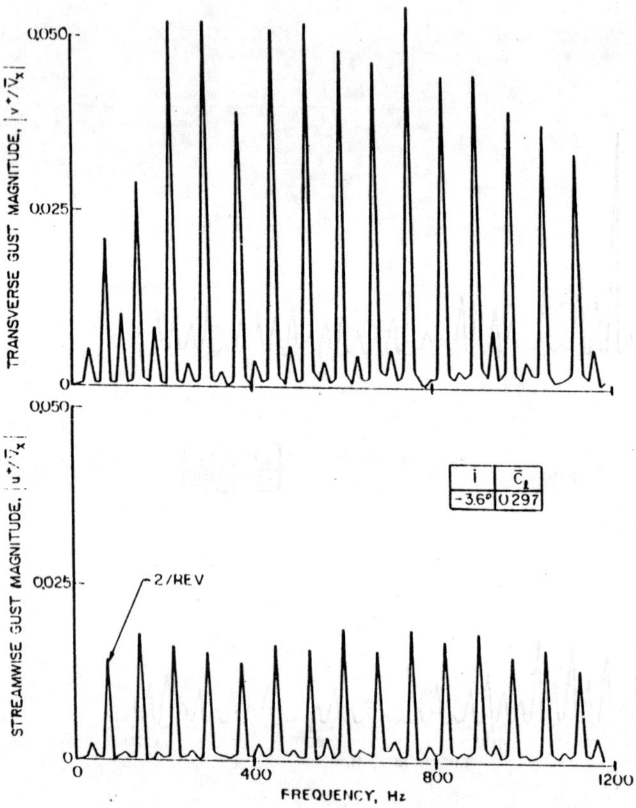

Figure 4. **Fourier decomposition of wake forcing function**

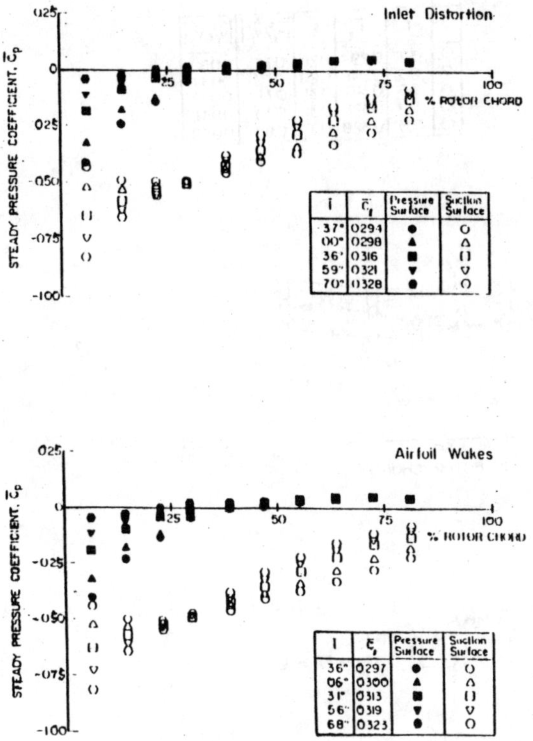

Figure 5. **Forcing function and steady loading effects on surface static pressure coefficient**

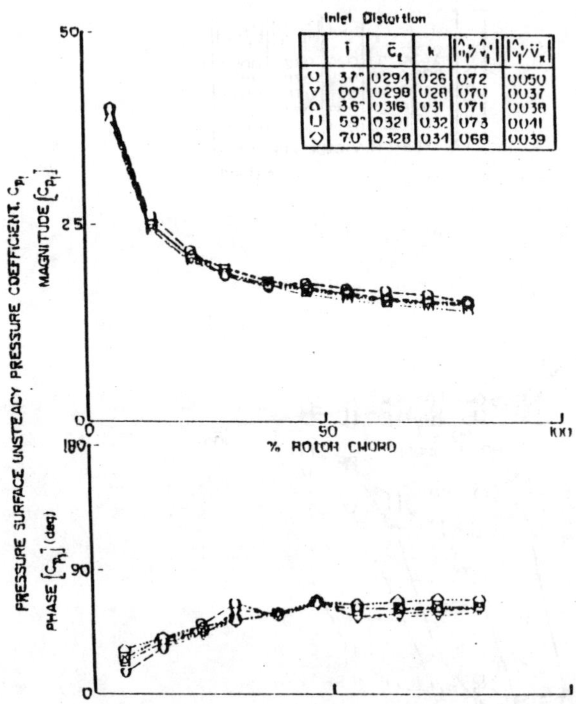

Figure 6. **Pressure surface steady loading effect on unsteady pressure**

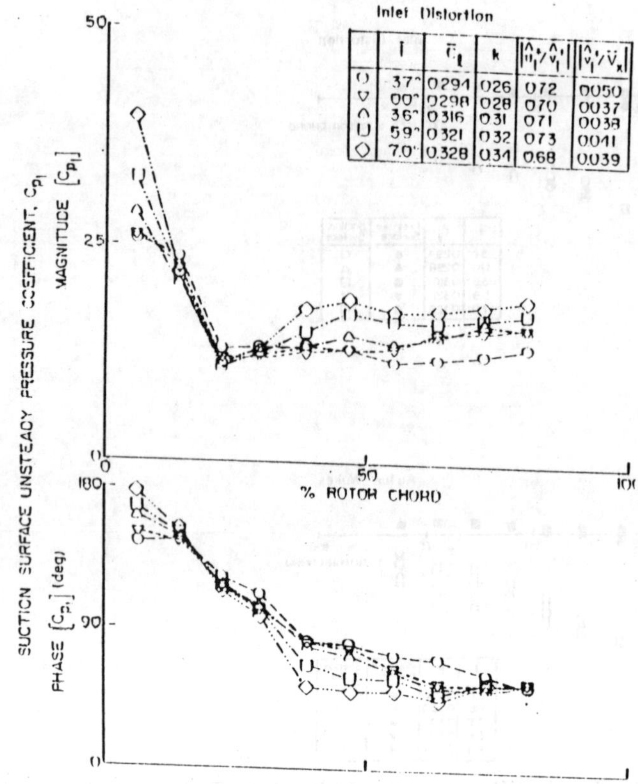

Figure 7. Suction surface steady loading effect on unsteady pressure

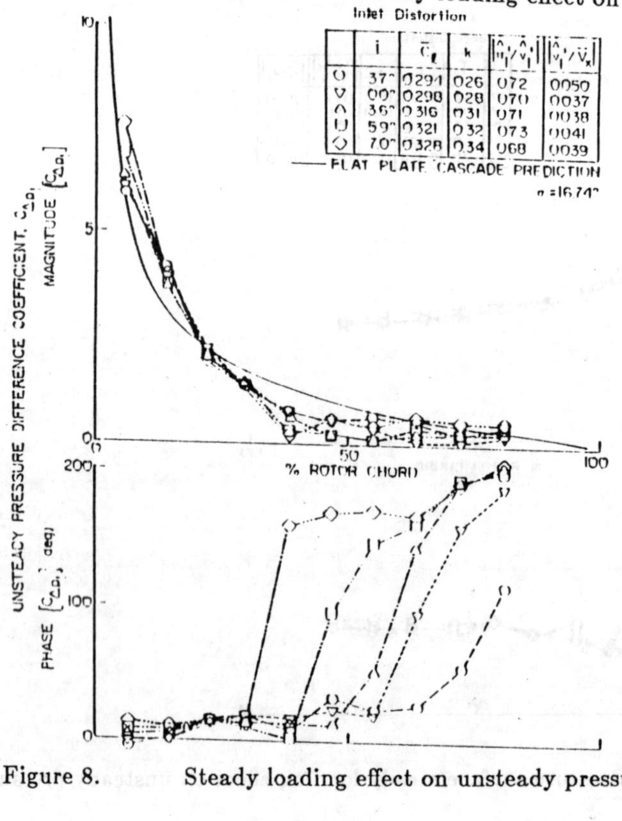

Figure 8. Steady loading effect on unsteady pressure difference

A STUDY ON THE ROTATING STALL OF CENTRIFUGAL COMPRESSORS.

Hideo Nishida, Hiromi Kobayashi, Takeo Takagi, Yasuo Fukushima
(Hitachi Ltd., Ibaraki, Japan)

ABSTRACT

Rotating stalls of centrifugal compressors, which often cause low frequency shaft vibration in high pressure compressors, is experimentally examined. The pressure fluctuations of 4 low specific speed centrifugal compressors, with vaneless diffusers, are measured by semiconductor pressure transducers. Rotating stall is caused by vaneless diffusers, where the stall cell number is 1 to 5 and its angular velocity to impeller tip speed ratio is in the range of 0.07~0.21. Critical inlet flow angle for the rotating stall depends not only on width ratio b_d/r_2, but also on contraction ratio b_d/b_2 and inlet curvature radius ρ_i. An empirical prediction formula for critical flow angle based on the Senoo method is derived.

INTRODUCTION

Multi-stage centrifugal compressors have recently been used in high pressure fields such as gas injection or various gas synthesis plants, according to progress in manufacturing techniques and reliability. However, low frequency shaft vibration problems[1~3] have often occurred, which probably are induced by aerodynamic force.

Fig. 1 shows one example[4] of low frequency shaft vibration which occurred in a carbon dioxide centrifugal compressor for a urea synthesis plant. It was thought that the shaft vibration was caused by a rotating stall of the centrifugal compressor. This was because the vibration appeared when suction volume flow rate dropped below a certain value and the frequency, which was about 42 Hz, was much lower than the rotor critical speed.

In regards to rotating stall many studies for centrifugal blowers, especially vaneless diffusers, have been reported.

Senoo and Kinoshita[5,6] repoted that the critical inlet flow angle of a vaneless diffuser for rotating stall could be predicted by the calculated angle, based on an axi-symmetric boundary layer analysis, at which the flow began to reverse. Detail flow measurements of rotating stalls of vaneless diffusers were conducted by Tsurusaki et al[7]. Frigne[8] reported that the rotating stall was either caused by the impeller or the diffuser.

There are many reports about the low frequency shaft vibration prob-

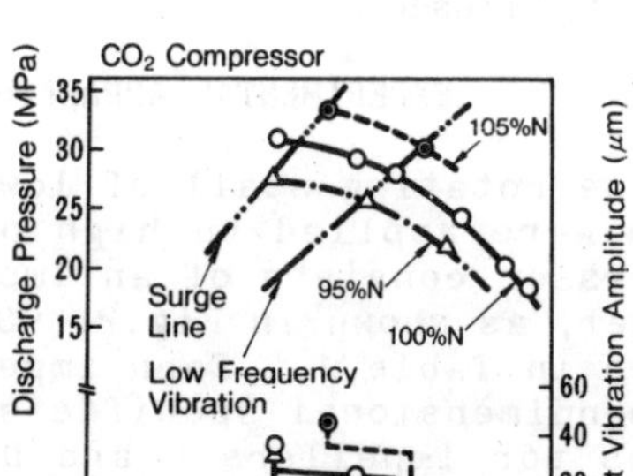

Fig.1 An example of low frequency shaft vibration for a high pressure centrifugal compressor.

lems of centrifugal compressors, which may be due to rotating stall. However, there are few examples that the relation between vibration and rotaing stall is clarified by measuring flow in compressors.

It is doubtful whether the abovementioned studies for centrifugal blowers can be applied to centrifugal compressors or not. This is because the impeller tip speed of the blowers is fairly lower than that of the compressors and blowers aren't equipped with a return-channel. So, the rotating stall of centrifugal compressors was examined experimentally with the conditions as close as possible to multi-stage compressors, except pressure.

NOMENCLATURE

b	:	flow passage width
f_p	:	pressure fluctuation peak frequency
H_{ad}	:	adiabatic head rise
Δh	:	head based on fluctuating pressure
i	:	stall cell number
N	:	impeller rotational speed
ns	:	specific speed
Q_s	:	suction volume flow rate
r	:	radius
t	:	blade thickness
u	:	impeller peripheral speed
α	:	absolute flow angle from tangential
ω_s	:	stall cell angular velocity
ω_{imp}	:	impeller angular velocity

Subscript

2	:	impeller outlet
i	:	diffuser inlet
crit	:	critical (stall onset)
d	:	diffuser

Table 1. Specifications and geometric parameters of each impeller.

Impeller	A	B	C	D
D_2 mm	300	300	300	300
b_2 mm	5.5	7.3	10.0	3.5
t_2 mm	3.0	10.0	3.0	2.4
u_2 m/s	300	300	300	270
ns	0.28	0.28	0.42	0.23

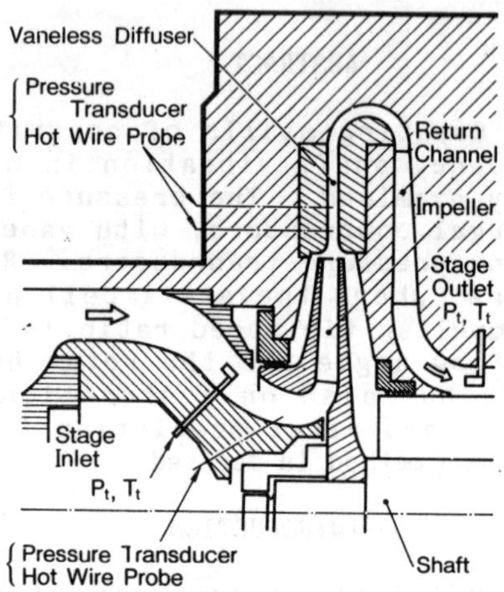

Fig.2 Tested compressor cross-sectional view.

EXPERIMENTAL APPARATUS AND MEASUREMENT TECHNIQUES

The rotating stall of low specific speed centrifugal compressors, which were applied to high pressure machines, was examined. A tested compressor consists of an impeller, a vaneless diffuser, and a return channel, as shown in Fig.2. Specifications and geometric parameters are listed in Table 1. Each impeller has several two-dimensional blabes. The nondimensional specific speed, ns, for impellers A and B is 0.28. The ns for impellers C and D is 0.42 and 0.23, respectively. Outlet blade thickness for impeller B is 10mm, which is larger than that of the other impellers.

Various parallel wall vaneless diffusers with an outlet-to-inlet radius ratio of 1.45 were examined. The figure for the diffuser title gives diffuser width in mm.

Return channels were designed for diffusers VL6.7T and VL3.8T and one or the other return channel was used in each experiment.

The impellers were driven by a 550kw DC-motor which rotational speed could be varied from 3,000 to 20,000 rpm. The compressor inlet was at atmospheric condition. The tested compressor flow rate was controlled by a throttle valve set in the delivery pipe.

Overall performance was measured by total pressure and total temperature probes mounted at both stage inlet and outlet. Flow rate was measured by an orifice flowmeter. Unsteady flow phenomena were measured by semiconductor pressure transducers and hot wire probes installed at the diffuser and the impeller inlets. The probes were installed at two different points on the circumference of the same radius to evaluate rotational speed and number of stall cells. Probe signals were fed through an amplifier or linealizer and recorded on tape by a data recorder. Signals were monitored by a FFT analyzer or an oscilloscope during the experiments.

EXPERIMENTAL PESULTS AND DISCUSSION

1. Overall performance and rotating stall

Experiments for each compressor were made at 3 or 4 different rotational speeds. Performance curves of the compressor equipped with impeller A and diffuser VL3.8T are shown in Fig.3. Diffuser pressure transducer signals and the frequency spectra achieved after processing them with a FFT analyzer are shown in Fig.4. Pressure fluctuation was slight at operating point 1 of Fig.3, and only impeller rotational frequency peak appeared. Periodic fluctuation caused by rotating stall appeared at operating point 2, and the flow rate was a slightly smaller than that of point 1. Peak frequency f_p is equal to 75Hz, which is about one-fourth as high as impeller rotational frequency, and stall cell number i, which was evaluated from the time lag between two pressure transducer signals, is 2. Amplitude of the pressure fluctuation increased and f_p became rather high when flow rate decreased.

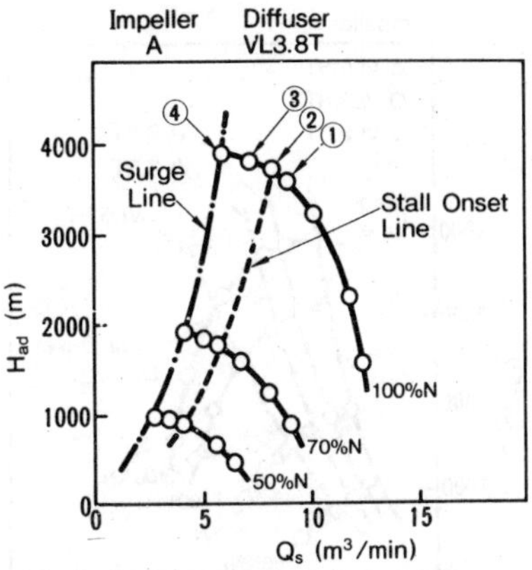

Fig.3 Performance curves and rotating stall onset line for a compressor.

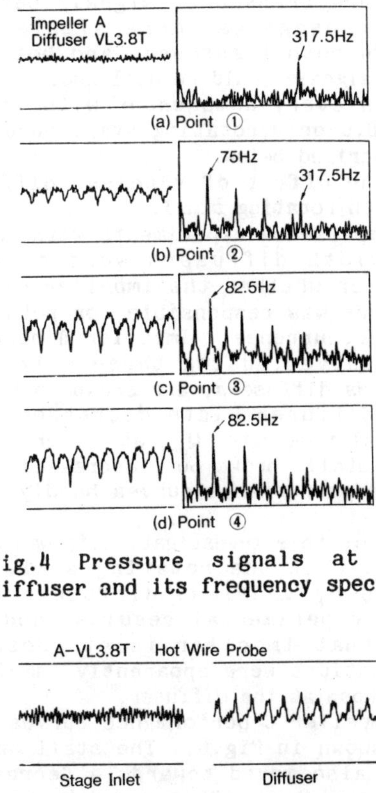

Fig.4 Pressure signals at the diffuser and its frequency spectra.

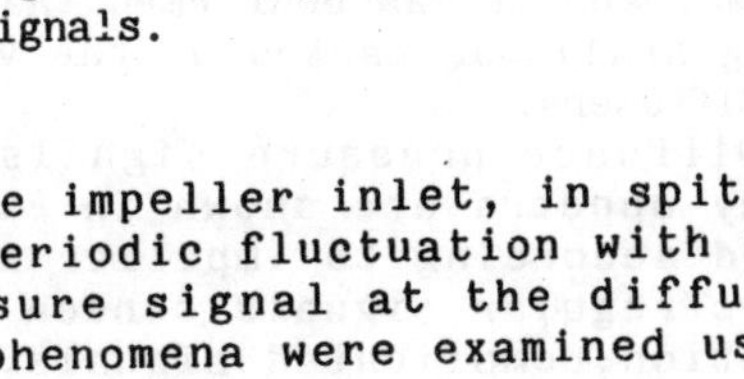

Fig.5 An example of hot wire probe signals.

Hot wire probe signals at operaing point2 are shown in Fig.5. Periodic fluctuation did not appear at the impeller inlet, in spite of rotating stall occurrence. In contrast periodic fluctuation with the same peak frequecy as that of the pressure signal at the diffuser appeared in the signal. Rotating stall phenomena were examined using

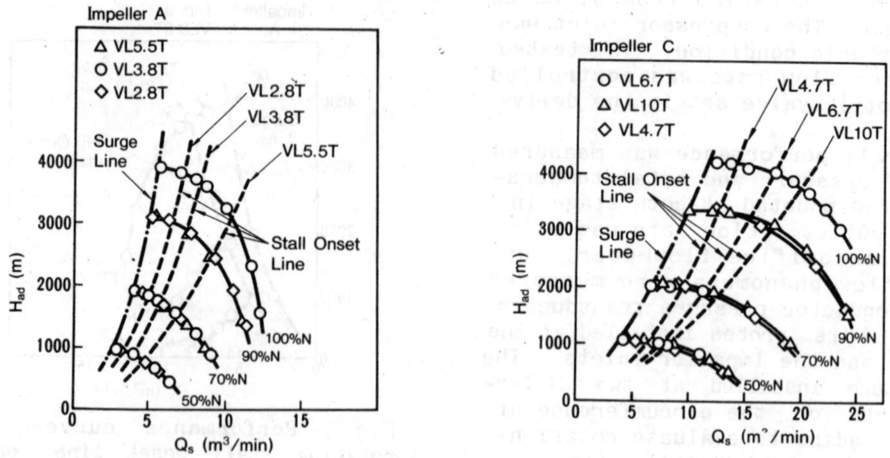

Fig.6 Effect of diffuser width on rotating stall onset.

pressure transducer signals because probes usage was more simple,even though both transducer and hot wire probe signals could be utilized.

In every compressor with impellers B,C or D,rotating stall occurred as described below.

2. The effect of vaneless diffuser width on rotating stall.

Impeller A experiments with various width diffusers were made to discover whether the impeller or the diffuser was responsible for rotating stall occurrence. Impeller A performance curves, using three different vaneless diffusers, are shown in Fig.6. When diffuser width decreased, the critical flow rate, Q_{crit}, at which rotating stall broke out, decreased, even though performance curves hardly changed.

In this investigation it was concluded that the rotating stall was caused by vaneless diffusers from these experimental results and the fact that impeller inlet pressure fluctuations were apparently smaller than those at the diffuser.

Impeller C performance curves are also shown in Fig.6. The stall onset lines also moved towards a decreased flow rate when diffuser width was decreased, and it was confirmed that rotating stall was caused by the vaneless diffusers.

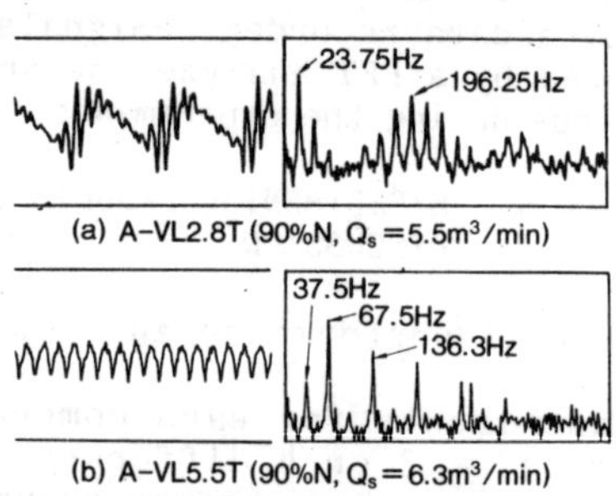

Fig.7 Impeller A pressure signals.

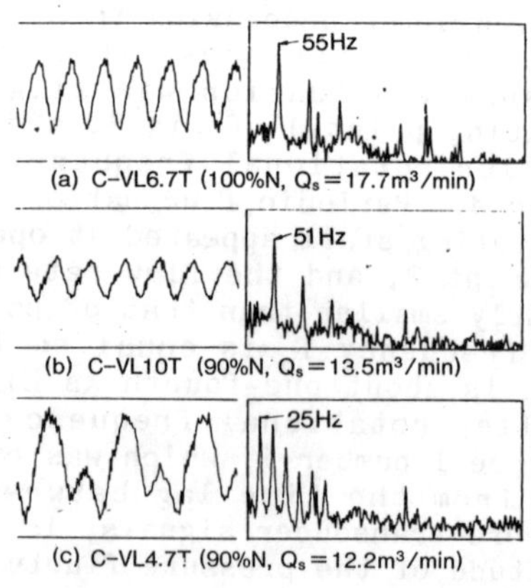

Fig.8 Impeller C pressure signals.

Diffuser pressure signals for various compressors and their frequency spectra are shown in Fig.7 and 8. Various pressure signals appeared according to impeller and diffuser combinations. In wide diffusers regular signals tended to appear. However, with diffuser width reduction,complicated signals, that were composed of low and high frequency components, appeared and fluctuation amplitude increased.

3. Features of rotating stall phenomena

Peak frequency ratio, $f_p/(\omega_{imp}/2\pi)$, rotating stall angular velocity ratio, ω_s/ω_{imp}, nondimensional pressure fluctuation amplitude, $\Delta h/(u_2^2/g)$, for ns=0.28 impellers are plotted versus the diffuser inlet flow angle, α_i, in Fig.9. Here, α_i was evaluated from the measured flow rate and the measured impeller temperature rise.

In every case f_p and ω_s increased and Δh also increased as α_i became small, which means flow rate decreased. When diffuser width decreased, Δh tended to increase, but the impeller effect on Δh seems to be small. However, f_p of impeller B differs greatly from that of impeller A. While $f_p/(\omega_{imp}/2\pi)=0.09\sim0.26$ for impeller A, $f_p/(\omega_{imp}/2\pi)$ for impeller B is in the range of $0.21\sim0.79$. Stall cell number, i, is 1 or 2 for impeller A, but i is 3 or 4 for impeller B. The reason why i of the two impellers differs from each other is not clear, but is assumed as below. The wake at the impeller B exit is considered to be larger than that of impeller A because the blade thickness at the impeller B outlet is larger than that of impeller A. The difference of this blade-to-blade flow distribution seems to affect the stall cell number.

The rotating stall features for impeller D are shown in Fig.10. Rotating stall, i=1, occurred in every case. From this result and the fact that i=1 in the case of impeller A with diffuser VL2.8T, it was found that the rotating stall, i=1, which led to a large force vibrating the compressor shaft, tended to occur if b_d/r_2 decreased below a certain value.

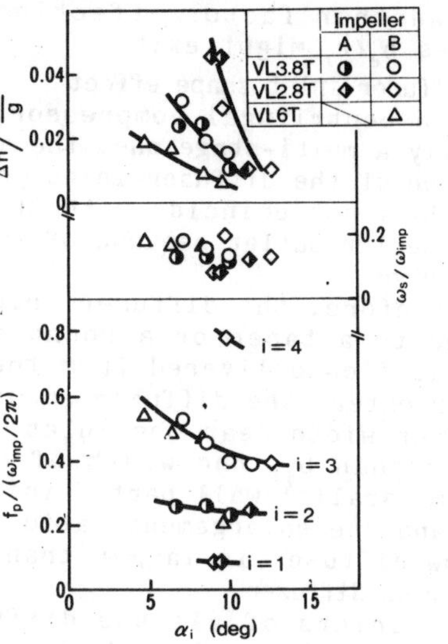

Fig.9 Rotating stall features for ns=0.28 impellers.

PREDICTION METHOD FOR ROTATING STALL ONSET POINT

It was ascertained that rotating stall was caused by vaneless diffusers, so critical conditions for rotating stall were examined.

The Senoo method is often quoted as a prediction method for the rotating stall of vaneless diffusers. Senoo and Kinoshita reported that rotating stall broke out when inlet flow angle became equal to a critical angle. This angle was 0.88 times as large as the angle at which the flow began to reverse based on axi-symmetric boundary layer analysis. This critical angle, α_{crit}, depends mainly on diffuser width to impeller outlet radius ratio, b_d/r_2.

The experimental values of VLXT, where X denotes diffuser width in mm, and the predicted values by the Senoo method of α_{crit}, are compared in Fig. 11. With a decrease in b_d/r_2, the pre-

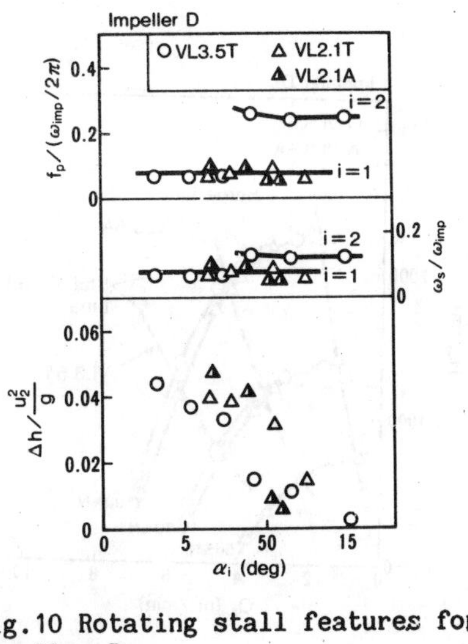

Fig.10 Rotating stall features for impeller D.

dicted values became smaller,but the experimental values of each impeller differed from one another and almost all the data were larger than the predicted values. The data of three out of four impellers showed a different tendency for b_d/r_2 than the predicted values.

It was considered that the difference between experimental and predicted data might have been due to flow distribution in an axial direction. Senoo and Kinoshita[9] reported that α_{crit}, in the case of distorted flow distribution at the inlet, was larger than that of uniform distribution. So flow distribution was measured by Pitot-tube and it was found that the flow at the diffuser inlet was almost uniform in an axial direction. Therefore it was believed that another factor affecting α_{crit} besides b_d/r_2 might exit.

1. Diffuser inlet shape effect

In a centrifugal compressor, especially a multi-stage one, the axial position of the diffuser inlet generally does not coincide with that of the impeller outlet because of assembling error.

Therefore, the diffuser inlet is formed to a taper or a round shape so that flow delivered from the impeller enters the diffuser smoothly. Diffuser width near the inlet is larger than b_d, the width of downstream parallel wall part, in this case and the enlargement ratio for a narrow diffuser is larger than that for a wide diffuser.

The inlets of all the diffusers in this paper were formed to a round shape as shown in Fig.12. The curvature radius, ρ_i, of the diffusers VLXT is 5mm. For example, in the case

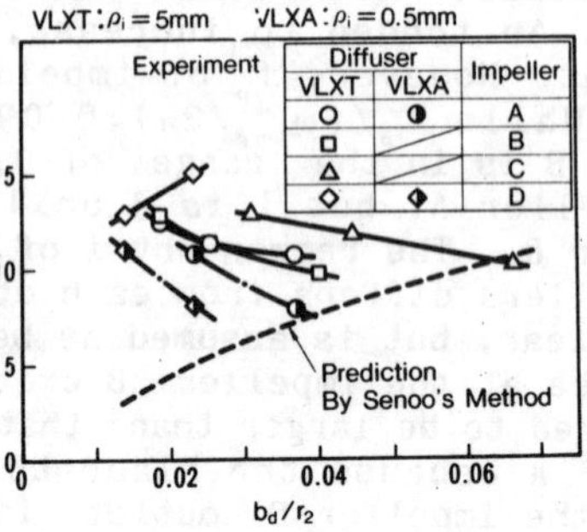

Fig.11 Comparison of experimental values and predicted values for inlet critical angle.

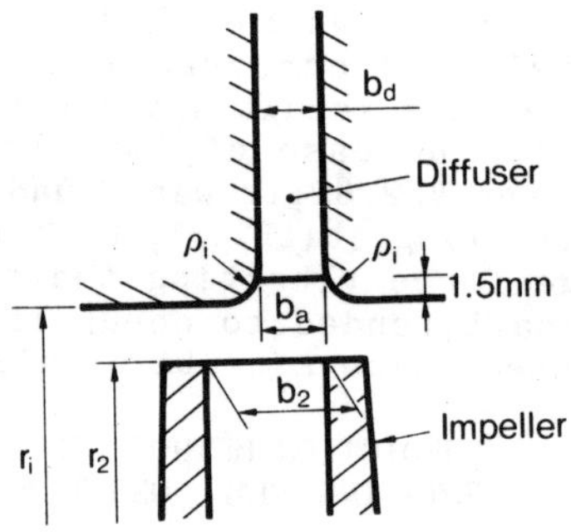

Fig.12 Diffuser inlet shape

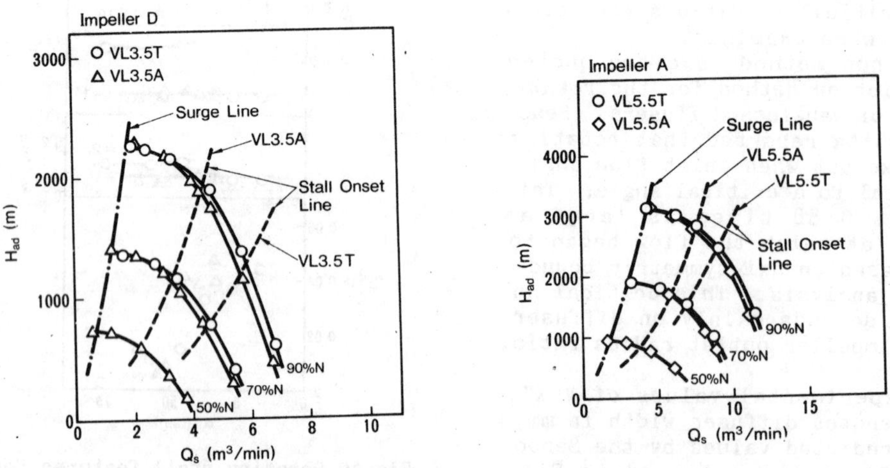

Fig.13 The effect of inlet curvature radius on rotating stall onset

where ρ_i=5mm, the width at radius which is 1.5mm larger than the inlet radius, b_a, is 2.35 times as large as b_d if b_d=2.1mm, while b_a/b_d=1.29 if b_d=10mm. Therefore flow angle in the vicinity of these diffuser inlets becomes smaller than that of a diffuser whose inlet is not formed to a round shape, and rotating stall may be caused early. So, inlet shape effect on rotating stall was experimentally examined.

The performance curves of impellers A and D with different ρ_i diffusers are shown in Fig.13. The diffuser, ρ_i=0.5mm, is titled VLXA.

In both cases, Q_{crit} for VLXA was less than that for VLXT, so it became evident that the effect of diffuser inlet shape on the rotating stall was very large. The difference in Q_{crit} between VLXA and VLXT, for a wide diffuser, is less than that for a narrow diffuser.

Fig.14 shows the reverse flow limit calculated by axi-symmetric boundary layer analysis[6] for various width diffusers. The left side of each curve indicates a reverse flow region. For example, in the case of b_d/r_2=0.0447 flow reversal does not appear at any place where α_i=10.5°. However, flow begins to reverse at r/r_2=1.175 where α_i=9.5°, and the reversal region spreads to r/r_2=1.10~1.27 when α_i=8.7°.

The radius ratio at which flow begins to reverse, $(r/r_2)_r$, is low in the case of a narrow diffuser. So, it is considered that the effect of inlet shape on rotating stall is large for small width diffusers.

On the other hand, as $(r/r_2)_r$ is large in wide diffusers, the effect of inlet shape is thought to be small.

For this reason and the fact that the enlargement ratio for a narrow diffuser is higher than that for a wide diffuser, it is believed that the difference of Q_{crit} for a small width diffuser is much more than that for a large width diffuser, as shown in Fig.13.

2. Contraction ratio effect on rotating stall

Experimental values of α_{crit} for diffuser VLXA are also plotted in Fig.11. The VLXA experimental values are smaller than those for diffusers VLXT, and approach to the predicted values. However, experimental values for one impeller show an opposite tendency for b_d/r_2 than the predicted values. The experimental values of the widest diffuser for each impeller agree fairly well with the predicted values and the experimental values deviate from predicted values, as the diffuser width decreases.

This fact indicates that α_{crit} is seriously affected by contraction ratio b_d/b_2. The experimental values of α_{crit} for VLXA (Impellers A and D) are plotted versus the contraction ratio in Fig.15. The data

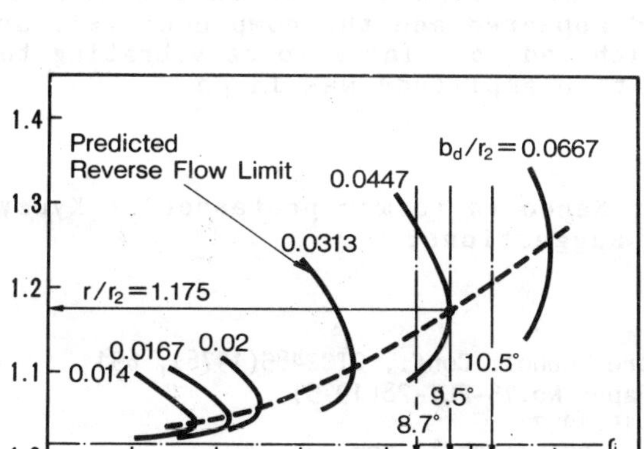

Fig.14 Reverse flow limit for various b_d/b_2 diffusers.

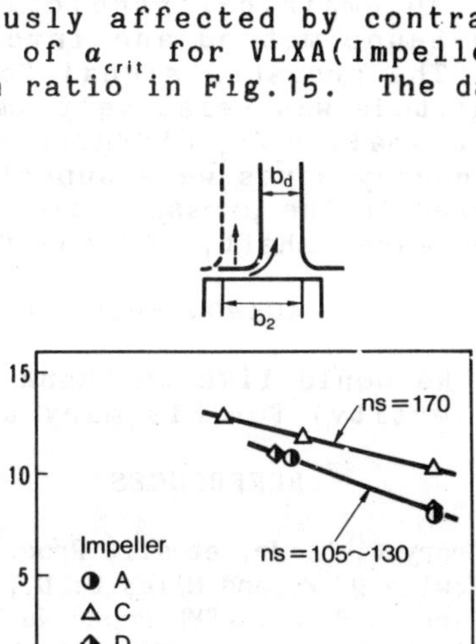

Fig.15 Contraction ratio effect on α_{crit}.

for VLXT(Impeller C) are also plotted in Fig.15, as in the case of a wide diffuser the effect of ρ_i on α_{crit} is considered to be small because of the above mentioned reason. Data for ns=0.23~0.28 impellers and ns=0.42 impeller approach each other. Though some differences still are seen between the two sets of data, this is considered to be due to the net effect of b_d/r_2.

In the case of a large b_d/r_2, the difference of α_{crit} between the two sets of data is large. On the other hand, in the case of a small b_d/b_2, the difference is small, so contraction ratio effect is very large.

The empirical prediction formula for α_{crit} was derived from these sets of experimental data, in consideration of the contraction ratio.

$$\alpha_{crit}=\alpha_{crit,\ senoo}+(17.02 - 74.20\cdot b_2/r_2)\cdot(1 - b_d/b_2)$$

The reason why α_{crit} varies with b_d/b_2 is assumed as below:in the case b_d/b_2=1.0, flow delivered from an impeller enters the diffuser radially as the dotted line shows in Fig.15. In the case of a small b_d/b_2, the flow enters the diffuser inclined to an axial direction as the solid line shows, so flow reversal may occur early.

It may be neccesary to clarify this point by flow analysis in considerataion of inlet shapes or detail flow measurements.

CONCLUSION

The following conclusions were drawn from this experimental investigation of rotating stall in low specific speed centrifugal compressors.

1. The rotating stall of centrifugal compressors was caused by vaneless diffusers.
2. Stall cell number was 1 to 5, and the relative angular velocity, ω_s/ω_{imp}, and the nondimensional peak frequency, $f_p/(\omega_{imp}/2\pi)$, of the stall cells was in the range of 0.07~0.21 and 0.07~1.01, respectively. The maximum value of nondimensional pressure fluctuation, $\Delta h/(u_2^2/g)$, was about 0.06.
3. The critical inlet flow angle, α_{crit}, depends not only on width ratio b_d/r_2, but also on contraction ratio b_d/b_2 and inlet curvature radius ρ_i. As b_d/b_2 decreases and ρ_i increases, α_{crit} becomes large.
4. An empirical prediction formula for α_{crit} was derived based on the Senoo method and involving the contraction effect.
5. The pressure signal for a large b_d/r_2 diffuser was regular, the amplitude was relatively small, and i was equal to 2. But, in the case of a small b_d/r_2 diffuser a complicated signal in which low and high frequency waves were superimposed appeared and the component i=1, appeared in the pressure signal, which led to a large force vibrating the compressor shaft, and the fluctuation amplitude was large.

ACKNOWLEDGEMENT

We would like to thank Dr. Y. Senoo (a former professor of Kyusyu University) for his many useful suggestions.

REFERENCES

1. Geary C. H. Jr. et al., Proc. offshore Techol. Conf., OTC2485(1976), 651
2. Fowlie D. W. and Miles D. D., ASME Paper No.75-PET-28(1975)
3. Ferrara P. L., ASME Paper No.77-DET-15(1977)
4. Kaneki T. and Eino T., Hitachi review, 28(6)(1979), 327
5. Kinoshita Y. and Senoo Y., ASME Paper No.84-GT-203(1984)
6. Senoo Y. and Kinoshita Y., ASME Paper No.78-GT-19(1977)
7. Tsurusaki H. et al., Turbomachinery (Japanese), 12(6)(1984), 323
8. Frigne P., ASME Paper No.83-GT-61(1983)
9. Senoo Y. and Kinoshita Y., Turbomachinery (Japanese), 5(2)(1977), 31

Measurements of Simulated Wake/Rotor Interaction Phenomena in Turbomachinery

K.Funazaki (Iwate Univ. Japan)
T.Nishiyama (Ishinomaki-Senshu Univ. Japan)

Abstract

Experimental investigations on simulated wake/rotor interaction phenomena were conducted by use of moving bars ahead of the stationary airfoil cascade. In this paper, results of overall measurements on the unsteady flow field, including static pressure on the airfoil and outlet total pressure, give some insight into these complex phenomena. And then comparisons between the measurements and the predictionshave verified the ability of the prediction code to some extent, whereas for unsteady total pressure, such code brings out only qualitative agreement with the measurements. Moreover, time-variant total pressure loss distributions are shown to give some views about the mechanism of aerodynamic performance deterioration in turbomachines by upstream wakes.

Nomenclature

<Symbol>

C : airfoil chordlength
C_P : unsteady pressure coefficient ($= 2p/\rho W_1^2 u_g/W_1$)
f_{BP} : bar passing frequency ($= U_C/t_S$)
K_n : reduced frequency of n-th harmonic
m : index of an airfoil
n : order of harmonics
P_1, P_2 : inlet,outlet total pressure
p : unsteady static pressure
T : period of inlet fluctuation ($= 1/f_{BP}$)
t_R : pitch of moving bars
t_S : pitch of airfoils
U_1 : inlet axial velocity
U_C : running speed of bars
u_g : amplitude of 1st harmonic of inlet velocity fluctuation
W_1 : inlet velocity
x : chordwise distance
y_t : circumferential distance

<Subscript>

1,2 : inlet,outlet

<Superscript>

$-$: steady state
$\sim$: unsteady state

1. Introduction

Until now, much efforts have been devoted to study of stator/rotor interactions in turbomachines, especially wake interactions, because they cause blade failure, noise, performance deterioration or shortfall of blade cooling. Owing to recent great progresses in computer performance and numerical schemes, unsteady flow simulation in turbomachines can be possible to some extent, where Hodson[1], Giles[2] Rai[3] have made great contributions to the development. Unfortunately, however, such calculations need much amount of cpu time and memory size even in the framework of two-dimensional inviscid flow. This situation could be worsened in the case of non-equal stator/rotor blade number configurations which are usually encountered in turbomachines, while several approaches were introduced in order to overcome this difficulty by first Erdos[4], then Giles(using inclined computational grids)[2] or Fourmaux and Le Meur[5] (using kind of approximation to stator/rotor blade number ratio). In addition, because of still insufficient turbulence model in the very complex flowfield, such codes are not so sophisticated for the purpose of estimations of the aerodynamic loss or increased heat transfer induced by the unsteady phenomena. In this sence, there seems much room for improvement in the numerical simulations of the wake interaction.

On the other hand, there are relatively limited number of experimental studies on the wake-interaction problem, which are indispensable to verification and improvement of numerical codes. In this area, Dring and Joslyn[6][7], Hodson[1][8] have made improtant works on turbine rotor, Gallus[10][11], Fleeter[12] on compressor cascade, for instance. Recently, mainly concerning with unsteady heat transfer on turbine blades, Doorly and Oldfield[13] or Blair et al[14] have done excellent studies. These experimental works do give us insights into the very complicated flowfield, but not a few previous researches mentioned above have concentrated considerable efforts upon measurements of unsteady pressure and/or unsteady boundary layer on blade surfaces, rather than measurements of the whole unsteady flowfields in turbomachines. The authors believe that such overall measurements are also needed for gaining more information on wake/rotor interaction mechanism(See a paper of Binder et al[16] for example).

In this paper, the authors present some results from their detailed measurements about a wake/rotor interaction simulated by the use of moving bars at a distance upstream from stationary turbine airfoils, where not only unsteady pressure distributions on the airfoil but also unsteady inlet/outlet total pressure distributions are measured for two different rotor/stator (moving bars) pitch ratio cases. This paper also shows some comparison between these measurements and numerical predictions which are obtained with the code developed by the authors[17]. Within the very limited range of test conditions covered this time, however, it is found that the predicted values of unsteady pressure distributions on the turbine airfoil agree with the experimental data fairly well. As for outlet total pressure, only qualitive agreement are obtained ,maybe, due to viscous effect. Furthermore, some noticable results are observed concerning with the total outlet pressure disributions, which are not reported in any othe rstudies as far as concerned.

2. Experimental Facilities and Instruments

2.1 Facilities

Fig.1 shows the experimental facility used to simulate a wake/rotor interaction, where moving bars of ϕ3.0mm diameter and 100 mm length are located at a distance 30mm ahead of the stationary turbine cascade. These bars are easily removed, which makes it possible to examine two test cases of rotor/stator (moving bar) pitch ratio, i.e., 0.5 and 1.0. The cascade configuration and inlet velocity triangle are shown schematically in Fig.2. The experiment was conducted with Reynolds number based on the airfoil chordlength $8.2 \times 10^4 \sim 1.3 \times 10^5$.

2.2 Instruments

Measurements were made on the flow properties as follows;

(1)unsteady velocity and total pressure at the station 1

(2)unsteady pressure distribution on the reference airfoil(m=0)

(3)unsteady total pressure distribution at the station 2

A block diagram of the measuring system is also shown in Fig.1. All transient signals from hot wire probe or pressure transducer were sampled with 50 μsec sampling time and then phase-locked averaged for 64 times using a photo-pickup system in order to improve S/N ratio of these signals.

Inlet flow field including wakes of upstream bars was measured using a conventional I-type hot wire probe which position was fixed at the center of the locus between two-neighboring blade leading edges. A custom-made total pressure probe with a pressure transducer was also used. Detected signals of wakes (velocity defects) were all FFT-analyzed as shown in Fig.4 and frequency of each 1st harmonic,i.e. BPF f_{BP} was used to determine the speed of those bars (U_c) from the relation $f_{BP} = U_c / t_S$. Reduced frequency (K_n) was calculated as follows;

$$K_n = 2\pi n f_{BP} C / U_1 , \tag{1}$$

where U_1 is an axial velocity and C is an actual blade chord.

On the wind tunnel-centered airfoil which was fabricated with 20 pressure tubes inside, unsteady pressure distribution was measured using a pressure transducer. This transducer had been previously calibrated not only statically and but also dynamically accompanied with the pressure transmission system. All analyzed data were phase-shifted by the amount of the time lag between the trigger pcint of data acquisition and the wake peak point, so that it can be considered that data acquisition started at the moment when the peak of the velocity deficit reached around the leading edge of the reference blade.

3. Prediction Code

Only brief explanation on the prediction code[14] are given in the following, accompanied by some comparison with Fleeter's experiment[10] for the sake of verification of its validity.

When dealing with a wake/rotor interaction problem, one of the most important and difficult aspects to be predicted by any prediction code is of course the wake transport phenomena through blade-to-blade passages. Wake itself has very complex structure, for example, high intense turbulence and various scale of eddies involved, but from an assumption that such internal structure of the wake doesn't play significant roles in the macro aspect of wake transport, wake can be considered to be convected downstream by the mean flowfiled and then the wake transport can be specified by solving the linearized vorticity transport equation for 2-D incompressible inviscid flow[14]. This approach was at first used by Adamczyk[15] and it was extended to compressible flow by Caruthers[16]. The authors' code also incorporates the same approach, combined with the surface singularity method.

In order to show how this code works in wake/rotor interaction problems, comparison was made between the measurement of unsteady pressure distributions on the compressor stator vane and the predictions, one of which is presented in Fig.3. Though slight discrepancies appear, the predictions agree with the measurements fairly well almost at the design operating condition, and these results gives the authors great confidence in applying the prediction code to the the turbine cascade case described in the following section.

4. Measurements

4.1 Unsteady Pressure Distribution on the Turbine Airfoil

Fig.5 shows typical data of pressure fluctuations on several measuring points, in which only AC components of the phase-locked averaged unsteady pressure (not phase-shifted) are presented. Clear view is obtained from this figure that sawtooth-like large pressure fluctuations occur at the blade leading edge. It should be noted that positive peaks appear at some measuring points on the suction surface when wakes reach these points, while negative peaks appear on the pressure surface. Another aspects of the unsteady pressure distributions are, for example, apparent decrease in amplitude towards the trailing edge, and very small pressure fluctuation near the trailing edge on the suction surface, maybe due to flow seperation which was identified from the steady pressure distribution.

Comparison between measured unsteady pressure distributions and predicted values are shown in Fig.6 through Fig.9, where only first harmonics of those data are presented.

Figs.6 and 7 are for the case of t_r/t_s=0.5. Due to the relatively low reduced frequencies, very slight chordwise variation in amplitude of both real and imaginary parts of the measured unsteady pressure are identified mainly on the pressure surface. This means in-phase fluctuation almost over the entire pressure surface. Meanwhile, the predictions show good agreement with measured data, although some discrepacies are seen in the imaginary parts, especially near the leading edge.

Figs.8 and 9 show the case of t_r/t_s=1.0, where it is clear that amplitudes of measured pressure fluctuations are smaller that those of the case of t_r/t_S=0.5. In addition, propagation of pressure fluctuation downstream to the trailing edge can be identified by looking into chordwise variation of the real and imaginary parts. Predictions, on the other hand, give fairly good results. Some difference between measured and predicted values in the figures above might be attributed to poor performance in prediction of wake transport around the blunt blade leading edge where the linearizing approach for unsteady terms no longer holds. Moreover, the wake decay, which is neglected in the prediction code, seems to have some effect on the accuracy.

4.2 Unsteady Total Pressure

Fig.10 shows transient total pressure measured at several points on the plane about one-pitch downstream from the locus of the trailing edge, where vertical dotted lines represent time standard of data analyzing. It should be noted that the direction of the probe's sensing face was keep constant (perpendicular to the exit flow direction) during the pressure traverse. Also note that one period (T) for the case of t_R/t_S=0.5 is two times the period of the case of t_R/t_S=1.0. Fig.11 shows calcaulated positions of the wake centerline at four different phases during one period for two cases, which helps the readers understand the relationship between the time-variant outlet total pressure fluctuations and wake positions on the measuring plane.

Seeing measured signals of the case of t_R/t_S=0.5 in Fig.10, one can notice that incoming wakes detected at the pressure side of the blade wake ⓓⓔ and those detected at the suction side ⓐⓑ are apparently different in terms of wake depth and width, i.e. measured total pressure defect and its time span. This difference might be attributed to wake chopping by the airfoils and its wake segment deformation by the mean non-uniform flow, which situations are represented in Fig.12 with calculated unsteady rotational velocities. In the case of t_R/t_S=1.0 in Fig.10, similar results are obtained except at the center point of the passage ⓒ. Note again that one period (T) in this case is a half of that of t_R/t_S=0.5. It is very hard to find out the wakes from the measured data, in contrast with the case of t_R/t_S=0.5. This result is very unbelievable at first glance, but after FFT-analyzing these data, a mechanism hidden behind appears as described later. Figs.13 and 14 show amplitude and phase distributions of first (●) and and second harmonics (○) of the outlet total pressure fluctuations in the case of t_R/t_S=0.5 and 1.0 respectively, where amplitudes of the first and second harmonics are both normalized by the first harmonic's amplitude of the inlet total pressure and dotted lines represent position of stagnation streamlines. Also shown are corresponding predicted outlet total pressures P are which are obtained by integrating linearized Euler equation along a steady streamline;

$$\tilde{P}(s) = \tilde{P}_I - j\omega \int_{so}^{s} u_s ds \ ; \ u_s = \vec{u} \cdot \vec{s} \qquad (2)$$

$\tilde{P}_I$: inlet unsteady total pressure
$\vec{u}$: unsteady velocity , $\vec{s}$: tangential unit vector on s

where an assumption is made that upstream static pressure fluctuation can be neglected in the equation (2).

One can easily notice that transverse amplitude variations of second harmonics in the case t_R/t_S=0.5 (Fig.13) and first/second harmonics in the case of t_R/t_S=1.0 (Fig.14) are very similar, in which minimum amplitudes appear almost at the center of the passage and the maximum amplitudes occur around the blade wakes. The reason for this similarity might not be far beyond understanding by casting a light on the point that second harmonics of the inlet total pressure in the case of t_R/t_S=0.5 and the first harmonics for the t_R/t_S=1.0 have almost the same characteristic in amplitude and phase, And then, such similarity means that each decomposed acts respectively to its corresponding component of the inlet total pressure, which implies the verification of the assumed linearity of the unsteady flowfield in the prediction code.

From the above discussion, also paying an attention on the point that in the case of t_R/t_S=1.0 inter-blade phase angles between two neighboring airfoils for all harmonics are zero, it can be concluded that pressure fluctuations with very small amplitude at the center of the exit plane are not special phenomena for the case t_R/t_S=1.0, but common when their inter-blade phase angles are equal to zero, in other words, when all blade-to-blade passages are at the identical flow condition.

Predictions for the fluctuation's amplitude give only qualitive agreement with the measurements, probably because of viscous effect on moving bar wakes and interaction between the bar wakes and blade wakes. Moreover, the assumption in the equation (2) can cause some errors in the estimation of the outlet total pressure. As for the phase delay, the predictions show slight overestimation at the suction side of the blade wakes. The authors think that flow acceleration due to the displacement effect of the blade wake is one of the main reasons for this overestimation.

4.3 Time-variant Total Pressure Loss

Fig.15 shows time-variant total pressure loss distributions during one period of each case, i.e. t_R/t_S = 0.5 and 1.0 accompanied by measured steady total pressure loss distributions. Note that outlet total pressure $\check{\tilde{P}}_2$ was not obtained direct from the signals of the pressure transducer, but calculated by summing up the steady total pressure and AC components of time-variant total pressure, and for the sake of simplicity, inlet total pressures in the loss coefficient (P_1) don't include the loss due to incoming wakes. Fig.15, with Fig.12 gives the readers clear feature about how the outlet total pressure distributions change according to the incoming wake movement. In the case of t_R/t_S=0.5, loss increase appears in the flow passage at t/T=0, which can be directly attributed to the bar wake as mentioned above. And it might be the case when the bar wake gets in touch with the blade wake at t/T=1/4 or 3/4 as shown in Fig.12.

In the case of t_R/t_S=1.0, drastic variation in loss coefficients can be seen around the blade wakes. At the center of the passage, on the other hand, only small variation in loss coefficients can be recognized during one cycle of the wake passing, accompanied by the appearance of negative loss at the suction side of the blade-to-blade passage when t/T=1/4 and at the pressure side when t/T=3/4.

The authors now must confess that any clear idears about appearance of the small amplitudes and the negative loss portions are still out of their reach. They also feel great necessity of further investigation of the relationship between incoming wakes and the induced total pressure loss.

5.Conclusions

The simulated wake/rotor interaction phenomena were investigated thoroughly, which have turned out to give the following major findings.

(1) Measured unsteady static pressure on the turbine airfoil shows positive peaks at the suction surface and negative peaks at the pressure surface, and also unsteady pressure amplitude decreases downstream along the blade surfaces.
(2) Predictions for first harmonics of unsteady pressure distributions on the blade surface agree with the measured values fairly well. Slight differences appear mainly in the imaginary(out-of-phase) parts of the pressure fluctuations, which can be attributed to poor performance in prediction of wake transport around the blunt blade leading edge.
(3) Amplitudes of outlet unsteady total pressure vary along the exit measuring plane. In general, larger amplitudes are detected around the blade wakes rather than amplitudes at the center of the blade-to -blade passage. In the case of $t_R/t_S=1.0$, especially, very small values of the total pressure amplitude appear at the center of the passage.
(4) As for the outlet unsteady total pressure, only qualitative agreement between the predictions and the measurements is obtained, which is probably due to viscous effect, i.e. wake decay.
(5) Time-variant total pressure loss distributions exhibit clear relationship with upstream wakes.

References

(1) Hodson,H.P.,"An Inviscid Blade-to-Blade Prediction of a Wake-Generated Unsteady Flow,"ASME Paper 84-GT-43,1984
(2) Giles,M.B.,"Stator/Rotor Interaction in a Transonic Turbine,"AIAA Paper 88-3093,1988
(3) Rai,M.M.,"Navier-Stokes Simulations of Rotor-Stator Interaction Using Patched and Overlaid Grids,"AIAA Paper 85-1519,1985
(4) Erdos,J.I. and Alzner,E.,"Numerical Solution of Periodic Transonic Flow throug a Fan Stage,"AIAA Journal, Vol.15, No.11,pp.1559-1568, 1977
(5) Fourmaux,A.,"Unsteady Flow Calculation in Cascade,"ASME Paper 86-GT-178,1986
(6) Dring,R.P. and Joslyn,H.D.,"Turbine Rotor-Stator Interaction," Trans.ASME J.Eng. for Power, Vol.104, pp.729-742, 1982.
(7) Hodson,H.P.,"Boundary Layer and Loss Measurements on the Rotor of an Axial-Flow Tu.bine,"ASME Paper 83-GT-4,1983
(8) Gallus,H.E. Grollius,H. and Lambertz,J.,"The Influence of Blade

Number Ratio and Blade Row Spacing on Axial-Flow Compressor Stator Dynamic Load and Stage Sound Pressure Level,"ASME Paper 81-GT-165, 1981
(9) Gallus,H.E. Hoenen,H.,"Measurement of 3-D Unsteady Flow Downstream of Rotor and Stator Blades in Axial-Flow Compressor,"1983 Tokyo Int'l Gas Turbine Congress,83-TOKYO-IGTC-68, 1983
(10) Fleeter,S. Jay,R.L. and Bennet,W.A.,"Wake Induced Time-Variant Aerodynamics Including Rotor-Stator Axial Spacing Effects," Trans. ASME J.Fluids Eng., Vol.103, No.1, pp.59-66, 1981
(11) Doorly,D.J. Oldfiled,M.L.G.,"Simulation of the Effects of Shock Wave Passing on a Turbine Rotor Blade,",ASME Paper 85-GT-112, 1985
(12) Blair,M.F. Dring,R.P. and Joslyn,H.D.,"The Effects of Turbulence and Stator/Rotor Interactions on Turbine Heat Transfer(Part I: Design Operating Conditions),"ASME Paper 88-GT-125,1988
(13) Binder,A. Forster,W. Kruse,H. and Rogge,H.,"An Experimental Investigation into the Effect of Wakes on the Unsteady Turbine Rotor Flow,"Trans.ASME J.Eng.Gas Turbines and Power, Vol.107, pp. 458-465, 1985
(14) Nishiyama,T. Funazaki,K.,"Aerodynamic Responses of Turbine Rotor Blade to Sinusoidal Gust of Large Vorticity," Int'l Gas Turbine Congress,83-TOKYO-IGTC-83, 1983
(15) Adamczyk,J.J.,"The Passage of a Distorted Velocity Field through a Cascade of Airfoils,"AGARD-CP-117(Unsteady Phenomena in Turbomachinery), 1975
(16) Caruthers,J.E.,"Theoretical Analysis of Cascaded Airfoils in Unsteady Rotational Flow,"Proceedings of the 2nd Int'l Symposium on Aeroelasticity in Turbomachines, pp.31-64, 1980

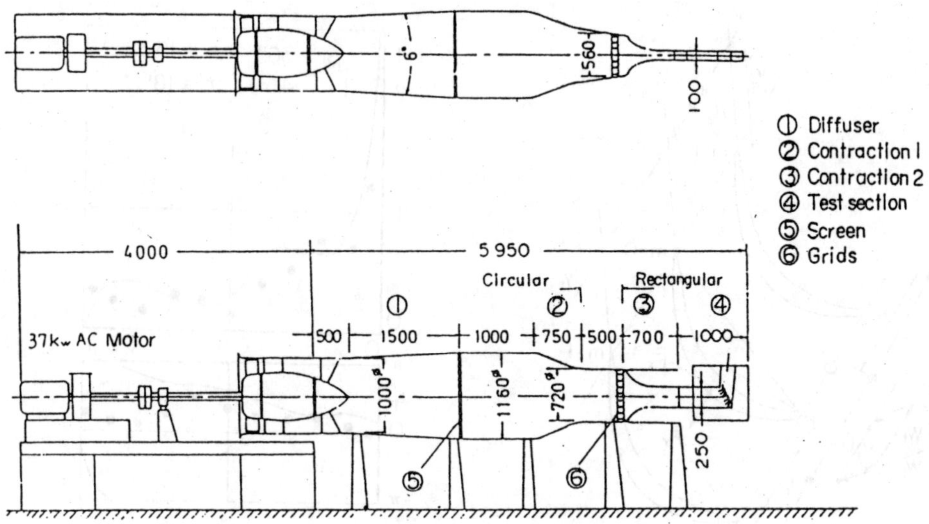

(a) Schematic View of the Cascade Tunnel

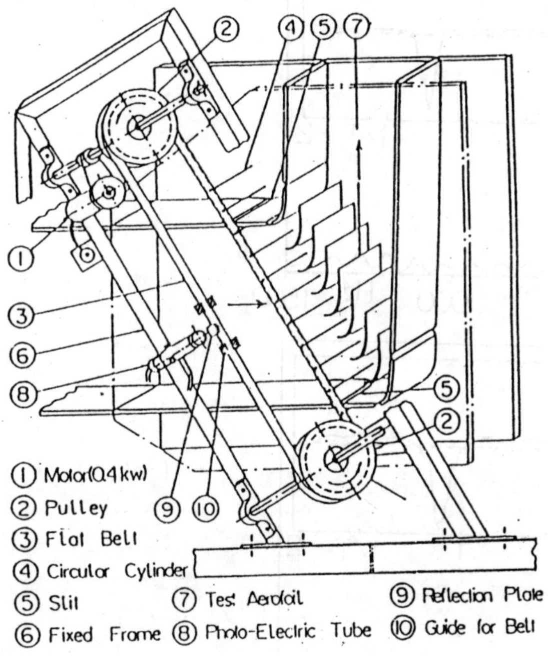

(b) Test Section

(c) Measuring System

Fig.1 Experimetal Facilities

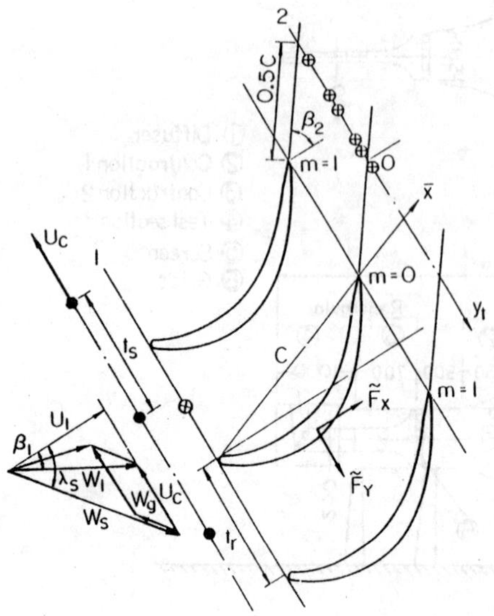

Fig.2 Cascade Configuration

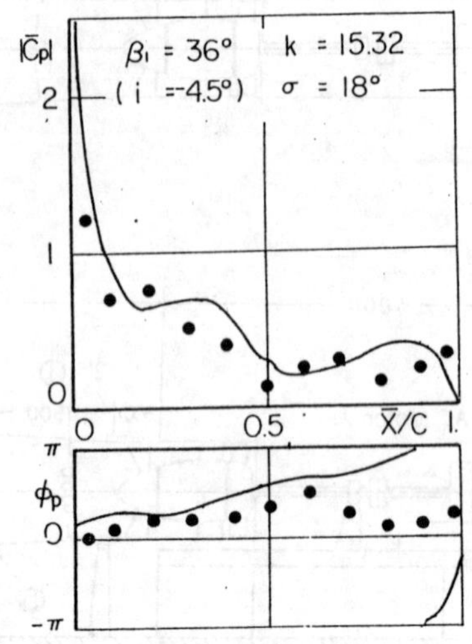

Fig.3 Comparison of Usteady Pressure

(● : Fleeter et al[10] / —— : Prediction)

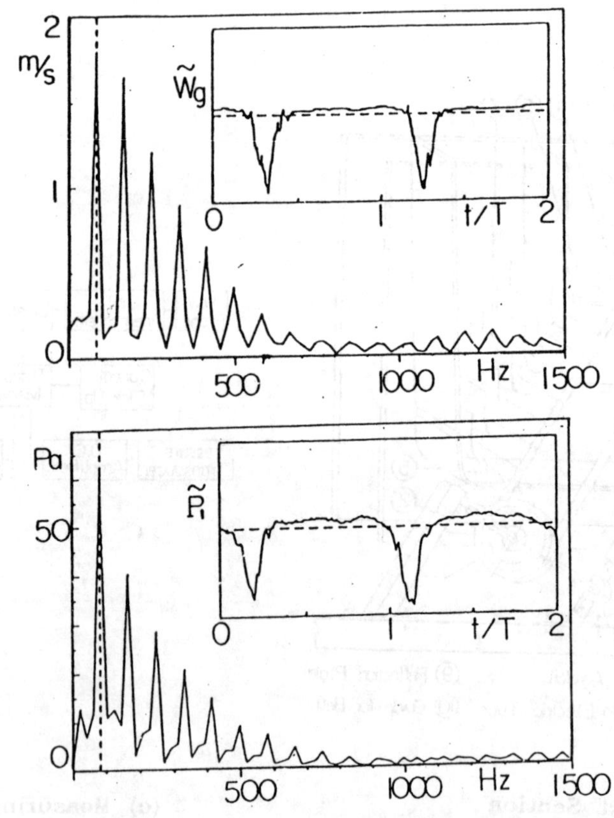

Fig.4 (upper) Measured Inlet Unsteady Velocity

(lower) Measured Inlet Unsteady Total Pressure

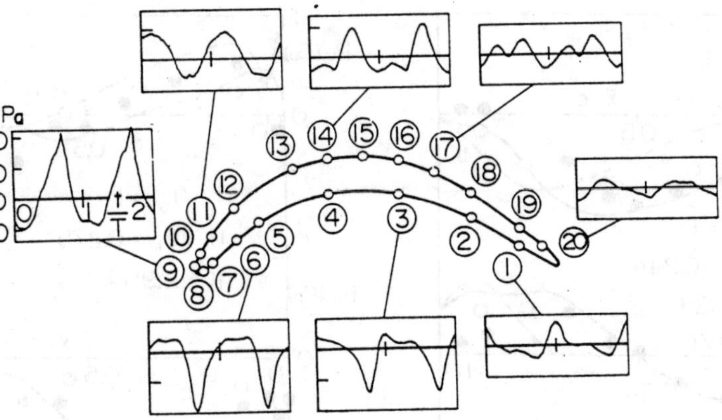

Fig.5 Pressure Fluctuations on the Reference Airfoil

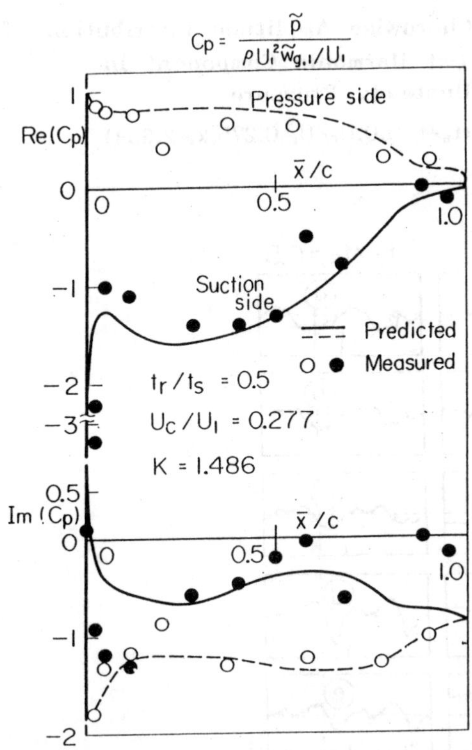

Fig.6 Chordwise Amplitude Distribution of 1-st Harmonic Component in Unsteady Pressure t_R/t_S=0.5, U_C/U_1=0.277, k_1=1.486)

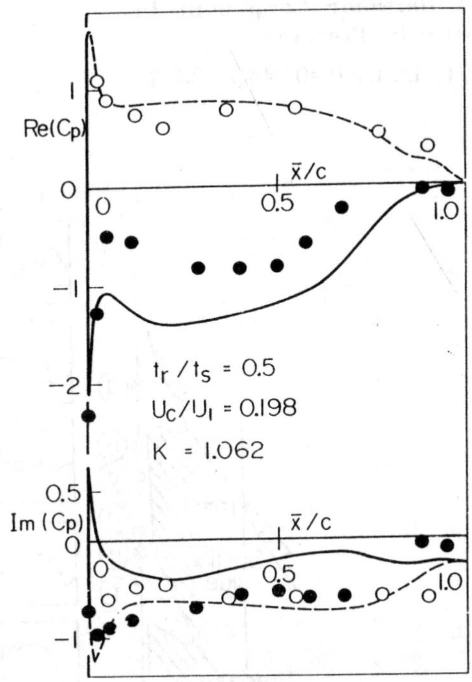

Fig.7 Chordwise Amplitude Distribution of 1-st Harmonic Component in Unsteady Pressure (t_R/t_S=0.5, U_C/U_1=0.198, k_1=1.062)

● : Suction Side , ○ : Pressure Side

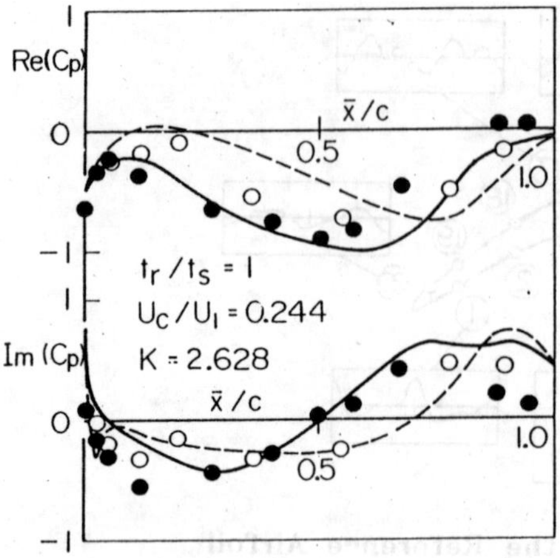

Fig.8 Chordwise Amplitude Distribution of 1-st Harmonic Component in Unsteady Pressure

(t_R/t_S=1.0,U_C/U_1=0.244,k_1=2.628)

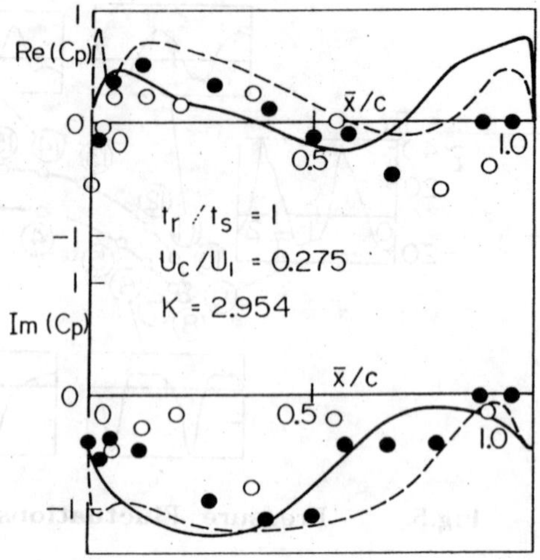

Fig.9 Chordwise Amplitude Distribution of 1-st Harmonic Component in Unsteady Pressure

(t_R/t_S=1.0,U_C/U_1=0.275,k_1=2.954)

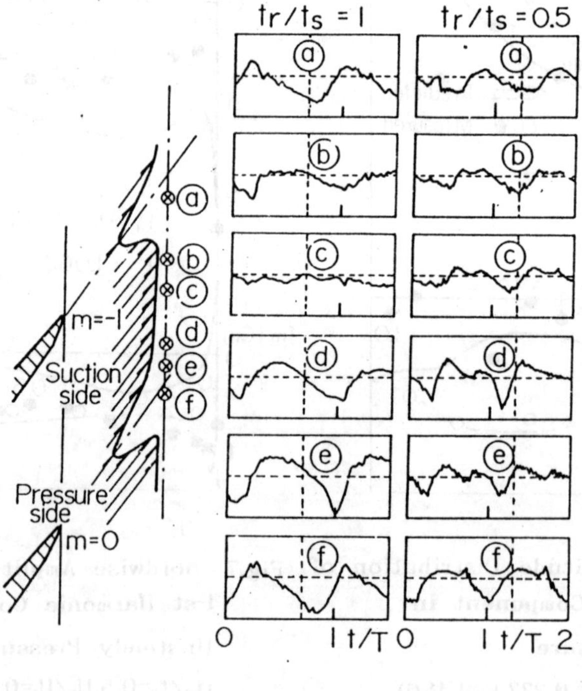

Fig.10 Signals of Transient Total Pressure at the Exit Plane and Measuring Points

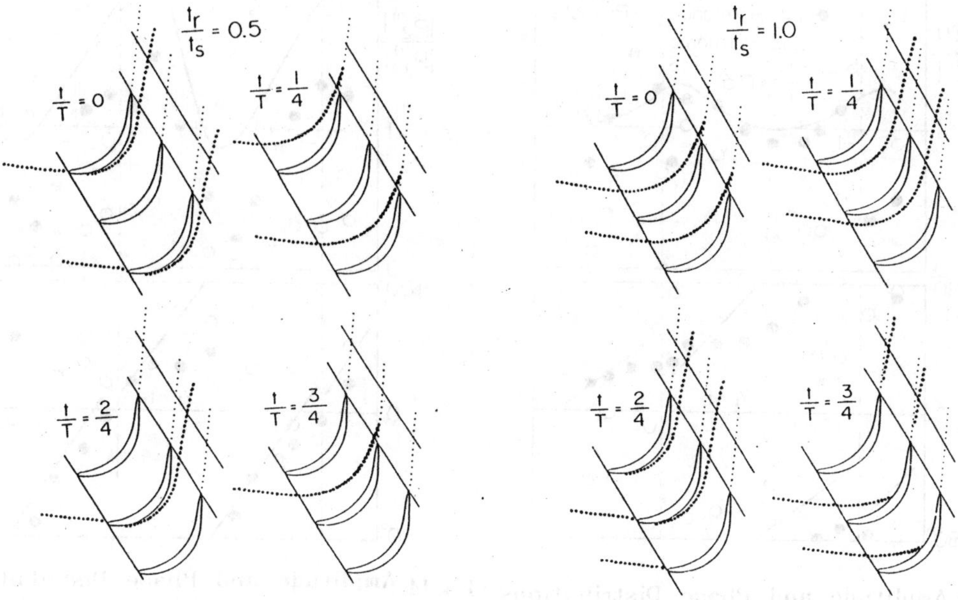

Fig.11 Calculated Wake Positions at Four Phases (U_c/U_1=0.277)

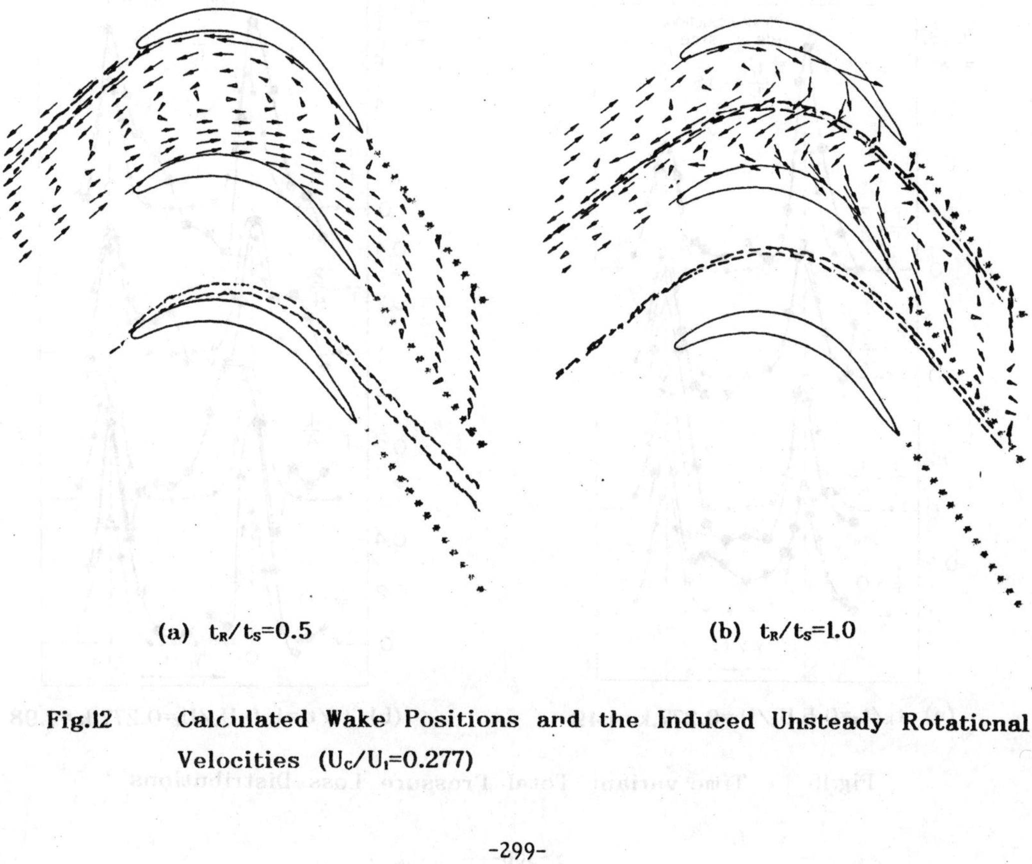

(a) t_R/t_S=0.5 (b) t_R/t_S=1.0

Fig.12 Calculated Wake Positions and the Induced Unsteady Rotational Velocities (U_c/U_1=0.277)

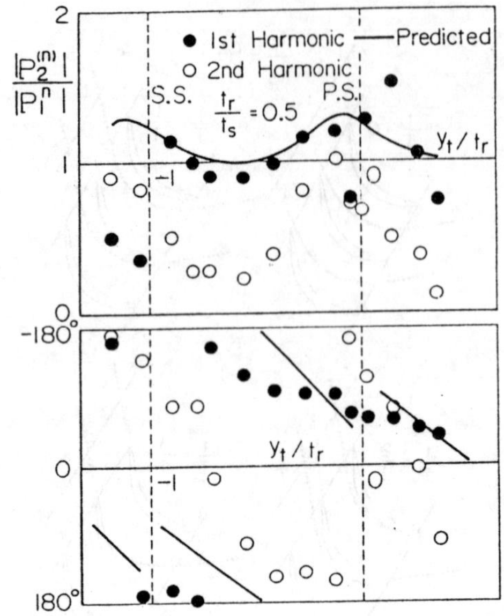

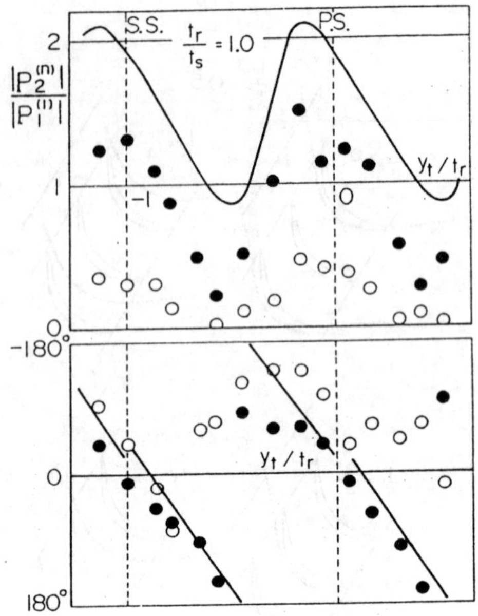

Fig.13 Amplitude and Phase Distributions of Outlet Total Pressure (t_R/t_S=0.5, U_C/U_1=0.277, k_1=1.49)

Fig.14 Amplitude and Phase Distributions of Outlet Total Pressure (t_R/t_S=1.0, U_C/U_1=0.277, k_1=2.98)

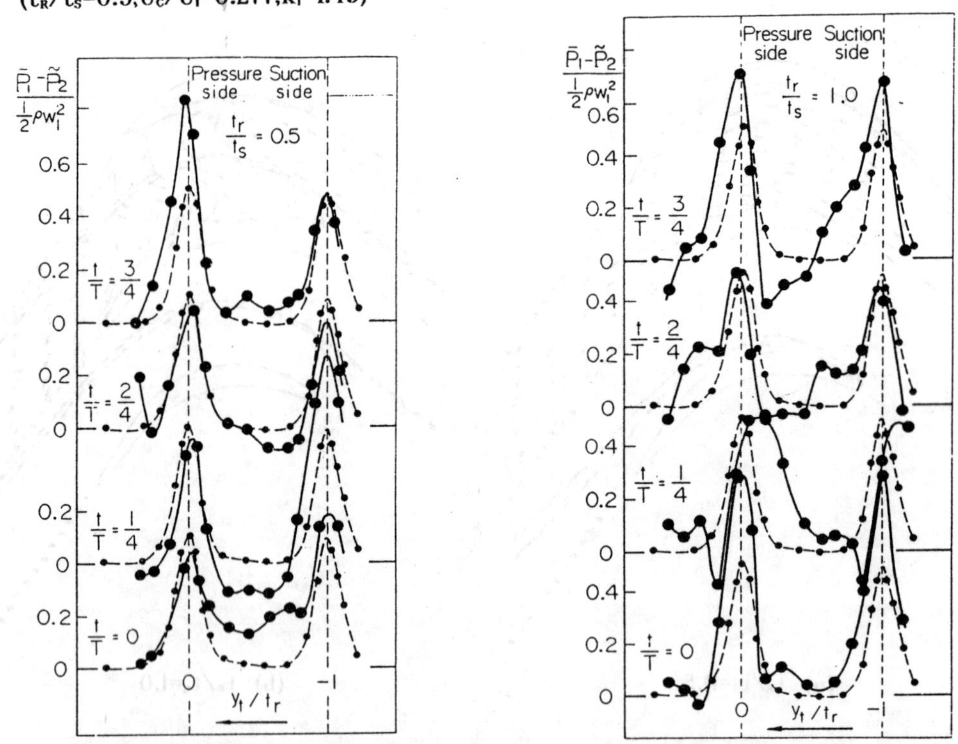

(a) t_R/t_S=0.5, U_C/U_1=0.277, k_1=1.49 (b) t_R/t_S=1.0, U_C/U_1=0.277, k_1=2.98

Fig.15 Time-variant Total Pressure Loss Distributions

THE EXPERIMENTAL RESEARCH IN THE FLUTTER CHARACTERISTICS OF THE BF–1 SERIES OF ROTORS

Song Zhao–Hong, Feng Yu–Cheng, Wang Yu–Wei
Beijing University of Aeronautics and Astronautics

ABSTRACT

This paper is a summary of the experiment in the flutter characteristics of the BF–1 series of rotors. The results of the experimental research in the factors, the blade radial clearance of the blade, the blade airfoil, the blade stagger, the mistuning pattern of the blade and the inlet flow condition, which affect the flutter characteristics, are introduced in this paper. Some useful test phenomena and methods of controlling flutter have been obtained. In addition, the acoustic phenomena of the stalling flutter and the rotating stall onset have been studied in the experimental procedures and the method of acoustic diagnostics in the flutter happening.

INTRODUCTION

Today, more and more attention has been paid to the research on the blade flutter. Although considerable progress has been made in that field from the aeroelastic stability point of view, the mechanism of the flutter onset has not been understood clearly because of the complexity. A lot of factors which affect the stability of blade system, such as the blade–disk coupling, the effect of the three dimension flow, the effect of secondary flow, large scale seperation region and the phase coupling, etc., have not been considered fully in numerical computation of predicting flutter boundary. And it is difficult to get satisfactory results by computation for some engineering methods of controlling flutter. Hence, the experimental research on the characteristics of the blade flutter is indispensable and very inportant work in the research on that subject.

Hitherto, most of the experimental researches on the blade flutter are either non–rotating cascade tests in wind channel or the flutter tests carried on the full scale turbomachine. But the specially designed modal rotor tests are very few.

In order to coordinate with the theoretic research in the blade flutter, the series of blade flutter rotors have been designed specially. The following tests and researches have been carried out.

(1) measuring the flutter boundary of the modal rotors. The testing results have been contrasted with that of the numerical results to verify the correctness of the computation subsystem of predicting flutter boundary.

(2) investigating the mechanism of flutter onset and observing the phenomena of flutter.

(3) investigating the effect of the radical clearance of blade on the flutter characteristics.

(4) investigating the effect of the geometric parameters of airfoil and the locations of the two centres on flutter characteristics.

(5) investigating the mechanism, effect and regularity of the mistuned rotor on controlling flutter.

(6) investigating the effect of the inlet flow field on the flutter characteristics.

(7) investigating the flutter acoustics and acoustic monitoring technique etc.

By analyzing the test results , not only some theoretic models have been vertified and some flow induced vibratory phenomena, which had not been reported in published literature, have been observed, but also some important change regularities and control parameters which affect the onset of the blade flutter have been obtained. Besides, the technique to measure the characteristics of rotating stall and the flutter of the blade by means of the acoustic monitoring method has been applied to the experiment.

THE DESIGN OF THE TEST BLADES

The design of the test blades must satisfy the requirement that the blade–disk system could flutter in the given opearating condition. The aerodynamic design is conducted on the basis of the principle of zero aerodynamic damping and by means of the constant circulation torsional law of simplified radial equilibrium. The gravity centres of the blade are modeled according to the linear super–imposition alone the span. The blade modelling is controlled by means of the reduced velocity parameter:

$$V_R = \frac{W}{\pi b f} \geqslant 8.0 \sim 8.5$$

Where W is the relative speed of the inlet flow

b is the blade chord

f is the first bending vibratory frequency

There are three schemes for the blade airfoil design, i.e. (1)double–cambered airfoil with double–overlaping–center in which the gravity centre is on the top of the torsional centre; (2) multi–cambered airfoil with ahead torsional centre; (3) multi–cambered airfoil with backward torsional centre. Three different stagger angles are designed, i.e. (1) the design stagger angle $\beta = 37.8°$, (2) entended stagger angle $\beta = 47.8°$ and (3) reduced stagger angle $\beta = 27.8°$.

Nine types of basic testing schemes, which can be gained by combining the different airfoil blades with different staggers, make the series of blade flutter testing rotors. If the various mistuning patterns or some other attached construction units to control flutter are considered, the research with more testing schemes can be carried out.

The main aerodynamic design parameters of the rotor are shown in Table1. The main construction parameters of the airfoil are shown in Table 2. The reduced velocities and reduced frequences are shown in Table 3.

Table 1. The main aerodynamic design parameters

Mame	Data	Name	Data
rate of flow	12.93	design incident angle	3.36
ratio of pressure	1.25	relative Mach number of blade tip	1.08
design rotating speed	16500	external diameter of rotor	355.8
efficiency	0.80	internal diameter of rotor	201.8
density of blade tip	1.035	wheel hub ratio	0.567
aspect ratio	2.64		

Table 2. The main construction parameters of the airfoil

Name	Date	Name	Data
number of blades	36	first bending frequency	363
length of blade	77	second bending frequency	1286
chord of blade tip	31	first torsion frequency	1185
chord of blade foot	26	second torsion frequency	4680
ratio of chords	1.2		

Table 3. The reduced velocity and reduced frequency

Mode of Vibration	Reduced velocity	Data	Reduced Frequency	Data
first bending	$V_{RF} = \frac{W}{\pi b f_1}$	9.71	$K_{RF} = \frac{1}{V_{RF}}$	0.103
first torsion	$V_{RT} = \frac{W}{\pi b T_1}$	3.04	$K_{RT} = \frac{1}{V_{RT}}$	0.328

TESTING INSTRUMENT AND MEASUREMENT SYSTEM

1. Testing apparatus

The testing apparatus is a single or double–stage transonic compresser facility whose rated power Ne = 2000~2500 horse power, rate of flow G = 0~ 20kg / sec., maximum rotating speed n = 22000 rpm. as shown in Fig. 1. The testing rotor is driven by a free turbine, which is powered by the turbopropeller engine Au–20A, through a gearing–up box. The testing rotor speed is controlled by means of changing the engine rotating speed and adjusting the bypass valve. The aerodynamic operating condition of the testing rotor is controlled by means of adjusting the exhaust butterfly valve.

2. Measurement System

The following measurement systems have been used in the experimental research. as is shown in Fig. 2.

(1) Measurement of the vibratory stress of blade: The vibratory stress signals received by means of the resistance wire strain foils and the piezoelectric crystal foils are sent to the multi–pass signal modulator through a mercury slip ring assembly. Then, the signals are recorded on a magnetic tape for further analysis. The strain foils and the piezoelectric crystal foils are striken on ten blades whose numbers are 2, 3, 8, 9, 14, 19, 20, 25, 26, and 31 respecially. Each foil is on the suction surface of the leading edge of the corresponding blade and is apart from the blade foot about 18mm, at which the bending vibratory stress is at its maximum by computation analysis.

(2) Measurement of the steady aerodynamic parameters: The pressure pulse sensing device, which is installed on the casing in the rotating plane of the blade, is used to measure the pressure fluctuation behind the blades. The results are analyzed by the pressure pulse auto–detector. The pressure pulse is also recorded by the water platoon of high pressure at the same time.

(3) Measurement of the dynamic pressure: The induction pressure conductor is used to monitor the fluctuatal pressure.

(4) Acoustic monitoring system: As the blade flutter is a kind of aeroelastic unstable phenomenon, the rate of flows does not change abruptly when the flutter occurs. Hence, some measured parameters, such as pressure, tempreture etc. do not change dearly. Since an abnormal sound can be heard when the stall flutter occurs, an

acoustic measure system is used to monitor that phenomenon.

Two high sensitive acoustic detectors are mounted on the casing in rotating plane of the blades and are 22 ° apart from each other. The third detector is put far away from the bellmouth. The signals detected are recorded on a magnetic tape for further analysis also.

(5) Measurement of the blade tip displacement: An unattached measuring instrument is used. The pulse vibration pick– up is installed on the casing in the rotating plane of the blades to pick up vibratory signals of the blade. Then the signals are sent to the amplitude display to show vibratory value curve of the blade on the fluoroscope. By using that instrument, the blade amplitude values can be measured and the flutter condition can be determined exactly.

EXPERIMENT RESULTS AND ANALYSES

The design that the flutter can occur near the designed rotating speed easily has been proved to be successful. The following are the tests and the testing phenomena.

1. The characteristic tests of the flutter rotor

The characteristic test involve two–respect contents, the aerodynamic response and aeroelastic stability of the blade. Three kinds of rotors, each with one of the three types of blade airfoils, were tested. The measurement was taken in four operating conditions. The following are the main phenomena.

(1). The phenomena of the aerodynamic response

The wake–induced blade resonance and the blade resonance caused by the rotating stall occured in the testing procedure.

For the rotor with the blade of the double–overlapingcontre, the phenomena oi the wake–induced blade resonance were observed at the rotating speed of 9000rpm. When the flow passes through the three prop staies, which are mounted on the internal wall of the front casing, the exciting frequency produced is:

$$f_B = Kn = 3 \times 9000 / 60 = 450Hz$$

That frequency is close to the kinetic frequency of the blade, $f_D \approx 460Hz$. Thus the blade resonance is excited by the aerodynamic force with the frequency f_B. As shown in Fig.3, the other rotors had also the kind of resonant phenomenon at low rotating speed.

The rotating stall phenomena occured under the condition of small throttle at several rotating speeds in the test. By measuring the rotating stall, the aerodynamic unstable boundary of the stage can be determined. The blade resonauce caused by the rotating stall had been recorded only at few rotating speeds in the testing procedure, because that resonance, a kind of aerodynamic response phenomenon, only occurs when the rotating frequency caused by the stall lumps is equal to the intrinsic frequeucy of the blade.

The rotating stall and the rotating stall resonance were observed at the rotating speed of 11530rpm. At this moment, the measured number of the stall lumps, K_S was 2, and the pulse frequency n_S, which was gained by conversion from the measured data, was equal to 85Hz. Hence, the rotating frequency

$$f_S = K_S n_S = 2 \times 85 = 170Hz.$$

The kinetic frequency of the blade was 510 Hz.

It is thus clear, that the third resonance of the blade was excited by the stall lumps at the rotating speed of 11530rpm.

Both those phenomena were also observed at the rotating speed of 11960 rpm.

The pressures in front of and behind the stage and the stage efficiency decrease clearly at the time that the rotating stall of the blade occurs. A large vibratory stress wave pattern can be measured and certain resonant sound can be heard when the rotating stall resonance occurs.

(2) The stall flutter phenomena

The stall flutter phenomena of blade, a kind of aeroelastic unstable vibration, had been observed many times in the testing procedure. For example, the BF–1–1 Type rotor fluttered at the rotating speed of 16000 rpm near the rotating stall boundary. At this moment, the incident angle of inlet flow $\alpha = 11.87°$, the reduced velocity $V_R = 8.92$. The measured vibrotory stress signals of the blades showed strong vibratory wave forms, and an intense whistling sound could be heard. Comparing the fluttering rotor with the nonfluttering one, it can be seen that the stage pressure ratio and the stage efficiency do not change obviously, for the blade flutter is a kind of aeroelastic unstable phenomenon. The vibratory stress wave forms measured at the time of fluttering, can be divided into two kinds, one is the stable sine wave, the other is the unstable 'beat' wave, as shown in Fig. 4.

The maximum vibratory stress σ_{max} which is converted from the vibratory stresses measured at the fluttering moment approximately equals 3600 kg / cm^2, which is close to the conventional fixed value.

The various ranges of the rotating speed at which the blade fluttered are shown in Table 4 for the five testing rotors.

Table 4.

Number of rotors	BF–1–1	BF–1–2	BF–1–3	FB–1–4	BF–1–5
Fluttering rotating speed $\bar{n}$	0.97	0.73–1	0.97–1	0.85–1	

2. The location of the two centres of the blade

There are three types of airfoil shapes, the airfoil with double–overlaping–centre in which the gravity centre is on the top of the torsional centre, the airfoil with ahead torsional centre and the airfoil with back ward torsional centre, in the designed blades. The geometric parameters of the airfoils are shown in Table 5.

Table 5. Geometric parameters of the airfoils

Parameters \ Airfoils		Double–overlapping–centre	Ahead torsional centre	Backward torsional centre
Area	A	36.75	36.81	37.07
Gravity	X_0	13.59	13.16	15.21
Centre	Y_0	0.058	0.090	0.037
torsional	X_T	13.26	11.77	16.99
centre	Y_T	0.173	0.229	0.089

Both theoretic and experimental results show that the relative location of the gravity centre and the torsional centre has a clarity effect on the flutter stability of the blade. The unstable characteristic figure of the three kinds of rotors, each of them with one of the three types of airfoil blades, is shown in Fig.5. It can be observed clearly that the flutter stability of the rotor with ahead torsional centre blade is the worst. The flutter boundary of the rotor with ahead gravity centre blade is close to or goes beyond the unstable characteristic boundary of the rotor, which means that it is difficult, or even impossible, to flutter for this kind of rotor. The stability of the rotor with double–overlapping–centre blade is situated between the above two rotors.

3. The effect of the mistuning on the flutter characteristics

Two kinds of mistuned blade rotors, type I rotor with a small frequency–difference of blades and type II rotor with a large frequency–difference of blades, are designed to investigate the effect of the mistuning on blade flutter characteristics. The mistuning patterns are shown in Fig.6 and Fig.7. The experiment results are shown in Fig.8.

Although the rotating stall phenomena were observed in the condition of turning down the throttle, the two kinds of rotors did not flutter at the small rotating speed. At the rotating speed of 12000 rpm, the stall flutter occured in type I rotor, which was with a small frequency–difference pattern. Along with turning down the throttle further, a typical rotating stall phenomenon was recorded also. But type II rotor, with a large frequency–difference, had not any stall flutter phenomena of that rotating speed. During the time that the rotating speed increased further, the flutter onset points of the rotor with a large frequency–difference pattern were always behind that of the rotor with a small frequency–difference pattern. Thus it can be seen that the flutter onset can be checked efficiently by increasing the amount of the mistuning.

4. The effect of the radial clearance of blade on the stall flutter.

Two kinds of tests about the radial clearance had been carried out for the BF–1–2 type rotor. The radial clearances are 1.6mm and 0.5mm respectively. The corresponding stall flutter characteristics are shown in Fig.9. It can be seen from Fig.9 that the stall flutter onset points of the rotors with different radial clearances varies greatly. The small radial clearance enlarges the fluttering range.

There are very complex flow phenomena between the blade tip and the annular wall. The boundary layers attached on the annular wall and the blade tip, the dead zone which makes the boundary layer separate easily and is consisted of the flow passage whirl, the annular wall and the suction surface of blade, the leaking flow and the scrape whirl increase the loss of the flow and affect the workdoing ability of the blade. It is the decrease of the blade tip clearance that makes the loss of the secondary flow decrease and the flow speed increase. As the mechanical damping does not change with the change of the flow speed and the aerodynamic exciting force increases with the increase of the flow speed, i.e., the aerodynamic work done on the blade by the aerodynamic force in a vibration circle increases, the small radial clearance makes the flutter occur easily.

Some literatures have pointed out that the best radial clearance at the blade tip is about 1~1.5%, represented by the ratio of clearance to chord. In this experiment, the ratioes of clearance to chord are 0.4% and 1.6% respectively.

5. The effect of total pressure distortion of the inlet flow on the stall flutter

In order to adjust the total inlet pressure of the compressor in the testing procedure, a propped net, which is 1 hole per inch and is weaved with the wire whose diameter is 2mm, was mounted in front of the inlet and was 0.5 times the inlet diameter from the inlet. The propped net was covered with the net which is used to disturb the inlet flow and cause the distortion. The net to disturb flow is ten holes per inch and is fixed by use of the stainless steel wire. Three kinds of tests were carried out

(1) The radial distortion at the blade tip. the covering area is one third of the total area.

(2) The radial distortion at the blade foot. The covering area is one third of the total area.

(3) The distortion along the circumference. The covering area is nine tenths of the total area.

The tests were made on the rotor with the ahead torsional centre blade, which is the most unstable one. The testing results are shown in Fig.10. It is found from the initial results that all the three kinds of inlet distortions defer the flutter onset and improve the aeroelastic stability of the rotors.

The testing rotors with the inlet distortion operated smoothly and the flutter phenomena were not observed in all the period of turning down the throttle from the full gate to the onset of the rotating stall. But the flutter phenomena occured in the testing rotors without the inlet distortion.

The following are the explanation about these phenomena. Since the tip section of the blade is blocked with the net to disturb flow, the flow field in the compressor changes along the radial direction, i. e. , both the total pressure of the inlet flow and the absolute velocity of the flow near the blade tip range reduce, which make the incident angle at the blade tip range increase and the separation on the back side of the blade serious. In addition, as the flow speed increases, the reduced–velocity V_R reduces. All of those defer the onset of the flutter.

SUMMARY

1. The aeroelastic response characteristics of the blade, which involve the blade resonance induced by the wake, the rotating stall and the blade resonance caused by the rotating stall lumps, have been studied in the experiment. The characteristics of the stall flutter and four factors which affect the characteristics of the stall flutter, i.e. the blade clearance, the airfoil shape of the blade, the mistuning and the distortion of the inlet flow, have been studied. Some beneficial test phenomena have been observed in the testing procedure.

2. All the designed blades, which are with three kinds of airfoils, fluttered in varying degrees under the designed operating condition, which has proved the reliablity of the design, and provides important basis for the further design of the nonfluttering blade.

3. It has been proved by the experiment that the flutter can be checked by changing the airfoil shape of the blade or by using the mistuned blade rotor. Some regularities about those controlling methods have been found from the test results, which is very valuable for the design and manufacture of the aeroengine.

4. The conclusions that the aeroelastic stability of the rotor can be improved by changing the radial clearance or by varying the distortion condition of the inlet flow are very useful.

5. Several kinds of measuring and monitoring methods have been used and studied in the experiment.

REFERENCES

1. Zhou sheng, etc., Aeroelasticity of Turbomachines.

2. Feng Yu–Cheng , Hu Zhong–An, ACTA AERONAUTICA ET ASTRONAUTICA SINICA. 1985

3. Song Tao, Song Zhaohong "The Theoretical Analysis of the Mistuned Blade Rotor for Controlling Flutter" ACTA AERONAUTICA ET ASTRONAUTIACA SINICA. 1988

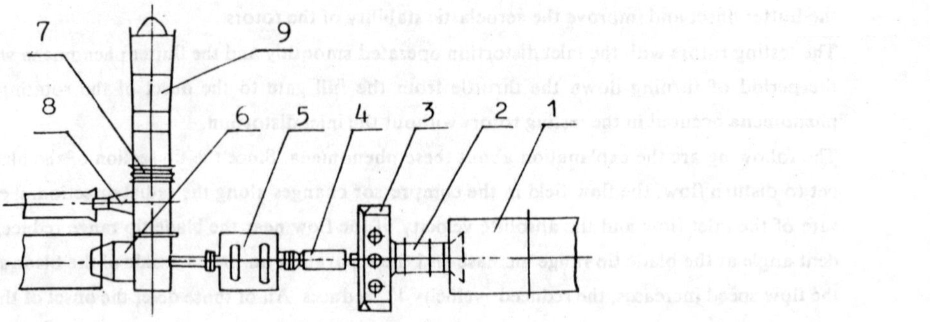

1. constant pressure box 2. working section 3. air evacuation throttle
4. detector 5. accelerator 6. free turbine
7. expansion segment 8. exhaust duct 9. AH-20 gas generator

Fig.1 Experimental set-up

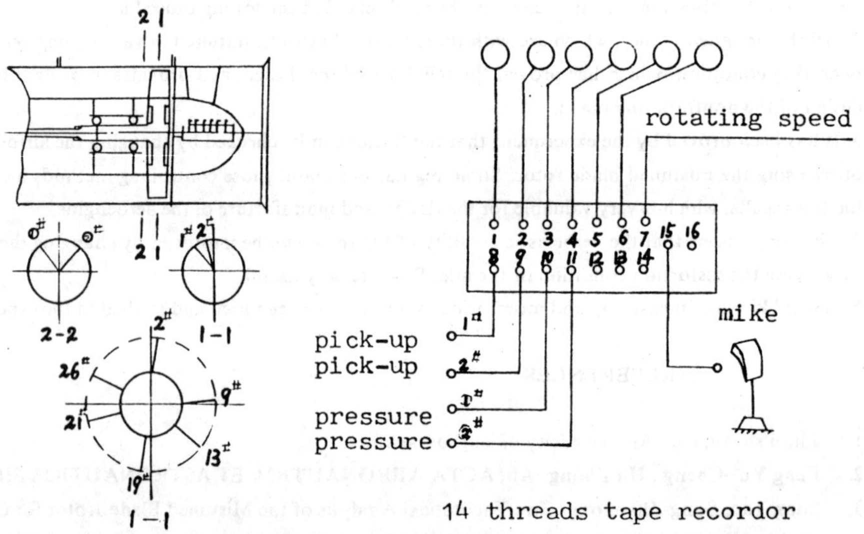

1--1 locating pick-up
2--2 locating baroreceptor

Fig.2 Testing system

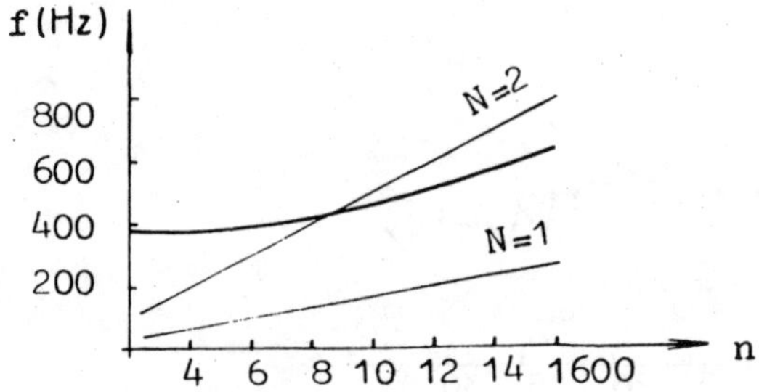

Fig.3 Resonance rotating speed

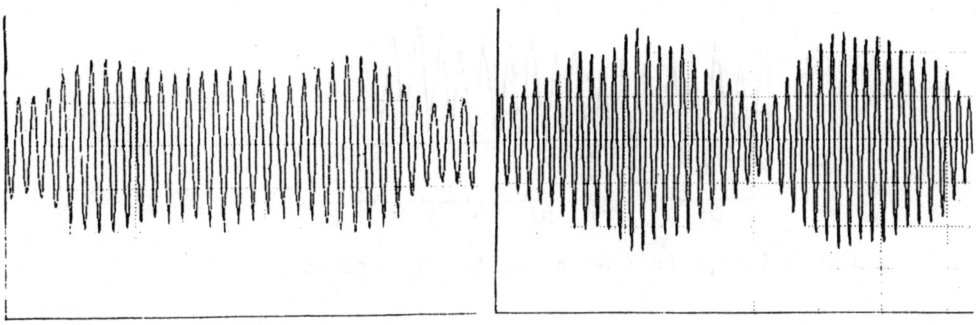

(a) sine wave (b) beat wave

Fig.4 Stall flutter wave-form

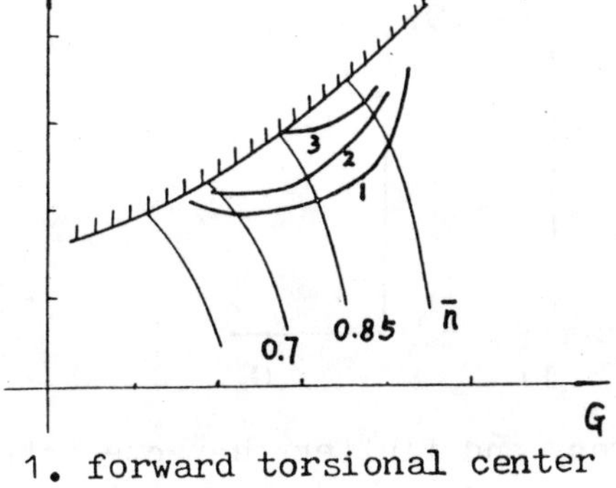

1. forward torsional center
2. double-overlapping-center
3. backward torsional center

Fig.5 Three centers and flutter boundary

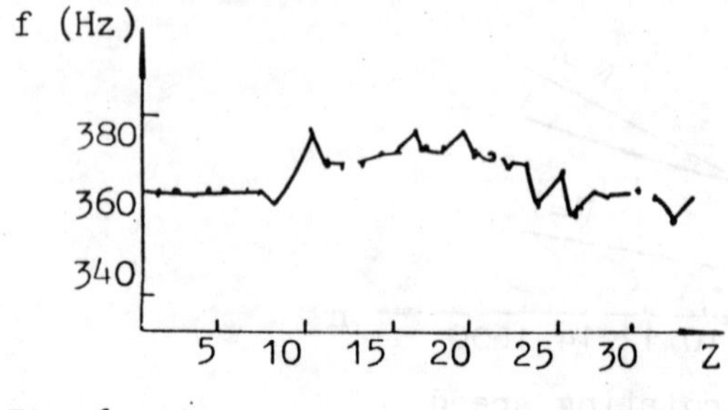

Fig.6 Little frequency difference scheme

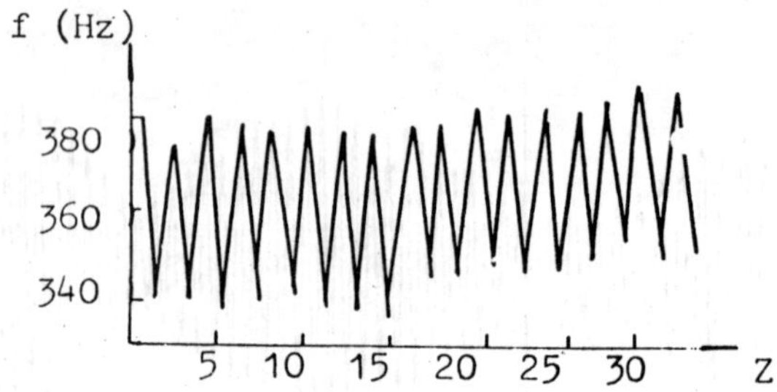

Fig.7 Large frequency difference scheme

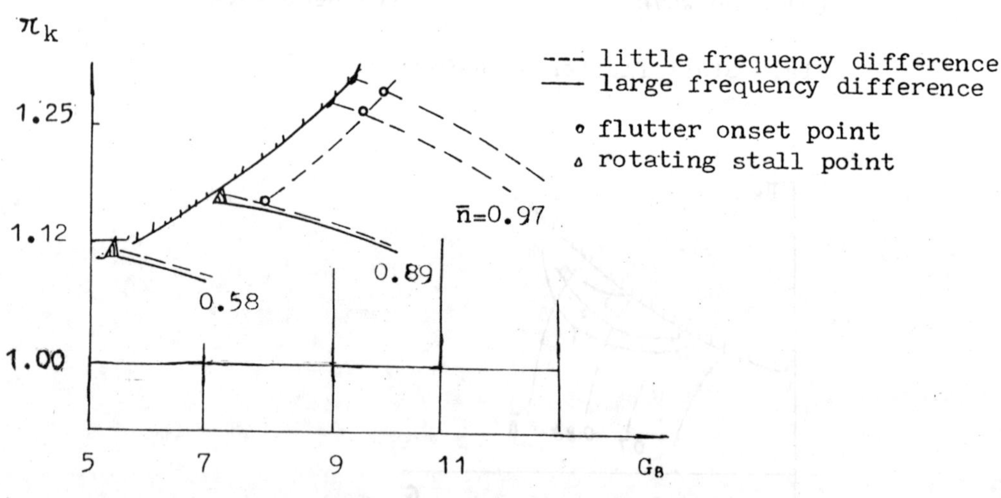

Fig.8 Mistuning frequency and flutter characteristic

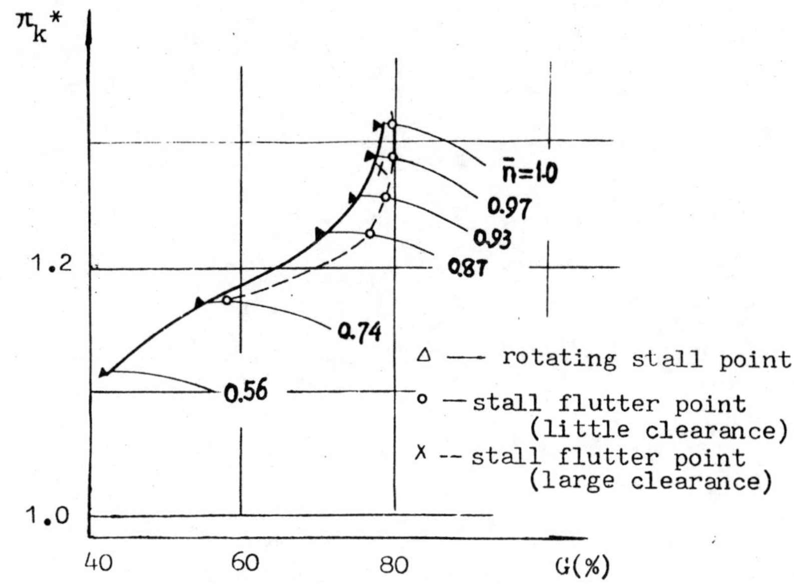

Fig.9 Flutter characteristics of two kinds of clearances

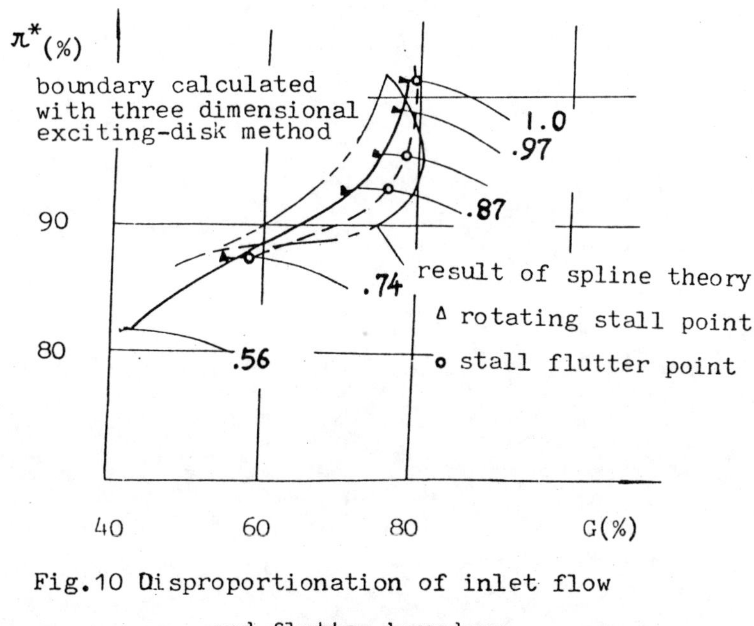

Fig.10 Disproportionation of inlet flow and flutter boundary

Part Three
Structural Dynamics

AEROELASTIC VIBRATIONS OF TURBOMACHINE BLADES AND PROPELLERS

Yu. S. Vorobiev, M.L.Korsunskiy
(Ukr. SSR Academy of Sciences, Institute for Problems in Machinery, Kharkov , USSR)

ABSTRACT

Vibrations of the blade system are considered accounting the interaction of blades through the flow and elastic bonds. Applicability of different mathematical models, excitation of different vibration modes mistuning effect and the other factors are discussed.

INTRODUCTION

The blade system in flow is the common system. It is convenient to describe the flow effect on the blade cascade with the help of disturbing polyharmonic loads and forces of aerodynamic damping. Depending on geometric configuration the blade is considered on the basis of the refined theory of twisted rods or as tree–dimensional body I. At aeroelastic vibration of burbomachine blades and propellers the lower modes usually are excited. So the fields of application of rod models usually are enough broad. However, even with these modes of blade vibration the deformation of the elongated cross–section in its plane is possible including also with the blades with intermediate bonds (Fig. 1). In these cases rod models can not be used I.

MATHEMATICAL MODELS AND RESEARCH METHODS

The blades are considered in the rectangular right coordinate system xyz with the origin 0 in the root cross–section, rotating together with the blade with the angular velocity Ω . The axis z is directed along a radius from the axis of rotation, the axis x is parallel to the axis of rotation (Fig. I). While using the theory of twisted rods, the local coordinate system xy are central, and $\xi\eta$ are main central ones in any cross–section. The corresponding systems x_1y_1 and $\xi_1\eta_1$ have an origin in the center of bending of the cross–section. An angle α between $x$and ξ is the angle of pretwisting of the cross–section. Further, dashes will denote the derivative along the coordinate z and points over functions will denote the derivative along the time t. Displacements of the elementary volume of the rod in directions x,y,xbased on the refined theory of rods will be

$$U = u_1 + u_2 - \theta y_1;$$
$$V = v_1 + v_2 + \theta x_1; \qquad (1)$$
$$W = w - u_1'x - v_1'y + u_2'x + v_2'\psi + \theta'\omega,$$

where $u_1(z,t), v_1(z,t)$ are displacements of bending, $u_2(z,t), v_2(z,t)$ are displacements of the shift in the directions x and y; $w(x,t)$ is the displacement in the direction z; $\theta(z,t)$ is the angular displacement with respect to the center of bending; $\varphi(\xi,\eta,z)$ is the function of torsion; $x,\psi(\xi,\eta,z)$ is the function of warping of the shift.

In the practical problems warping of the cross–section is accounted for by the definition of tangent deformations and stresses 2. Deformations of the rod finally are taken as 1,2:

$$\varepsilon_z = w' - u_1''x - v_1''y + u_2'' + v_2\psi + \theta''\varphi + \theta'\tau_o r_1^2;$$

$$\gamma_{xz} = \theta'\left(\frac{\partial\varphi}{\partial x} - y_1\right) + u_2^1\left(1 + \frac{\partial x}{\partial x}\right) + v_2^1\frac{\partial\psi}{\partial x} \quad ; \tag{2}$$

$$\gamma_{yz} = \theta'\left(\frac{\partial\varphi}{\partial y} + x_1\right) + u_2^1\frac{\partial x}{\partial y} + v_2^1\left(1 + \frac{\partial\psi}{\partial y}\right); \qquad \varepsilon_x = \varepsilon_y = \gamma_{xy} = 0,$$

where $\tau_0 = \alpha'$ is pretwisting; $r_1^2 = \xi_1^2 + \eta_1^2$

The expressions (1) represent the inertial interconnection of bending and torsion due to mismatch of the center of mass and the center of the cross–section bending and the expression (2) represents deformation interconnection of different types of displacements. While defining stresses, the internal damping in the material is taken into account 2. Variations of the work of internal forces at deformation of the rod holds:

$$\delta\Pi = \int_0^l \int_F [\sigma_z\delta\varepsilon_z + \tau_{xy}\delta\gamma_{xy} + \tau_{yz}\delta\gamma_{yz}]dFdz,$$

where l is blade length, F is the area of the cross–section. The kinetic energy of the rotating rod is the sum of the kinetic energy of vibrations and works of centtrfugal and the Koriolis forces:

$$T = \frac{1}{2}\int_0^l \int_F \rho\{[\dot{U}^2 + \dot{V}^2 + (\dot{W}^2 - \dot{\zeta}^2)] + \frac{\Omega^2}{2}[(R + z + W - \zeta)^2 + (V + y)^2] + \Omega[(V + y)(\dot{W} - \dot{\zeta}) - (R + z + W - \zeta)\dot{V}]\}dFdz,$$

where ρ is the density of blade material, ζ is the change of projection of longitudinai fiber on the axis z at its deformation.

Interblading bonds are considered as rods, displacements of ends of which are defined by displacements of the corresponding cross–secti–ons of the blades. Bonds experience space bending, torsion and tension. Forces of structural damping are supposed to be proportional to the coefficients of structural damping and relative movements of the corresponding blade cross–sections I. A disc is simulated as a round plate of variable thickness executing bending–torsion–radial vibrations in the general case, rim and bush are simulated as curved beams. The rotor of disc–drum construction may be presented as a set of round plates, shells of rotation and curved beams.

Each harmonic of the disturbing load of the flow is defined by the frequency w_k and variable amplitude $q_K(z)$ and phase $\varphi_K(z)$ in blade length, where K is harmonic number 2. The aerodynamic interaction of profiles in the cascade are defined by the forces, that are expressed by the aerodynamic effect coefficients K_{jK}, e.g.:

$$Q_{\eta n} = Q_0\sum_K [K_{\xi k}\frac{U_{\xi,n+K}}{b} + K_{\eta K}\frac{V_{\eta,n+K}}{b} + K_{\theta K}\theta_{n+K}],$$

where b is cross–section chord, U_ξ, V_η are displacements in the direction of axes ξ,η,η+K are numbers of interaction blades 3. In the practical cases it is sufficient to take the account of the interaction of neighbouring blades (K = 1,0,–1).

The problem for nonconservative system is solved by the variational method that differs by the fact, that not displacements, but moments and forces are the values being approximated. The basic functions satisfy force boundary conditions. Displacements and their derivatives are determined by integration, e.g.:

$$U_1' = [l_\eta M_\eta \cos\alpha - l_\xi M_\xi \sin\alpha]_{z=0} + \int_0^z \left(\frac{M_\eta}{EJ_\eta}\cos\alpha - \frac{M_\xi}{EJ_\xi}\sin\alpha\right)dz,$$

where M_ξ, M_η are bending moments, J_ξ, J_η are moments of inertial, E is elasticity modulus, l_ξ, l_η are the coefficients of blade root compliance.

So all the boundary conditions are met and at calculations the numerical integration is mainly carried out, that allows to solve non–conser vative problems and to improve the convergency of the methods. Besides, the consequent complication of the system is used by choosing the solutions of the partial problems for the basic functions. It results in decreasing a number of variable parameters at each stage of the calculation to a significant extent and convergency and stability of the computational process are greatly increased. If all the blades in the system are supposed to be equal, then the assumptions on turning assymetry of vibration modes of the system may be used and the problem may be simplified. If small technological deviations of blade sizes are taken into account, intrapackage vibrations are considered or assumptions on turning symmetry of vibration modes of the system in flow are not introduced, then each blade has own set of basic funcitions and the problem is greatly complicated. In this case the calculation of the system in parts plays a great role. To make the choice of the mathematical model of the blade at the concret calculations easier, the sets of the boundaries of the application fields for these models depending on nondimensional parameters, representing the main geometric configurations of the object and the account of the different factors (pretwisting, non–symmetry of the cross–section, shift, warping of the cross–section, etc).

Application fields of rod models for problems of aeroelasticity are enough broad. However, they are restricted by the parameters of the blades, at which the "shell" vibration modes occur or discrepancy of theoretical and experimental results takes place due to cross–section deformation, even at "rod" modes of vibrations. So, in Fig. 2 the boundaries of application fields of the rod models are shown depending on the angle of twist of species α at $\frac{b}{l} = 0.25$ and relations $\frac{b}{l}$ for species with the linear change of thickness.

The three–dimensional body is the most suitable mathematical model for a number of blades. An isoparametric hexahedral curvilinear element HK 60 with square–law for a function of displacements has been chosen that is important for the description of stresses (Fig. I). The element has 20 nodes and belong to the Serendip family. The finite element used allows to describe blades of the more complicated configuration with a bevel, flange, blade rotor and cooling channels. For some blades it is necessary to account blade root deformation. It should be noted, that the given method is convenient while calculating thinwall compressor blades. In a number of cases the calculation of blades for vibrations follows the determination of its temperature field. These results are used to calculate the given geometric performances of the blades for rod modes or the account of inhomogeneity of the material using the finite element method. To account flow effect on the finite element model of the blade the aerodynamic coefficients of effect and data on the distribution of the nonstationary pressure on the blade may be uses 4.

SOME INVESTIGATION RESULTS

The methods developed, open the broad possibilities to investigate vibrations of turbomachine blade system. The comparison of theoretical and experimental results for the blade and the blade system has been carried out I. As an example in Fig. 3 are given theoretical (446.6Hz and experimental (442.5 Hz) modes of displacements v a··1 stresses σ at the leading edge for the model package of five blades with the given mistunig. The data of the experiment is maked by points.

For comparison the modes of packages of identical blades (425.2 Hz) are shown by a dashed line. The great effect of mistuning and a good correspondence of theoretical and experimental results are seen even in such a complicated case.

The calculation of vibration of blading with cricular flange bond is carried out on the basis of the experimental data on the circumferential nonuniformity of the flow I. Distributions of displacements and stresses in blade length at the leading edge are shown by solid lines in Fig. 4 and their phases at vibrations with three nodal diameters are shown by dash–and–dot lines. The location of circular bond is shown. Displacements relate to the value $v(l)$, and the stress relates to $\sigma(0)$. The data of strain measurement correspond to vibration mode, that is practically resonant. The phase shift between bending and torsion and also in blade length, and also the effect of the bond on vibration mode are seen.

Sets of such calculations allow to construct amplitude–frequency performances of the system, that show comparative excitation of different modes of vibration and the possiblility of self–vibration. Figure 5 gives amplitude–frequency performances of a package of five blades (a solid line) and a package with mistuning (dashed lines) within the given range of frequencies.

Aeroelastic bibration modes, as a rule, correspond to the lower rod modes of blade vibration. The vibration of thin–wall peripheral section close to local shell modes may be excited in flow of blades of powerful turbomachine last stages, the cross–section of which varies greatly from root to periphery. The peripheral section of these bladings is the cascade of small solidity and sometimes their calculation of unit profiles in flow is possible. For aeroelastic vibration of blades in flow the phase shift plays the large role between displacement and incidence angle of profile. It defines the instantaneous angle of incidence, that effects aerodynamic forces at the given moment. The Koriolis forces, arising at vibration of rotating rod effect also phase shift between these displacements 1,2. Their effect may be clearly demonstrated by an example of unit rod (propeller type) vibration in the nondisturbing axial–symmetric flow. For the rod of two–symmetrical cross–section the Koriolis forces cause the phase shift between bending and torsion $\varphi = \pi / 2$. With increasing the parameter $\rho_{\xi} = \xi_s / r_p$ (ξ_s is the distance between the center of mass and the center of bending, γ_p is a radius of cross–section inertia) the phase shift both for the first (φ_1) and for the second modes (φ_2) decreases that influences on the critical velocity of flutter (Fig. 6).

Centrifugal forces increase the critical velocity of flutter V_k (V_{k0} is the critical velocity of the nonrotating rod). For non–twisted rod the Koriolis forces increase, and for the twisted rod ($\alpha = \pi / 6$) decrease the critical velocity. The curve I refers to the case of large increase of parameters of nonsymmetry of the rod(polar–axial moments of intertial 1,2) by five fold.

The calculation results show, that at vibration of the blade system in flow the role both of the inertial and deformation connections between different vibration modes is sufficient; mistuning of the system appreciably influences on excitation of vibration modes; some factors(such as e.g., the Koriolis forces, that may be neglected while investigating the free vibration should be taken into account.

REFERENCES

1. ВоРобеВ Ю.С. КолебаНИЯ ЛоПаТоЧНоГо аППаРата Турбомащин. Киев: Наук. дма(1988), 224.

2. Воробев Ю.С., Щорр Б.Х. Теория закрученных Стержней. Киев: Наук. думка(1983), 188.

3. Горобьев Д.Н., Курзин В.Э., Сарен В.Э. Аэродинамика Решеток В Нестационарном Потоке, Новосибирск: Наука(1971), 272.

4. Соколокийя Г.А., Гнесин В.И. Нестационарные Трансзвуковые и Вязкие Течения В Турбомацин ах. Киев: Наук > думка(1986),264.

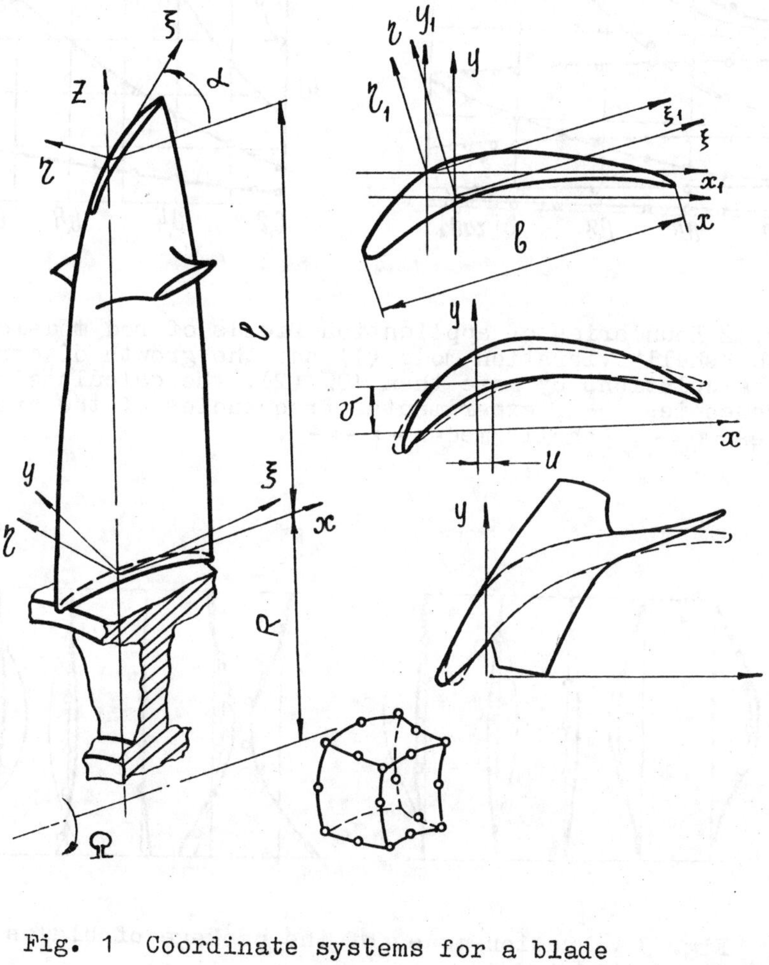

Fig. 1 Coordinate systems for a blade

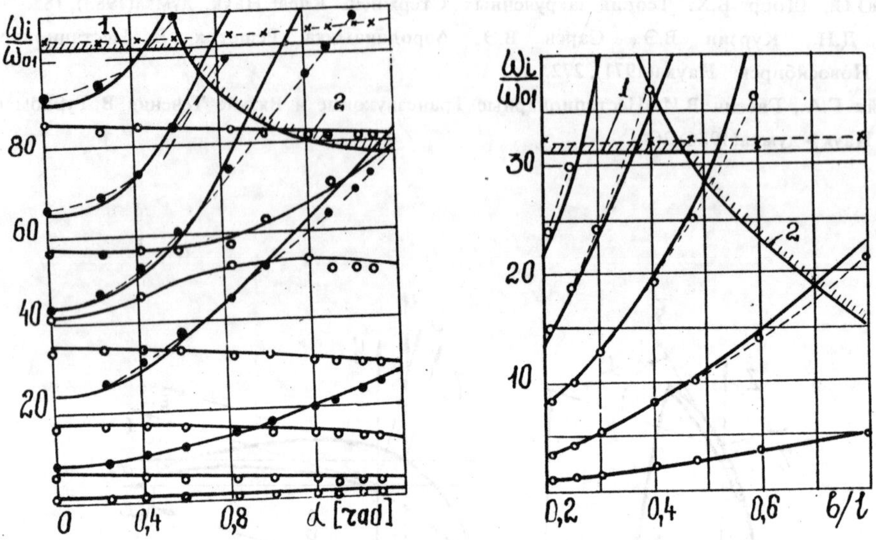

Fig. 2 Boundaries of application fields of rod models with "shell" vibration mode (1) and the growth of errors of calculations by more than 10% (2), the calculated frequencies – – ; experimental frequencies: of the "rod" modes -o-o- ; "shell" modes -x-x- .

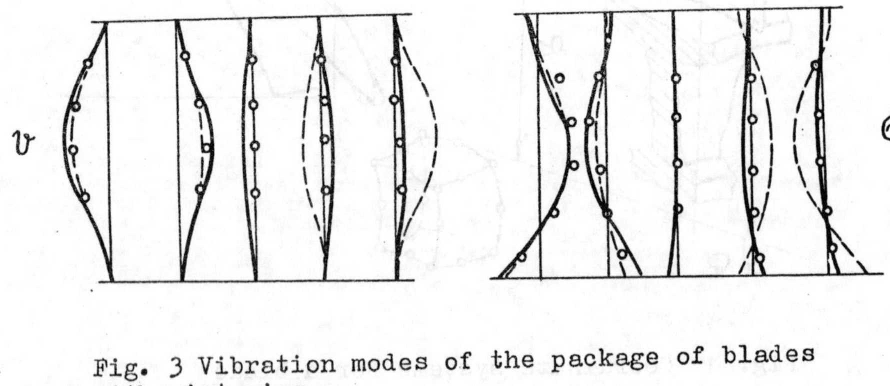

Fig. 3 Vibration modes of the package of blades with mistuning.

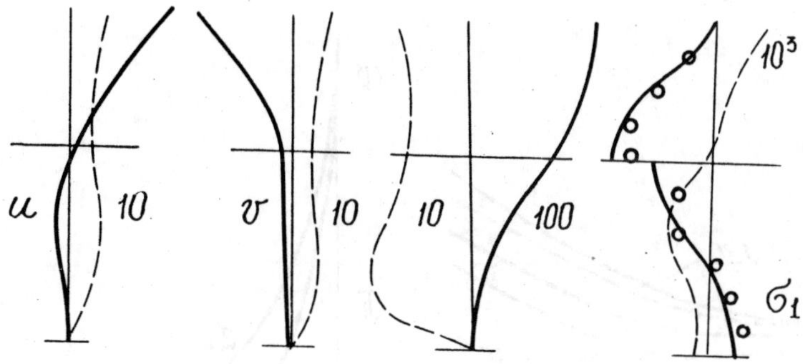

Fig. 4 Displacements, stresses and their phases at vibration of blades with circular connection in flow

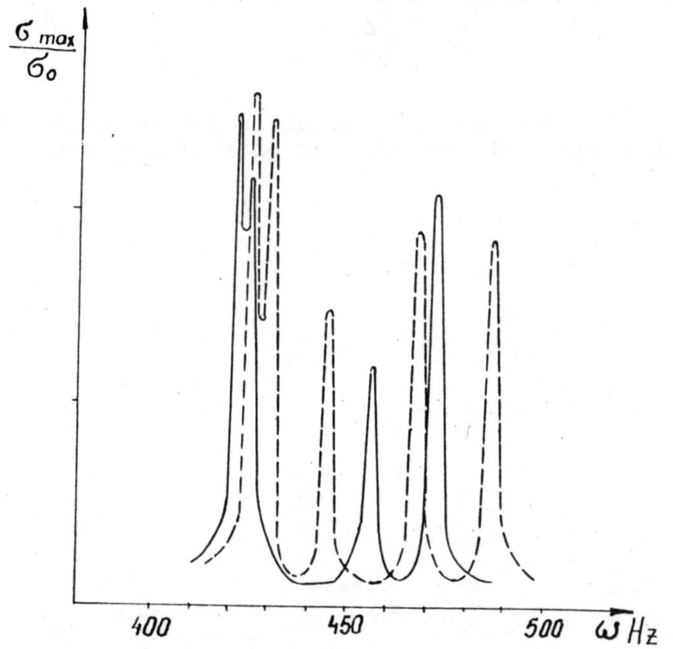

Fig. 5 Amplitude-frequency performances of the package of five blades.

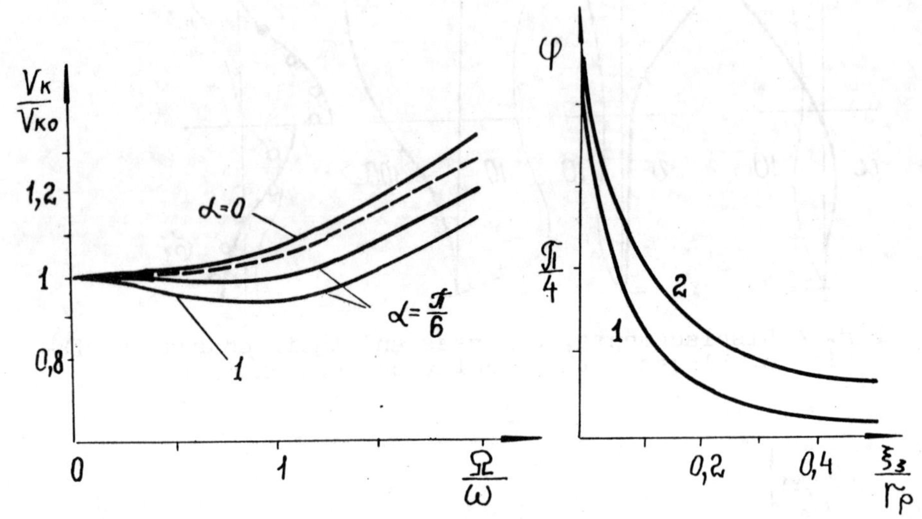

Fig. 6 Effect of Koriolis forces on phase shift (φ) and critical velocity V_K of flutter of rotating rod.

AEROELASTIC TAILORING ANALYSIS FOR PRELIMINARY DESIGN OF ADVANCED TURBO PROPELLERS WITH COMPOSITE BLADES

Takashi Yamane
(Mechanical Engineering Laboratory, Tsukuba, 305 Japan)
Peretz P. Friedmann
(University of California, Los Angeles, CA 90024, U.S.A.)

ABSTRACT

An analytical study of aeroelastic tailoring has been conducted to determine the flutter characteristics of Advanced Turbo Propellers for preliminary design purposes.

The aeroelastic tailoring technique, which utilizes the anisotropy of composites, has the potential for eliminating flutter from Advanced Turbo Propellers, without any weight penalties. Though several experiments indicated the validity of the technique, a fairly simple analysis, useful for preliminary design purposes, is not available at present.

In this paper the structural dynamic model for composite swept propeller blades developed by Kosmatka and Friedmann is combined with unsteady cascade theory to produce an aeroelastic analysis tool. The blade is modeled as a set of rotating beam elements with moderate deflections and section warping, including the geometric nonlinearity. For unsteady aerodynamic analysis, Smith's unsteady compressible cascade theory is used and combined with the sweep corrections and finite span corrections based on modified strip theory.

First, the predicted steady aerodynamic loads on the SR-3 propeller model were compared with the results of wind tunnel experiments. By the aeroelastic analysis for the SR-3 propeller model made of graphite/epoxy, it was found that the natural frequencies of the blade can be changed from 100% to 35% and that the corresponding modes or sign of pitch-flap coupling can be modified only through changing composite fiber orientation. Using a p-k type of flutter analysis it was found that the flutter velocity is sensitive to fiber orientation.

INTRODUCTION

The Advanced Turbo Propeller (ATP) represents a new propulsion system which maintains high efficiency, similar to that of conventional turboprop engines, at cruising speeds similar to that in turbofan engines. Characteristics of the ATP blades are (1) high sweep angles to reduce the effective Mach number, (2) low aspect ratio to generate high power with a limited size, and (3) low thickness ratio to reduce noise and drag.

Wind tunnel experiments done by Mehmed et al.[1,2] revealed that the ATP blades have potential for bending-torsion flutter. The importance of cascade effects was also noted. To improve flutter characteristics of the ATP, several ideas such as aeroelastic tailoring[3,4], blade mistuning[5], mass balance, and geometry modification[4] have been examined. Aeroelastic tailoring, which utilizes the anisotropy of composite materials to improve the aeroelastic properties of structure, was examined qualitatively[3,4], and is expected to be applied to current ATP's under development.

Attempts at sophisticated flutter analyses can be found in Refs. 6 and 7. However, the purpose of this analysis is to provide a fairly simple analysis tool useful for the preliminary design of the ATP blades and to examine the usefulness of aeroelastic tailoring for the ATP. Since the numerical code may be used by any designer, the code should be as simple as possible even if the estimated flutter velocity might include slight errors.

In this paper the structural dynamic model for composite swept propeller blades developed by Kosmatka and Friedmann[8,9] is combined with unsteady cascade theory to produce an aeroelastic analysis tool. The blade is modeled as a set of rotating beam elements with moderate deflections and section warping, including the geometric nonlinearity. For unsteady aerodynamic analysis, Smith's unsteady subsonic cascade theory[8] is used combined with the sweep corrections and finite span corrections based on modified strip theory. The present method is aimed at including, for example, the compressibility at Mach number of 0.85, the cascade effect of an 8- to 10-bladed rotor, the effect of sweep up to 45 degrees, the finite span effect due to aspect ratio less than 10.

First, the variations of the section stiffness, the spatial elastic axis location, the natural modes and frequencies for the SR-3 propeller model were studied by changing the fiber orientation as a parameter. The predicted steady aerodynamic performance was compared with the results of a wind tunnel experiment, and the importance of elastic deformation is examined. Finally, the effects of fiber orientation on the flutter velocity are clarified by applying the p-k method on the reduced system using a modal representation.

1. Structural Dynamic Analysis

A. Rotating Anisotropic Beams with Pretwist and Sweep

The ATP blades are regarded as rotating anisotropic beams with pretwist and sweep. Kosmatka and Friedmann[8,9] derived the following curvilinear strain expressions in terms of shear center displacements in a Cartesian coordinate system defined for each element, (u, v, w, ϕ), and section curvilinear coordinates, (η, ζ):

$$\begin{aligned}
\varepsilon_{xx} &= u_{,x} + (v_{,x}^2 + w_{,x}^2)/2 + (\phi_{,x}^2/2 + \beta_{,x}\phi_{,x})(\eta^2 + \zeta^2) + (\psi\phi_{,x})_{,x} \\
&\quad - v_{,xx}\{\eta\cos(\beta+\phi) - \zeta\sin(\beta+\phi)\} - w_{,xx}\{\zeta\cos(\beta+\phi) + \eta\sin(\beta+\phi)\} \\
\gamma_{x\eta} &= (\psi_{,\eta} - \zeta)(\phi_{,x} + \phi_0) \\
\gamma_{x\zeta} &= (\psi_{,\zeta} + \eta)(\phi_{,x} + \phi_0) \qquad (1) \\
\phi_0 &= \{w_{,x}\cos\beta - v_{,x}\sin\beta\}\{v_{,xx}\cos\beta + w_{,xx}\sin\beta\}
\end{aligned}$$

where u is taken along the elastic axis, v is taken in the plane determined by the elastic axis and the chord at 75% radius, and w is taken normal to the plane. The curvilinear coordinate system (x,η,ζ) rotates about the elastic axis by the amount of pretwist, $\beta(x)$. The total system equations of motion can be obtained from discretized Hamilton's principle:

$$\int_{t_1}^{t_2} \sum_{e=1}^{n} (\delta U - \delta T - \delta W_E)\, dt = 0 \tag{2}$$

This implies that the total dynamic potential, $\Sigma_e(U - T - W_E)$, is an extremum over the time integral from t_1 to t_2. The variation of strain energy can be obtained by

$$\delta U = \int_V \begin{Bmatrix} \delta\varepsilon_{xx} \\ \delta\gamma_{x\zeta} \\ \delta\gamma_{x\eta} \end{Bmatrix}^T \begin{bmatrix} Q_{11} & Q_{15} & Q_{16} \\ Q_{15} & Q_{55} & Q_{56} \\ Q_{16} & Q_{56} & Q_{66} \end{bmatrix} \begin{Bmatrix} \varepsilon_{xx} \\ \gamma_{x\zeta} \\ \gamma_{x\eta} \end{Bmatrix} dV \tag{3}$$

where $[Q_{ij}]$ reflects anisotropy of the material and is condensed from the 6x6 stiffness matrix $[C_{ij}]$ based on the assumption as follows:

$$\sigma_{\eta\eta} = \sigma_{\zeta\zeta} = \tau_{\eta\zeta} = 0 \tag{4}$$

The matrix $[C_{ij}]$ is given in Ref. 11 in terms of elastic moduli, E_L, E_T and G_{LT}, and Poisson's ratios, ν_{LT} and ν_{TT}. The variation of kinetic energy, δT, is described in terms of shear center displacements, section warping, and curvilinear coordinates. The virtual work, δW_E, is composed of steady and unsteady aerodynamic forces. The complete expressions for δU, δT and δW_E are given in Ref. 9.

In the case of a cross section close to an ellipse $(\eta^2/a^2 + \zeta^2/b^2 = 1)$, the warping function can be given as follows[13]:

$$\psi(\eta,\zeta) = \frac{S_{45}\, a^2}{S_{44}\, b^2 + S_{55}\, a^2}\left(\zeta^2 - \frac{b^2\, \eta^2}{a^2}\right) + \frac{S_{44}\, b^2 - S_{55}\, a^2}{S_{44}\, b^2 + S_{55}\, a^2}\cdot\eta\zeta \tag{5}$$

where $[S_{ij}] = [C_{ij}]^{-1}$ is a compliance matrix. In the case of a thin airfoil section with homogeneous anisotropy, an approximate equation for the shear center location can be derived through a slight modification to Ref. 13 as follows:

$$X_{SC} = \left[\left(1 - \frac{4S_{13}}{S_{55}}\right) \int_{LE}^{TE} \tau^3\, x\, dx - \left(\frac{3}{4}\, \frac{S_{36}+S_{45}}{S_{55}}\right) \int_{LE}^{TE} \tau^4\, dx\right] \Big/ \int_{LE}^{TE} \tau^3\, dx \tag{6}$$

where the section contour is assumed as $y = \pm\tau(x)$, and $x = 0$, LE, and TE correspond to the section area centroid, the leading edge, and the trailing edge, respectively. Complete analysis for arbitrarily laminated composites can be obtained from the analysis described in Refs. 9 and 12.

B. Finite Element Analysis

The shear center displacements are described by Hermite interpolation polynomials within j-th beam element as follows:

$$
\begin{aligned}
v &= \{\Phi_3(x)\}\ \{v^{j-1},\ v_{,x}{}^{j-1},\ v^{j},\ v_{,x}{}^{j}\}^T \\
w &= \{\Phi_3(x)\}\ \{w^{j-1},\ w_{,x}{}^{j-1},\ w^{j},\ w_{,x}{}^{j}\}^T \\
\phi &= \{\Phi_2(x)\}\ \{\phi^{j-1},\ \phi^{j},\ \phi^{j-0.5}\}^T \\
u &= \{\Phi_2(x)\}\ \{u^{j-1},\ u^{j},\ u^{j-0.5}\}^T
\end{aligned} \tag{7}
$$

where $\{\Phi_2(x)\}$ and $\{\Phi_3(x)\}$ denote a set of quadratic and cubic Hermite interpolation polynomials, respectively, and $(\ )^{j-0.5}$ shows mid-element value. The following boundary condition is applied at the root cutoff:

$$\{v^0,\ v_{,x}{}^0,\ w^0,\ w_{,x}{}^0,\ \phi^0,\ u^0\} = 0 \tag{8}$$

The generalized equations of motion for 8n degrees of freedom are obtained as follows:

$$[M]\mathbf{q}_{,tt} + [D]\mathbf{q}_{,t} + [K_L+K_{NL}(\mathbf{q})]\mathbf{q} + \{F_T\} = \{0\} \tag{9}$$

where

$$\mathbf{q} = \{\phi^{0.5},\ u^{0.5};\ v^1,\ v_{,x}{}^1,\ w^1,\ w_{,x}{}^1,\ \phi^1,\ u^1;\ \ldots\ldots,\ \phi^n,\ u^n\}^T \tag{10}$$

The matrix $[K_{NL}]$ represents nonlinear stiffness due to moderate deflections, warping, and steady load effects, and $\{F_T\}$ refers to steady/unsteady aerodynamic forces and steady centrifugal forces.

2. Aerodynamic Analysis

A. Steady Momentum Analysis

Steady aerodynamic loads are calculated based on Glauert's general momentum theory[14] including axial and angular momentum in combination with Prandtl's finite span correction[14]. When the axial induced velocity and the angular induced velocity at the disk plane are denoted by aV and $br\Omega$, respectively, they can be determined by

$$
\begin{aligned}
4Fa(1+a) &= (Bc/2\pi r)\ C_L\lambda(1-b)\ [((1+a)^2+\lambda^2(1-b)^2]^{1/2} \\
4F\lambda b &= (Bc/2\pi r)\ C_L\ [((1+a)^2+\lambda^2(1-b)^2]^{1/2}
\end{aligned} \tag{11}
$$

where $\lambda=r\Omega/V$, and B denotes number of blades. The quantity F is Prandtl's finite span correction factor for a B-bladed rotor:

$$F(a,b,r) = \frac{2}{\pi}\cos^{-1}\exp\left[-\frac{B}{2}\ (1-\frac{r}{R})\ \{1+(\frac{r\Omega(1-b)}{V(1+a)})^2\}^{1/2}\right] \tag{12}$$

The sweep effects and effective Mach number are included in lift curve slope. Finally, the power coefficient distribution, $dC_P/dx = \pi^3(dP/dr)/4\rho\Omega^3R^4$, can be obtained from the induced velocities, where P denotes the propeller power. The finite span correction factor, F, will be also used for unsteady aerodynamic loads.

B. Unsteady Subsonic Cascade Analysis

Smith's unsteady subsonic cascade theory[10] is used in combination with sweep corrections and finite span corrections based on modified strip theory. A brief description of methodology is provided next.

The cascade airfoil geometry is specified by chord length c, stagger angle θ_s, and blade spacing s. The pressure and two velocity components are expanded in Fourier series:

$$\begin{Bmatrix}\Delta p\\ \Delta u\\ \Delta v\end{Bmatrix} = \sum_{r=-\infty}^{\infty} \begin{Bmatrix}p_r\\ u_r\\ v_r\end{Bmatrix} \exp i(\omega t+\alpha_r x+\beta_r y) \tag{13}$$

where x and y are the orthogonal coordinates in the directions of the axial flow and the tangential flow, respectively. Each velocity/pressure field is called a cascade wave which corresponds to a fundamental solution for isentropic inviscid flow. Since the cascade airfoils are regarded a system where cascade waves are superposed. The normal induced velocity on the reference airfoil, w, can be determined by summing up all the cascade waves and integrating over the chord:

$$w(z,t) = \int_0^c \sum_{r=-\infty}^{\infty} \sum_{k=1}^{3} w^{(k)} \Gamma' dz \exp i(\omega t+\alpha_r x+\beta_r y)^{(k)}$$

$$= \exp(i\omega t) \int_0^c \Gamma'(z_0)\, K\left(\frac{z-z_0}{c}\right) \frac{dz_0}{c} \tag{14}$$

where Γ' is unit-chord circulation amplitude when the free vortex component being subtracted, and K is a kernel function[10], whose singularity can be removed by the method described in Ref. 10.

The boundary conditions are expressed in terms of leading edge heave, h, and pitching, δ, as follows:

$$\frac{w}{W} = 2ik\left(\frac{h}{c} + \delta\frac{z}{c}\right) + \delta; \qquad k = \frac{\omega c}{2W} \tag{15}$$

where W denotes the resultant velocity, and k is the reduced frequency. In the case of a propeller, h is defined as the elastic displacement normal to the helical stream lines, and δ is defined as the elastic twist angle about the radius. The structural variables which are defined along the line of shear centers will be transformed to these variables. The phase of airfoil motion have been assumed as

$$\exp i(\omega t+m\sigma); \qquad m=0,\pm 1,\pm 2,\ldots,\pm\infty$$

where σ is called interblade phase angle. Finally, the 2-D lift and moment about leading edge is obtained by integrating Γ' along the chord. This yields

$$L = L_h(k,\sigma)\, h + L_\delta(k,\sigma)\, \delta$$

$$M = M_h(k,\sigma)\, h + M_\delta(k,\sigma)\, \delta \tag{16}$$

where L_h, L_δ, M_h, M_δ are complex, and k is represented by the reduced frequency of a reference section. Reduced frequency for j-th element can be given by $k_j = k(Wc_j/cW_j)$. Compressibility effects on complex value of L_h is presented in Fig. 1 for the case of an isolated airfoil. The equivalent lift deficiency function shown in Fig. 1 is the unsteady-to-steady lift ratio excluding the virtual mass term. Since the virtual mass term cannot be separated in compressible flow cases, it is denoted as "equivalent". In the calculation, the quantities

$$\gamma_i = \frac{\Gamma_i'}{W}\frac{\sin\psi_i}{2}\Delta\psi_i ; \quad \text{where} \quad \cos\psi_i = 1 - \frac{2z_0}{c} \ (i=1,2,\ldots N) \tag{17}$$

have been selected as unknown variables because they have a finite value at the leading edge. It can be shown that the computation time is almost proportional to the following index:

$$\frac{\text{Blade spacing}}{\text{Chord length}}\frac{[\text{Chordwise segment number}]^2}{\text{COSINE (Stagger angle)}} \tag{18}$$

C. Finite Span Correction and Sweep Correction

In order to count the 3-D finite span effects, the 2-D unsteady subsonic aerodynamic loads should be multiplied by the finite span correction factor, F, which was also used in the steady aerodynamic analysis. This can be regarded the simplest modified strip theory[15].

On the other hand, when sweep angle at quarter chord is denoted by Λ and the effect of spanwise flow is neglected, the sweep correction for airfoils specified along the stream lines can be obtained by multiplying by $\cos\Lambda$ the 2-D lift and moment[16]. Furthermore, the effective Mach number is reduced to $M = M_0\cos\Lambda$, where M_0 is helical Mach number. Unit-span drag is assumed to be unaffected by sweep angle. Finally unit-span lift, drag and moment can be obtained as follows:

$$\begin{aligned} L_{3D} &= F\cos\Lambda\ L_{2D}(M=M_0\cos\Lambda) \\ D_{3D} &= D_{2D} \\ M_{3D} &= F\cos\Lambda\ M_{2D}(M=M_0\cos\Lambda) \end{aligned} \tag{19}$$

3. Modal Flutter Analysis Using the p-k Method

A. Steady deformation

The steady blade deformations due to steady aerodynamic loads and centrifugal forces are determined first. The static system considered is as follows:

$$[K_L + K_{NL}(\mathbf{q})]\ \mathbf{q} + \{F_T(\mathbf{q})\} = \{0\} \tag{20}$$

The equilibrium position for $\mathbf{q}$ is obtained by the Newton-Raphson method. The general coordinates for the (n+1)th iteration, $\mathbf{q}^{(n+1)}$, can be obtained by the the following formula[9].

$$\mathbf{q}^{(n+1)} = \mathbf{q}^{(n)} - [J^{(n)}]^{-1}\{[K_L+K_{NL}{}^{(n)}]\mathbf{q}^{(n)} + F_T{}^{(n)}\} \tag{21}$$

where

$$[J^{(n)}] = \{\partial/\partial\mathbf{q}^{(n)}\}\{[K_L+K_{NL}{}^{(n)}]\mathbf{q}^{(n)} + F_T{}^{(n)}\} \tag{22}$$

The converged nonlinear solution has to be generated carefully to avoid spurious solutions. Therefore the following calculation scheme was adopted:

1) The rotational velocity is increased step by step to avoid erroneous convergence to inadequate equilibrium positions. A converged solution of the previous step is used as the first approximation for the next step.

2) Direct substitution method as well as the Newton-Raphson method are used alternatively because solutions given by the Newton-Raphson method diverge rapidly when convergence is not obtained.

B. Modal Representation

After the nonlinear equilibrium position is obtained, linearization of Eq. (9) leads to the following equations:

$$[M]q_{,tt} + [D]q_{,t} + [K_L+K_{NL}(q_0)]q + [A_R(k,\sigma)+iA_I(k,\sigma)]q = \{0\} \qquad (23)$$

where the dimension of matrices is 8nx8n (n: number of elements). It is well known that modal representation is convenient for reducing matrix size and for distinguishing the flutter mode. When a transformation $q = [T]y$ is introduced and [T] is composed of the first m eigenvectors for the undamped system for each rotational velocity, the above 8nx8n system will be reduced to an mxm system:

$$[m]y_{,tt} + [d]y_{,t} + [k_{NL}]y + [a_R + ia_I]y = \{0\} \qquad (24)$$

where

$$[m] = [T]^T [M] [T]$$
$$[d] = [T]^T [D] [T]$$
$$[k_{NL}] = [T]^T [K_L + K_{NL}] [T]$$
$$[a_R + ia_I] = [T]^T [A_R + iA_I] [T]$$

These equations give correct eigenvalues for the undamped system, and a good approximation for weakly damped systems. The eigenvectors for transformation [T] are given by the eigenvalue analysis with the QZ method.

C. The p-k Method of Flutter Analysis

Though the above aerodynamic matrices $[A_R + iA_I]$ or $[a_R + ia_I]$ are based on a simple harmonic motion, they give a good approximation for weakly damped systems since simple harmonic motion is achieved at the flutter points. In this sense, Laplace operator, $p = \lambda+i\omega$, and Fourier operator, $k=c\omega/2W$, (physically reduced frequency at the reference section as was mentioned previously) are used simultaneously in the p-k method. Since p tends to $i\omega = i2kW/c$ near flutter points, imaginary unit $i=\sqrt{-1}$ may be approximated by $i \approx pc/2kW$. Therefore, the following eigenvalue problem can be solved iteratively[17,18]

$$\left(\begin{bmatrix} k' & d' \\ 0 & I \end{bmatrix} - p \begin{bmatrix} 0 & -m \\ I & 0 \end{bmatrix} \right) \begin{Bmatrix} y \\ y_{,t} \end{Bmatrix} = \begin{Bmatrix} 0 \\ 0 \end{Bmatrix} \qquad (25)$$

where

$$[d'] = [d] + \frac{c}{2kW}[a_I(k,\sigma)]$$
$$[k'] = [k_{NL}] + [a_R(k,\sigma)]$$

until the reduced frequency, k, converges to $k \approx Im(p)\ c/2W$ of the mode of interest. The above equations are exact for flutter points but give an approximation for other points.

RESULTS AND DISCUSSION

1. Discretization by a Finite Element Model

Assuming the SR-3 blade model[19], made of high-strength graphite/epoxy, the 3-D elastic axis (line of shear centers) was determined first. The line through shear centers of the initial section profiles, which were defined perpendicular to the pitch change axis, was taken as the first approximation. The nodal sections were replaced by those perpendicular to the approximate line of shear centers, and thus the elastic axis was obtained after a few iterations. The elastic axis geometry and the finite element grid for the structural dynamic analysis are shown in Fig. 2. On the other hand, the grid for the aerodynamic analysis is shown in Fig. 3. The aerodynamic grid differs from the structural dynamic grid due to the reasons provided below.

For the structural dynamic grid, the nodal sections should be perpendicular to the elastic axis based on the definition of the curvilinear coordinates used in each element. An exception is the tip section which has to coincide with the actual tip. This is because the mass property of the tip element has a major effect on the system characteristics whereas precise structural modeling at the tip is less important.

For the aerodynamic grid, the nodal sections should have the same radius or, approximately, should be perpendicular to the pitch change axis because the dynamic pressure should be constant along each chord. Otherwise, two-dimensional aerodynamic theory cannot be applied.

When the two grids were generated, the nodal points on the elastic axis were kept identical for the two grids and furthermore the total aerodynamic forces and moments were left unchanged.

2. Blade Deformation Effects on Steady Aerodynamic Loads

Using the SR-3 geometry shown in Ref. 19, the aerodynamic load distribution after deformation was predicted. Assuming the material to be titanium, the result was compared with the experimental results of Ref. 19, as is shown in Fig. 4. The curve denoted by "Elastic Analysis" represents the results obtained from our analysis. It is evident from Fig. 4 that the elastic deformation does not have a major influence on aerodynamic performance, for this case. The discrepancies between experiment and analysis can be probably attributed more to the aerodynamics being used than to the elastic deformation of the blade. These discrepancies were not explained in a satisfactory manner in Ref. 19 either.

The results for the finite span correction factor are shown in Fig. 5. These results will be also used for introducing finite span corrections in the unsteady analysis.

3. Tailoring Effects on Blade Natural Modes and Frequencies

The effects of fiber orientation on the nonrotating natural frequencies and corresponding modes of the SR-3 blade, made of unidirectional high-strength graphite/epoxy, were examined. A twelve element model was used in the analysis. The fiber orientation was assumed uniform over the blade and the section was assumed to have homogeneous anisotropy.

The fiber orientation, Λ_F, is defined to be zero when the orientation coincides with the pitch change axis and to be positive when swept back.

The results shown in Fig. 6 indicate that the natural frequencies of graphite/epoxy blade can be controlled from 100% to 35% by changing the fiber orientation. The maximum frequency is obtained when fibers are oriented along the blade span, which corresponds to 15 - 30 degrees in the figure.

The corresponding natural modes are shown in Fig. 7. The third mode, for example, clearly show the change of the pitch-flap (ϕ,w) coupling. Though there is negative pitch-flap coupling for fiber sweep of -30 degrees, coupling of different distribution can be seen for fiber sweep of 15 degrees.

4. Aeroelastic Tailoring Effects on Flutter Velocity

Throughout the flutter analysis, advance ratio, J (= 3.0), and blade pitch setting, $\beta_{0.75}$(=60.7 deg), were kept constant and flight Mach number M was changed proportionally with the rotor speed of rotation Ω, assuming the interblade phase angle σ as a parameter. The blade material is the same as the previous section. The geometry was the SR-3 model with a diameter of 0.622 meters and the number of blades was taken to be eight. Six modes were used in the modal p-k analysis.

First, blade natural frequencies vs. rotational velocity was checked in vacuo, as is shown in Fig. 8. In order to see the effects of interblade phase angle, the p-k flutter analysis was conducted up to Mach number 0.4, varying interblade phase angle as a parameter. The real parts of eigenvalues are shown in Fig. 9, and it is suggested that interblade phase angle of zero seems least stable in this case.

Next, the p-k flutter analyses were conducted, varying Mach number and rotational velocity and changing the fiber orientation as a parameter. Some typical results are presented in Figs. 10 and 11 which illustrate the effect of fiber orientation on flutter. Flutter does not occur for zero fiber sweep, namely when fibers are aligned along the pitch change axis, since all the real parts of the converged eigenvalues are negative. However a fairly small change in fiber orientation to only 15 degrees of sweep back, is sufficient to cause flutter as shown in Fig. 11, where it is evident that the second mode has low damping and flutter occurs at M = 0.66. This clearly illustrates the sensitivity of the flutter velocity to fiber orientation.

CONCLUDING REMARKS

This paper presents a fairly simple aeroelastic analysis tool useful for the preliminary design of the ATP blades, and the usefulness of aeroelastic tailoring for the ATP has been examined. The aeroelastic model consists of a structural dynamic model of rotating beam elements with moderate deflections and section warping for composite swept blades, combined with unsteady subsonic cascade theory modified by sweep corrections and finite span corrections based on modified strip theory. Flutter analysis is carried out by applying the p-k method on the system after reducing the number of degrees of freedom by a normal mode transformation.

Using this tool, the aeroelastic analysis for the SR-3 propeller model, made of graphite/epoxy, was carried out. It was found that the natural frequencies for the blade can be changed from 100% to 35% and that the corresponding modes or sign of pitch-flap coupling can be modified by changing fiber orientation. It was also shown that flutter velocity is sensitive to fiber orientation. These indicate the effectiveness of the aeroelastic tailoring technique and demonstrate the convenience of the present analysis technique for Advanced Turbo Propellers.

ACKNOWLEDGEMENT

The authors would like to express their gratitude to Dr. John B. Kosmatka, Virginia Polytechnic Institute and State University, for his advice and to Dr. Oral Mehmed, NASA Lewis Research Center, for his useful suggestions.

REFERENCES

1. Mehmed, O., Kaza, K.R.V. et al.: NASA TM-82975 (1982).
2. Mehmed, O. and Kaza, K.R.V.: NASA TM-88792 (1986).
3. Turnberg, J.E.: NASA CR-179512 (1986).
4. Ducharm, E.H. and Crawley, E.F.: MIT GTL Report 191 (1987) 139-231.
5. Kaza, K.R.V., Mehmed, O. et al.: AIAA Paper 87-0739 (1987).
6. Elchuri, V. and Smith, G.C.C.: NASA CR-167926 (1982).
7. Kaza, K.R.V., Mehmed, O. et al.:AIAA Paper 87-0738 (1987).
8. Kosmatka, J.B. and Friedmann, P.: AIAA Paper 87-0740 (1987).
9. Kosmatka, J.B.: Ph.D. Dissertation, Univ. Calif. Los Angeles (1986).
10. Smith, S.N.: ARC R&M 3709 (1972).
11. Jones, R.M.: Mechanics of composite materials, Scripta (1975) 31-41.
12. Kosmatka, J.B.: AHS Specialist Meeting on Advanced Rotorcraft Structures (1988).
13. Lekhnitskii, S.G.: Theory of elasticity of an anisotropic elastic body, Holden-Day (1963) 191-197, 313-316.
14. Glauert, H.: Aerodynamic Theory, Vol.4, Dover (1963) 236, 268.
15. Yates, E.C., Jr.: J. Aircraft, 3-1 (1966) 25-29.
16. Bispringhoff, R.L., Ashley, H. et al.: Aeroelasticity, Addison-Wesley (1955) 250-251.
17. Hassig, H.J.: J. Aircraft, 8-11 (1971) 885-889.
18. Turnberg. J.: AIAA Paper 83-0847 (1983).
19. Rohrbach, C., Metzger, F.B. et al.: NASA CR-3505 (1982) 52-53, 103.

APPENDIX A - List of Symbols

$[A_R]$, $[A_I]$ = real and imaginary parts of aerodynamic stiffness matrix

a = nondimensional axial induced velocity

B = number of blades

b = nondimensional angular induced velocity

$[C_{ij}]$ = material stiffness matrix

C_L, C_D = section lift and drag coefficients, normalized by $\rho W^2 c/2$

c = chord length

[D] = system damping matrix[9]

dC_p/dx = power coefficient distribution

F = Prandtl's finite span correction factor, Eq.(12)

$\{F_T\}$ = generalized forces due to aerodynamic and steady centrifugal forces
h = leading edge heave, normal to the helical wake surface
J = advance ratio, $\pi V/\Omega R$
$[J]$ = Jacobian matrix for the Newton-Raphson method
K = kernel function
$[K_L], [K_{NL}]$ = linear and nonlinear stiffness matrices
k = reduced frequency at the reference section
M = axial Mach number, V/a_0, where a_0 denotes sound velocity
$[M]$ = system mass matrix[9]
p = Laplace operator in the p-k analysis
Δp = pressure perturbation
p_r = pressure amplitude of the r-th cascade wave
$[Q_{ij}]$ = condensed material stiffness matrix
$\mathbf{q}$ = general coordinates of the beam, Eq.(10)
R = rotor radius
r = radius from the center of rotation
s = blade spacing, namely leading edge distance
$[S_{ij}]$ = material compliance matrix, $[S_{ij}] = [C_{ij}]^{-1}$
T = kinetic energy
$[T]$ = transformation matrix for size reduction
t = time
U = strain energy
u,v,w = axial, lead, and flap displacements of the shear center
$\Delta u, \Delta v$ = axial and tangential velocity perturbations
u_r, v_r = velocity amplitudes of the r-th cascade
V = axial flow velocity at a blade section
W = resultant velocity at the cascade
w = velocity normal to the chord
W_E = work of nonconservative forces
x,y = orthogonal coordinates of the cascade
x,η,ζ = curvilinear coordinates of the beam
x_{sc} = shear center distance from the area centroid, positive aft
z, z_0 = coordinates along the chord, origin: leading edge
α_r, β_r = r-th wave numbers in the x- and y-directions
β = pretwist angle
$\beta_{0.75}$ = blade pitch angle at 75% radius
Γ_i' = circulation per unit chord, except for free vortex component
δ = section pitch angle around the leading edge, positive trailing edge up
$\varepsilon_{xx}, \gamma_{x\eta}, \gamma_{x\zeta}$ = strain components
θ_s = stagger angle, namely angle between resultant velocity and axial velocity
Λ = aerodynamic sweep angle, positive sweep back
Λ_F = fiber sweep angle from the pitch change axis
λ = local speed ratio, $r\Omega/V$
ρ = air density
σ = interblade phase angle
$\sigma_{\eta\eta}, \sigma_{\zeta\zeta}, \tau_{\eta\zeta}$ = stress components
τ = chordwise distribution of the airfoil thickness (=2τ)
$[\Phi_2], [\Phi_3]$ = Hermite interpolation polynomials
ϕ = elastic twist angle
ϕ_0 = second order terms for elastic twist
ψ = section warping function, positive convex
Ω = angular velocity of the rotor, in rad/s
ω = system frequency
$(\ )^{j-0.5}$ = mid-element value
$(\ )^{(n)}$ = value of the n-th iteration
$(\ )_r$ = value of the r-th cascade wave
$[\ ]^T$ = transpose
$(\ )_{,x}$ = partial derivative with respect to x

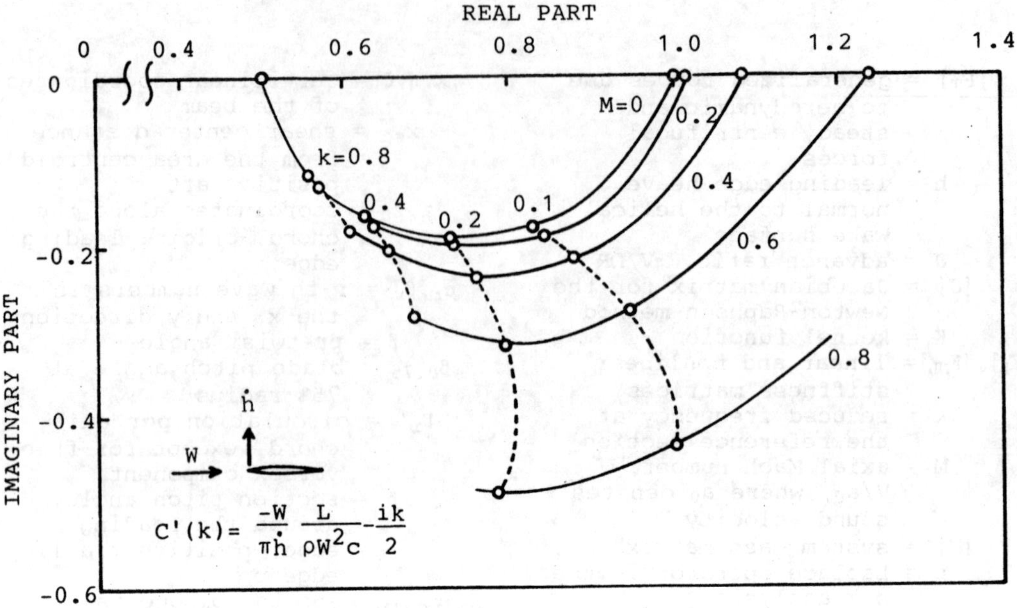

Fig. 1 Compressibility Effect on Equivalent Unsteady Lift Deficiency Function (σ=0, θ_s=0, s/c=200)

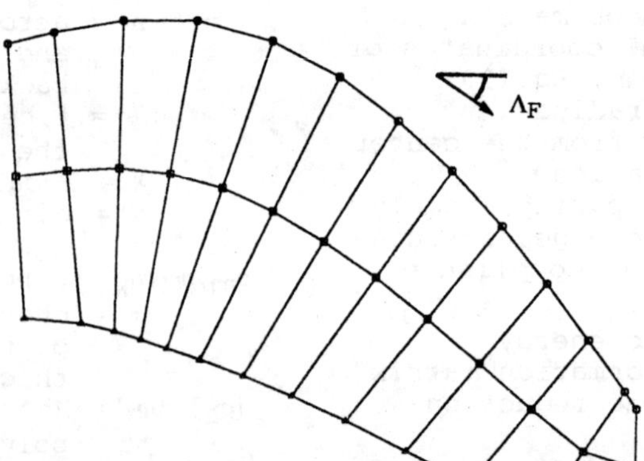

Fig. 2 Structural Dynamic Grid

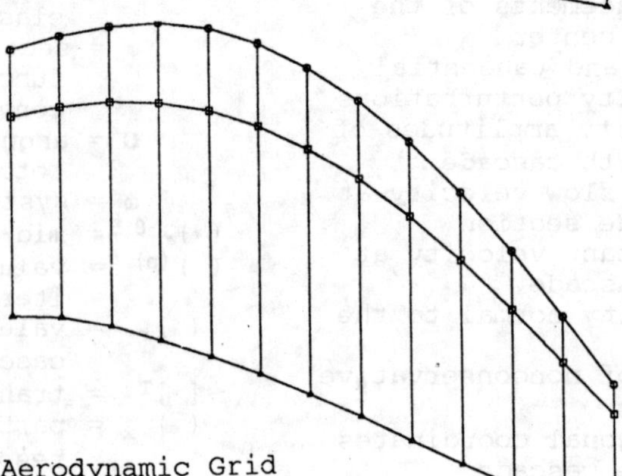

Fig. 3 Aerodynamic Grid

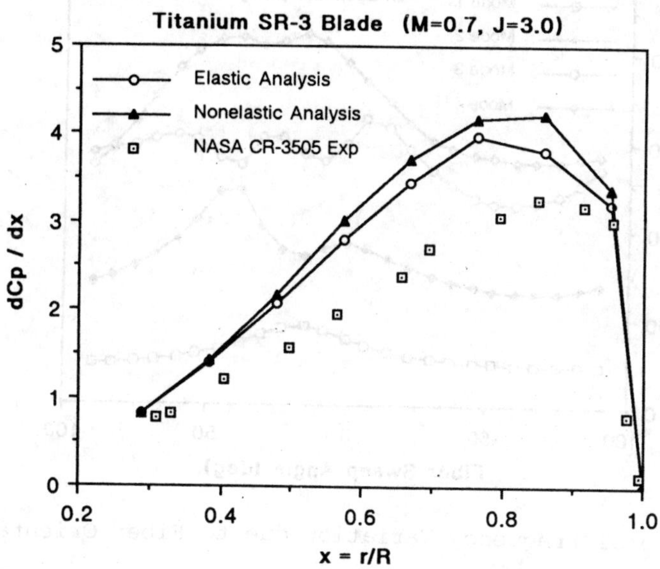

Fig. 4 Blade Deformation Effect on Steady Aerodynamic Load

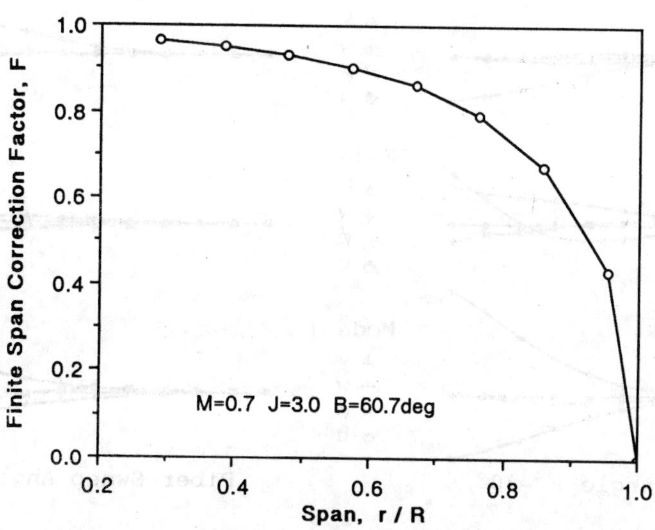

Fig. 5 Prandtl's Finite Span Correction Factor

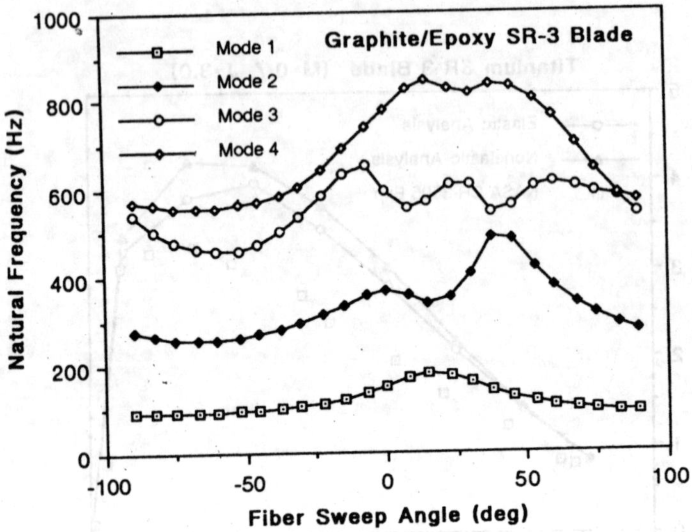

Fig. 6 Natural Frequency Variation due to Fiber Orientation

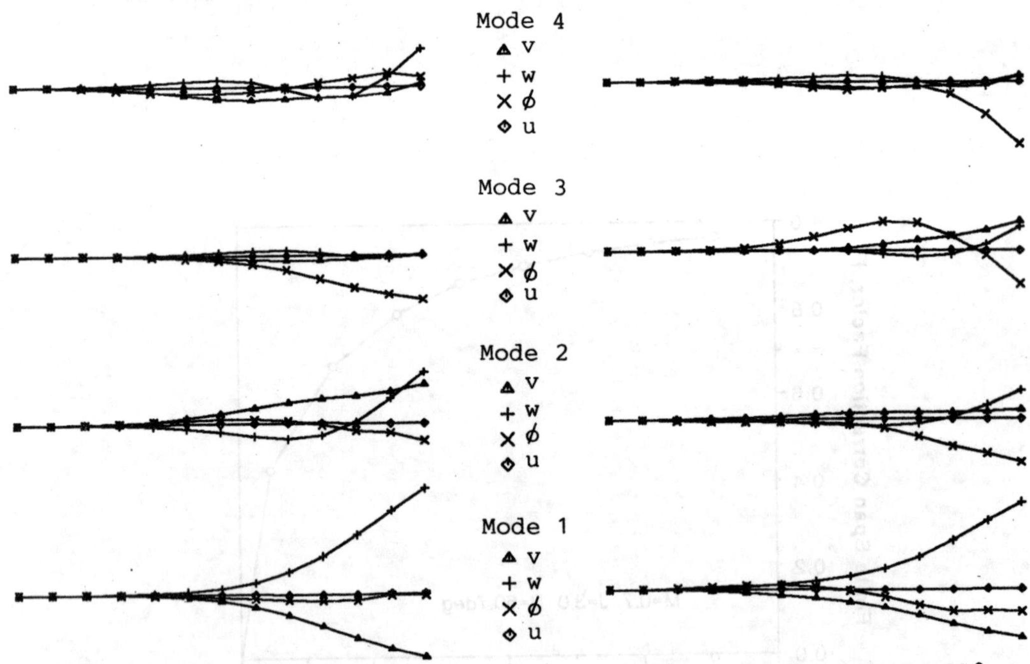

Fig. 7 Natural Mode Variation due to Fiber Orientation

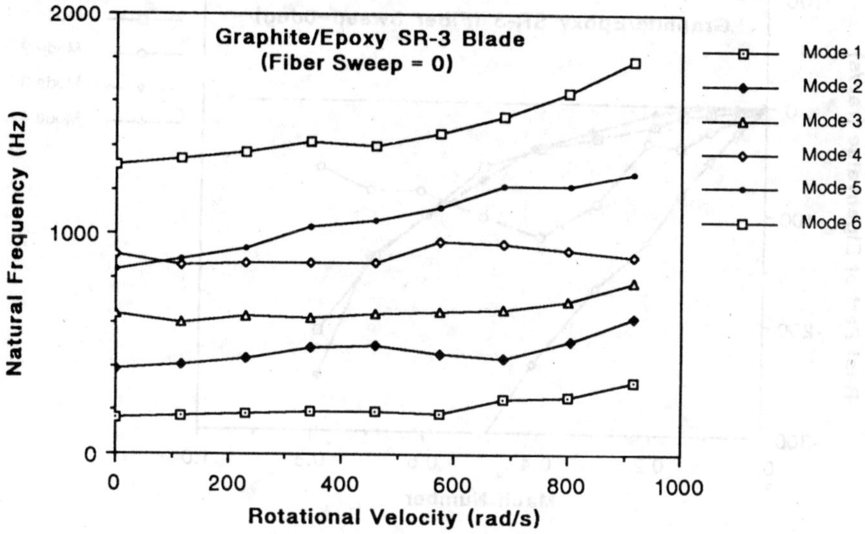

Fig. 8 Blade Natural Frequency vs. Rotor Speed

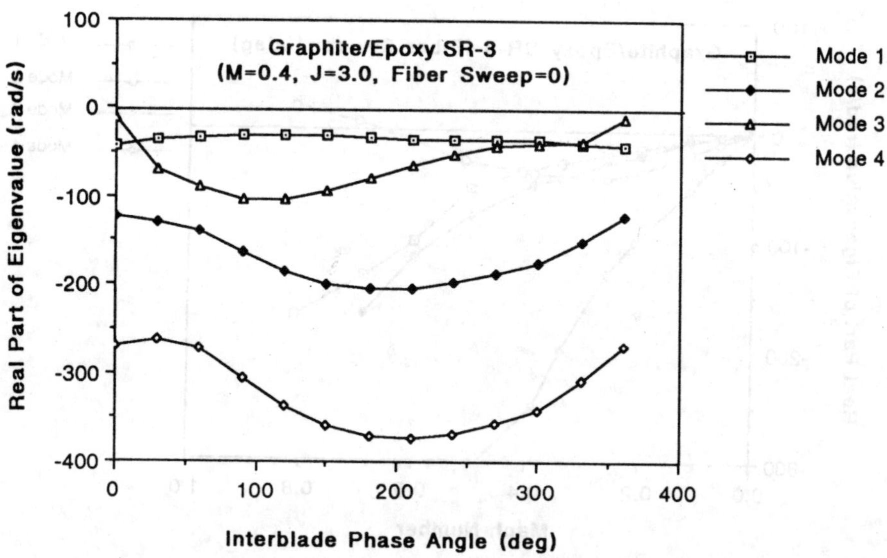

Fig. 9 Modal Damping Variation due to Interblade Phase Angle

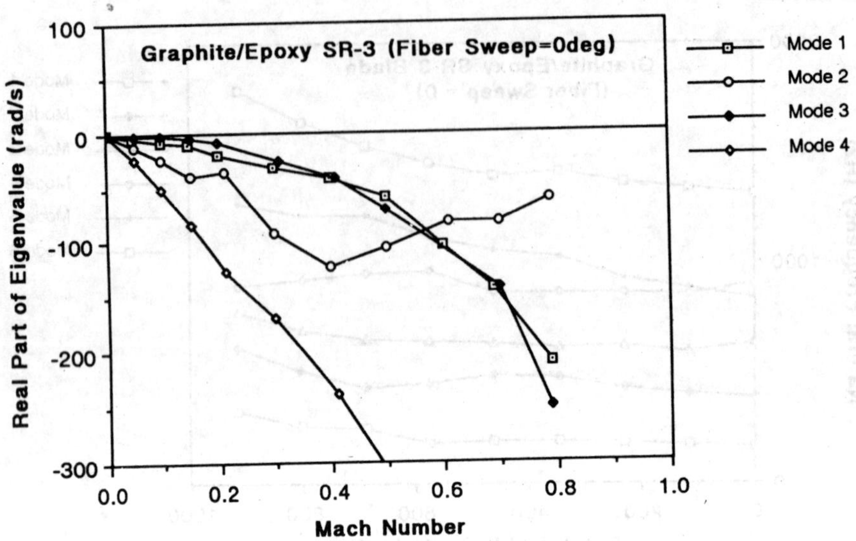

Fig. 10 Modal p-k Flutter Analysis (Fiber Sweep = 0 deg)

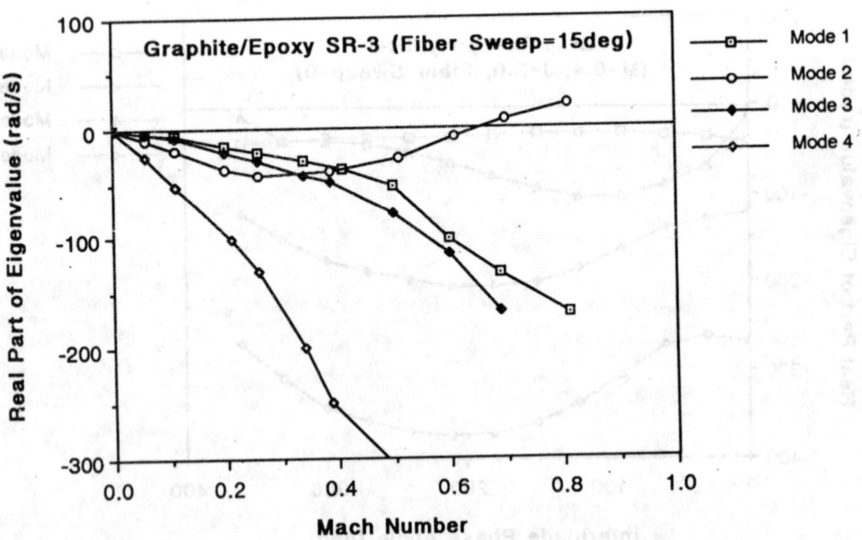

Fig. 11 Modal p-k Flutter Analysis (Fiber Sweep = 15 deg)

EFFECT OF BLADE MISTUNING AND COUPLED DISK-BLADE ON CASCADE FLUTTER BOUNDARIES

Qun Chen, Jingbo Chen and Feida Fan
(Northwestern Polytechnical University, China)

ABSTRACT

This paper deals with effects of structure factors, such as blade mistuning, coupled disk–blade and location of elastic axis etc., on the flutter boundaries of a cascade in incompressible flow. The flutter boundaries of cascades are evaluated with the eigenvalue method . Flutter boundaries of three mistuning orders have been studied, taking into account the effects of coupled disk–blade and the location of elastic axis. Three mistuning orders are as the follwing:

1. First order bending and torsional frequencies are all mistuning;
2. First order bending frequency is tuning and first order torsional frequency mistuning;
3. First order bending frequency is mistuning and first order torisional frequency tuning. Results show that the stability of the above three mistuned cascades is more beneficially improved than that of tuned cascade. The larger the mistuning amplitude, the better the stability of cascade. When the effects of the coupled disk–blade are considered, the flutter boundaries can be expected to move forwards and their stability to become worse. The results also indicate that different forms of mistuning distribution, namely the arrangement of requencies of blades on the disk, have effect of different degrees even though the mistuning amplitudes, mistuning orders and location of elastic axis are the same. Calculations show that the stability of cascade with alternating distribution is better than that with cosine distribution. The existence of an optimal frequency distribution is possible which can suppress the flutter more effectively. The centrifugal force makes the flutter boundaries move backward.

MECHANICAL MODELS AND ELEMENTARY EQUATIONS

In this paper, a real blade is simplified as a model of the characteristic section at 80% height of the blade, as shown in Fig.1.

A real cascade is represented by a two–dimensional annular cascade with N airfoil, as shown in Fig.2. The aeroelastic stability of a real blade is represented by that of the characteristic section.

THE SELF–EXCITED VIBRATION DIFFERENTIAL EQUATION OF MISTUNED CASCADE

Assume that all blades in each mode of the tuned cascade simply harmonically vibrate with the same frequency and amplitude. The phase angle between adjacent blades is constant, whose value is

$$\beta_r = \frac{2\pi r}{N} \qquad (r = 0.1,\cdots,N-1)$$

Where N is the number of blades, r is the order of mode. At a given moment, the displacement of sth blade can be written as:

$$\begin{bmatrix} \frac{h_z}{b} \\ \alpha_z \end{bmatrix} e^{i\omega t} = \begin{bmatrix} \frac{h_{az}}{b} \\ \alpha_{ar} \end{bmatrix} e^{i\omega t + \beta_r e)} \tag{1}$$

When the cascade is mistuned, frequencies and amplitudes of the blades are different, and phase difference of the adjacent blade is not constant any longer, either but variable. Considering spatial peiodicity, the displacement of the mistuned cascade can be decomposed into the superposition of all modes of the tuned cascade by mens of Fouries Series. Therefore, the displacement of the sth blade of the mistuned cascade can be written as:

$$\begin{bmatrix} \frac{h_z}{b} \\ \alpha_z \end{bmatrix} e^{i\omega t} = \sum_{r=0}^{N-1} \begin{bmatrix} \frac{h_{av}}{b} \\ \alpha_{av} \end{bmatrix} e^{i(\omega t + \beta_r e)} \qquad (s = 0.1,\cdots,N-1) \tag{2}$$

Combining these N equations above, we can obtain displacements of all blades of the whole cascade in matrices:

$$\{X\}e^{i\omega t} = [E]\{Y\}e^{i\omega t} \tag{3}$$

where:

$$\{X\} = \left\{ \frac{h_0}{b} \quad \alpha_0 \quad \frac{h_1}{b} \quad \alpha_1 \quad \cdots \quad \frac{h_{N-1}}{b} \quad \alpha_{N-1} \right\}^T$$

$$\{X\} = \left\{ \frac{h_{a0}}{b} \quad \alpha_{a0} \quad \frac{h_{a1}}{b} \quad \alpha_{a1} \quad \cdots \quad \frac{h_{a(N-1)}}{b} \quad \alpha_{a(N-1)} \right\}^T$$

matrix [E] is a square matrix and its elements are

$$e(s,r) = e^{\frac{12\pi s r}{N}} \tag{4}$$

Regardless of coupled disk–blade, the self-excited vibration quation of the sth blade is

$$\begin{bmatrix} M_s & S_{as} \\ S_{as} & I_{as} \end{bmatrix} \begin{bmatrix} \frac{d^2}{dt^2}(h_s e^{i\omega t}) \\ \frac{d^2}{dt^2}(\alpha_s e^{i\omega t}) \end{bmatrix} + \begin{bmatrix} (1+2i\zeta_{hs})m_s\omega_{hs}^2 & 0 \\ 0 & (1+2i\zeta_{as})I_{as}\omega_{as}^2 \end{bmatrix} \begin{bmatrix} h_s e^{i\omega t} \\ \alpha_s e^{i\omega t} \end{bmatrix} = \begin{bmatrix} -L_s \\ M_s \end{bmatrix} \tag{5}$$

where s = 0,1,2,....,N-1

Because blade flutter belongs to self–excited vibration, the aerodynamic force L_s and moment M_s are function of blade states, and they can be calculated as follows:

$$L_s = -\pi \rho b^3 \omega^2 \sum_{r=0}^{N-1} \left(\frac{l_{hhr} \cdot h_{ar}}{b} + l_{har} \cdot \alpha_{ar} \right) e^{i(\omega t + \beta_r s)} \tag{6a}$$

$$M_s = \pi \rho b^4 \omega^2 \sum_{r=0}^{N-1} \left(\frac{l_{ahr} \cdot h_{ar}}{b} + l_{aar} \cdot \alpha_{ar} \right) e^{i(\omega t + \beta_r s)} \tag{6b}$$

$$\left.\begin{aligned} l_{hhr} &= \frac{2i}{k}(c_{Fa})_\eta \\ l_{aar} &= \frac{4}{k^2}(c_{Ma})_\eta \\ l_{har} &= \frac{2}{k^2}(c_{Fa})_\eta \\ l_{ahr} &= \frac{4i}{k}(c_{Ma})_\eta \\ k &= \frac{\omega_h}{V} \end{aligned}\right\} \tag{7}$$

When k,b,s / c,θ and are given, we can calculate the following coeficients $(c_{Fq})_\eta$,$(c_{Fa})_\eta$,$(c_{Mq})_\eta$, and $(c_{Ma})_\eta$. Details are seen in Ref [6]. Combining equations(5), we will obtain N nondimensional equations of self–excited vibration in matrices.

$$\{[P]-\gamma[L]\}\{[Y]\}=\{0\} \tag{8}$$

where

$$[P]=[E]^{-1}[D][E]+[E]^{-1}[G][E][A]$$

$$[D]=\begin{bmatrix} [d_0] & & & & & \\ & [d_1] & & & & \\ & & \cdot & & & \\ & & & \cdot & & \\ & & & & \cdot & \\ & & & & & [d_{N-1}] \end{bmatrix}$$

$$[d_s]=\mu_s\begin{bmatrix} \frac{1}{G_{khs}} & \frac{X_{as}}{G_{khs}} \\ \frac{X_{as}}{G_{khs}} & \frac{r_{as}^2}{G_{kas}} \end{bmatrix}$$

$$\left.\begin{aligned} G_{khs} &= \mu_s\gamma_{hs}^2(1+2i\xi_{hs}) \\ G_{kas} &= \mu_s r_{as}^2 \\ \gamma_{hs} &= \frac{\omega_{hs}}{\omega_0} \\ \gamma_{as} &= \frac{\omega_{as}}{\omega_0} \\ \mu_s &= \frac{m_s}{\pi\rho b^2} \\ r_{as}^2 &= \frac{I_{as}}{m_s b^2} \\ X_{as} &= \frac{S_{as}}{m_s b} \\ \gamma &= \left(\frac{\omega_0}{\omega}\right)^2 \end{aligned}\right\} \tag{9}$$

$$[G]=\begin{bmatrix}[g_0] & & & & \\ & [g_1] & & & \\ & & \cdot & & \\ & & & \cdot & \\ & & & & \cdot \\ & & & & & [g_{N-1}]\end{bmatrix}$$

$$[g_r]=\begin{bmatrix}\frac{1}{G_{khs}} & 0 \\ 0 & \frac{1}{G_{kas}}\end{bmatrix}$$

$$[A]=\begin{bmatrix}[a_0] & & & & \\ & [a_1] & & & \\ & & \cdot & & \\ & & & \cdot & \\ & & & & \cdot \\ & & & & & [a_{N-1}]\end{bmatrix}$$

$$[a_r]=\begin{bmatrix}l_{hhr} & l_{har} \\ l_{ahr} & l_{aar}\end{bmatrix}$$

The model of the characteristic section of the coupled disk–blade is show in Fig.3.

Because frequency and damping change with the number r of node diameters, frequencies ω^2_{hs}, $\omega^2_{as}\xi_{as}$ and ξ_{as} in equation [5] represent are by symbols ω^2_{hav}, ω^2_{hav}, ξ^2_{hav}, and ξ^2_{hav}, where the footnote r respensents number of node diameter. In addition, suppose thate lastical restoring force and moment are proportional to the modal displacement$[h_{ar}\alpha_{ar}]^T$, but not to physical displacement $[h_s,\alpha_s]^T$ of the blade. According the description mensioned above. The differential equation of self–excited vibration of sth blade for coupled disk–blade system is:

$$\begin{bmatrix}M_s & S_{as} \\ S_{as} & L_{as}\end{bmatrix}\begin{bmatrix}\frac{d^2}{dt^2}(h_s e^{i\omega t}) \\ \frac{d^2}{dt^2}(\alpha_s e^{i\omega t})\end{bmatrix}+\sum_{r=0}^{N-1}\begin{bmatrix}(1+2i\xi_{har})m_s\omega^2_{har} & 0 \\ 0 & (1+2i\zeta_{har})l_{as}\omega^2_{har}\end{bmatrix}\begin{bmatrix}\frac{h_{ar}}{b} \\ \alpha_{ar}\end{bmatrix}e^{i(\omega t+\lambda_r s)}=\begin{bmatrix}-L_s \\ M_s\end{bmatrix} \tag{10}$$

Therefore from equation [10] we can obtain nondimensional self–excited vibration equation of the cascade of coupled disk–blade.

$$\{[P']-\gamma[Q]\}\{Y\}=\{0\} \tag{11}$$

$$[P']=[G'][E][A]+[D'][E] \tag{12}$$

$$[d'_s]=\mu_s\begin{bmatrix}\frac{1}{G_{khs0}} & \frac{X_{as}}{G_{khs0}} \\ \frac{X_{as}}{G_{kas0}} & \frac{r^2_{as}}{G_{kas0}}\end{bmatrix}$$

$$[g'_s]=\begin{bmatrix}\dfrac{1}{G_{khs0}} & 0\\ 0 & \dfrac{1}{G_{khs0}}\end{bmatrix}$$

$$G_{khsr}=\mu_s\gamma^2_{har}(1+2i\xi_{hsr})$$

$$G_{khsr}=\mu_s r^2_{as}\gamma^2_{har}(1+2i\xi_{asr})$$

$$\gamma_{hsr}=\frac{\omega_{hsr}}{\omega_0}$$

$$\gamma=(\frac{\omega_0}{\omega})^2$$

$$[Q]=\begin{bmatrix}1 & 0 & \dfrac{(1+2i\xi_{h01})\gamma^2_{h01}}{(1+2i\zeta_{h00})\gamma^2_{h00}}e(0,1) & 0 & \dfrac{(1+2i\xi_{h02})\gamma^2_{h02}}{(1+2i\zeta_{h00})\gamma^2_{h00}}e(0,2) & \cdots\\ 0 & 1 & 0 & \dfrac{(1+2i\xi_{a01})\gamma^2_{a01}}{(1+2i\zeta_{a00})\gamma^2_{a00}}e(0,1) & 0 & \cdots\\ 1 & 0 & \dfrac{(1+2i\xi_{h11})\gamma^2_{h11}}{(1+2i\zeta_{h10})\gamma^2_{h10}}e(0,2) & 0 & \dfrac{(1+2i\xi_{h12})\gamma^2_{h12}}{(1+2i\zeta_{h10})\gamma^2_{h10}}e(0,2) & \cdots\\ \cdot & \cdot & \cdot & \cdot & \cdot & \cdots\\ \cdot & \cdot & \cdot & \cdot & \cdot & \cdots\\ \cdot & \cdot & \cdot & \cdot & \cdot & \cdots\end{bmatrix} \quad (13)$$

See Ref[4] to determine the values of $\gamma_{hsr}, \gamma_{asr}$.

DETERMINATION OF FLUTTER BOUNDARIES

Flutter boundaries are predicted by eigenvalue , which is obtained from soluting the standard complex eigenvalue problm [equation (8)] without regard to the coupled disk–blade. The relationship between and complex frequency of the system can be expressed as

$$\frac{i\omega}{\omega_0}=\frac{i}{\sqrt[2]{\gamma}}=\bar{\mu}+i\bar{v} \quad (14)$$

Where the real part $\bar{\mu}$ represents total damping of the system. When $\bar{\mu}>0$, flutter occurs and the system is unstable; when $\bar{\mu}<0$, the system is stable; when $\bar{\mu}=0$ the system is in the critical state. When $\bar{\mu}=0$, let $k=k_F$, $\gamma=\gamma_F$ and $V=V_F$, the nondimensional critical flutter speed can be written as

$$\frac{V_F}{b\omega_0}=\bar{v}/k_F$$

In regard to coupled disk–blade, eigenvalue will be determined by the generalized eigenvalue equation [11].

EXAMPLES AND ANALYSIS

Suppose that frequencies of blades are distributed in the form of cosine arrangement on the circumference. The parameters of the three calculated cascades are in tab.1 Futter bounderies of three mistuning forms under five mistuning amplitudes are calculated for above three cascades. Here three mistuning forms are respectively:

(1) Torsional frequency of the first order is mistuned, bending frequency of the first order is tuned which is called the T order mistunning.

(2) Bending frequency of the first order is mistuned and torsional frequency of the first order is tuned, which is called the B order mistuning.

(3) Both are mistuning, which is called the BT order mistuning.

Table.1. paramenters of casecade

	fore torsional center	after torsional center	torsional and gravity center are conincide
N	12	12	12
s / c	1	1	1
μ_s	225	225	225
η	0.25	0.75	0.50
X_{as}	−0.5	0.5	0
r_{as}	0.5774	0.5774	0.5774
θ	45 °	45 °	45 °
$\omega_{hs} / \omega_{as}$	0.239	0.239	0.636

Calculated results are shown in Fig. 4–6.

It will be seen from the above figures that mistuning can increase the speed of the critical flutter and make it larger with increasing of the mistuning amplitude. The effect of different mistuning orders on flutter boundaries is not the same. The BT and T order mistuning makes the critical flutter speed increase greatly, but B order mistuning increases very little. That means that effect of mistuning of torsional frequency on control ling the flutter is more significant than that of mistuning of bending frequrncy. This can be explained by the following. The calculated results indicate that the flutters of the former two cascades in tab.1 are combined bending–torsional vibration in which the torsional vibration is the dominat one. Consequently, the above conclusion can be obtained. The flutter of the third cascade in tab.1 (the torsional center coincides with the gravity one) is a pure torsional vibration, so B order mistuning of order B hardly affects the velocity. It is only under this mistuning order close to the flutter mode that the flutter can be controlled effectively.

Ref [7] indicates that mistuning amplitude may increase the respones amplitude of forced vibration of some blades. And the effect of mistuning on the response amplitude is very small only when the exciting frequencies close to the mode which is tuning and other orders are mistuned. We can think from above that different mistuning orders can be obtained by frequency–choosing in operation.

As long as the mistuning amplitude of the mistuning mode whose frequency is close to the exciting frequency is as small as possible, while the mistuning mode whose frequency is close to the flutter frequency is as large as possible. In this way not only flutter can be controlled effectivly, but also the increment of the response amplitude of the forced vibration can be limited.

in the above cases,the effect of coupled disk–blade is not considered. the results with the coupled disk–blade are showing in Fig. 7.

It is clear from this figure that the basic rules of the critical flutter velocity are similar to those rules disscussed previously and its value is much smaller, but still larger than that of tuning. With regard to coupled disk–blade, every blade of mistuned cascades will vibrate with the same frequency and different amplitude. It may regard the case as that between of mistuning and tuning regardless of coupled disk–blade between in both. So stability is between mistuning and tuning.

With the aid of figure [8]. the problem can be discribed more clearly. The system is unstable when $\bar{\mu} > 0$ And the greater $|\bar{\mu}|$ is, the greater the negative damping are, the faster the amplitude increaes. On the contrary, the system is stable when $\bar{\mu} < 0$ and the greater $|\bar{\mu}|$ is, the greater the positive damping is, the faster the amplitude decreases, It can be seen from fig. 8 ($\bar{\mu} < 0$) that the minimum positive damping $|\bar{\mu}|$increases (the structural damping keeps constant so that the minimum positive damping increases) and maximum positive damping decreases after the cascade is mistuned. Because of the fact that the stability of the system only depends on the minimum positive damping, more precisely, the reason why mistuning can restrain flutter lies in increasing of the minimum aerodynamic damping of the cascade.

The different effect of the different orders on the flutter boundaries also lies in different sensitivityy of the minimum aerodyndamic positive damping to the different mistuning order (Fig.8)

Effect of frequency distribution forms are showing in Fig. 10.

Effect of centrifugal force are showing in Fig.11.

CONCLUSION

1. Mistuning of blade can delay flutter. The flutter boundary will be defered more and more with increasing of mistuning amplitude within the certain range of the amplitude.
2. Effect of different mistuning order on flutter boundaries is different. The effect deponds on the order of vibration when flutter occurs.
3. Coupled blade–disk makes the flutter boundaries of mistuned cascade move forwards.
4. Comparatively, stability of cascade with the rear torsional center is the best, then the second is one with the front torsional certer, and the third the considerd center.

REFERENCE

1. Kaza, K.R.V. and Kielb, R.E., Flutter and Response of a Mistuned Cascade in Incompressible Flow, AIAAJ, Vol.20, No.8,1981.
2. Kaza, K.R.V. and KIelb, R.E., Coupled Bending–Bending–Torsion flutter of Mistuned Cascade With Nonuniform Blades, AIAA paper 82–0726, 198.
3. Jinming, Xu. and Zhaohong, Shong. Stability Analysis of Rotor Stagger Frequency Blade. Beijing Institute of Aeronautics and Astronautics.
4. Kieilb, R.E. and KAZA, K.R.V.,Effects of Structural Coupling on Mistuned Cascade Flutter and Ronsponse, ASMEA, VOL.106, No.1,1984.
5. Bendiksen, O.Q, Flutter of Mistuned Turbomachinery Rotors,ASME.A VOL. 106, NO.1, 1984.
6. Whitehend, D.S., Forced and Momrnt Coefficients for Vibrating Airfois in Cascades, A.R.C., RM3254,1960.
7. Huitao, Fan and Dakuan, Shen Effect of Blade Mistuning on Coupled Vibration of Disk–blade. Norsthwersten Polytechnical University. 12, 1985.

NOMENCLATURE

$[A_r]$	aerodynomic matrix dut to vibration in rth mode(r= 1,2,3···N−1)
b	semichord
c	chord
r	order of mod
β_r	interblade phase anger of rth mode($2\pi r/N$)
h_s	amplitude of bending deflection of sth blade
α_s	amplitude of torsional deflection of sth blade
h_s	amplitude of bending deflection of blade in r mode of tuned cascade
α_s	amplitude of to~sional deflection of blade in rth mode of tuned cascade
ω	frequency
{Y}	columu matrix amplitude of all modes of sth blade for tuned cascade
{X}	columu matrix amplitude of all modes of sth blade for msituned cascade
m_s	massper unit span of sth blade
I_{as}	mass mokment of intertia of sth blade about ellasticaxis per unit span($m_s r_{as}^2 b^2$)
S_{as}	static mass moment of sth blade per uniit span about elastic axis when center of gravity is after elastic axis
ζ_{hs}	damping ratio of sth blade in bending
ζ_{as}	damping ratio of sth blade in torsion
L_s, M_s	unsteady aerodynomic lift and moment of sth blade per unit span
ρ	fluid density
K	reduced flutter frequency($\frac{\omega b}{V}$)
V	freestream speed relative to the blade
θ	stagger angle (Fig.1)
η	locatio of elastic exis measured from leading edge
γ	nondimensional eigenvalue($\frac{\omega_0}{\omega}$)^2
μ_s	mass ratio of sth blad($\frac{m_s}{\pi\rho b^2}$)
ω_{hs}	uncoupled bending frenquency of sth blade
ω_{as}	uncouple torsional frenquency of sth blade
γ_{hs}	nondimensional uncoupled bending frequency of sth blade
γ_{as}	nondimensional uncoupled torsioalg frequency of sth blade
k_{hs}, k_{as}	bending and torsional stiffness respectively of sth mode
r_{as}^2	radius of gyration of sth blade nondimensionalized with respect to b
x_{as}	dimensionless static unblance of sth blade($\frac{S_{as}}{m_s b}$)

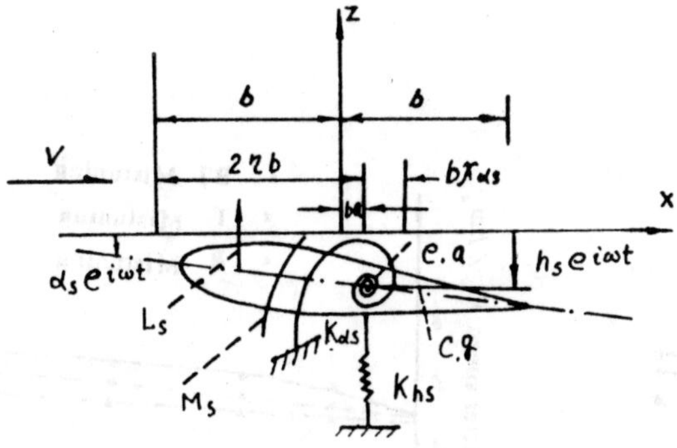

Fig.1. Model of the sth blade of a mistuned cascade

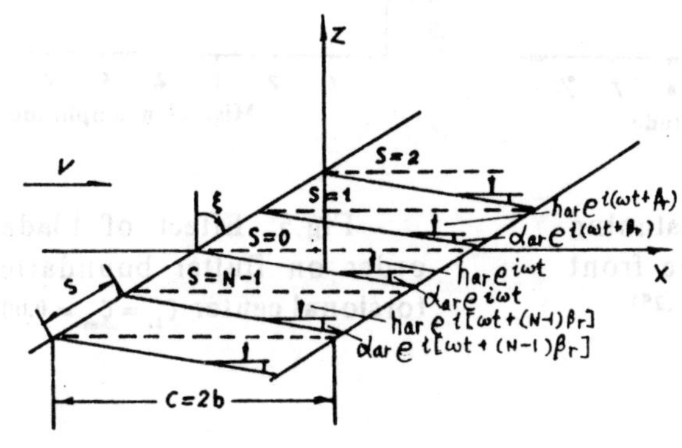

Fig.2. Model of a tuned cascaed in rth mode

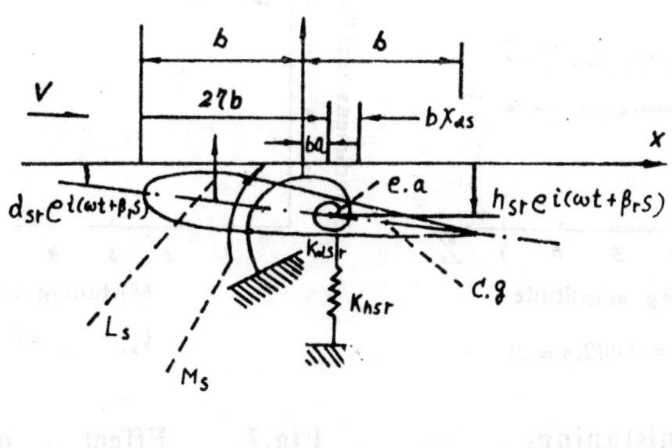

Fig.3. Model of The sth blade of a mistuning cascade (coupled disk–blade).

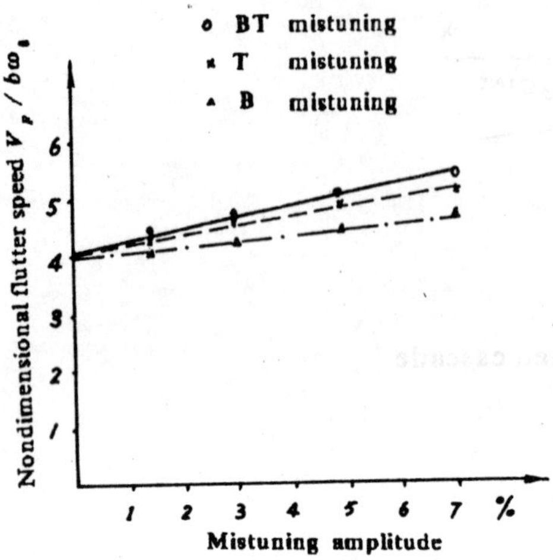

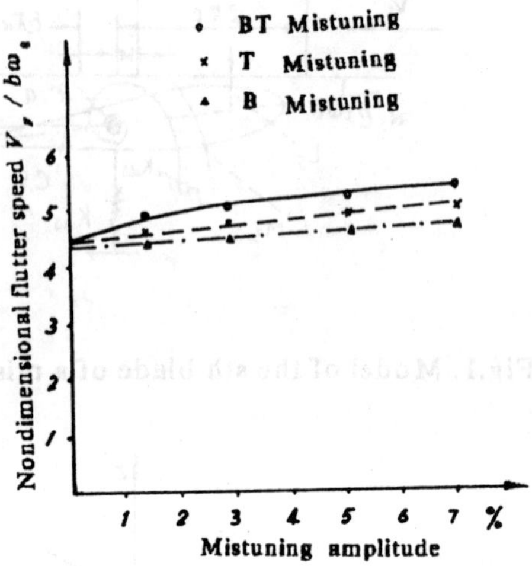

Fig.4. Effect of blade mistuning order on flutter boundaries (the front torsional center $\zeta_{hr} = \zeta_{ar} = 0.002, \eta = .25$)

Fig.5. Effect of blade mistuning order on flutter boundaries (the rear torsional center $\zeta_{hr} = \zeta_{ar} = 0.002, \eta = .75$)

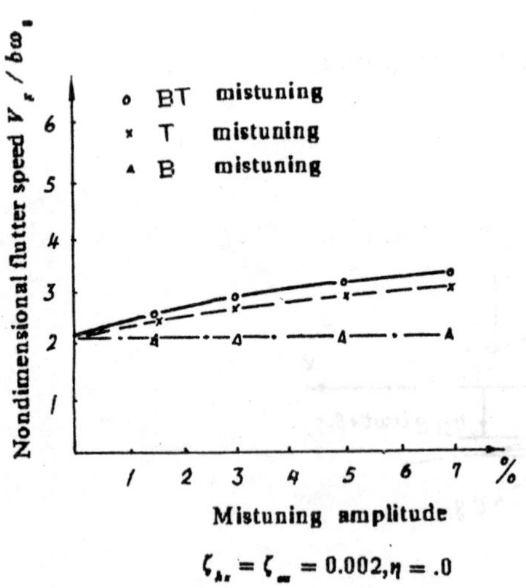

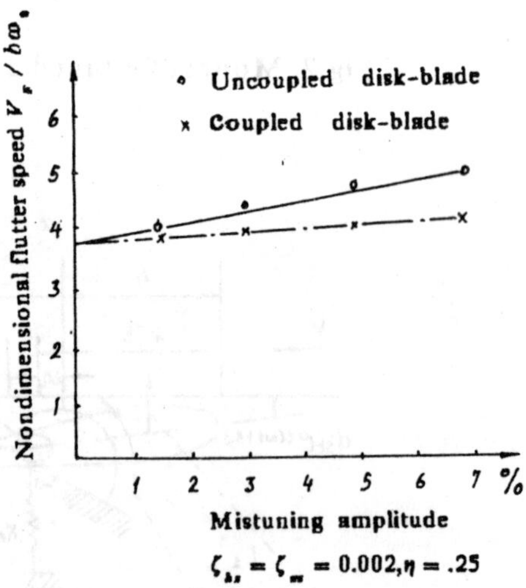

Fig.6. Effect of blade mistuning order on flutter boundaries (torsional and gravity center are coincided

Fig.7. Effect of coupled disk–blade on flutter boundaries

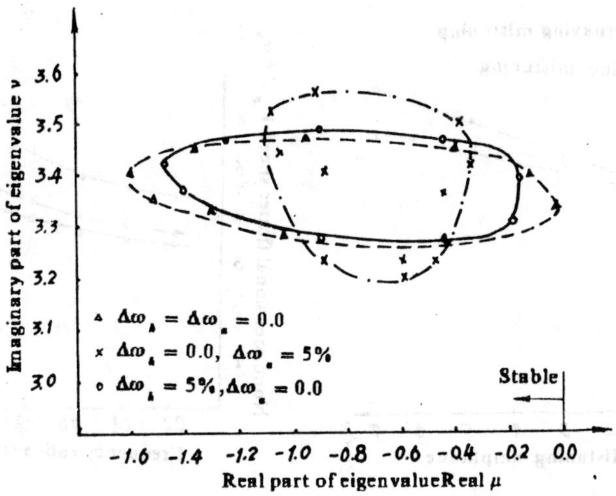

Fig.8. Eigenvalue distribution of a uncoupled disk–blade system

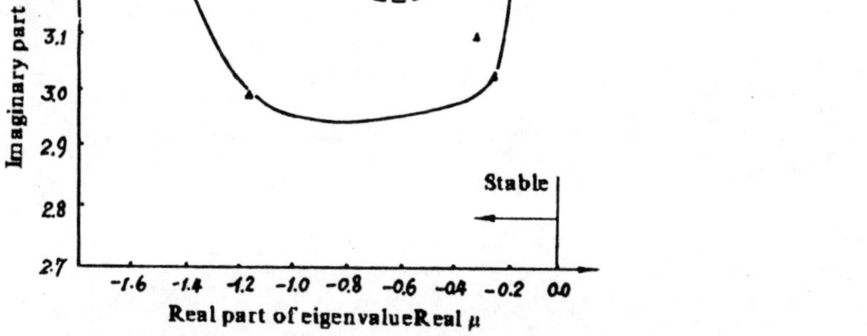

Fig.9.Eigenvalue distribution of a coupled disk–blade system

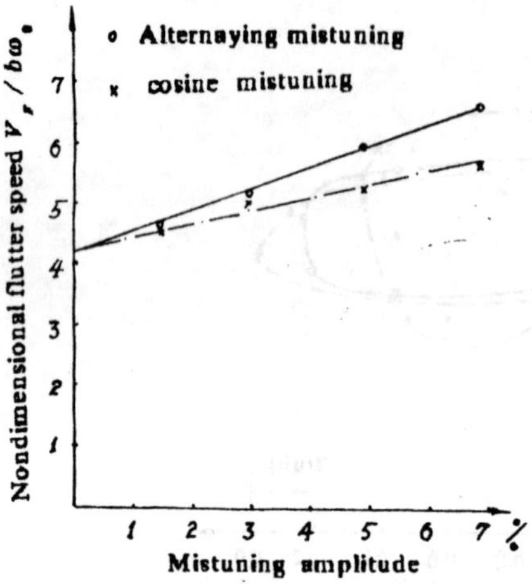

Fig.10. Effect of frequency distribution on flutter boundaries

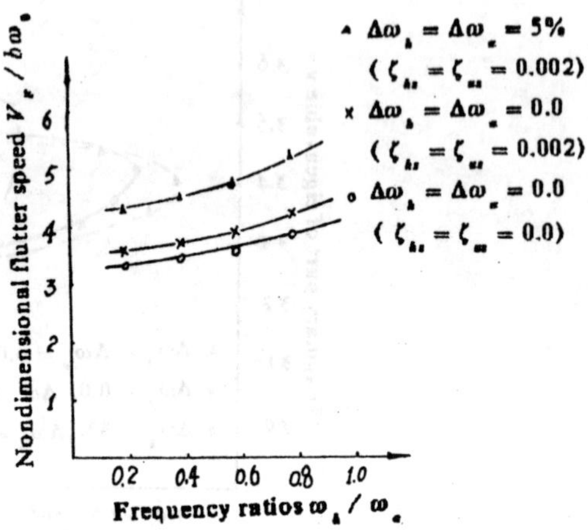

Fig.11. Effect of centrifugal force on flutter boundaries

THE RESEARCH OF BLADE FLUTTER CHARACTERISTICS UNDER QUASI–THREE DIMENSION UNSTEADY AERODYNAMICS LOAD

Jun Yue, Rui–Lian Kong, Zhao–Hong Song
Beijing University of Aeronautics And Astronautics

ABSTRACT

This paper presents an investigation of the flutter characteristics of rotating blades. The method is based on quasi–three–dimension unsteady cascade theory combined with a structural finite element for rotating blades. The aerodynamic load is assumed as a function of modal coordinates. The eigenpairs solved from the equation of motion of the blade have been used to judge whether the system is stable or not. The effects of interblade phase angles,modal coupling, and structural damping, etc. are considered in the calculation. The results show modal coupling effect is detrimental to stability.

INTRODUCTION

In recent years with the rapid development of aero–engine performances,it makes pressure ratio high, mass flow large and stage load increasing very quickly. Those means the further worse for the aeroelastic stability of aero–engine blades. Since the sixty years, a number of theoretical analysis and experimental research have been performed about the above problems.

In the early sixty D.S.Whitehead etc.[1] built an analysing model and derivated a series of formulae for cascade blade using the Theodorsen theory for airfoil. In his theory, the aerodynamic force is considered as a linear function of displacements. From then on, there are many calculating methods being developed on the base of Whitehead's theory by D.S.Whitehead, S.N.Smith[6],M.E.Goldstans,O.O.Bendikson[4] and V.Kaza R.Kielb etc..J.M.Verdon[11],F.Sisto[10],He Li and Zhou Sheng[7] etc. have developed numerical methods of two–dimension cascade subsonic unsteady flow to predict the choke and stalling flutter of compressor blades. In their works, the equivalent cross section assumption have been used. Only one cross section is used substituting the whole blade to predict flutter. So the above mentioned theories all belong to two–dimension methods and only have two modal shapes bending and torsion. The modal shape of flutter is either pure bending or pure torsion. The vibrational modal shape of turbo–blade is very complex. It is not very reasonable using only two–dimension methods to solve flutter problem of actual blade,especially for low aspect ratio blade and prop fan. In order to solve those problems, O.O.Bendikson and J.F.White[5] tried to do something,as follows:flat blade modal analysis connecting with aerodynamics force distribution of whole blade surface is used to research the aeroelasticity of blade. The unsteady pressure distribution is assumed as a function of generalized coordinates. that is :

$$[p(x,y)]_{total} = \sum_{i=1}^{n} [p_o(x,y)]_i q_i$$

This method's result shows the aeroclastic stability of blade is very sensitive with aspect ratio and thickness of blade. Xu Jianmin and Song Zhaohong[8] used two dimensional quasi–steady method to simulate three dimensional effect of blade and used three dimensional method for structural dynamic analysis. Their calculating results got good agreement with experimental data. K.R.V.Kaza and R.E.Mehmed etc.[2] have used two dimensional analysis and three dimensional analysis both for fan blade cascade structures predicting calculation. By a comparison with these two results, it shows the flutter speed predicted by two dimensional analysis is less accurate than by three dimensional analysis. Those above mentioned researches got some successful conclusions,but it still need further research work to do.

Based on advantages of above research works, an aeroelasticity analysis for actual,rotating compressor blades have been made in this paper. The finite element method is adopted in structural vibration analysis. In the calculation of unsteady aerodynamic pressure,many vertical sections along blade are searched. The unsteady numerical calculation of cascade is used to every section,then the unsteady pressure distribution of blade surface can be got through interpolation. Assuming the unstable pressure distribution as a function of modal coordinates. The aeroelastic problem is transferred to solve a vibration differential equation, then the stability of the first several modes can be predicted together.

FORMULA DERIVATION

A. Basic Assumption

a) The air flow is in nonviscous and nonrotating condition.

b) The vibration of blade is in small turbulence condition.

c) When flutter occurs, blades have harmonic vibration.

In general condition,the vibration mode may be expressed as follows:

$$u(x,y,z,t) = \sum_{i=1}^{n} \varphi_i(x,y,z) q_i(t) \qquad (1-1)$$

where q_i —— the blade generalized modal coordinates

(x,y,z) —— blade mode

According to the basic assumption, the linear theory has been used in the paper. Then the unstable pressure distribution on blade surface can be expressed approximately by superposition.

$$p(x,y,z,t) = \sum_{j=1}^{n} p_j(x,y,z) q_j(t) \qquad (1-2)$$

where $p_j(x,y,z)$ —— unsteady pressure distribution of the jth mode of blade.

B. Basic Equation of Flutter

About blade flutter the basic differential equation can be written as :

$$[M]\{\ddot{u}\} + [K]\{u\} = \{F\} \qquad (1-3)$$

where {F} —— unsteady aerodynamic force vectors.

[M] —— mass matrix

[K] —— stiffness matrix

After [M] and [K] are uncoupled,the equation can be expressed by generalized coordinates:

$$[M]\{\ddot{q}\} + [K]\{q\} = \{Q\} \qquad (1-4)$$

where {Q} —— the generalized unsteady aerodynamic force. It is a function of time,flow parameters etc.
The key to solve flutter problem is to solve the {Q}.
C. Generalized Aerodynamic Force :
From equation (1–2)

$$\{p\} = [p_1 \ \ p_2 \ \ p_3 \ \cdots\cdots \ p_n]\{q\} = [p]\{q\} \tag{1-5}$$

$p_1(x,y,z),\cdots\cdots,p_n(x,y,z)$——the unsteady pressure distribution in the first n modal modes.
Equation (1–1) is written as matrix form,that is :

$$\{u\} = [\Phi]\{q\} \tag{1-6}$$

According to virtual work principle:

$$\delta W_{ext} = \sum_{i=1}^{n} Q_i \delta q_i = \iint_s [p]\delta u ds = \iint_s \{p\}[\Phi] ds \{\delta q\} \tag{1-7}$$

then,

$$\{Q_i\} = [A]\{q\} \tag{1-8}$$

$$A_{ij} = \iint_s p_i \varphi_j \, ds \tag{1-9}$$

So,the generalized aerodynamic force vector is expressed as a function of generalized coordinates.
Where A_{ij} —— coefficient of aerodynamics load. Because the unsteady aerodynamic force is a complex, the A_{ij} is as complex form, and $[A]$ is a complex matrix. From equation (1–9) it is obvious that A_{ij} connects with the modal parameters of structural system directly. Substituting Eq. (1–8) into the basic equation of flutter Eq. (1–4) leads to :

$$[M]\{\ddot{q}\} + [K]\{q\} = \{Q\} = [A]\{q\} \tag{1-10}$$

THE CALCULATION OF DYNAMIC CHARACTERISTICS OF STRUCTURE

In order to analyse the flutter of blade,it is necessary to study the dynamic characteristics of blade. The calculation of dynamic characteristics of blade has close relationship with the analysis of flutter.
In rotating condition the equation of motion of blade can be derivated as follows:
Coordinate system shown in Fig.1, XYZ —— static Cartesian coordinates,xyz —— a rotating coordinates that rotate at angle velocity with blade. The position of a point p at blade is.

$$\bar{r} = \begin{Bmatrix} x \\ y \\ z \end{Bmatrix} = \bar{r}_0 + \bar{r}' = \begin{Bmatrix} x_0 \\ y_0 \\ z_0 \end{Bmatrix} + \begin{Bmatrix} u \\ v \\ w \end{Bmatrix} \tag{2-1}$$

the transient velocity of point p is

$$\begin{Bmatrix} \dot{x} \\ \dot{y} \\ \dot{z} \end{Bmatrix} = \begin{Bmatrix} \dot{u} \\ \dot{v} \\ \dot{w} \end{Bmatrix} + \vec{\Omega} \times \begin{Bmatrix} x + R_0 \\ y \\ z \end{Bmatrix} = \begin{Bmatrix} \dot{u} \\ \dot{v} \\ \dot{w} \end{Bmatrix} + \Omega[S]\begin{Bmatrix} x + R_0 \\ y \\ z \end{Bmatrix} \quad (2-2)$$

$$[S] = \begin{bmatrix} 0 & -sin\theta & -cos\theta \\ sin\theta & 0 & 0 \\ cos\theta & 0 & 0 \end{bmatrix}$$

So the kinetic energy of blade can be written as

$$T = \frac{1}{2}\int \rho(\dot{r}' + \vec{\Omega} \times \vec{r}) \cdot (\dot{r}' + \vec{\Omega} \times \vec{r})dv = \frac{1}{2}\int \rho\begin{Bmatrix} \dot{u} \\ \dot{v} \\ \dot{w} \end{Bmatrix}^T \begin{Bmatrix} \dot{u} \\ \dot{v} \\ \dot{w} \end{Bmatrix} dv + \frac{1}{2}\Omega^2 \int \rho\begin{Bmatrix} x + R_0 \\ y \\ z \end{Bmatrix}^T [S]^T[S]\begin{Bmatrix} x + R_0 \\ y \\ z \end{Bmatrix} dv$$

$$+ \Omega \int \rho\begin{Bmatrix} \dot{u} \\ \dot{v} \\ \dot{w} \end{Bmatrix}^T [S]\begin{Bmatrix} x + R_0 \\ y \\ z \end{Bmatrix} dv \quad (2-3)$$

the strain potential energy is

$$V = \frac{1}{2}\int \{\varepsilon\}^T[D]\{\varepsilon\}dv = V_l + V_{nl} \quad (2-4)$$

Using the finite element method to analyse the rotating frequencies of blade, assuming the shape function as N ,that is:

$$\begin{Bmatrix} u \\ v \\ w \end{Bmatrix} = [N]\{\delta\}^e \quad (2-5)$$

Lagrangian Formulation is

$$\frac{d}{dt}\frac{\partial T}{\partial\{\dot{\delta}\}^e} - \frac{\partial T}{\partial\{\delta\}^e} - \frac{\partial V}{\partial\{\delta\}^e} = \{Q\} \quad (2-6)$$

Substituting equation (2–5) into (2–3), then into equation (2–6) leads to:

$$[M]^e\{\ddot{\delta}\}^e + [M_G]^e\{\dot{\delta}\}^e + [K_c]\{\delta\}^e + \int_e [\bar{B}]^T\{\delta\}W = \{Q_c\}^e \quad (2-7)$$

where

$[M]^e = \int_e \rho[N]^T[N]dv$—— element mass matrix

$[M_G]^e = 2\Omega\int_e \rho[N]^T[S][N]dv$—— element Coriolis matrix

$[K_c]^e = \Omega^2\int_e \rho[N]^T[S]^T[S][N]dv$—— centrifugal stiff matrix of element

$$[Q_c]^e = \Omega^2 \int_e \rho [N]^T [S]^T [S] \begin{Bmatrix} x_0 + R_0 \\ y_0 \\ z_0 \end{Bmatrix} dv$$ —— centrifugal force vector of element

$$\int_e [B]^T \{\delta\} dN$$ —— element strain energy

Because centrifugal force acts on the twisted blade,the deflection of blade appears small geometric non–linear. The transform matrix $[\bar{B}]$ may be divided into two part $[B_0]+[B(\delta)]$, the first linear term and the second nonlinear term.

The linear stiff matrix of element is:

$$[K_0]^e = \int_e [B_0]^T [D][B_0] dv \tag{2-8}$$

Because nonlinear effects are not very large,it is neglect to the high term of strain energy and the quadratics term of stress. The non–linearty is considered to be caused mainly by the quadric term of deflection.

$$V^e_{nl} = \int_e \{\delta\}^T \{\varepsilon_2\}^e dv$$

$$\{\varepsilon_2\}^e = \begin{Bmatrix} \frac{1}{2}\left[\left(\frac{\partial u}{\partial x}\right)^2 + \left(\frac{\partial v}{\partial x}\right)^2 + \left(\frac{\partial w}{\partial x}\right)^2\right] \\ \frac{1}{2}\left[\left(\frac{\partial u}{\partial y}\right)^2 + \left(\frac{\partial v}{\partial y}\right)^2 + \left(\frac{\partial w}{\partial y}\right)^2\right] \\ \frac{1}{2}\left[\left(\frac{\partial u}{\partial z}\right)^2 + \left(\frac{\partial v}{\partial z}\right)^2 + \left(\frac{\partial w}{\partial z}\right)^2\right] \\ \frac{\partial u}{\partial x}\cdot\frac{\partial u}{\partial y} + \frac{\partial v}{\partial x}\cdot\frac{\partial v}{\partial y} + \frac{\partial w}{\partial x}\cdot\frac{\partial w}{\partial y} \\ \frac{\partial u}{\partial z}\cdot\frac{\partial u}{\partial x} + \frac{\partial v}{\partial z}\cdot\frac{\partial v}{\partial x} + \frac{\partial w}{\partial z}\cdot\frac{\partial w}{\partial x} \\ \frac{\partial u}{\partial z}\cdot\frac{\partial u}{\partial y} + \frac{\partial v}{\partial z}\cdot\frac{\partial v}{\partial y} + \frac{\partial w}{\partial z}\cdot\frac{\partial w}{\partial y} \end{Bmatrix} \tag{2-9}$$

Nonlinear stiff matrix

$$[K_{nl}]^e = \partial^2 V^e_{nl} / \partial\{\delta\}^e \partial\{\delta\}^e \tag{2-10}$$

Substitute equation (2–5)

$$[K_{nl}]^e = \int_e \left([H]_u^T [\sigma'][H]_u + [H]_v^T [\sigma'][H]_v + [H]_w^T [\sigma'][H]_w \right) dv \tag{2-11}$$

where σ' is the initial stress causing by centrifugal force.

In order to obtain matrix $[K_{nl}]$, at first the initial stress should be solved. It is necessary to know the stiff matrix to solve the initial stress. Therefore,iteration is used to solve $[K_{nl}]$, such as

(1)giving initial displacement,that is $\{\delta_0\} = [K_0]^{-1}\{Q_c\}$

(2)according to $\{\delta_0\}$, $[\sigma_0]$ can be solved,then $[K_{nl}]$ can be obtained.

(3)from the equation: $\{\delta_1\} = ([K_0] + [K_{nl}] - [K_c])^{-1}\{Q_c\}$, the new displacement can be obtained.

(4)repeat these steps.

After several iteration times ,an approaching result can be obtained.

Generally,the matrix $[M_G]$ and $[K_c]$ in the equation of motion of blade have small effect on the several low frequencies of blade, it is feasible to neglect the two terms. So the free avibration differential equation of blade can be simplified as follows :

$$[M]\{\ddot{\delta}\} + [K]\{\delta\} = 0 \tag{2-12}$$

$$[K] = [K_0] + [K_{nl}] \tag{2-13}$$

THE CALCULATION OF UNSTEADY AERODYNAMIC FORCE

In the formula (1–9), P_i is the unsteady pressure distribution of blade surface in the ith nodal. In order to get the matrix $[A]$, it is neccessary to solve the unsteady flow–field. Since the limitation of strorage capacity of computer and its calculating speed,it is very difficult to resolve three–dimentional unsteady flow field directly. In this paper, the method to be used is taking many sections along the vertical direction of blade and solving the two dimensional unsteady flow–field of each section in each modal , then forming the unsteady areodynamics pressure distribution of all blade surface through interpolation. Duan's [9] method is used to calculate two dimensional aerodynmic force in blade sections.

THE SOLUTION OF FLUTTER PROBLEM

According to the basic equation of flutter problem (1–10)

$$[M]\{\ddot{q}\} + [K]\{q\} = [A]\{q\} \tag{3-1}$$

From (1–9), it can be seen that A_{ij} is directly correlated with the nodal parameters of structural system, so the equation (3–1) is nonlinear.

In actual analysis, the ratio of flow density where blades exist to mass density of blade is very small and the effect of mass couple is very small. So,flutter frequencies are always very close to natural frequencies. The dynamic load coefficient A_{ij} are obtained by using the natural frequencies and modal shapes directly,not by using iteration method.

The flutter problem is converted into a known differential equation after solving out $[A]$. In general,according to the assumption above,as the blade flutter onset, it is always in harmonic vibration. Assume λ to be the flutter frequency, the generalized modal coordinate can be written as:

$$q = \bar{q}e^{i\lambda t} \tag{3-2}$$

Substituting q into vibration equation, the flutter problem is transformed into a solving complex eigenpairs problem.

$$([\bar{K}] - [A])\bar{q} = \lambda^2[\bar{M}]\bar{q} \tag{3-3}$$

Non–dimensional equation is as follow:

$$\left.\begin{aligned} \lambda^* &= (\lambda / \omega_{ref})^2 \\ k^*_{ii} &= \bar{k}_{ii} / (m_{ii}\omega^2_{ref}) \\ A^*_{ij} &= A_{ij} / (m_{ii}\omega^2_{ref}) \end{aligned}\right\} \qquad (3-4)$$

then,

$$([K^*] - [A^*])\bar{q} = \lambda^* \bar{q} \qquad (3-5)$$

Where ω_{ref} ---- reference frequency, in general,it is fundamental torsion frequency.
Solving (3–5),then the complex roots are obtained.

$$\lambda^* = \lambda^*_R + i\lambda^*_I \qquad (3-6)$$

Right now,the vibrational frequency of blade is

$$\omega' = \sqrt{\lambda^*_R}\,\omega_{ref} \qquad (3-7)$$

The coefficient of aerodynamic damping is

$$\delta_A = \frac{\lambda^*_I}{2\lambda^*_R} \qquad (3-8)$$

For the kth generalized displacement $q^{(k)}$ can be written as

$$q^{(k)} = \bar{q}^{(k)} e^{-\delta^{(k)}_A + i\omega'^{(k)}}$$

and when $\delta^{(k)}_A > 0$, then vibration is convergent, the flutter of blade doesn't occur; when $\delta^{(k)}_A = 0$, vibration is in the critical state of flutter; when $\delta^{(k)}_A < 0$, vibration is divergent and the flutter of blade occurs.

CALCULATING EXAMPLE

The first stage and second stage compressor blade of a jet engine have even appeared flutter fault. Some actual analyses of these two stages compressor blades are performed in this paper.
The first stage rotor consists of 31 blades, the stagger angle of blade is 5 ° .
The second stage rotor consists of 53 blades, the stagger angle of blade is 24 ° .

RESULTS AND DISCUSSION

There are many factors that affect flutter. At present,the investigation of flutter mechanism is being made. It is necessary to investigate the characteristics of flutter from different aspects and with different methods. The effects of blade flutter of modal coupling among modes, the phase angle between blades and structural damping etc. are emphasized to analyse in this paper.
(1) The effect of modal coupling between vibrational modes
Before the flutter occurs,the motion of blade concludes small vibration of various order frequencies. So there are unsteady aerodynamic coupling forces among modes. The aerodynamic force caused by any order mode will af-

fect other orders vibration. The mutual action in a certain condition can make it negative for the aerodynamic damping of a mode, the mode will become unstable and be dominant vibration mode. In general, lower modes are easy invalid, so in this calculation only the first three modes are considered for emphasizing major contradiction.

The coupling effects among modes are shown in Fig.3 and Fig.4. It can be seen that the first–bending mode is affected very small with the modal coupling ,but large change occurs at the first torsion. For the second–bending mode of the first–stage blades,the coupling effect among modes is not very large, but for the second–bending mode of the second–stage blades, the effct becomes terrible. It is similarly to the first–torsion mode. From the frequency characteristics of the two stage blades(Fig.2), it can be seen that the frequency of second–bending of the second stage blades is nearly to the frequency of first–torsion mode and much more nearly with the increasing of rotational speed, so the coupling effect between two modes is very serious. The flutter characteristics of second–bending is affected seriously with first–torsion mode and the flutter characteristics of first–bending mode is hardly affected.

In general, the first–bending vibration dominants over blade vibration,and it is very difficult that the other orders vibration affect it largely. When it comes to all, the coupling effect makes the dynamic damping amplitude of every order mode larger,especially makes the minus dynamic damping larger, and this is very harmful to restrain flutter.

(2) The effect of phase angle among blades

Referring Fig.5,it can be seen that phase angle strongly affects the blade flutter. In certain condition, the change of phase angle can makes the blades unstable when the other conditions have no changes. Bendiksen,Li He and Jianmin Xu etc. got the same conclusion. The change of phase angle among blades directly changes the fluid field of cascade and lead to the changes of aerodynamic work on blades. The effect of phase angle have a rule: the unsteady aerodynamic damping coefficient is a sinusoid with the change of phase angle.

(3) The effect of structural damping

For a general structure mode of solid material,when material slide or glide occurs between two layers of material which includes friction, it produces structural damping. So in general, structural damping hinders vibration.

It can be seen from the results (shown in Fig.5) of this paper that when the effects of structural damping are considered, it makes aerodynamic damping to increase a little toward positive direction. So,the existence of structural damping is benefit for restraining the occurrence of flutter.

CONCLUSION

On the foundation of pioneers,this paper carried on some inquiry for the mechanism of the blade flutter onset combined with the finite element dynamic analysis of rotating shell and quasi–three–dimension unsteady dynamic calculation. Some conclusions are as follows:

(1) The coupling effect among modes of blade makes the aerodynamic damping increasing in every mode, and it has bad effect for flutter. The smaller difference of frequencies among various modes, the larger coupling effect will be obtained. The modal coupling effect of first–torsion mode is the largest.

(2) The phase angle among blades has a great effest on flutter onset and its effect is a sinusoid.

(3) The structural damping is favorable to restrain the onset of flutter.

The subtransonic stall flutter of blade is a very complicated aeroelastic problem, the discussion mentioned above is just the basic idea, and further improvements are expected.

REFERENCE

1. Whitehead D.S.
   Force and Moment Coefficients of Vibrating Aerofoils in Cascade
   A.R.C. R&M 3254, 1961.
2. Kaza K.R.V., Mchmed O., Narayanan G.V. and Murthy O.V.
   Flutter Investigation of a Composite Advanced Propfan Modal.
   AIAA paper 87–0738

3. Kaza K. R. V., August V.
   Vibration and Flutter Characteristics of the Sr–7L Large–Scale Propfan
   N88–18036191 XAD
4. Bendikson O. O. and Fridman P. P.
   Coupled Bending–torsional Flutter in Cascade
   AAIA Journal Vol.18,NO.2,1980
5. White J. F. and Bendikson O. O.
   Aeroelastic Behaviour of Low–Aspect Ratio Metal and Composite Blades.
   ASME 86–GT–243
6. Elchuri V. and Smith C. C. C.
   Flutter Analysis of Advanced Turbopropellers
   AIAA 83–0846
7. He Li
   Essay for Master
   BUAA 1984
   8. Xu Jianmin, Song Zhaohong
   Prediction of Blade Flutter in a Tuned Rotor.
   ASME 1985
9. Duan Shizhong
   Essay for Master
   BUAA 1986
10. Sisto F.
    Stall Flutter
    N87–24404
11. Verdon J.M.
    The Unsteady Aerodynamic of a Finite Supersonic Cascade with Subsonic Axial Flow
    Tran. Applied Mechanics Vol.40

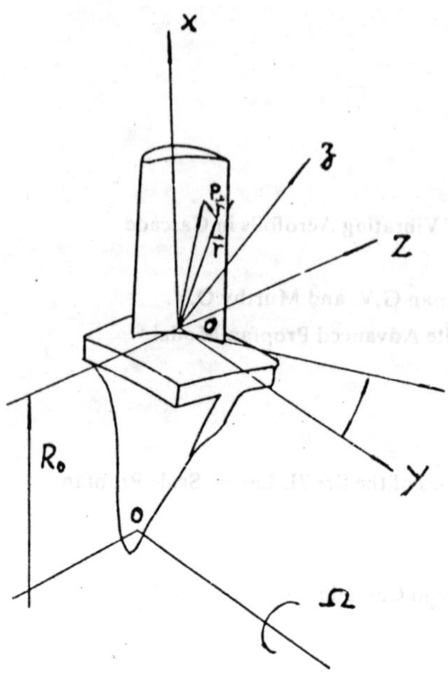

Fig.1 Coordiate system of rotating blade

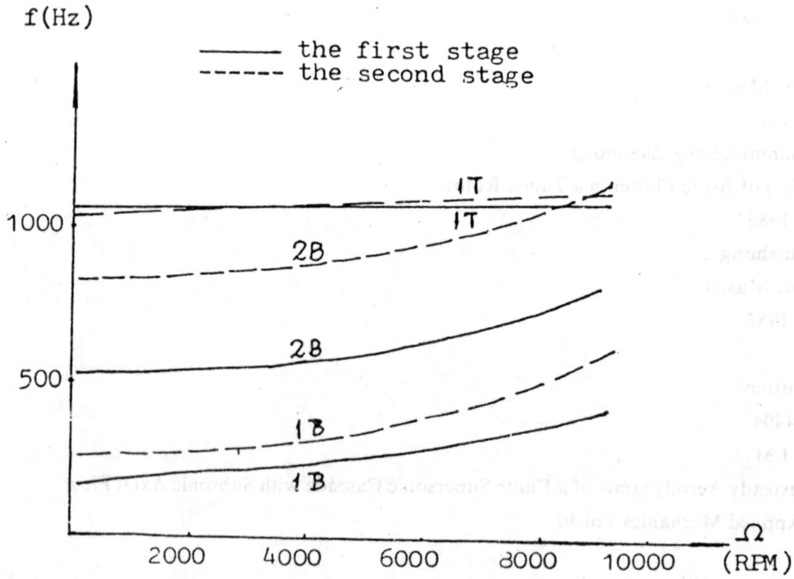

Fig.2 Variation of blade natural frequencies with rotational speed

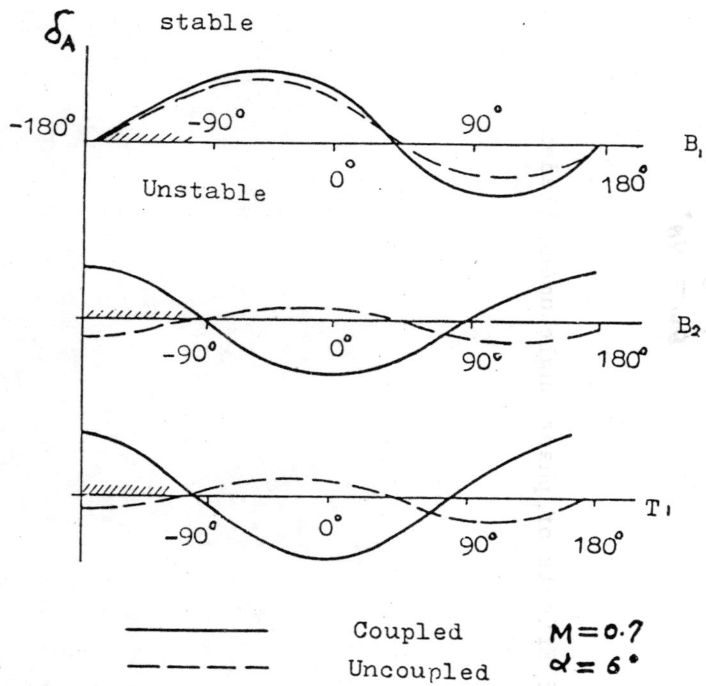

Fig.3 Effects of modal couplingbetween vibrational modes of the first-stage blades

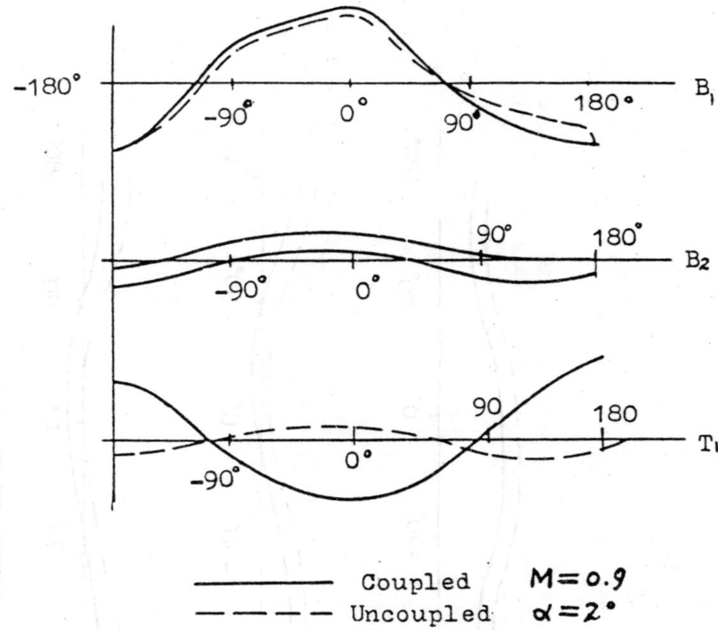

Fig.4 Effects of modal coupling between vibrational modes of the second-stage blades

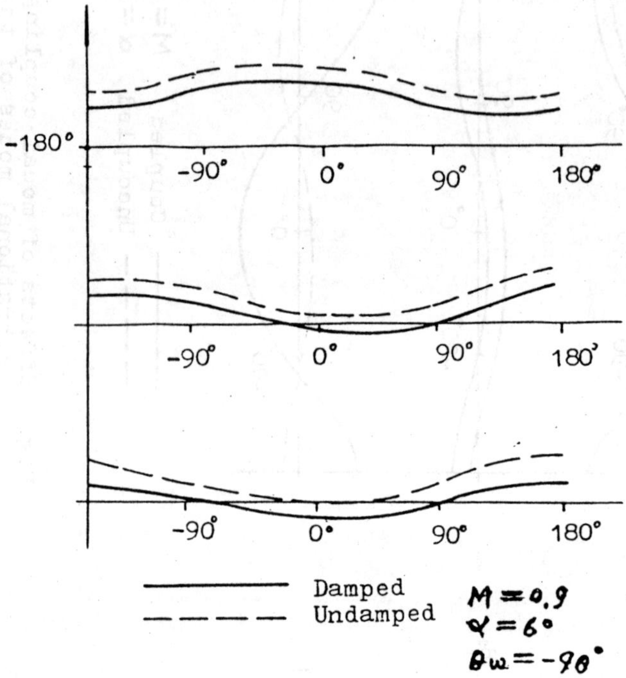

Fig.5 Effects of phase angle among blades

MISTUNING EFFECTS ON FULLY STALLED CASCADE FLUTTER

Ichiro Fujimoto
(Takushoku University, Tokyo, Japan)
Hiroshi Nagai
(Nihon University, Chiba, Japan)
Hideo Tanaka
(Tokai University, Tokyo, Japan)
Takanori Hirano
(Takushoku University, Tokyo, Japan)
Susumu Ishii
(Nihon University, Chiba, Japan)

ABSTRACT

Mistuning effects on fully stalled cascade flutter in torsional mode are surveyed by changing the blade natural frequency in the case of the alternating blade arrangement, which presents a significant flutter suppression effect for unstalled cascades. The flutter suppression effect is confirmed for stalled cascades as well as for unstalled cascades. Furthermore, an emphasis is to be placed on the fact that more remarkable flutter suppression effect is expected by slight mistuning for stalled cascades than for unstalled cascades. There is no discontinuous change in flutter mode in the case of stalled cascades throughout the whole range of mistuning concerned except in extremely close vicinity of tuned condition, while the discontinuous modal change is seen in unstalled cascade flutter. Mistuning has more beneficial effect in turbines than in compressors.

1. INTRODUCTION

The flutter of cascade blades is one of various problems awaiting solution about unsteady internal flow of turbomachines. The operation in a stall region or its vicinities is unavoidable when starting turbomachines or when running them with a partial load, and the stall flutter must hence be taken into due account. Since the flutter velocity becomes noticeably low on stalled condition as compared with the case on unstalled condition, its improvement is being demanded.
There is an attempt by the non-uniform design of cascade blades called mistuning or detuning as a measure of suppressing cascade flutter. The concept of mistuning was treated by Whitehead(1) in 1964. Bendiksen(2) reviewed recent research on passive flutter control including mistuning. On unstalled condition, it has been found that the vibration having a uniform amplitude and a uniform interblade phase angle, as seen in tuned cascades, is suppressed when mistuning is applied by changing the elastic condition without changing any geometric condition of cascades(3)(4). By positively utilizing the above aspect from the view point of flutter suppression, the effect of mistuning on unstalled cascade flutter has been analyzed paying attention to the influences of a change in blade natural frequency, arrangement of blades, etc.. It

has been numerically proved that the flutter velocity becomes the highest when two kinds of blades having different natural frequencies are alternately arranged so long as the surveyed range is concerned[8]. However, the mistuning effect on cascade flutter under stalled condition has not been made clear up to the present.

This paper deals with the fully stalled cascade where the flow around blades shows the leading edge separation. Torsional flutter of mistuned cascades is examined by changing the natural frequency of blades in the case of the alternating blade arrangement, as a typical example, which presents a significant flutter suppression effect for unstalled cascades.

2. THEORETICAL FOUNDATION

Let's consider a model cascade where a blade group consists of N blades and is cyclically arranged by taking the cyclicity of flutter in annular cascades into consideration as shown in Fig.1. The vibration of each blade is assumed to be in a one degree of freedom torsional mode which is important for stall flutter. Only the elastic condition is changed to accomplish mistuning of the cascade without degrading any steady cascade specifications. In this paper, the equations of motion are considered as follows by simply changing the natural frequency, neglecting any friction term.

$$\begin{aligned}
\alpha_1 M_1 + \alpha_2 M_N e^{-i\tau} + \cdots\cdots + \alpha_L M_{N+2-L} e^{-i\tau} + \cdots\cdots + \alpha_N M_2 e^{-i\tau} &= I_1 \ddot{\alpha}_1 + K_1 \alpha_1 \\
\alpha_1 M_2 + \alpha_2 M_1 + \alpha_3 M_N e^{-i\tau} \qquad + \alpha_L M_{N+3-L} e^{-i\tau} + \qquad + \alpha_N M_3 e^{-i\tau} &= I_2 \ddot{\alpha}_2 + K_2 \alpha_2 \\
\vdots \qquad\qquad\qquad\qquad & \qquad \vdots \\
\alpha_1 M_K + \alpha_2 M_{K-1} + \cdots\cdots + \alpha_L M_{N+K-L+1} e^{-i\tau} + \cdots + \alpha_N M_{K+1} e^{-i\tau} &= I_K \ddot{\alpha}_K + K_K \alpha_K \qquad (1) \\
\alpha_1 M_{L-1} + \alpha_2 M_{L-2} + \cdots + \alpha_{L-1} M_1 + \alpha_L M_N e^{-i\tau} + \cdots\cdots + \alpha_N M_L e^{-i\tau} &= I_{L-1} \ddot{\alpha}_{L-1} + K_{L-1} \alpha_{L-1} \\
\alpha_1 M_L + \alpha_2 M_{L-1} + \cdots\cdots + \alpha_L M_1 + \alpha_{L+1} M_N e^{-i\tau} + \cdots + \alpha_N M_{L+1} e^{-i\tau} &= I_L \ddot{\alpha}_L + K_L \alpha_L \\
\vdots \qquad\qquad\qquad\qquad & \qquad \vdots \\
\alpha_1 M_N + \alpha_2 M_{N-1} + \cdots\cdots + \alpha_L M_{N+1-L} + \cdots\cdots + \alpha_N M_1 &= I_N \ddot{\alpha}_N + K_N \alpha_N
\end{aligned}$$

Here, $I_i = mr^2$, $K_i = I_i \omega_i^2$, and $\alpha_i e^{i\omega t}$ are the moment of inertia, torsional stiffness and angular displacement of vibration of the i-th blade respectively, and Mi represents the unsteady aerodynamic moment coefficient induced on the i-th blade by all 1st blades vibrating with the intergroup phase angle τ. Equations (1) can be written as

$$[\, A_{ij} \,][\, \alpha_i \,] = 0 \qquad (i,j = 1,2,\cdots,N) \qquad (2)$$

where

$$A_{ij} = \begin{cases} M_{i-j} & (i>j) \\ M_{N+i-j+1} e^{-i\tau} & (i<j) \\ M_i + I_i(\omega^2 - \omega_i^2) & (i=j) \end{cases} \qquad (3)$$

then ω is a blade vibration frequency and ω_j is a natural frequency of the j-th blade. Given the unsteady aerodynamic moment M_i, the eigenvalue ω/ω_1 and eigenvector $\{\alpha_j\}$ are obtained from Eq.(2) as the flutter frequency and flutter mode respectively. The flutter velocity $U/c\omega_1$ is determined from equation (4).

$$\frac{U}{c\omega_1} = \frac{U}{c\omega}\,\frac{\omega}{\omega_1} = \frac{1}{k}\,\frac{\omega}{\omega_1} \qquad (4)$$

where k is the reduced frequency ($=c\omega/U$, c : blade chord length, U : upstream velocity relative to blades).
Unsteady aerodynamic moment Mi due to the vibration of the 1st blade in each group on stalled condition is calculated[7] by using the unsteady moment in the case of all-blades vibration keeping the interblade phase angle β constant. In order to obtain the latter unsteady moment, the theoretical analysis using the method of singularity by Yashima[6] is utilized since his analytical results qualitatively coincide with experimental results. Mistuning effects on unstalled cascade flutter with an attack angle of $\alpha=4°$ are also surveyed in this paper for comparison. For details, refer to the reference (7).

3. RESULTS AND DISCUSSION

The cascade conditions are fixed such that the attack angle of $\alpha=30°$, torsional axis location of Pc=0.5(midchord position), solidity (chord/pitch ratio) of $\sigma=1.0$ and stagger angles of G=± 30°, 20°, 10°. Mistuning effects of blade arrangements or its combinations on flutter suppression should be surveyed in detail. However, the authors pay attention to the blade group where the blades having two kinds of natural frequencies are alternately arranged, since this particular arrangement noticeably increases the flutter velocity on unstalled conditions[5]. In this paper, the reference blades(1st blades) and adjacent blades(2nd blades) having a different natural frequency are alternately arranged, i.e., two kinds of blades are arranged next to each other to have every other blade identical. The authors presume a group composed of four blades in total, as an actual calculation model, where two sets of the 1st and 2nd blades are arranged and the groups of such blade arrangement are infinitely repeated.

3.1 AERODYNAMIC DAMPING

For basic understanding of flutter of tuned cascade with leading edge flow separation, the aerodynamic dampings obtained from the imaginary part of the unsteady moment are illustrated in Fig.2. The solid line in each figure represents the flutter boundary with $Im(C_M)=0$. Inside the boundary with positive value of $Im(C_M)$ is an unstable area and outside the boundary with negative value of $Im(C_M)$ is a stable area.
Figures 2(A)-(C) show the cases for turbines at stagger angle G= -30°, -20°, and -10°. The maximum reduced frequency kmax on the flutter boundary increases from 1.0 to 1.20 as the stagger angle increases from -30° to -10°. Results for compressors at G=10°,20°, and 30° are shown in Fig.2(D)-(F). The profile of flutter boundary, which extends in the range of interblade phase angle from $\beta=40°$ to 2π, largely differs from that of turbines. However, the maximum value kmax increases from 1.55 to 1.69 with an increase in stagger angle as well as the case for turbine cascades.
The aerodynamic dampings of unstalled cascades, for comparison, are shown in Fig.3 on corresponding cascade conditions. The unstalled cascade is composed of double circular arc blades having a camber angle of 16°. As the stagger angle G increases from -30° to 30°, kmax decreases once, but it increases on the contrary in and after G=10°. This can easily be understood from the geometry of the cascade on the physical plane. This flow field is almost symmetrical with respect to G=0°, and kmax changes correspondingly. The flow around cascade blades is found to be similar to each other at G=30° and G=-30°, for example. On the contrary, kmax continuously increases as G increases from -30° to 30° for stalled cascades as shown in Fig.2. On this point, referring to Fig.4 which illustrates the flow with the wake behind blades of the

stalled cascade used in the present analysis, it is recognized that the flow field is quite different between G=30° and G=-30°. In other words, as the stagger angle increases from -30° to 30°, the wake region in the flow field develops and kmax changes according to this development.

3.2 FLUTTER VELOCITY AND FLUTTER MODE

The flutter velocity and flutter mode such as amplitudes and interblade phase angles are obtained through the eigen equation(3) keeping all cascade conditions unchanged except the blade natural frequency. The mass ratio μ $(=m/1/2\rho c^2)=300$ and radius of gyration $r=\sqrt{0.073}$ during calculations. Mistuning is accomplished by means of arranging two kinds of blades next to each other and changing the natural frequency ratio(ω_2/ω_1) of adjacent two blades, i.e., normalizing the natural frequency ω_2 of the 2nd blade by the natural frequency ω_1 of the 1st one(the reference blade).
The relation between the flutter velocity $U/c\omega_1$ and the intergroup phase angle τ of the tuned cascade($\omega_2/\omega_1=1.0$) when $\alpha=30°$ and G=-30° is shown in Fig.5(A). Figure 5(B) shows the same results from a different view point such as the relationship between $U/c\omega_1$ and the interblade phase angle β. Points a, b, and c in the figures correspond to three points a, b, c on the flutter boundary in Fig.2(A), and the symbol ↑ shows the critical flutter velocity position. The flutter velocity is $U/c\omega_1=4.10$ at point a($\tau=40°$, $\beta=190°$) on the curve in each figure. The flutter velocity decreases as τ and/or β increases until it reaches the point ↑ where $U/c\omega_1$ becomes the lowest(1.002). From this point the flutter velocity continuously increases as τ and/or β increases through the point b($U/c\omega_1=1.085$) to the point c($U/c\omega_1=2.109$, $\tau=340°$, $\beta=355°$). The relation between intergroup phase angle τ and interblade phase angle β of the tuned cascade is as follows.

$$\beta = \frac{2\pi n + \tau}{N} \qquad (n = 0,1,2,\cdots,N-1) \tag{5}$$

where N is a number of blades in a blade group. Accordingly, a change of β from π to 2π in Fig.2(A) or Fig.5(B) corresponds to variation of τ within the span of 4π in Fig.5(A).
By changing the natural frequency ratio ω_2/ω_1, the effects of mistuning on $U/c\omega_1$-τ relationship are shown in Fig.6. The vibration amplitude ratios of A2/A1 between the 1st and 2nd blade are indicated in the figures together with the interblade phase angles β_1 between the 1st and 2nd blade and with β_2 between the 2nd and the successive 1st blade in the adjacent blade group. What is evident from the figures is the critical flutter velocity increases in correspondence to an increase in ω_2/ω_1, which confirms the flutter suppression effect by means of mistuning in blade natural frequency. As might be expected, when the degree of mistuning is extremely low as shown in Fig.6(A), the curves are similar to those of the tuned cascade in Fig.5(A). Accordingly, the flutter mode becomes close to the tuned mode of $A_1=A_2$ or a mode derived from that. However, when ω_2/ω_1 is increased, the curve is divided into two parts as shown in Fig.6(B)($\omega_2/\omega_1=1.05$). One is of the flutter mode of $A_1 < A_2$ where the amplitudes of higher natural frequency blades(2nd blades) are larger than those of lower ones(1st blades). The other is of the flutter mode of $A_1 \gg A_2$ where the amplitudes of lower natural frequency blades(1st blades) are dominant. The critical flutter velocity(the point↑) is obtained on the curve of the latter one. Though the blade natural frequency ratio is increased further as shown in Fig.6(C), the tendency of flutter modes does not change.
The variations of the critical flutter velocity $U/c\omega_1$ and interblade

phase angles β_1, β_2 versus the natural frequency ratio are shown in Fig.7 and Fig.8 respectively. Results for unstalled cascades(α =4°) are also illustrated for comparison in both figures. As the degree of mistuning is gradually developed from the tuned condition(ω_2/ω_1=1.0), the flutter velocity increases rapidly as well as β_1 and β_2 while the corresponding change for the unstalled case is moderate. This implies the possibility of quite effective flutter suppression by slight mistuning. When the natural frequency ratio ω_2/ω_1 exceeds about 1.05 for the stalled case, the flutter velocity rise becomes dull and both interblade phase angles remain almost unchanged. Consequently, there is no discontinuous change both in flutter velocity and in flutter mode in the case of stalled cascades throughout the whole range of ω_2/ω_1 concerned, although for the unstalled case, the discontinuous change in β_1 and/or β_2 is found at the point near ω_2/ω_1=1.35 as shown in Fig.8, which means the discontinuous change in flutter mode.

3.3 EFFECT OF STAGGER ANGLE

As an example for a different stagger angle, the relation between the flutter velocity $U/c\omega_1$ and the intergroup phase angle τ of the tuned cascade(ω_2/ω_1=1.0) with G=30° is shown in Fig.9(A). Figure 9(B) shows the same results in a different way using the relation between $U/c\omega_1$ and the interblade phase angle β. On the compressor side of stall cascades, β varies within the span of about $3\pi/2$ as shown in Fig.9(B) corresponding to Fig.2(F), while it varies within the span of about π on the turbine side. As is evident from equation(5), the span of variation in τ accordingly grows to 6π as shown in Fig.9(A), whereas it is about 4π on the turbine side. In spite of such difference mentioned above, the aspect of flutter mode on the compressor side is similar to that on the turbine side(G<0°), as shown below.
The variations of the amplitude ratio A_2/A_1, interblade phase angles β_1 and β_2 between adjacent blades, and critical flutter velocity $U/c\omega_1$ of stalled cascades for various stagger angles versus the blade natural frequency ratio ω_2/ω_1 are shown in Figs.10, 11, and 12 respectively. As ω_2/ω_1 increases from 1.0, A_2/A_1 for each stagger angle rapidly decreases and approaches to 0 where the amplitudes of lower natural frequency blades(1st blades) are dominant as shown in Fig.10. Both β_1 and β_2 accordingly show sharp changes and after that they keep almost constant indicating no significant change in the flutter mode as shown in Fig.11. Consequently, regardless of the stagger angle, the flutter velocity rapidly increases to show a considerable flutter suppression effect according to an increase in mistuning such as the imperfection of blade natural frequency up to around 5% as shown in Fig.12. Beyond this region of mistuning, the flutter velocity remains almost unchanged accompanied by no change in flutter mode as mentioned above, which means such amount of mistuning brings no improvement on flutter suppression. It is also understood from Fig.12 that the flutter velocity overall increase as the stagger angle decreases.
Figure 13 shows the flutter velocity ratio which is obtained by normalizing the flutter velocity of the mistuned cascade with that of the tuned one. The relevant results for unstalled cascades are shown in Fig.14 with stagger angles of G=± 30° and ± 20°. In Fig.14, the symmetrical aspect is seen, which in Section 3.1 is explained as the geometric symmetry of the flow field with respect to the stagger angle of G=0° for unstalled cascades. In this figure, a similarity can be observed between the results of G=-30° and 30°, and the curve for G=-20° is similar to that for G=20°. On the other hand, for the case of stalled cascades shown in Fig.13, the flutter velocity ratio generally

increases as the stagger angle decreases.
In addition, the flutter velocity ratio for stalled cascades ranges from 1.0 to around 1.7, and it ranges from 1.0 to around 2.0 for unstalled cascades. Thus, no significant difference is seen between stalled and unstalled case from the view point of the rate of flutter velocity rise, although the flutter velocity itself noticeably differs between two cases as shown in Fig.7. Further, the flutter velocity ratio in the vicinity of $\omega_2/\omega_1=1.0$ rapidly increases according to an increase in natural frequency ratio in Fig.13, while it moderately increases in Fig.14. Special emphasis is to be placed on such fact that more remarkable flutter suppression effect is expected by slight mistuning for stalled cascades than for unstalled ones.

4. CONCLUSIONS

Mistuning effects on fully stalled cascade flutter in torsional mode have been investigated by changing the natural frequency of blades in the case of the alternating arrangement which presents a significant flutter suppression effect for unstalled cascades. The conclusions of the present study are summarized as follows:

1. The flutter suppression effect by mistuning in blade natural frequency is confirmed for stalled cascades as well as for unstalled ones. Furthermore, an emphasis is to be placed on the fact that more remarkable flutter suppression effect is expected by slight mistuning for stalled cascades than for unstalled ones.
2. There is no discontinuous change in flutter mode in the case of stalled cascades throughout the whole range of mistuning concerned except in extremely close vicinity of tuned condition, while the discontinuous modal change is seen in unstalled cascade flutter.
3. Mistuning has more beneficial effect on flutter in turbines than in compressors.

REFERENCES

(1)Whitehead,D.S.,Great Britain A.R.C. R&M 3429,1964.
(2)Bendiksen,O.O.,J.AIAA,Vol.4,No.2,1988,164.
(3)Tanaka,H.,Hanamura,Y.and Hattori,S.,Trans.JSME,41-345,1975,1437.
(4)Hanamura,Y.and Tanaka,H.,Proc.'77 Tokyo Joint Gas Turbine Congress, 1977,418.
(5)Tanaka,H.,Hanamura,Y.and Fujimoto,I.,Proc.'83 Tokyo International Gas Turbine Congress,Vol.2,1984,651
(6)Yashima,S.and Tanaka,H.,Trans.ASME,J.Engg.Power,100-2,1978,317
(7)Tanaka,H.and Fujimoto,I.,Proc.4th International Symposium on Unsteady Aerodynamics and Aeroelasticity of Turbomachines and Propellers,(held in Aachen), 1987,547.

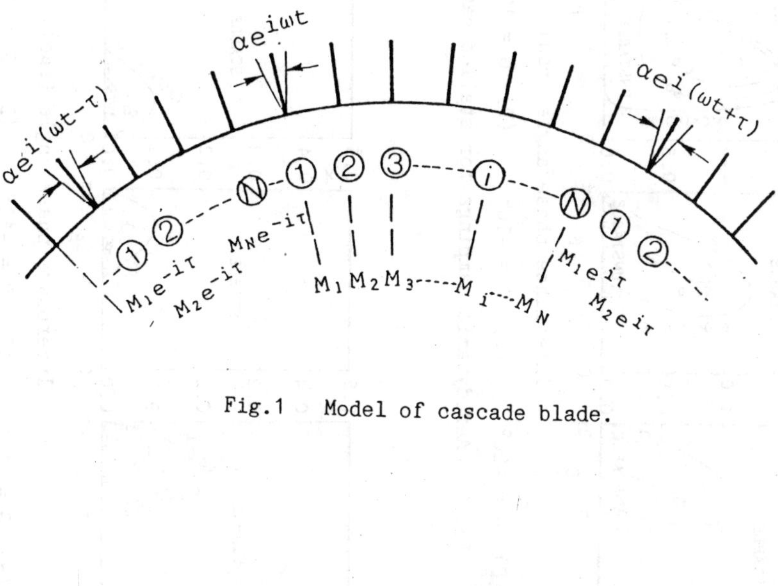

Fig.1 Model of cascade blade.

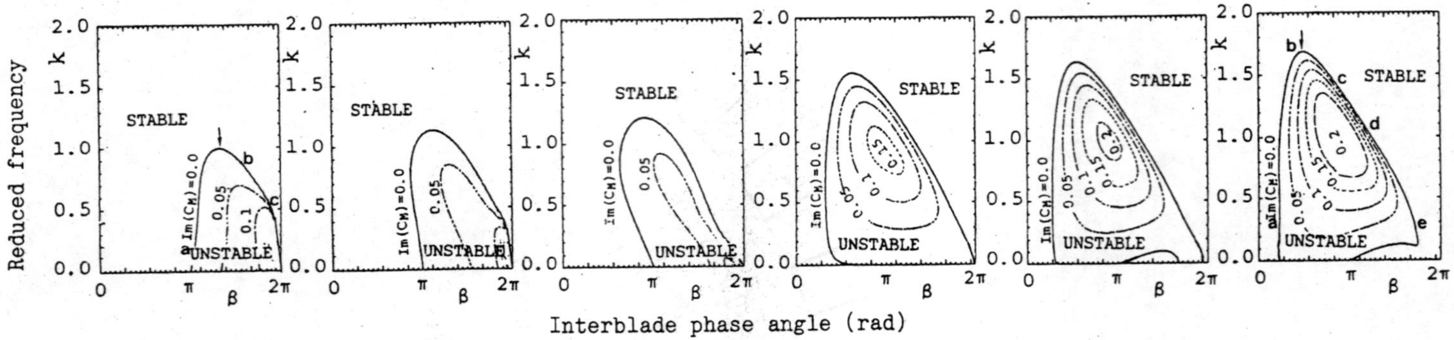

(A) G = -30° (B) G = -20° (C) G = -10° (D) G = 10° (E) G = 20° (F) G = 30°

Fig.2 Aerodynamic dampings for stalled cascades.

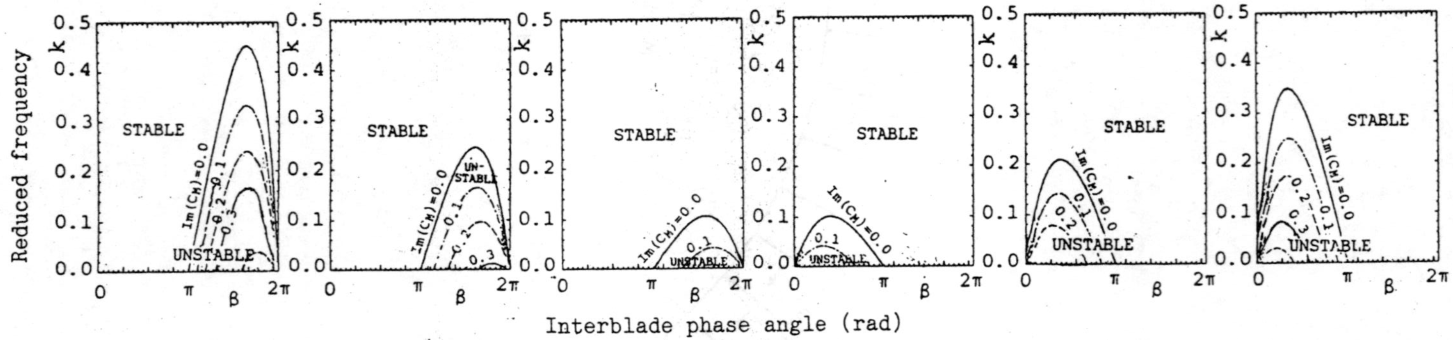

(A) G = -30° (B) G = -20° (C) G = -10° (D) G = 10° (E) G = 20° (F) G = 30°

Fig.3 Aerodynamic dampings for unstalled cascades.

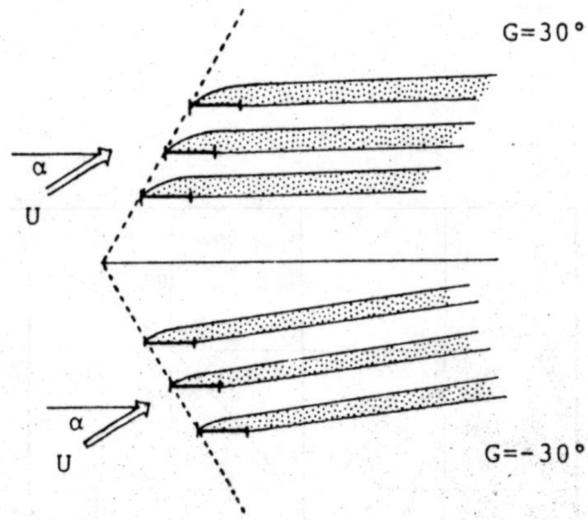

Fig.4 Flow field with the wake behind blades of stalled cascades.

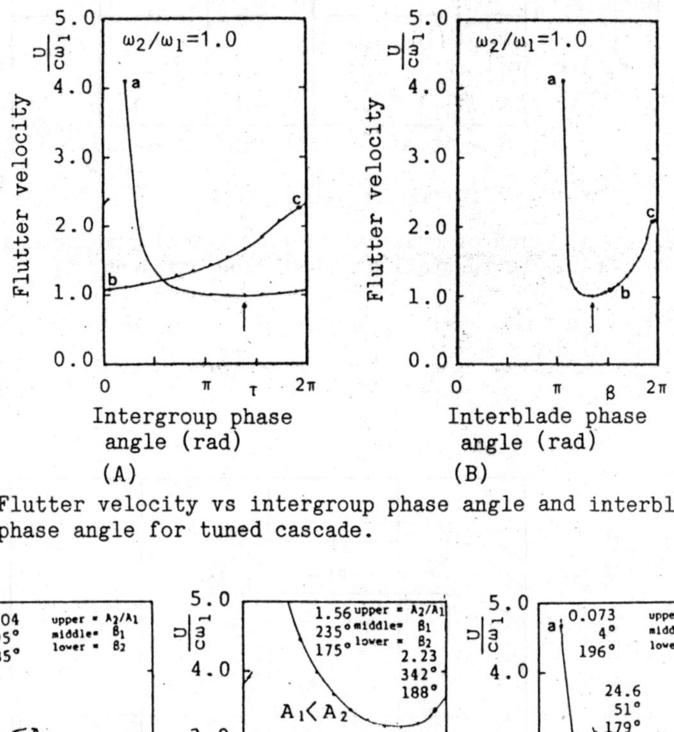

Fig.5 Flutter velocity vs intergroup phase angle and interblade phase angle for tuned cascade.

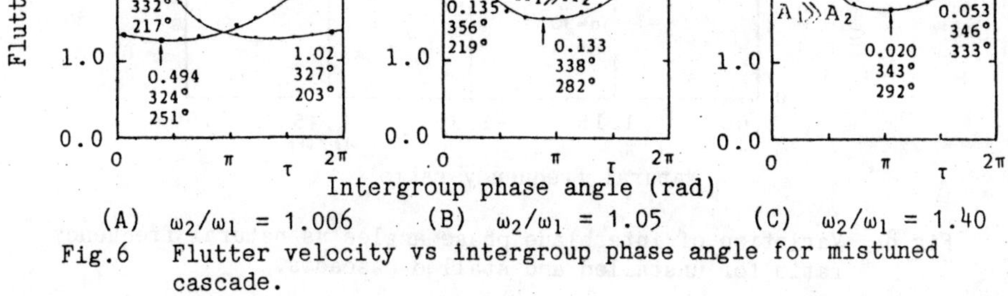

(A) ω_2/ω_1 = 1.006 (B) ω_2/ω_1 = 1.05 (C) ω_2/ω_1 = 1.40

Fig.6 Flutter velocity vs intergroup phase angle for mistuned cascade.

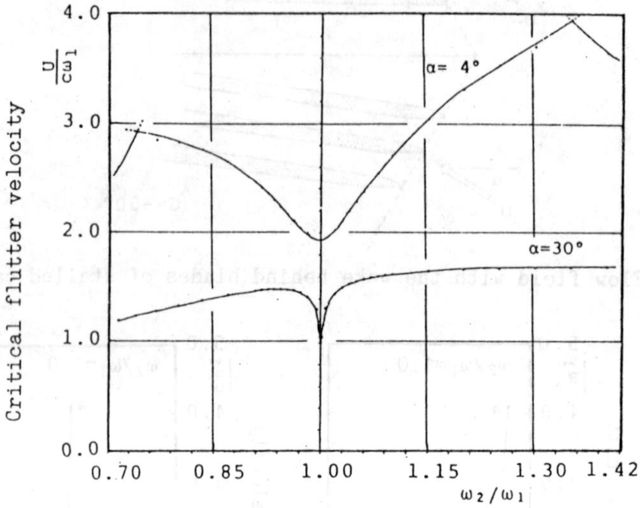

Fig.7 Variation of critical flutter velocity vs natural frequency ratio for unstalled and stalled cascades.

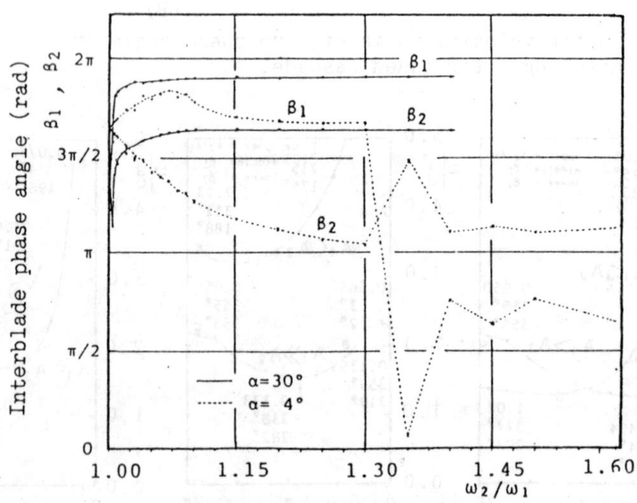

Fig.8 Variation of interblade phase angles vs natural frequency ratio for unstalled and stalled cascades.

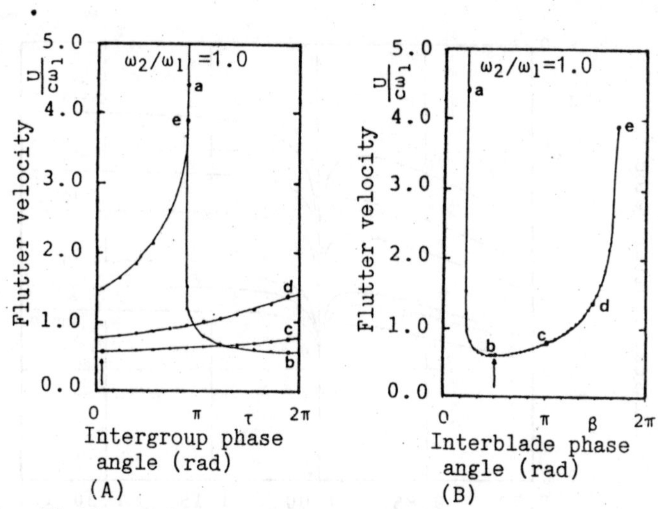

Fig.9 Flutter velocity vs intergroup phase angle and interblade phase angle for a tuned cascade.

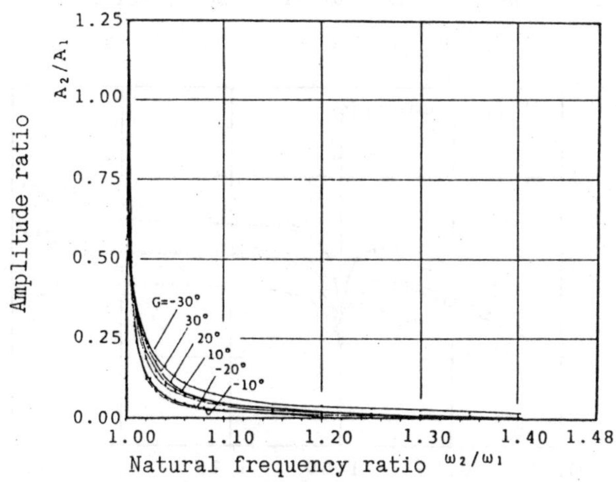

Fig.10 Variation of amplitude ratio vs natural frequency ratio.

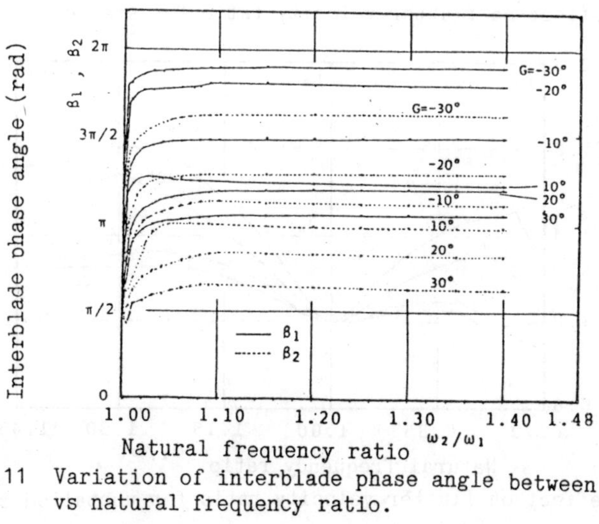

Fig.11 Variation of interblade phase angle between adjacent blades vs natural frequency ratio.

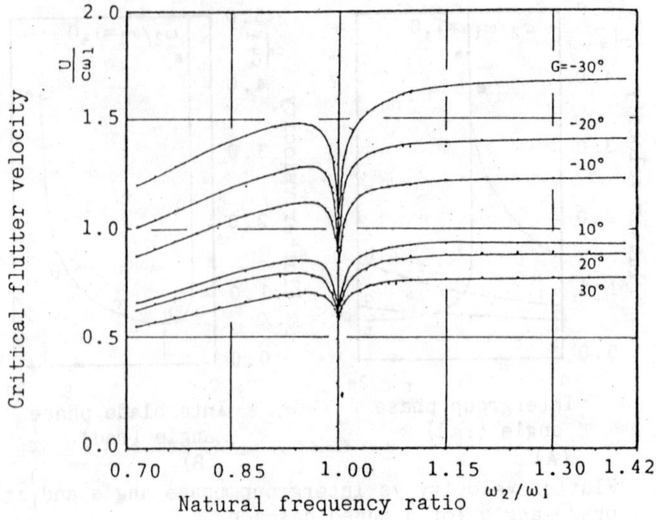

Fig.12 Mistuning effect on critical flutter velocity for stalled cascades.

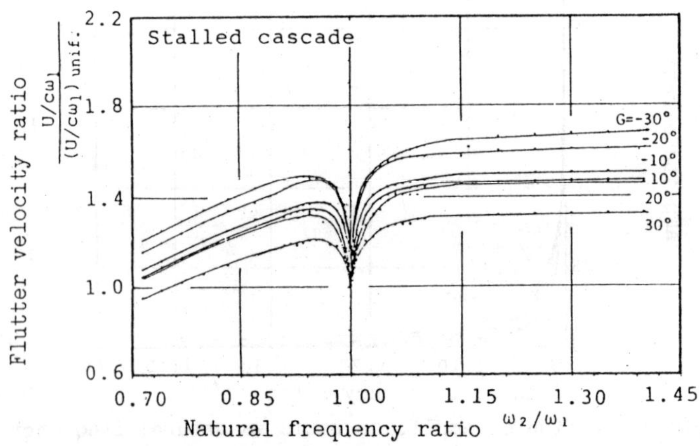

Fig.13 Mistuning effect on flutter velocity ratio for stalled cascades.

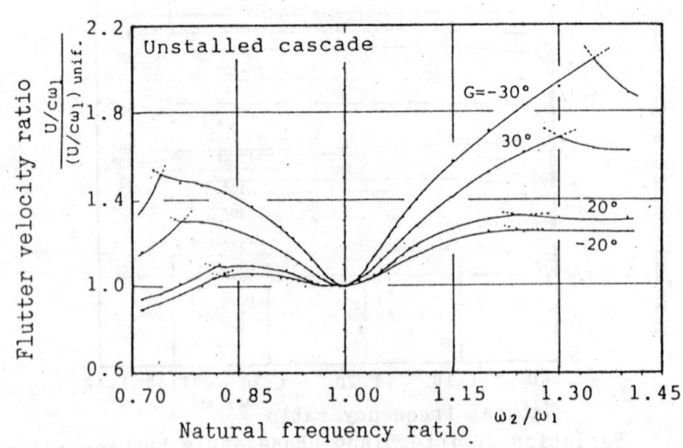

Fig.14 Mistuning effect on flutter velocity ratio for unstalled cascades.

Part Four
Aeroacoustics

AN ANALYSIS OF CAVITY RESONANCE IN THE AEROENGINE CASING DURING RIG TESTING

Wang Tongqing, Sheng Yuansheng
(Shenyang Institute of Aeronautical Engineering, China)

Xu Ejun
(Shenyang Aeroengine Research Institute, China)

ABSTRACT

A pure tone which occured in the prototype aeroengine of a turbojet during rig testing was studied in this paper. It has been pointed out that the pure tone arises from the Helmholtz resonance in a ring cavity outside of the compressor casing which is originally designed to reduce unstability of air flow during engine transient process in a way to match the high–and low–pressure compressor. When the frequency of cavity resonance is in coincidence with that of casing structure resonance such tone will be strongly radiated. For studying such a phenomenon, the formulas of acoustic radiation impedance of a circular hole on the infinitely rigid plate and the mathematic model of Helmholtz resonance in the ring cavity with multi holes were explored. The consequences of calculation are comformable to the results measured in rig testing.

PHENOMENA OBSERVED

During the rig testing of the prototype of a turbojet engine a strange sound was observed when the engine ran up to the speed about 93% of its maximum design speed and the sound was disappeared slowly when it passed over 94% of its maximum speed. Being considering that the sound might be, perhaps, due to the flutter of the compressor blades, the tester dared not shift the throttle further to increase the engine speed until such a sound was identified to be a pure tone with the frequency of 392.5HZ radiated from the compressor casing . It is seen that, the frequency component of 392.5HZ appeared in the 3–D sound power spectrum measured near the compressor, Fig 1, while the strange sound was radiated; meanwhile, the component of same frequency also appeared in the cross power spectrum between the fluctuating flow pressure measured and the sound pressure measured, as well as in the flow pressure spectrum, shown in Fig 2,3. The figures showed that the strange sound is a tonal component of 392.5HZ. After the hole on the liner of the ring cavity structure of the engine casing, shown in Fig 4, which were originally designed for the pheumatic matching between the high– and low–pressure compressor, were blocked the strange sound was disappeared. Thus, it was demonstrated that such a pure tone was produced from the ring cavity of the engine casing.

ANALYSIS & MATHEMATICAL SOLUTION

The ring cavity of the compressor casing is located at the conjunction of the high– and low–pressure compressor and there are 41 circular holes having the diameter of 16mm on its liner connecting the flow path with the cavity. The cavity noise is an important constituent part of airplane noise. Since 1950s many researchers have studied this subject theoretically and experimentally.In references 1–4 the studies on the cavity noise induced by the stream passing through a rectangular cavity and the empirical formulas were presented, which predict the cavity noise by the vortical–Acoustic oscillation model for shallow cavity and by standing wave model for deeper cavities, and the results given by which agree perfectly with that of experiments. In the case being analyzed, for the ring cavity with holes, the ratio of the surface area of holes to the volume of the cavity is larger when comparing with the rectangular cavity, it can be considered as a Helmholtz Resonantor.

The complex impedance Z for a Helmholtz Resonator with one hole, by definition, is

$$Z = \frac{P}{Q} \tag{1}$$

Where

P—the sound pressure in the hole

Q—the volume flow

And $Z = Z_A + Z_{ri} + Z_{ro}$ (2)

Where Z_A is the Helmholtz resonance impedance due to the inertial of air mass in the cylindrical volume of the hole and the elastisity of air in the cavity. Zri and Zro are the impedances due to sound radiation propogating in and out the cavity.

$$Z_A = \frac{\rho l \omega}{\pi a^2} i + \frac{\rho C^2}{\omega V i} \tag{3}$$

Where

ρ — the mass density of air

l — the length (i.e depth) of the hole

a — the radius of the hole

V — the volume of the cavity

C — the sound speed

ω— the angular frequency of the sound wave.

E.Q (3) is valid only if

(1) a. l and $\sqrt[3]{V} << \lambda$,where λ is wave length.

(2) $l \cdot \pi a^2 << V$

(3) the wall of the cavity is rigid

The average rediation impedance of a circular piston vibrating on a infinitely rigid plane can be calculated by a double integration. 6

$$Z_r = \frac{i\omega\rho}{2\pi^3 a} \iint_{s}\iint_{s'} \frac{e^{\frac{-ika}{\beta}}[D - M(x' - x)]}{D} dxdydx'dz' \tag{4}$$

where x,y; x′,y′ are the non–dimensional co–ordinates of the area elements on the piston surface, the x–axis is oriented in the direction of air flow, s and s′ are the areas of the piston surface,and

$$\begin{cases} K = \omega / c \\ \beta^2 = 1 - M^2 \\ D = \sqrt{(x - x')^2 + \beta^2 (y - y')^2} \end{cases} \tag{5}$$

By using special trick in integration associated with the Taylor expansion for the exponential fuction of the integrand, the following equation was obtained.

$$Z_r = \frac{16\rho F(M)}{3\pi^3 a}\omega i + \frac{1}{2\pi}\frac{\omega^2 \rho}{\beta^2 C} \tag{6}$$

The formula is of second order appoximation, when $\frac{ka}{\beta} << 1$.

Where F(M) is the elliptic integral of first kind.

$$F(M) = \int_0^{\frac{\pi}{2}} \frac{d\theta}{\sqrt{1 - M^2 \sin^2\theta}} \tag{7}$$

Inside the cavity, M = O, thus

$$Z_{RI} = \frac{8\rho\omega i}{3\pi^2 a} + \frac{\rho\omega^2}{2\pi C} \tag{8}$$

Outside the cavity, there is flow of Mach number M in X–direction, it is obtained that.

$$Z_{RO} = \frac{16\rho F(M)\omega i}{3\pi^3 a} + \frac{\rho\omega^2}{2\pi\beta^2 C} \tag{9}$$

Substituting (3), (8), (9) into (2), then

$$Z = \frac{\rho\omega^2}{2\pi\beta^2 C} + \frac{\rho\omega^2}{2\pi C} + i(\frac{16\rho F(M)}{3\pi^3 a} + 8\rho / 3\pi^2 a \quad + \frac{\rho l}{\pi a^2})\omega + \frac{\rho C^2}{V\omega i} \tag{10}$$

Let

$$\begin{cases} R_T = \frac{\rho\omega^2}{2\pi\beta^2 c} + \frac{\rho\omega^2}{2\pi c} \\ L_T = \frac{16PF(M)}{3\pi^3 a} + \frac{8\rho}{3\pi^2 a} + \frac{\rho l}{\pi a^2} \\ C_T = \frac{V}{\rho c^2} \end{cases} \tag{11}$$

Thus

$$Z = R_T + L_T\omega i + \frac{1}{C_T\omega i} \tag{12}$$

Where R_T,L_T and C_T are the electrical analogue acoustic resistance, inductance, capacitance respectively. For the ring cavity of the engine casing, there are 41 circular hole with the same diameter on the liner. With the assumption of equal sound pressure in all the hole, the total impedance of the ring cavity from the electrical analogue acoustic circuit, shown in Fig 5, is obtained as follows.

$$Z_T = \frac{1}{41}(R_T + i\omega L_T) + \frac{1}{iC_T\omega} \tag{13}$$

Thus,theHelmholtzresonantfrequencyofiheringcavityis

$$\omega_r = \sqrt{\frac{41}{L_T C_T}} \tag{14}$$

The pressure ratio of that in the cavity and outside the holes (i.e. in the compressor flow path) can be calculated as follows.

$$\frac{P_c}{P} = \frac{1}{\sqrt{(1-\frac{\omega^2}{41}L_T \cdot C_T)^2 + (\frac{\omega}{41}R_T \cdot C_T)^2}}$$

$$= \frac{1}{\sqrt{[1-\frac{V\omega^2}{41c^2}(\frac{16F(M)}{3\pi^3 a} + \frac{8}{3\pi^2 a} + \frac{l}{\pi a^2})]^2 + [\frac{V\omega^3}{82\pi C^3}(\frac{1}{\beta^2+1})]^2}} \tag{15}$$

When $\omega = \omega_m$ $(\omega_m = \omega_r)$,Pc / P reaches its maximum volume. Where Wm is the frequency at which the pressure oscillation in the cavity is the largest volume. If that frequency coincides with the natural frequency of the structure, then structure resonance will be excited and a strong resonant sound of pure tone will be generated.

RESULTS OF CALCULATION AND DISCUSSION

In order to prove the analytical results derived above, the structure frequency was measured by hammer excitation and the spectrum is shown in Fig 6. A peak component with the frequency of 393.5 HZ is found in coincidence with that of the strange sound.The caculation of the cavity resonant frequency is as follows. The geometrical parameters of the ring cavity area = 0.008m, V = 0.0058m, l = 0.003m.

Substituting into (15), then

$$\frac{P_c}{P} = \frac{19678200}{\sqrt{[19678200 - \frac{\omega^2}{41}(F(M) + 33.774 + 14.9)]^2 + [\frac{0.000468\omega^2}{41}(\frac{1}{\beta^2}+1)]^2}} \tag{16}$$

The Mach numver of air flow measured at the exit of the 3rd stage of the compressor is shown is Fig7, which is varied with engine speed. While the engine is operating at the 93% of its maximum speed, the M of the exit flow from Fig.7 is about 0.7. Thus from (16), when P / P takes its largest value of 20.2 the frequency obtained is W = 480HZ.When comparing the frequency calculated with that has been measured it shows an error of about 22%. The following discussion, perhaps, will be the reasons of the error.

(1) The gaps and clearance between the assembled parts of the cavity forms the leakage of air which cause volume increasing.

(2) Some holes was blocked by measuring sensors during testing, thus the actural number of hole was reduced.

(3) The Mach numeber employed in the calculation is n / t exactly that outside the holes.

(4) The approximation of real boundary conditions and flow also cause errors in calculation.

If the volume of the cavity is increased 10% and the 10 holes out of 41 holes are blocked, the resonance frequency calculated will be 395H.

CONCLUSTION

(1) The calculation of the resonance frequency of the cavity with holes derived by using Helmholtz resonance theory and acoustic radiation impedance formula of the air flow passing through holes can be used to explain the resonance phenomena occured in ring cavity of engine casing.

(2) The resonance frequence is ralated to the geometry configuration of the cavity and its holes as well as the Mach number of air flow. The Mach Number increases as the engine speed increases, and the resonance frequency decreased. While the resonaut frequency coincides with the structral mode frequency the structure resonance occours and a strong pure tone excites.

(3) The resonance frequency calculated by the derived formula is 22% lorger than the frequency measured during engine testing, the further study on exploring the differences theoretically and experimentally are needed.

REFERENCES

1. Patricia J. W. Block, Noise response of cavities of varying dimensions at subsonic speeds, NASA TN–D–8351
2. L..F.East, Aerodynamically induced resonance in rectagular cavities, J.S.V. (1966) 3(3) P277–287
3. Hanno H.Heller and Donald B.Bliss, The physical mechanism of flow–induced pressure fluctuations in cavities and concepts for their suppression, AIAA paper 75–491.
4. P.J.W.Block, measurement of the tonal component of cavity noise and comparison with theory, NASA TP 1013
5. A.P.Dowling and J.E.FFOWCS WILLIAMS, Sound and Sources of Sound, Ellis Horwood limited . Publishers. Chichester (1983) P130–131
6. Plumblee, H.E.Gibson. J.S. and Lassiter. L.W., A theoretical and experimental investigation of the acoustic response of cavities in an aerodynamic flow, WADD TR–61–75

AEROELASTIC VIBRATIONS OF TURBOMACHINE BLADES AND PROPELLERS

Yu. S. Vorobiev, M.L.Korsunskiy
(Ukr. SSR Academy of Sciences, Institute for Problems in Machinery, Kharkov , USSR)

ABSTRACT

Vibrations of the blade system are considered accounting the interaction of blades through the flow and elastic bonds. Applicability of different mathematical models, excitation of different vibration modes mistuning effect and the other factors are discussed

INTRODUCTION

The blade system in flow is the common system. It is convenient to describe the flow effect on the blade cascade with the help of disturbing polyharmonic loads and forces of aerodynamic damping. Depending on geometric configuration the blade is considered on the basis of the refined theory of twisted rods or as tree–dimensional body I. At aeroelastic vibration of burbomachine blades and propellers the lower modes usually are excited. So the fields of application of rod models usually are enough broad. However, even with these modes of blade vibration the deformation of the elongated cross–section in its plane is possible including also with the blades with intermediate bonds (Fig. 1). In these cases rod models can not be used I.

MATHEMATICAL MODELS AND RESEARCH METHODS

The blades are considered in the rectangular right coordinate system xyz with the origin 0 in the root cross–section, rotating together with the blade with the angular velocity Ω . The axis z is directed along a radius from the axis of rotation, the axis x is parallel to the axis of rotation (Fig. I). While using the theory of twisted rods, the local coordinate system xy are central, and $\xi\eta$ are main central ones in any cross–section. The corresponding systems x_1y_1 and $\xi_1\eta_1$ have an origin in the center of bending of the cross–section. An angle α between $x$and ξ is the angle of pretwisting of the cross–section. Further, dashes will denote the derivative along the coordinate z and points over functions will denote the derivative along the time t. Displacements of the elementary volume of the rod in directions x,y,xbased on the refined theory of rods will be

$$U = u_1 + u_2 - \theta y_1;$$
$$V = v_1 + v_2 + \theta x_1; \qquad (1)$$
$$W = w - u_1'x - v_1'y + u_2'x + v_2'\psi + \theta'\omega,$$

where $u_1(z,t)$,$v_1(z,t)$are displacements of bending, $u_2(z,t)$,$v_2(z,t)$are displacements of the shift in the directions x and y; $w(x,t)$is the displacement in the direction z; $\theta(z,t)$is the angular displacement with respect to the center of bending; $\varphi(\xi,\eta,z)$is the function of torsion; $x,\psi(\xi,\eta,z)$is the function of warping of the shift.

In the practical problems warping of the cross–section is accounted for by the definition of tangent deformations and stresses 2. Deformations of the rod finally are taken as 1,2:

$$\varepsilon_z = w' - u_1''x - v_1''y + u_2'' + v_2\psi + \theta''\varphi + \theta'\tau_o r_1^2;$$

$$\gamma_{xz} = \theta'\left(\frac{\partial\varphi}{\partial x} - y_1\right) + u_2^1\left(1 + \frac{\partial x}{\partial x}\right) + v_2^1\frac{\partial\psi}{\partial x}; \tag{2}$$

$$\gamma_{yz} = \theta'\left(\frac{\partial\varphi}{\partial y} + x_1\right) + u_2^1\frac{\partial x}{\partial y} + v_2^1\left(1 + \frac{\partial\psi}{\partial y}\right); \qquad \varepsilon_x = \varepsilon_y = \gamma_{xy} = 0,$$

where $\tau_0 = \alpha'$is pretwisting; $r_1^2 = \xi_1^2 + \eta_1^2$

The expressions (1) represent the inertial interconnection of bending and torsion due to mismatch of the center of mass and the center of the cross–section bending and the expression (2) represents deformation interconnection of different types of displacements. While defining stresses, the internal damping in the material is taken into account 2. Variations of the work of inertial forces at deformation of the rod holds:

$$\delta\Pi = \int_o^l \int_F \int [\sigma_z \delta\varepsilon_z + \tau_{xy}\delta\gamma_{xy} + \tau_{yz}\delta\gamma_{yz}]dFdz,$$

where l is blade length, F is the area of the cross–section. The kinetic energy of the rotating rod is the sum of the kinetic energy of vibrations and works of centtrfugal and the Koriolis forces:

$$T = \frac{1}{2}\int_o^l \int_F \int \rho\{[\dot{U}^2 + \dot{V}^2 + (\dot{W}^2 - \dot{\zeta}^2)] + \frac{\Omega^2}{2}[(R + z + W - \zeta)^2 + (V + y)^2] + \Omega[(V + y)(\dot{W} - \dot{\zeta}) - (R + z + W - \zeta)\dot{V}]\}dFdz,$$

where ρ is the density of blade material, ζ is the change of projection of longitudinal fiber on the axis z at its deformation.

Interblading bonds are considered as rods, displacements of ends of which are defined by displacements of the corresponding cross–secti–ons of the blades. Bonds experience space bending, torsion and tension. Forces of structural damping are supposed to be proportional to the coefficients of structural damping and relative movements of the corresponding blade cross–sections I. A disc is simulated as a round plate of variable thickness executing bending–torsion–radial vibrations in the general case, rim and bush are simulated as curved beams. The rotor of disc–drum construction may be presented as a set of round plates, shells of rotation and curved beams.

Each harmonic of the disturbing load of the flow is defined by the frequency w_k and variable amplitude $q_K(z)$ and phase $\varphi_K(z)$in blade length, wnere K is harmonic number 2. The aerodynamic interaction of profiles in the cascade are defined by the forces, that are expressed by the aerodynamic effect coefficients K_{jK},e.g.:

$$Q_{\eta n} = Q_0 \sum_K [K_{\xi k}\frac{U_{\xi,n+K}}{b} + K_{\eta K}\frac{V_{\eta,n+K}}{b} + K_{\theta K}\theta_{n+K}],$$

where b is cross–section chord, U_ξ,V_η are displacements in the direction of axes ξ,η,η+K are numbers of interaction blades 3. In the practical cases it is sufficient to take the account of the interaction of neighbouring blades (K = 1,0,–1).

The problem for nonconservative system is solved by the variational method that differs by the fact, that not displacements, but moments and forces are the values being approximated. The basic functions satisfy force boundary conditions. Displacements and their derivatives are determined by integration, e.g.:

$$U_1' = [l_\eta M_\eta \cos\alpha - l_\xi M_\xi \sin\alpha]_{z=0} + \int_0^z \left(\frac{M_\eta}{EJ_\eta}\cos\alpha - \frac{M_\xi}{EJ_\xi}\sin\alpha\right)dz,$$

where M_ξ, M_η are bending moments, J_ξ, J_η are moments of inertial, E is elasticity modulus, l_ξ, l_η are the coefficients of blade root compliance.

So all the boundary conditions are met and at calculations the numerical integration is mainly carried out, that allows to solve non–conser vative problems and to improve the convergency of the methods. Besides, the consequent complication of the system is used by choosing the solutions of the partial problems for the basic functions. It results in decreasing a number of variable parameters at each stage of the calculation to a significant extent and convergency and stability of the computational process are greatly increased. If all the blades in the system are supposed to be equal, then the assumptions on turning assymetry of vibration modes of the system may be used and the problem may be simplified. If small technological deviations of blade sizes are taken into account, intrapackage vibrations are considered or assumptions on turning symmetry of vibration modes of the system in flow are not introduced, then each blade has own set of basic funcitions and the problem is greatly complicated. In this case the calculation of the system in parts plays a great role. To make the choice of the mathematical model of the blade at the concret calculations easier, the sets of the boundaries of the application fields for these models depending on nondimensional parameters, representing the main geometric configurations of the object and the account of the different factors (pretwisting, non–symmetry of the cross–section, shift, warping of the cross–section, etc).

Application fields of rod models for problems of aeroelasticity are enough broad. However, they are restricted by the parameters of the blades, at which the "shell" vibration modes occur or discrepancy of theoretical and experimental results takes place due to cross–section deformation, even at "rod" modes of vibrations. So, in Fig. 2 the boundaries of application fields of the rod models are shown depending on the angle of twist of species α at $\frac{b}{l} = 0.25$ and relations $\frac{b}{l}$ for species with the linear change of thickness.

The three–dimensional body is the most suitable mathematical model for a number of blades. An isoparametric hexahedral curvilinear element HK 60 with square–law for a function of displacements has been chosen that is important for the description of stresses (Fig. I). The element has 20 nodes and belong to the Serendip family. The finite element used allows to describe blades of the more complicated configuration with a bevel, flange, blade rotor and cooling channels. For some blades it is necessary to account blade root deformation. It should be noted, that the given method is convenient while calculating thinwall compressor blades. In a number of cases the calculation of blades for vibrations follows the determination of its temperature field. These results are used to calculate the given geometric performances of the blades for rod modes or the account of inhomogeneity of the material using the finite element method. To account flow effect on the finite element model of the blade the aerodynamic coefficients of effect and data on the distribution of the nonstationary pressure on the blade may be uses 4.

SOME INVESTIGATION RESULTS

The methods developed, open the broad possibilities to investigate vibrations of turbomachine blade system. The comparison of theoretical and experimental results for the blade and the blade system has been carried out I. As an example in Fig. 3 are given theoretical (446.6Hz and experimental (442.5 Hz) modes of displacements v and stresses σ at the leading edge for the model package of five blades with the given mistunig. The data of the experiment is maked by points.

For comparison the modes of packages of identical blades (425.2 Hz) are shown by a dashed line. The great effect of mistuning and a good correspondence of theoretical and experimental results are seen even in such a complicated case.

The calculation of vibration of blading with cricular flange bond is carried out on the basis of the experimental data on the circumferential nonuniformity of the flow I. Distributions of displacements and stresses in blade length at the leading edge are shown by solid lines in Fig. 4 and their phases at vibrations with three nodal diameters are shown by dash–and–dot lines. The location of circular bond is shown. Displacements relate to the value $v(1)$, and the stress relates to $\sigma(0)$. The data of strain measurement correspond to vibration mode, that is practically resonant. The phase shift between bending and torsion and also in blade length, and also the effect of the bond on vibration mode are seen.

Sets of such calculations allow to construct amplitude–frequency performances of the system, that show comparative excitation of different modes of vibration and the possiblility of self–vibration. Figure 5 gives amplitude–frequency performances of a package of five blades (a solid line) and a package with mistuning (dashed lines) within the given range of frequencies.

Aeroelastic bibration modes, as a rule, correspond to the lower rod modes of blade vibration. The vibration of thin–wall peripheral section close to local shell modes may be excited in flow of blades of powerful turbomachine last stages, the cross–section of which varies greatly from root to periphery. The peripheral section of these bladings is the cascade of small solidity and sometimes their calculation of unit profiles in flow is possible. For aeroelastic vibration of blades in flow the phase shift plays the large role between displacement and incidence angle of profile. It defines the instantaneous angle of incidence, that effects aerodynamic forces at the given moment. The Koriolis forces, arising at vibration of rotating rod effect also phase shift between these displacements 1,2. Their effect may be clearly demonstrated by an example of unit rod (propeller type) vibration in the nondisturbing axial–symmetric flow. For the rod of two–symmetrical cross–section the Koriolis forces cause the phase shift between bending and torsion $\varphi = \pi / 2$. With increasing the parameter $\rho_{\xi} = \xi_s / r_p$ (ξ_s is the distance between the center of mass and the center of bending, γ_p is a radius of cross–section inertia) the phase shift both for the first (φ_1) and for the second modes (φ_2) decreases that influences on the critical velocity of flutter (Fig. 6).

Centrifugal forces increase the critical velocity of flutter V_k (V_{k0} is the critical velocity of the nonrotating rod). For non–twisted rod the Koriolis forces increase, and for the twisted rod ($\alpha = \pi / 6$) decrease the critical velocity.

The curve I refers to the case of large increase of parameters of nonsymmetry of the rod(polar–axial moments of intertial 1,2) by five fold.

The calculation results show, that at vibration of the blade system in flow the role both of the inertial and deformation connections between different vibration modes is sufficient; mistuning of the system appreciably influences on excitation of vibration modes; some factors(such as e.g., the Koriolis forces, that may be neglected while investigating the free vibration should be taken into account.

REFERENCES

1. ВоРобеВ Ю.С. КолебаНИЯ ЛоПаТоЧНоГо аППаРата Турбомащин. Киев: Наук. дма(1988), 224.

2. Воробев Ю.С., Щорр Б.Х. Теория закрученных Стержней. Киев: Наук. думка(1983), 188.

3. Горобьев Д.Н., Курзин В.Э., Сарен В.Э. Аэродинамика Решеток В Нестационарном Потоке, Новосибирск: Наука(1971), 272.

4. Соколокийя Г.А., Гнесин В.И. Нестационарные Трансзвуковые и Вязкие Течения В Турбомацин ах. Киев: Наук>думка(1986),264.

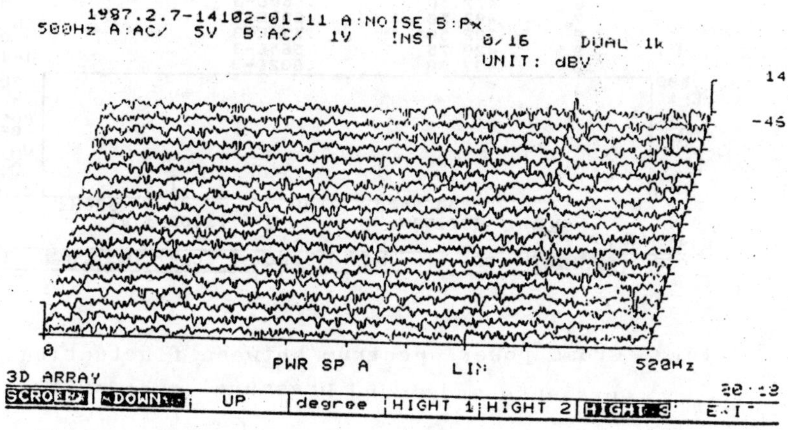

Fig1. 3-D Pweer spectrum of sound pressure, speed=93%

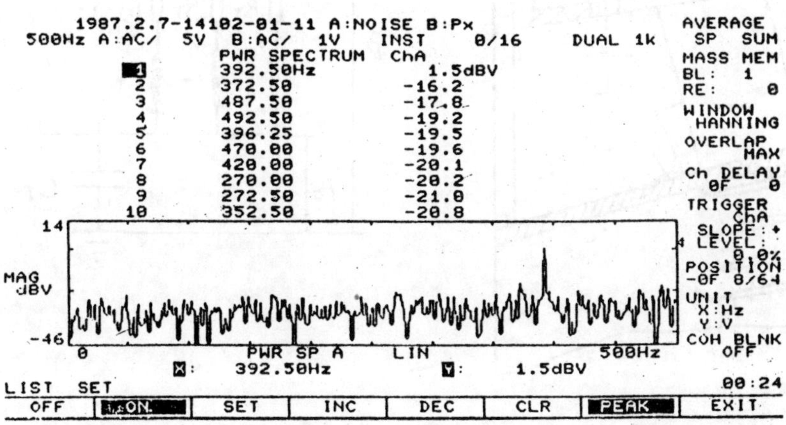

Fig2. Fluctuating flow pressure spectrum, speed=93%

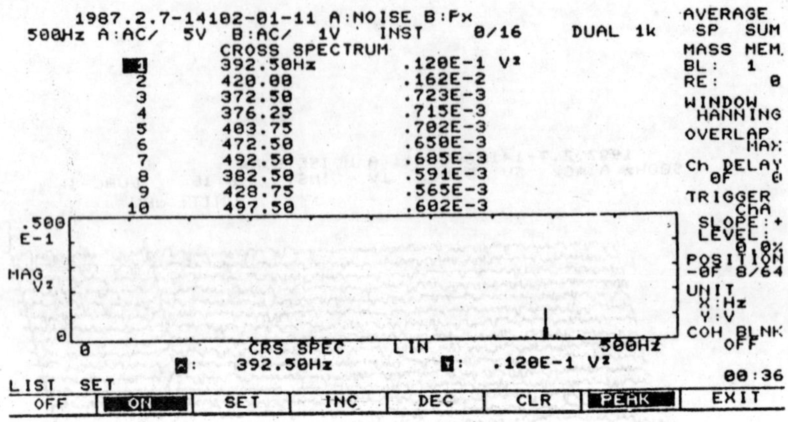

Fig3. Cross power spectrum between fluctuating flow pressure and sound pressure, speed=93%

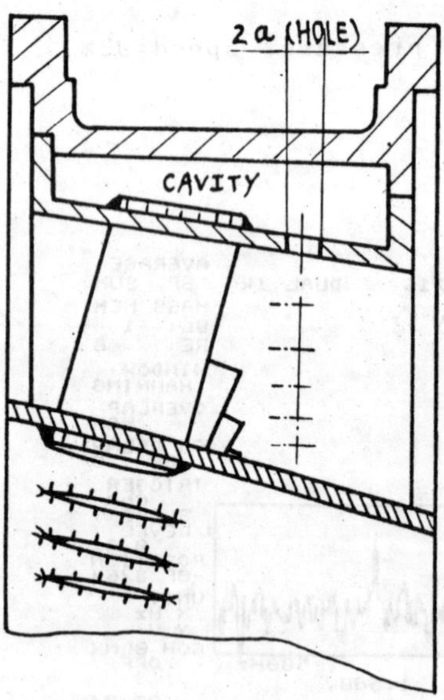

Fig4. Ring cavity and the holes on the liner

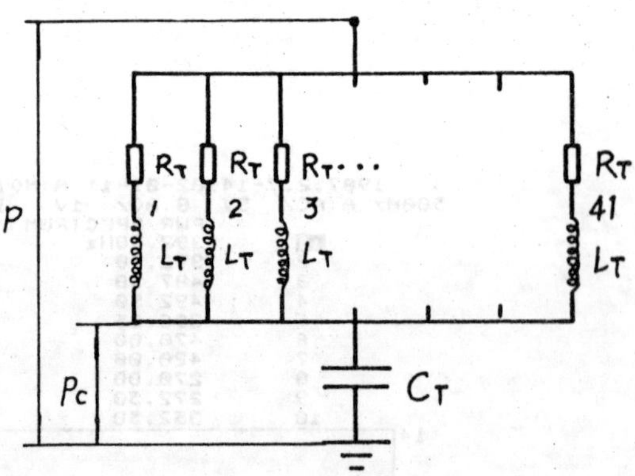

Fig5. Sound circuit scheme

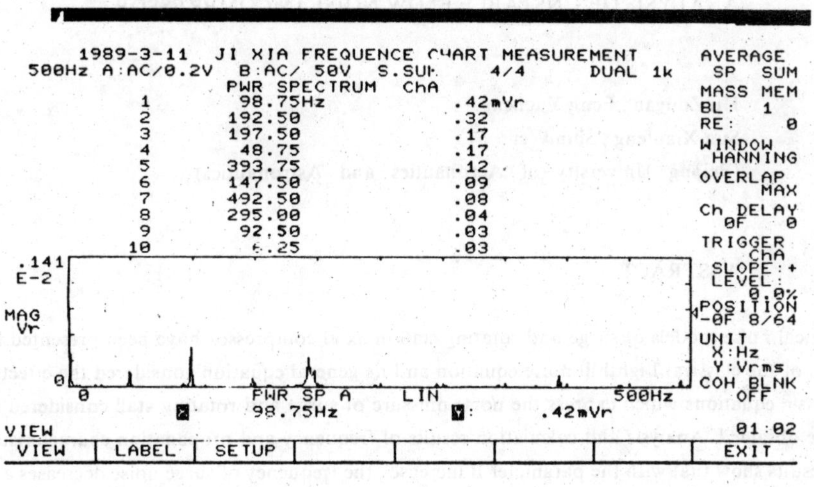

Fig6. Structure vibration spectrum by hammer excitation

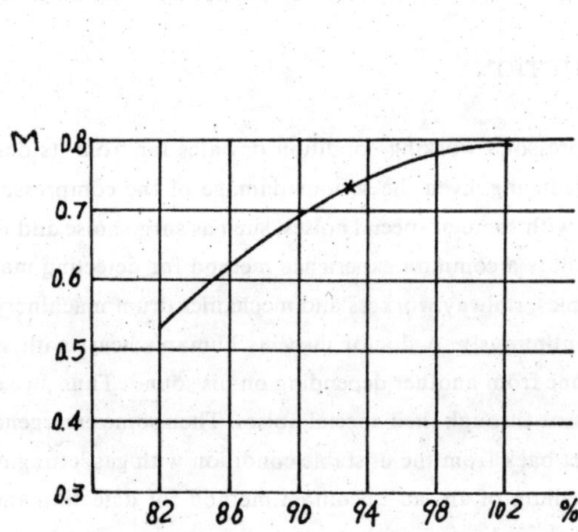

Fig7. The Mach number varing with the engine speed

ANALYSIS OF UNSTABLE FLOW NOISE IN COMPRESSOR

Hu Zongan , Feng Yucheng,
Sun Xiaofeng , Shin Wei
(Beijing University of Aeronautics and Astronautics)

ABSTRACT

Two theoretical noise models of surge and rotating stall in axial compressor have been presented in this paper. On the basis of the classical Lighthill noise equation and its general equation considered the effects of flow medium, the basic equations which express the noise pressure of surge and rotating stall considered the effects of pipe wall are obtained. Analysis and calculation results of frequency and propagation characteristics are made. And these results show that with the parameter B increase , the frequency of surge noise decreases and its intensity increases. They are corresponding to the phenomena from experimental results. Having studied the frequency characteristics of noise from rotating stall, a formula which can predict the propagation velocity of stall cell is obtioned. And the analytical result also shows that the noise of rotating stall can not propagate to a large scope when $Mr < 1$. In this paper, some experimental results of rotating stall noise are presented. These conclusions lay the foundation for diagnosis of unsteady flow onset in compressor by using the noise measurement methods.

INTRODUCTION

Unstability occurs when compressor's working condition deviates far from its design point. This will result in pressure ratio and efficiency reducing, even the serious damage of the compressor. With compressor unstable flow occurs, it will accompany with some of special noises, such as surge noise and rotating stall noise etc.

Noise or vibration measurement is a common experience method for detecting machine's (or structure's) working condition. For some examples, railway workers and mechanics drum machinery parts with small hammer to see whether they can work continuously. A doctor inspects human's heart with stethoscope. With computer's development, we can identify one from another depending on his sound. Thus , we can determine which kinds of unstable flow onset in compressor through their special noises. Then some emergence methods can be taken. For instance, a compressor can kept back from the unstable condition with gas letting off. Because the sound propagation spreads through the medium of air, an uncontact method for detecting and diagnosis can be obtained. Without question, it is very simple and convenient.

Based on the principle suggested by E. M. Greitzer in Ref.[1], two kinds of noise theoretical model of compressor surge and rotating stall are presented . Having soluted the Lighthill acoustic equation (2) and general acoustic equation considered the effect of moving medium, the sound pressure equations of surge and rotating stall in compressor are deduced. And the characteristics of frequency, propagation and their influence factors have discussed respectively. In this paper, an experimental method is also used to test them in a high speed axial compressor. They will lay the foundation of acoustic diagnosis in running compressor (including air–blower).

THE FLOW UNSTABILITY IN DIFFUSE TURBOMACHINE AND ITS CHARACTERISTIC PARAMETERS

In order to describe unstable flow in flow field, the value K ($K = \omega L / V$, L, V where L is characteristic length, V is flow velocity and $1/\omega$ is vibration frequency) is usually applied. The magnitude of value K shows the unstable degree of flow field. The flow is steady when $k << 1$, while unsteady flow dominates the whole flow field, when $K >> 1$.

There are two typical unstable flows in compressor . Ref. [1]points out that they are determined by a characteristic parameter B which is called Greitzer determination value. When value B is greater or equal to 0. 7, compressor is in surge state meanwhile the gas flows along axial direction of compressor, oscillates in low frequency(5–15HZ) and gives off a strong"rumble"noise. And the niose tune becomes lower , the frequency becomes lower and the noise intensity becomes stronger when the value of K becomes less. Compressor is in rotating stall state when B is less than 0. 7. In such case, unstable flow occurs along circumferential direction while the mean flow velocity along axial direction is constant. Being affected by rotating stall cells, the rotor blades produce a sort of special "rustle" noise which is different from the noise in good working condition. And it resembles (but any difference) as rotor / stator or stator / rotor interaction noise.

MECHANISM AND CALCULATION OF SURGE NOISE IN AXIAL COMPRESSOR

Just as experiment shows: When the compressor surge occurs, there are not only large amount flow oscillation along axial direction but also strong low frequency noise. Therefore the surge noise of compressor can be regarded as the result of a small piston oscillation along axial direction as show in Fig. 1.

The Lighthill noise equation is:

$$\frac{1}{C_0^2}\frac{\partial^2 p}{\partial t^2} - \nabla^2 p = \frac{\partial Q}{\partial t} - \nabla \cdot \bar{f} + \frac{\partial^2 T_{ij}}{\partial y_i \partial y_j} \tag{1}$$

Three items of the right side of above equation represent three different basic sound sources which are Single Polar, Dual Polar and Tetra Polar respectively. With regard to sound efficiency , the highest of these three kinds of sound source is Single Polar , Dual Polar is the second and Tetra Polar is the third. But at the point of sound power, they are in converse sequence. And their sound powers are proportional to the flow velocity's 4th, 6th and 8th power shown in Ref. [3]. When compressor's working condition is in surge, there are characteristics of low frequency oscillation along axial direction and low flow velocity . Thus , the physical model due to compressor surge noise can be simplified as changing mass flow item. It is to say tha. only the Single Polar results in the surge noise while neglects the Dual and Tetar Polar's effects. So equation (1) can be simplified as:

$$\frac{1}{C_0^2}\frac{\partial^2 p}{\partial t^2} - \nabla^2 p = \frac{\partial Q}{\partial t} \tag{2}$$

If the center of the oscillating piston can be regarded as the original point in a polar ordinate, then the instant sound pressure of any point in the space can c written as:

$$p = \frac{1}{4\pi r}\frac{\partial Q}{\partial t} e^{i\omega(t - \frac{r}{C_0})} \tag{3}$$

Where r is the distance from any point in the space to the original center, $\partial Q / \partial t$ is the variable rate of axial mass flow in compressor, ω is the circular frequency of flow oscillation along axial direction (or of piston oscillation); Co is the sound speed.

Equation (3) is the basic formula for predicting surge noise characteristics of compressor. From this equation , the surge noise frequency and sound wave propagation characteristics are mainly determined by the dimensions of compressor pipe system. Taking compressor center as the center of the polar coordinate, sound wave spreads all around space in the form of spherical surface. The sound pressure magnitude of any point in the sound field is in converse proportion to the distance r . With the oscillating model given by Ref. [1], the differential equation groups of fluid dynamic equation and continuous equation of the pipe system can be derived. Then solving these equations by means of the method of the fourth order of Rung–Kuta , the value $\partial Q / \partial t$ of can be obtained. Fig. 2 and Fig. 3 are the calculation results when B = 1.0 And 1.3 respectively.

From calculation results with formula (3), it is found that the sound pressure at the same point increases and oscillation frequency decreases a little with value B increase. This is correspond to the experimental phenomena when compressor is in surge state.

THE NOISE MECHANISM AND CALCULATE RESULTS OF COMPRESSOR IN ROTATING STALL STATE

1. Theoretical noise model of rotating stall

Rotating stall noise is mainly due to periodical interaction of rotating stall cells with rotor blades, as shown in Fig. 4. Therefore, aerodynamical noise caused by rotating stall is produced by a typical Dual Polar sound source. Just as Ref. [1] indicating, when rotating stall occurs in compressor, if neglect the influences of Single and Tetra Polar sound sources, calculation results would be a little error because the mean flow velocity along compressor axis is constant and its value is not much bigger. But because the rotating stall noise is mainly due to the interaction result of rotary stall cells with rotor blades, the influence of axial mean velocity of air medium on sound propagation can not be neglected. So the general Lighthill equation give by Ref. [4] is written as follows:

$$\frac{D_0}{Dt} p - C_0^2 \frac{\partial^2 p}{\partial y_i \partial_j} = \frac{\partial^2 T_{ij}}{\partial y_i \partial y_j} \tag{4}$$

where

$$\frac{D_0}{Dt} = \frac{\partial}{\partial t} + \overline{U}\frac{\partial}{\partial y_i}$$

In order to solve the general Lighthill equation conveniently, some hypotheses of the pipe boundary conditions of compressor are made as follows :

(1) The solution region is in a circular ring of infinite length pipe.

(2) The flow field is uniform and M < 1 .

(3) The pipe boundary surface is a hard wall without sound wave penetrating through.

The boundary condition is shown as Fig. 5 .

Solving equation (4), the basic formula which will predict the characteristics of rotating stall noise in compressor can be deduced as follows:

$$p(\vec{x},t)=\sum_{m=-\infty}^{+\infty}\sum_{n=1}^{\infty}\psi_m(K_{n,m}r)e^{i(m\Phi-r^{\pm}_{n,m}x_i)}\times e^{-i(NB\Omega-qB_{rs}\Omega_{rs})t}\,p_{m,n} \quad (5)$$

where

$$p_{m,n}=\frac{B}{2\Gamma_{n,m}K_{n,m}}\int_{s_0}\psi_m(K_{m,n}r)\left[\frac{NB-qB_{rs}}{r}f^0_D-\gamma_{n,m}f^0_T\right]$$

$$exp\,i[\gamma_{n,m}Y_i-(NB-qB_{rs})\varphi]ds$$

$$K_{n,m}=\sqrt{\left(\frac{NB\Omega-qB_{rs}\Omega_{rs}}{C_0}\right)^2-\beta^2K_{m,n}} \quad (6)$$

$$\gamma_{n,m}=\frac{1}{\beta^2}\left[\frac{M(NB\Omega-qB_{rs}\Omega_{rs})}{C_0}\pm K_{n,m}\right]$$

In formula (6) f^0_D and f^0_T are axis and circumferential direction forces of the pressure value Δp^0_q forced on the blades. where

$q=\pm 1, 2, \ldots\ldots\ldots$

B_{rs} : stall cell number

Ω : rotational velocity of blades

Ω_{rs} : absolute propagating velocity of stall cells

B: blade number

S_0 : surface area of blade

$\Phi=\overline{\Phi}-\Omega t$ (see Fig. 5)

$\beta^2=1-m^2$

$\varphi_m(K_{m,n}r)=AJ_m(K_{m,n}r)+By_m(K_{m,n}r)$

$J_m(K_{m,n}r), y_m(K_{m,n}r)$ represent the first and second class Beessel Function respectively. Km, n is its characteristic root (detail information is shown in Ref. [5]).

Formula (5) is pretty good for aero–acoustics calculation of far or near sound field.

2. Characteristics of frequency and propagation and its experimental result of sound wave due to rotating stall

From formula (5), the sound wave frequency due to rotating stall can be written as the following form:

$$f=\frac{1}{2\pi}(qB_{rs}\Omega_{rs}-NB\Omega) \quad (7)$$

where

N = 0, ± 1, ± 2, ± 3, ...

q = 0, ± 1, ± 2, ± 3, ...

For different composes of N and q , there would be different frequency components of sound wave. But these frequency components all can not satisfy the relation of multiple frequency range unless N is equal to 0. When N = 0, the sound wave frequency satisfying the relation of multiple frequency range is as follows:

$$f=\frac{1}{2\pi}qB_{rs}\Omega_{rs}$$

where

$$q=1,\ f_{1rs}=\frac{1}{2\pi}B_{rs}\Omega_{rs}$$

$$q=2,\ f_{2rs}=\frac{1}{2\pi}2B_{rs}\Omega_{rs}$$

$$q=3,f_{3rs}=\cdots\cdots\cdots$$

Thus , the frequency satisfying multiple frequency relation can be found in experiment, the frequency magnitude of the first order can be determined. Therefore, having known the stall cell's number (can be decided by experiment) , the propagating velocity of stall cells can be immediately decided, as follows:

$$\Omega_{rs}=\frac{2\pi f_{1rs}}{B_{rs}} \tag{8}$$

Because sound wave can spread through air medium, some important characteristic parameters of rotating stall, such as stall cell's number and its propagating speed can be easily obtianed by using such kind of uncontact acoustic measurement method. One of the characteristic parameters has been measured in a high speed compressor facility in BUAA by using above measurement method. The experimental apparatus is shown in Fig. 6. Detail information of this facility is presented by Ref. [5]. Two microphones (type CH16) separated from 90 ° are fixed on the wall behind a chord of the rotor. The third one is fixed behind the dustproof net for measuring the sound wave of far field . The signals coming from these microphones enter into an oscillgraph and are recorded on a tape recorder.

Fig. 7 are the time wave curves measured by two microphones at the downstream of rotor when rotary speed is equal to 12600rpm and compressor is not in rotating stall state. Fig. 8 are the time wave curves measured at the same positions when rotating stall occurs. Comparing Fig. 8 with Fig. 7, the time wave curve of rotating stall has greatly changed and it shows that there exists new frequency components . From the time difference of these two curves in Fig. 8, the numbers of the stall cell Brs can be calculated by the method presented by Ref. [6]. In this experiment Brs is approximately equal to 1.

Fig. 9 and Fig. 10 are the power density spectra of the noise measured by two microphones behind the rotor (near field) in rotating stall state, while Fig. 11 is the density spectrum of one of the two microphones in non–rotating stall state . From these , it is easy to see that rotating stall results in diffuse component of low frequency and this can illustrate why rotating stall produces a special tone.

From Fig. 9 and Fig. 10, the first frequency of the noise due to rotating stall is 126 HZ, ie. , f = 126HZ. The second and third are 252HZ and 382HZ respectively. Thus, the propagating speed of stall cell can be deduced from formula (8) when n = 12600rpm.

The flutter group's propagate speed Ω

$$\Omega_{rs}=\frac{2\pi f_{1rs}}{B_{rs}}=79128Au/s$$

From this , the relative propagation rate can be obtained as follows:

$$K=\frac{\Omega_{rs}}{\Omega}=0.60$$

where value K is in the range presented by Ref. [5].

3. Propagation characteristics and analysis of experimental results in rotating stall state

From formula (5), when

$$K_{n,m}^2 = \left(\frac{NV\Omega - qB_{rs}\Omega_{rs}}{C_o}\right)^2 - \beta^2 K_{n,m} < 0$$

While N = 0 , the tenable condition of above formula is:

$$\left(\frac{qB_{rs}\Omega_{rs}}{C_o}\right)^2 - \beta^2 K_{n,m} < 0 \qquad (9)$$

It means that the sound wave of rotating stall frequency which satisfies formula (9) will attenuate. Because the characteristic root $K > qB_{rs} / r$, (r is the outer radius of the rotor), an equal condition satisfying formula (9) can be deduced as:

$$M_r^2 = M_{rs}^2 + M^2 < 1$$

where $M_{rs} = \frac{\Omega_{rs} r_d}{C_0}$

It means that only the Mach number of stall cell relative to the rotor blades greater than 1, the sound wave of rotating stall can spread far away. Otherwise the sound wave will attenuate with distance increasing. Meanwhile it shows that the attenuation magnitude with distance increasing is dependent on value $|K_{n,m}|$. The larger of $|K_{n,m}|$, the faster of attenuation with distance increasing will be. And the magnitude of $|K_{n,m}|$ is determined by value m. The larger of m , the larger of $|K_{n,m}|$ will be. $m = NB - qB_{rs}$ when model (N, q) composition keeps less value m ($m = NB - qB_{rs}$) , the attenuation will be slower with distance increasing. Especially when N = 0 and m = $- qB_{rs}$, it would be. It means that the attenuation of the first frequency in rotating stall state is less than its higher order harmonic wave.

Fig. 12 is the charts of time wave for comparing the sounds of near field with far field. Fig. 13 is the relevant frequency spectrum of far field of the sound wave. From these charts, either time wave or frequency spectrum distribution, there are great differences between these message of far and near sound fields in rotating stall state of a compressor. And from formula (6), $K_{n,m}^2$ is less than zero when n = 12600rpm. Because the sound wave frequency in rotating stall state is a type of attenuation, the time wave form of far field sound is not the same like near one . And, the frequency spectrum of far field does not have a clear diffuse mountain value.

4. Sound pressure calculation and analysis in rotating stall state of a compressor

From formula (5), it is shown that $p(\bar{x},t)$ is also decided by the intervals acted by f_D and f_r and characteristic parameters of frequency . And it is very difficult to calculate the large scale seperating flow. Especially there is no use to calculate the sound pressure in rotating stall state of a compressor. So further analysis is not discussed in this paper. Ref. [5] has given some effects of the characteristic parameters of rotating stall, such as stall cell number, the width of stall region and its rotary speed on the rotating stall noise.

CONCLUSION

1) In compressor , unstable flow degree is determined by value K and the frequency among K is the main frequency of this unstable flow.

2) Compressor surge noise is Single Polar Sound source. The less of K, the less of surge noise frequency and

the larger of sound intersity will be.

3) Rotating stall noise is Dual Polar Sound source. The sound pressure in pipe is related to blade force, stall cell number and its rotary velocity. When Mr < 1, rotating stall noise can not spread to far distance in upstream.

4) Uncontact sound measurement method can be used to diagnose and identify the generation of unsteady flow in compressor.

REFERENCES

[1] E. M. Geritzer., "Review Axial Compressor Stall Phenomena", Journal of Fluids Engineering, June, 1980, Vol. 102, Transaction of the ASME

[2] R. D. Belavense., "Fluid Induce Vibration", Machinery Industry Publishing House, 1983

[3] Lixi. Huang., " Propeller Acro–acoustics ", Science Research Report of BUAA, BH–B2751, 1987

[4] M. Oldstein., "Unified Approach to Aerodynamic Sound Generation in the Presence of Sound Boundary", J. Acoust. Soc. Am., 56, 497–509, 1974

[5] Xiaofeng. Sun., "Some Aeroacoustic Problems' Research of Aeronautics Power Apparatus", BUAA, Doctor paper, 1988

[6] Yajun. Lu., "Turbomachinery Unsteady Flow", Teaching Materials, 1983

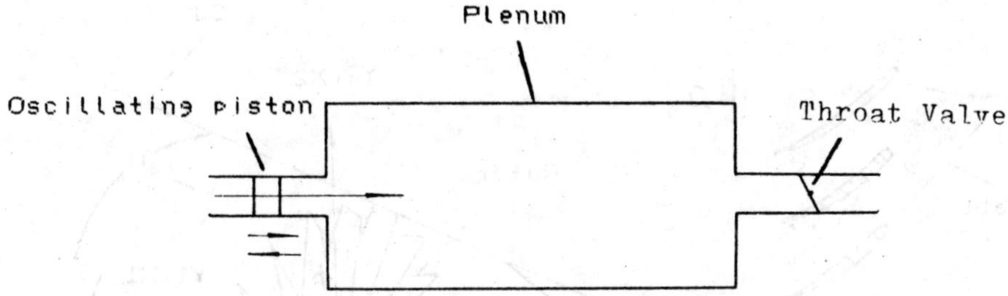

Fig.1 Surge noise model

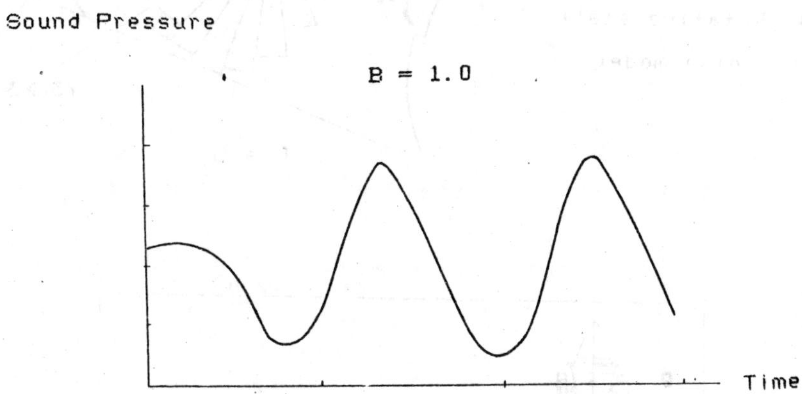

Fig.2 Time Wave Curve

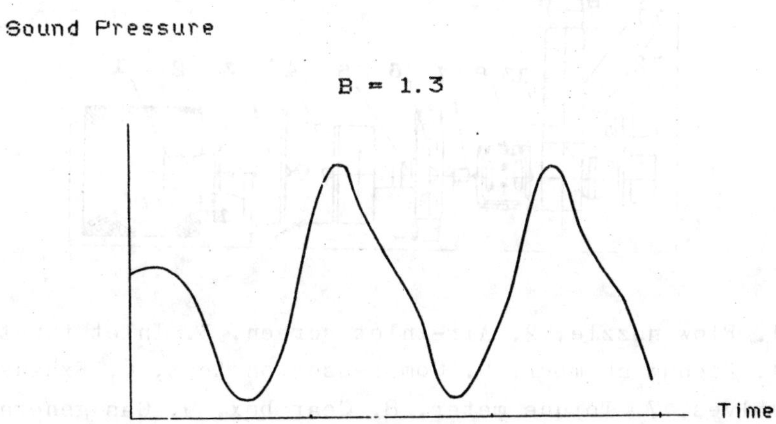

Fig.3 Time Wave Curve

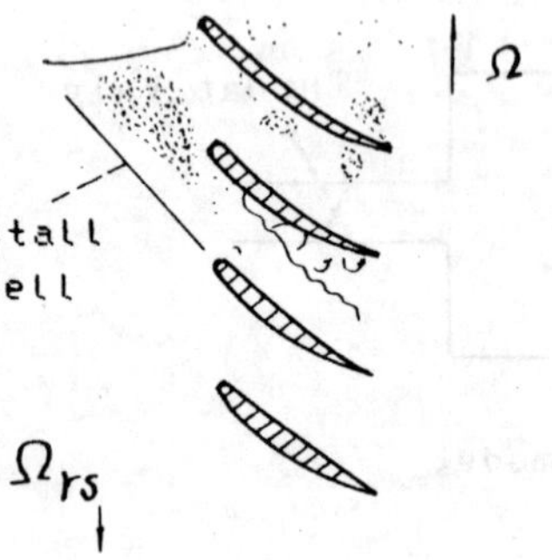

Fig.4 Rotating stall noise model

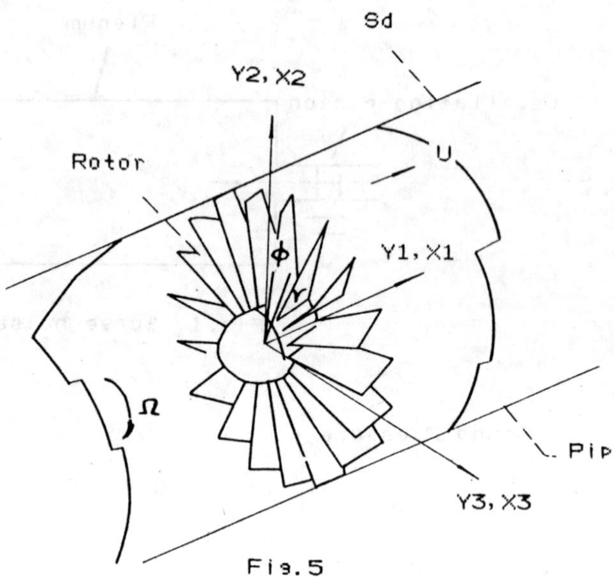

Fig.5

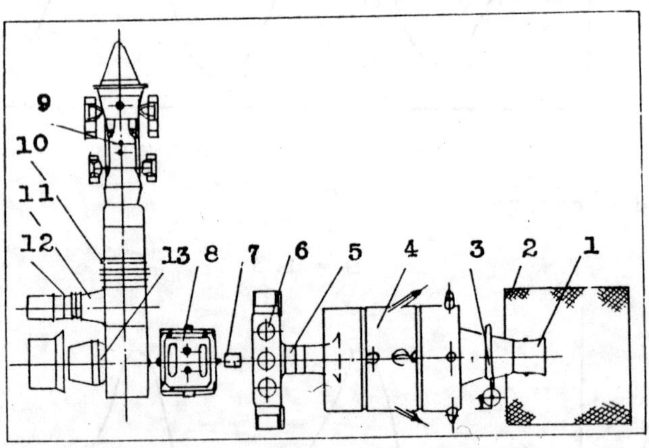

1. Flow nozzle, 2. Air-inlet screen, 3. Inlet throttle, 4. Plenum chember, 5. Compressor on test, 6. Exhaust valves, 7. Torque meter, 8. Gear box, 9. Gas generator, 10. Expanding ring, 11. By-pass exhaust system, 12. Regulating valves, 13. Free turbine.

Fig. 6 Transonic Compressor Test Facility

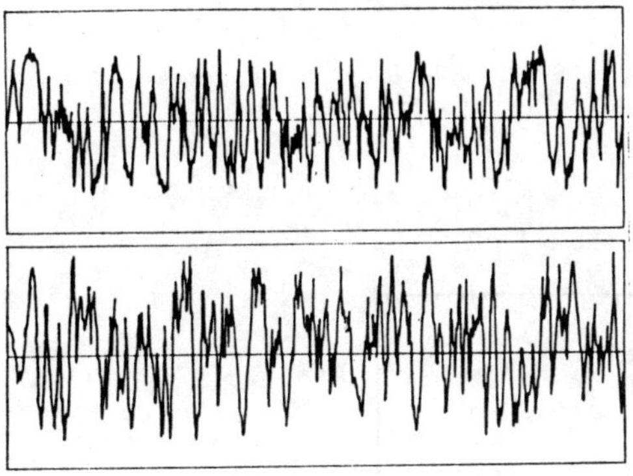

Fig.7 Time Wave Curves
(non-rotating stall state)

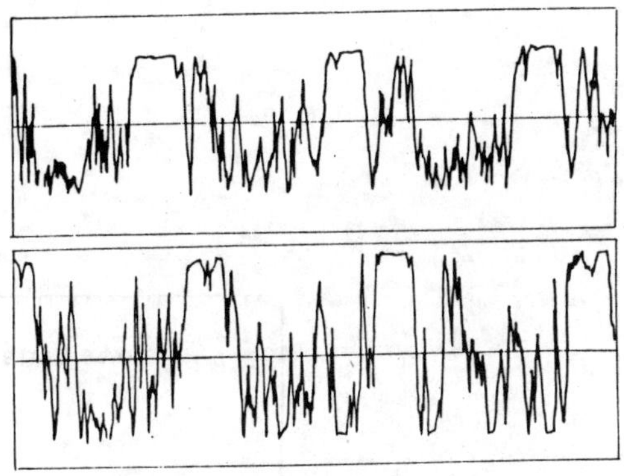

Fig.8 Time Wave Curves
(rotating stall state)

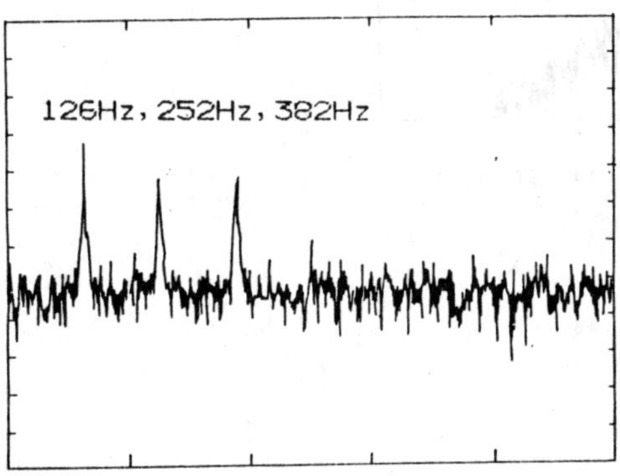

Fig.9 Power density spectrum

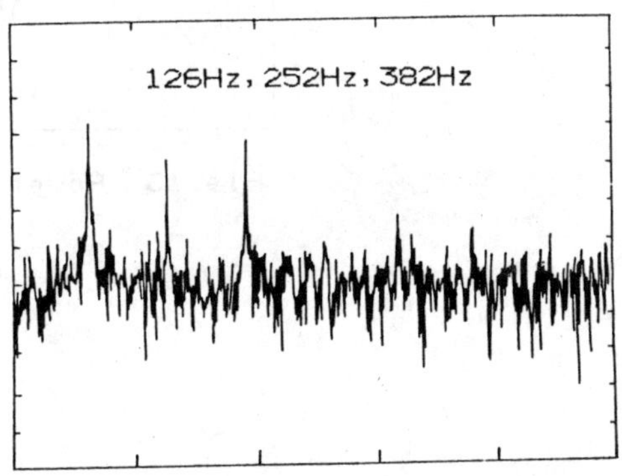

Fig.10 Power density spectrum

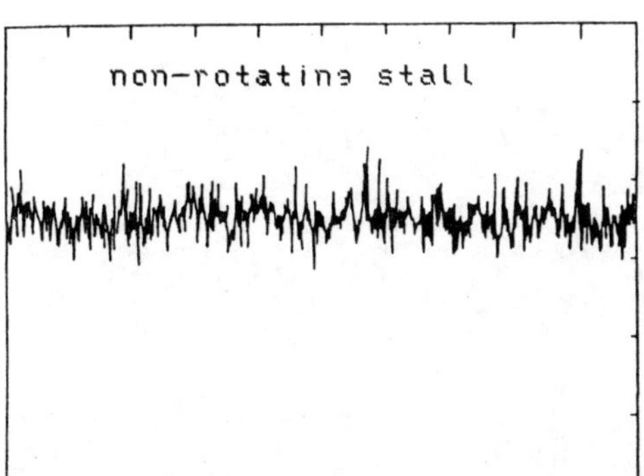

Fig.11 Power density spectra

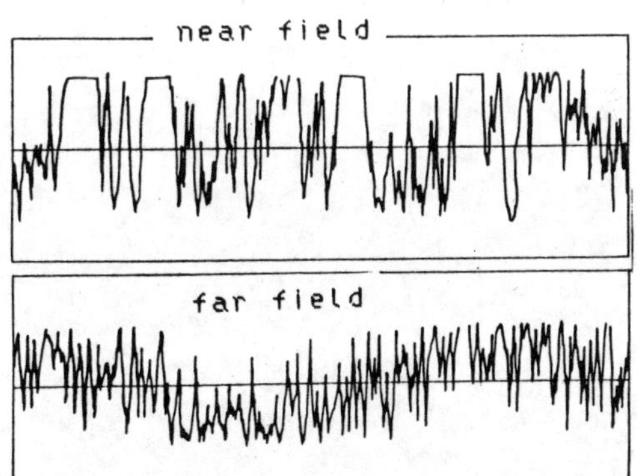

Fig.12 Time Wave Curves

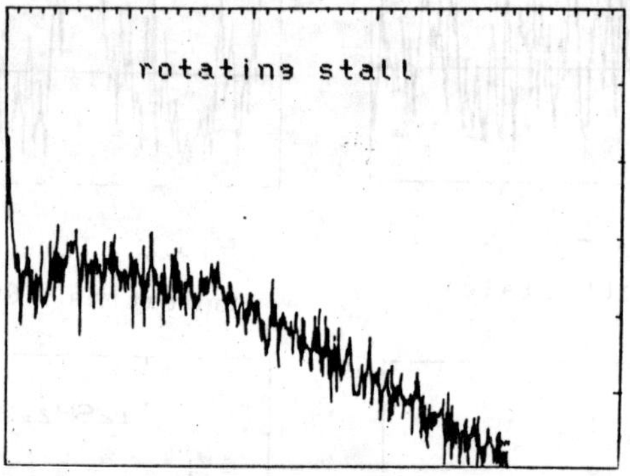

Fig.13 Power density spectra

On the unsteady loading noise of counter–rotating propeller

Sun Xiao–feng, Hu Zong–an, Zhou Sheng

(Beijing University of Aeronautics and Astronautics)

Abstract

The mechanism of noise generated by counter–rotating propeller or propfan is quite different from that by single–rotating one. The mechanism of unsteady loading noise generated by interaction of front rotor's wake with aft rotor in a counter–rotating (CR) propeller has been analyzed and the set of corresponding noise radiation formulas has been derived. The numerical predictions of noise generated by a CR propeller are given and compared with published experimental data. It looks like that the sound pressure level (SPL) in axial direction of rotor is fully determined by the component of unsteady loading noise. but in rotating plane the main contribution to SPL comes from the component of steady loading noise, and the effect of unsteady loading noise is just limited in the range of higher harmonics.

§ 1 Introduction

It has been demonstrated by theoretical analysis and test results that counter–rotating (CR) propeller posseses higher efficiency than single–rotating (SR) one. Therefore, most of next generation's commercial aircrafts pushed by propfans will be equipped with counter–rotating propfans. But the mechanism of noise generated by CR–propfan is more complicated than that by SR–propfan. Apart from steady loading noise just like that generated by SR–propfan, there is additional unsteady loading noise which is generated by interaction between front and aft rotors in CR–propfan. Then, the overal sound power level generated by CR–propfan will be higher than that generated by comparable SR one, if there were not any effective sound–suppression means. So, the models, methods and computer codes for predicting far field and near field of sound generated by CR propeller and propfan are quite necessary, because the noise and interrelated problems must be considered during the process of developing a commercial aircraft. Regarding the steady loading noise, there are lots of theories and methods [3] by means of which the predictons are usually consistent with test results. However, the situation on the unsteady loading noise is quite different. Besides some of test results [2,4,5], the published prediction methods for calculating unsteady loading noise generated by CR–propeller and propfan are not effective enough.

Recently, Hanson extended his compressible helicoptal surface theory of propeller aerodynamics and noise [6] into the analysis [7] of sound far–field generated by unsteady loading of CR propeller. By comparing the calculations with test results. Hason suggested that some new interaction models should be developed in order to improve the consistence between calculations and test results.

Some further explorations on predicting noise generated by CR–propeller are worth trying. The prediction of unsteady loading noise generated by interaction between viscous wakes from front rotor blades and aft rotor at a CR–propeller has been presented at this paper. At the next section, a set of sound radiation formulas describing unsteady loading noise generated by CR–propeller will be derived, which is an extension of the theories from Lowson[8] and Wright[9]. At the third section, the model of blade wakes and the model of unsteady aerodynamics used at this paper will be described. The results and discussions will be presented at the fourth section.

§ 2 The sound radiation formulas

According to Lighthill's basic aeroacoustic theory, Lowson[8]、 Wright[9] and White[10] derived the different forms on sound radiation formulas which described the sound far-field. The basic method from White[10] has been used at this paper in order to extend the sound radiation formulas of SR-propeller into the situation of CR-propeller.

The noise radiated from CR-propeller is mainly generated by steady aerodynamic load and by unsteady aerodynamic load which comes from interaction between front rotor and aft rotor. It means that the sound sources are the sound dipoles. Assume that ρ is the air pulse density , and C_0 is the sound speed, then the lighthill's equations can be written in to following form:

$$\frac{\partial^2 \rho}{\partial t^2} - C_0^2 \frac{\partial^2 \rho}{\partial x_i^2} = -\frac{\partial F_i}{\partial x_i} \qquad (i = 1,2,3) \tag{1}$$

Firstly, the sound pressure at observing point in free space which radiated by the force F located at point (y_1, y_2,y_3) is

$$G(\vec{y},\tau|\vec{x},t) = \frac{\delta\left(\tau - t + \frac{r}{C_0}\right)}{4\pi r}$$

where $\delta(\ \ )$ is the pulse function. Then the solution of Eq.(1) is

$$P(X_i,t) = -\frac{\partial}{\partial X_i}\left[\frac{F_i\left(t - \frac{r}{C_0}\right)}{4\pi r}\right] \tag{2}$$

Let us write Eq. (2) into the form of sphere's coordinates (reference Fig. (1)), threfore,

$$P(R,\theta_0,\Phi_0,t) = \frac{\partial}{\partial X}\left[\frac{L\left(t - \frac{r}{C_0}\right)}{4\pi r}\right] - \frac{1}{a}\frac{\partial}{\partial \Phi_0}\left[\frac{D\left(t - \frac{r}{C_0}\right)}{4\pi r}\right] \tag{3}$$

where L(t) the force component parallel to the axis of rotor, and D(t) the circumstantial force component.

The intensity of sound source at a space-fixed point will be also time-dependent, for the sound radiated from aerodynamic load on rotor blades is equivalent to that radiated from continuously distributing sound sources fixed at rotating plane, and the aft rotor is rotating with angular velocity $\Omega_2 \neq \Omega_1$. For a space-fixed point (Φ,a) at rotating plane, the intensity of sound source at this point will be equal to blade load while every blade passes this point, otherwise the intensity will be equal to zero while no blade passes that point. Therefore, the sound pressure radiated from the sound sources at rotating plane is

$$dp = \frac{\partial}{\partial X}\left[\frac{L(\Phi,t)}{4\pi r}\right]\sum_{m=-\infty}^{\infty}\delta\left[\Phi + \frac{2\pi m}{B_2} - \Omega_2\left(t - \frac{r}{C_0}\right)\right] - \frac{1}{a}\frac{\partial}{\partial \Phi_0}\left[\frac{D(\Phi,t)}{4\pi r}\right]\sum_{m=-\infty}^{\infty}\delta\left[\Phi + \frac{2\pi m}{B_2} - \Omega_2\left(t - \frac{r}{C_0}\right)\right] \tag{4}$$

At the coordinates fixed at ground, the wakes of front rotor will rotate with angular Ω_1.S_0 the unsteady forces caused by the wakes of front rotor will be expanded into following Fourier series:

$$L(\Phi,t) = L(\Phi + \Omega_1 t) = \sum_{n=-\infty}^{\infty} A_n exp[-i(nB_1\Phi + nB_1\Omega_1 t)]$$

$$D(\Phi,t) = \sum_{n=-\infty}^{\infty} B_n exp[-i(nB_1\Phi + nB_1\Omega_1 t)]$$

$$dp = \frac{B_2}{8\pi_2} \sum_{n=-\infty}^{\infty} \sum_{m=-\infty}^{\infty} e^{-i(mB_2\Omega_2 + nB_1\Omega_1)t} \left[A_n \frac{\partial E}{\partial X} - B_n \frac{\partial E}{a\partial \Phi_0} \right]$$

where

$$E = \left[\frac{exp[i(mB_2 - nB_1)\Phi + \frac{(mB_2\Omega_2 + nB_1\Omega_1)C_0}{r}}{r} \right]$$

In this way, the sound pressure radiated from whole rotationg plane is

$$P(R,\theta_0,\Phi_0,t) = \frac{B_2}{8\pi^2} \sum_{n=-\infty}^{\infty} \sum_{n=-\infty}^{\infty} e^{-i(mB_2\Omega_2 + nB_1\Omega_1)t} \int_{r_H}^{r_0} \int_{-\pi}^{\pi} \left[A_n \frac{\partial E_r}{\partial X} - B_n \frac{\partial E_r}{a\partial \Phi_0} \right] d\Phi dr$$

As shown from Fig.1, for far–field, we can use $r \approx R - acos(\Phi - \Phi_0)sin\theta$.

$$\frac{1}{F} \approx \frac{1}{R}$$

and define

$$M_s = \frac{\Omega_2 a}{C_0}$$

$$\Omega_{12} = \frac{\Omega_2}{\Omega_1}$$

$$\frac{\partial(\ )}{\partial x} = cos\theta_0 \frac{\partial(\ )}{\partial R}$$

then

$$P(R,\theta_0,\Phi_0,t) = \frac{B^2}{4\pi Ra} \sum_{n=-\infty}^{\infty} \sum_{n=-\infty}^{\infty} i^{[(mB_2 - nB_1)+1]} \times e^{i\left[(mB_2 - nB_1)\Phi_0 - (mB_2\Omega_2 + nB_1\Omega_1)\left(t - \frac{R}{C_0}\right)\right]}$$
$$\times \int_{r_H}^{r_0} \left[\frac{(mB_2\Omega_2 + nB_1\Omega_1)acos\theta_0}{C_0} A_n - (mB_2 - nB_1)B_n \right] J_{mB_2}$$
$$- nB_1[(mB_2 + nB_1\Omega_{12}) \times M_s sin\theta_0]dr \qquad (5)$$

where J_{mB2} is the first kind's Bessel function. Eq. (5) is the final sound radiation formula derived for CR–propeller. When $\Omega_1 = 0$, the formula is consistent with the results of ref. [8,9]. Moreover, the formula is also consistent with the result of ref. [7] if we don't consider the forward velocity effects.

The further simplifications have been made during calculation process. For unsteady loads, the effects of blade thickness, camber and incidence are neglected, but for blade steady loads, the above–mentioned effects have been included.

§ 3 Blade wake model and unsteady aerodynamic model

Some of our works on axial fan / compressor rotor–stator interaction noise generation has been shown at ref[12]. Ref[13] is the extension of ref[12] into the noise generation from swept cascade interaction. At ref[12] and ref[13] , rotor exit velocity is decomposed into the mean wake velocity component and the pulse velocity component. According to the test results , empirical distributing rule of pulse velocity could be used. By adding all the effect of wake from each blade of a rotor, the whole pulse velocity field after a rotor could be obtained. Finally, by solving unsteady linearinged Euler eguations, the unsteady aerodynamic load on blade could be calculated. Then, we can get the rotor / stator interaction noise.

The sound generation mechanism of unsteady loading noise by CR–propeller is rather similar to that of rotor / stator interaction noise of axial fan / compressor. Howercr, the situation of CR–propeller is more complicated than fan / compressor, because the angular velocity of aft rotor is different to that of front rotor . So , the wake model and unsteady load model used in this paper are the extension of ref.[12,13]. The velocity distribution in wakes of front rotor can be expanded into a Fourier series as follows:

$$W = \sum_{n=-\infty}^{+\infty} W_n exp\left\{ i\left[Kz + 2\pi nm\frac{B_1}{B_2} - nB_1(\Omega_1 + \Omega_2)\tau \right]\right\} \qquad (m = 0, \pm 1, \pm 2, \cdots\cdots) \qquad (6)$$

where z is the chordwise variable , Wn are the velocity coefficitents, which are given by means of the method of ref.[11]. The unsteady aerodynamic forces on blade could be calculated by means of the method of ref.[13]. Then, the effects of blade swept could be also include.

§ 4 Results and analysis

The numerical example and corresponding geometric and aeroacoustic data are taken from ref.[5,14], where the blade numbers of front and aft propeller$B_1 = B_2 = 3$, the r.p.m. of front and aft propeller $N_1 = N_2 = 9000$ rpm, and the swept angles are both 27° . The unsteady loading noise and steady loading noise at different observing points aue shown at Table I,II,III, respectively, where the tested results are taken from Fig.3 of ref.[5]. The values of steady loading noise are the sums of the noise from front rotor and that from aft rotor.

From Table I we can see that the steady loading noise at rotating axis is egual to zero. it means that the contribution to sound pressure entirely comes from unsteady loading noise. In Table II , where $\theta_0 = 45°$, the peak of steady loading noise just corresponds to the rotor blade–passing–frequency (B.P.F) and in the higher frequencies (1350–2700 Hz), and the unsteady loading noises are higher than the steady loading noises obviously. In Table III, where $\theta_0 = 90°$, at B.P.F and its second harmonic the steady loading noises are higher than the unsteady loading noises obviously, and the effect of unsteady loading noise are mainly at higher harmonics of B.P.F. It means that the primary contribution to the overall sound pressure level at rotating plane comes from the steady loading noise. Above–mentioned numerical results are consistent with the results of SR–propellers.

Table I (R = 2.5m, $\theta_0 = 0°$)

frequency (Hz)	450	900	1350	1800	2250	2700
steady loading noise (dB)	0	0	0	0	0	0
unsteady loading noise (dB)	0	118.91	0	123.25	0	125.37

Table Ⅱ (R = 2.5m, θ_0 = 45 °)

frequency (Hz)	450	900	1350	1800	2250	2700
steady loading noise (dB)	97.75	73.34	61.47	46.05	25.76	8.14
unsteady loading noise (dB)	81.87	108.67	111.03	108.00	117.08	110.66

Table Ⅲ (R = 2.5m, θ_0 = 90 °)

frequency (Hz)	450	900	1350	1800	2250	2700
steady loading noise (dB)	112.12	101.98	91.83	81.84	72.01	62.28
unsteady loading noise (dB)	86.89	76.17	89.57	89.37	86.56	92.90
OASPL by prediction (dB)	112.12	101.98	94.03	91.67	86.74	92.91
OASPL by test[4] (dB)	109.5	98.5	98.0	95.0	91.5	92.5

§ 5 Concluding remarks

The mechanism of unsteady loading noise of counter–rotating propeller and propfan has been explored by means of a simplified model and corresponding sound radiation formulas. According to the calculated results compared with a few test examples, some remarks could be made as follows:

1. It seems that the sound pressure level in axial direction of CR–propeller was fully determined by the component of unsteady loading noise, but in rotating plane the main contribution to S.P.L. comes from the component of steady loading noise. The result is similar with that of SR–propeller. The difference between CR and SR propellers is that the steading loading noise should be the sum of noise from front rotor and that from aft rotor.

2. The calculated results have been compared with a few test data, and the comparison showed that their tendencies are qualitatively consistent each other.

References

[1]Strack, W.C., "Technology and Benefits of Aircraft Counter Rotation Propeller" NASA TM 83983, 1982.

[2]Korkan, K.D., "Experimental Study of Noise Generated by Counterrotating, Propeller Systems", AIAA 84–2264, 1984.

[3]Farassat, F., "A Collection of Formulas for Calculation of Rotating Blade Noise–Compact and Non–Compact Source Results". AIAA 80–0996, 1980.

[4]Block, P.J.W., "Noise Radiation Patterns of Counter–Rotation and Unsteady Loaded Single–Rotation Propellers" Journal of Aircraft Vol. 22, No.9, pp. 776–783, 1985.

[5]Fujii, S., "Noise and Performance of a Counter–Rotation Propeller" Jaurnal of Aircraft, vol. 23, No.9, pp. 719–724, 1986.

[6]Hanson, D.B., "Compressible Helicoidal Surface Theory of Propeller Aerodynamic and Noise" AIAA Journal Vol.21, No. 6. pp. 881–889. 1983.

[7]Hanson, D.B., "Noise of Counter–Rotation Propellers" Journal of Aircraft Vol.22, No.7. pp. 609–617, 1985.

[8]Lowson, M.V., "A Theoretical Study of Helicopter Rotor Noise" Journal of Sonnd and Vibration, Vol. 9, No.2, pp. 197–222, 1969.

[9]Wright, S.E., "Sound Radiation from a Lifting Rotor Generated by Asymmetric Disc Loading" Journal of Sound and Vibration, Vol.4.7, pp.371–385, 1970.

[10]White, R.G., "Noise and Vibration" John Wiloy Sons, pp. 439–462, 1982.

[11]Shaw. I.M., "Rotor–Stator In teraction Noise Generation" Journal of Aircraft. Vol.19 pp.954–962, 1982.

[12]Sun, X.F., Hu, Z.A., Zhou, S., "Noise Prediction of Rotor–Stator Interaction for Fan / Compressor", ACTA Aeronautica Et Astronautica Sinica, Ser. A, Vol.10, No.1, 1989.

[13]Sun, X.F., Hu, Z.A., Zhou, S., "Noise Generation by Swept Cascade" Journal of Beijing University of Aeronautics and Astronantics, Vol.49, No.1, pp.39–49, 1989.

[14]Fujii, S., "Aeroacoustics of an Advanced Propeller Design U–nder Takeoff and Landing Conditions", Journal of Aircraft, Vol.23, pp.136–141, 1986.

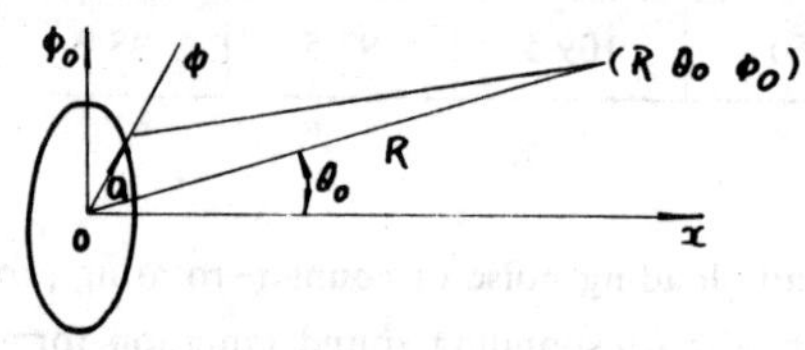

Fig.1 The coordinates of sound source point (o) and observing point (R,θ_0,Φ_0)

ON NUMERICAL PREDICTION OF SOUND FIELD GENERATED BY PROPELLER

By Huang Li–xi, Sun Xiao–feng and Zhou sheng
Beijing University of Aeronautics and Astronautics

Abstract

A numerical prediction method has been presented now, and free space field of propeller noise, mainly the near sound field radiated by noncompact sources, has calculated by a retarded time integration in time domain. The method distinguishes itself from many others by its features of no extra partial differencials nor singularities, therefore, it is accurate and efficient. Some comparisons of calculations with test results showed that the method and code could be used as an engineering tool.

§ 1. Introduction

A task of improving propellers of a kind of Chinese commercial aircraft has been received recently. The main technical objects of the new type of propellers include increasing efficiency and decreasing noise radiated by propeller. In order to realize those technical objects, the sub–system for prediction of sound field must be included among the whole designing system of propellers. It means that the acoustic performance of new type of propellers should be taken into account during every stage of developing the new type of propellers. Therefore, we need to possess a numerical method for predicting sound field radiated by subsonic propellers and the corresponding computer code. A numerical prediction method has been presented now. It is a kind of fast and accurate calculation method, and free space field of propeller noise, mainly the near sound field radiated by noncompact soureces, has calculated by a retarded time integration in time domain.

For solving the sound field generated by moving boundaries of solid body in flow field , the Ffowcs Williams–Hawkings equation has been used, in which the quadrupole term has been neglected and noncompact sources mean the thickness source and the load source, both of them have been included in the dipole term.

By usig strip assumption , the sound field generated by 3D shaped propeller could be cosidered as the sum of the terms radiated by every strip element along spanwise. Then, the pressure distribution along profile chordwise at every element should be known before solving the sound field. A code solving 2D compressible full velocity potential equation with body–fitting coordinates has been used which was developed by BUAA earlier.

It is well known that sound is a small distribution in the medium. Then, during numerical simulation process, the systematical errors of sound pressure will play more significant role on the accuracy of numerical results. The method distinguishes itself from similar others by the following improvement: i.e., the blade surface curve coordinates have been used to replace the chordwise coordinates which usually are used by some of similar methods. From this improvement, there is no any singularities in the integration at leading edge and trailing edge of blade.

§ 2. The retarded time integration in time domain

Firstly, the Ffowcs Williams–HawKings equation is

$$\Box^2 p' = \frac{\partial^2 T_{ij}}{\partial x_i \partial x_j} - \frac{\partial}{\partial x_i}[l_i|\nabla f|\delta(f)] + \frac{\partial}{\partial t}[\rho_0 V_n|\nabla f|\delta(f)] \tag{1}$$

where p′—— sound pressure ; $\Box^2$——$\dot{d}$Alember wave operator; T_{ij}—— Lighthill tensor; $l_i = P_a - P_o$, P_a is air pressure at surface and P_o the undisturbed pressure; δ——Dirac function ; ρ_0—— undisturbed air density; V_n—— velocity component of normal direction $\vec{n}$ at blade surface; t ——time at observing point , and equation

$$f(\vec{x},\ t) = 0 \tag{2}$$

describes the movement of blade surface; f<0—— outside surface; f<0 ——inside surface; $\vec{x}$——vector locating the observing point.

For subsonic propellers, the quadrupole term in eq.(1) could be neglected. Moreover, if a wave equation has the form

$$\Box^2 \Phi = Q(\vec{x},\ t) \cdot |\nabla f| \cdot \delta(f) \tag{3}$$

where Φ ——an unknown function ; Q——source, then, using Green function method, its solution is

$$4\pi\Phi = \int_{f=0} = \left[\frac{Q(\vec{y},\ \tau)}{r|1 - M_r|}\right]_{ret} ds \tag{4}$$

where $\vec{r}$—— a vector along sound transmiting direction and r is its modul; M_r—— $\vec{r}$ direction component of Mach number of blade movement; ret——retarded time; dS——element along blade surface. $\vec{y}$——vector locating source point.

By means of convolution formula:

$$\int \frac{\partial}{\partial y}[g_{(y)}] f_{(x-y)} \cdot dy = \frac{\partial}{\partial x}\int f_{(x-y)} g_{(y)} dy \tag{5}$$

we can get

$$4\pi p'_{(\vec{x},\ t)} = \frac{\partial}{\partial t}\int_{f=0}\left[\frac{\rho_0 V_n}{r|1 - M_r|}\right]_{ret} ds - \frac{\partial}{\partial x_i}\int_{f=0}\left[\frac{l_i}{r|1 - M_r|}\right]_{ret} ds \tag{6}$$

Because

$$-\frac{\partial}{\partial x_i}\left[\frac{\delta(g)}{r}\right] = \frac{1}{c}\frac{\partial}{\partial t}\left[\frac{\hat{r}_i \cdot \delta(g)}{r}\right] + \frac{\hat{r}_i \cdot \delta(g)}{r^2} \tag{7}$$

and while

$$g = \tau - t + \frac{r}{c} = 0 \tag{8}$$

we have

$$\frac{\partial}{\partial t}(\ \ )=\left[\frac{1}{1-M_r}\frac{\partial(\ \ )}{\partial \tau}\right]_{ret} \tag{9}$$

then, we can exchange into following form:

$$4\pi p'_{(\vec{x},\ t)}=\int_{f=0}\left[\frac{\rho_0 c^2 Q_A}{cr}+\frac{\rho_0 c^2 Q_B}{r^2}\right]_{ret}\cdot dS+\int_{f=0}\left[\frac{(p_a-p_0)Q_1}{cr}+\frac{(p_a-p_0)Q_2}{r^2}\right]_{ret}\cdot dS \tag{10}$$

where

$$Q_A=\frac{\Omega(n_1\hat{r}_2-n_2\hat{r}_1)}{(1-M_r)^2}+\frac{\Omega cos\theta(M_1\hat{r}_2-M_2\hat{r}_1)}{(1-M_r)^3} \tag{11}$$

$$Q_B=\frac{cos\theta(1+M_r)-M_n}{(1-M_r)^2}+\frac{cos\theta(M_r-M^2)}{(1-M_r)^3} \tag{12}$$

$$Q_1=\frac{\tilde{p}cos\theta}{(1-M_r)^2}+\frac{\Omega(n_1\hat{r}_2-n_2\hat{r}_1)}{(1-M_r)^2}+\frac{\Omega cos\theta(M_1\hat{r}_2-M_2\hat{r}_1)}{(1-M_r)^3} \tag{13}$$

$$Q_2=\frac{cos\theta(1+M_r)-M_n}{(1-M_r)^2}+\frac{cos\theta(M_r^2-M^2)}{(1-M_r)^3} \tag{14}$$

$$\tilde{p}=\frac{1}{(p_a-p_0)}\frac{\partial(p_a-p_0)}{\partial\tau} \tag{15}$$

where c——sound speed of undisturbed flow ; Ω –rotating angular speed of propeller; M_1 —— $\vec{r}_i$ component of Mach number of blade movement; $\vec{r}_i$ ——component of $\vec{r}$ at x_i direction; $\hat{r}_i$ —— $\hat{r}_i=\vec{r}_i/r$; M_n —— normal component of Mach number of blade movement; θ—— angle between $\vec{n}$ and $\vec{r}$; M —— Mach number of blade movement; τ——time of radiated sound by source .

The eg.(10) is the final integration solution for thickness noise and load noise radiated by propeller.

3. Numerical calculations

The integration is along the sound source surface i.e.the blade surface $f(\vec{y},\ \tau)=0$. Let p to represent an arbitrary point at blade surface. The vector $\vec{p}$ corresponds to point p, and determines on two parameters: radial coordinate x and non–dimensional curve length along blade surface l_0. So, the normal vector $\vec{N}$, unit normal vector $\vec{n}$ and element area ds, all of which corresponds to the point p, are as follows:

$$\vec{N}=\left(\frac{\partial\vec{p}}{\partial x}\right)\left(\frac{\partial\vec{p}}{\partial l}\right) \tag{16}$$

$$\vec{n}=\vec{N}/N \tag{17}$$

$$dS=N\cdot dx\cdot dl \tag{18}$$

respectively.
It means that the difference net is formed by means of two crossing curve families: x=constant and l=constant. Because aeroacoustic numerical simulation is very sensitive to the systematical errors, the accurate and efficient Gauss–Legendre integration is used for both directions .That is

$$\int_a^b f_{(x)}dx = \sum_{i=1}^{m} f_{(x_i)}\omega_i \tag{19}$$

and

$$x_i = \frac{1}{2}(b+a) + \frac{1}{2}(b-a)\xi_i \tag{20}$$

where ξ_i is position of zero point of Legendre polynome, and $\xi_i \in (-1,\ 1)$; ω_i——weight coefficient.The two groups $\{\xi_i\}$ and $\{\omega_i\}$ for a fixed class m can be found from standard mathematical tables.
The solution of retarded time τ is the key point of above mentioned numerical process.The retarded time equation is eq.(8). We use the Newton alternate method to find the solution of eq.(8), for

$$\frac{\partial g}{\partial \tau} = 1 - M_r > 0 \tag{21}$$

and g(τ) is a monotonous funtion for the case of subsonic propellers.
Therefore, there is no singularities in the integration eqution(10), and the calculating process doesn't contain any calculation of partial derivatives besides calculating normal vector $\vec{N}$.

§ 4. A numerical example

Some numerical examples have been made, and an example is described here.
The example is picked up from Ref.[1], and the main parameters of propeller are as follows: blade number——3.0; tip diameter——2.6m; shaft rotating speed——2145rpm; sound speed——345m / sec; flight velocity——40.14 m / sec. The observing point at the rotating disk plane moves with the aircraft together, and the distance between observing point and shaft axis is 7.25m. The spanwise distributions of three main geometric parameters have been presented at Fig.1, where CH is chord of profile; TH is thickness of profile; β is propeller's, blade angle, i.e.the angle between chord with rotating plane.
The comparison of numerical results with experimental data has been shown at the spectral distribution figure (Fig.2), and the numerical results from Ref.[1, 2] have been presented at the same figure too. The measured spectral distribution showed that the pure tones corresponding to blade passing frequency and its harmonics possess the predominant above the background broadband noise and the SPL of pure tones decrease monotonically with the number of harmonics.The prediction data, including ours and those from Ref.[1, 2], have caught the main feature of measured spectral distribution.
Fig.3 showed the individual contribution from thickness source term and from load source term to SPL.It is notable from Fig.2 that, for thickness source, the tendency of decreasing SPL with the number of harmonics is slower than the tendency for thickness source, and the peak isn't at the first harmonics.
In a word, some comparisons of calculations with test results showed that the method and code could be used as an engineering tool.During whole designing process of the new type of propellers, the near and far sound fields of different projects have been predicted by means of the code.The practice has shown that the code for predicting sound field is necessary of quaranting the combination of high efficiency with low noise.

Concluding Remarks

A numerical method for predicting free space aeroacoustic field generated by propeller has been presented in this paper. It is a retarded time integration in time domain .The method uses the strip simplification and considers noncompact sources, i.e. thickness source and load source.The method distinguishes itself from many others by its features of no extra partial differencials nor singularities.Thus it is accurate and efficient.
Some comparisons of calculations with test results showed that the method and code could be used as an engineering tool.During the designing process of a new type of propellers, the near and far sound fields of different projects have been predicted by means of the code.The practice of developing the new propellers has shown that the code for predicting sound field is necessary of guaranting the combination of high efficiency with low noise.

References

[1] Farassat, F.and Brown, T.J., A New Capability for Predicting Helicopter Rotor and Propeller Noise Including Forward Motion, NASA TM X–74037, 1977.

[2] Woan, C.J., and Gregorek, G.M., The Exact Numerical Calculation of Propeller Noise, AIAA 78–1122, 1978.

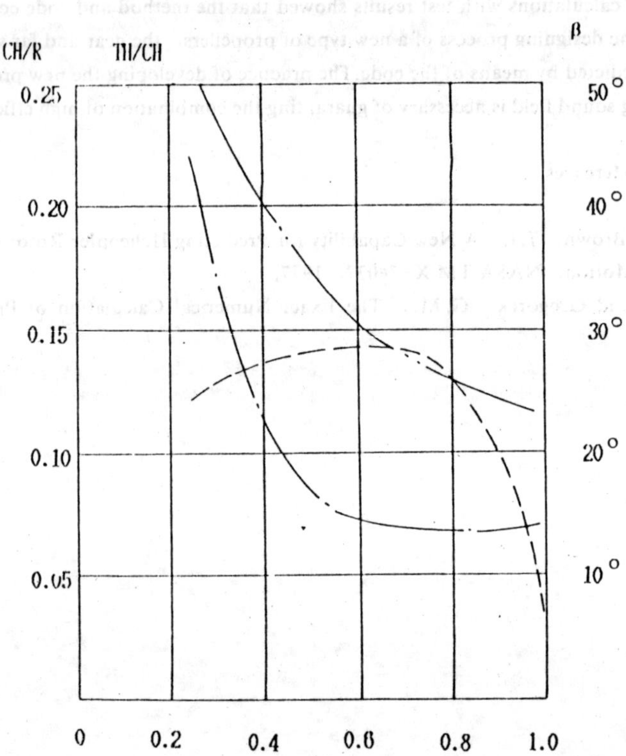

Fig.1 The spanwise distribution of three geometric parameters

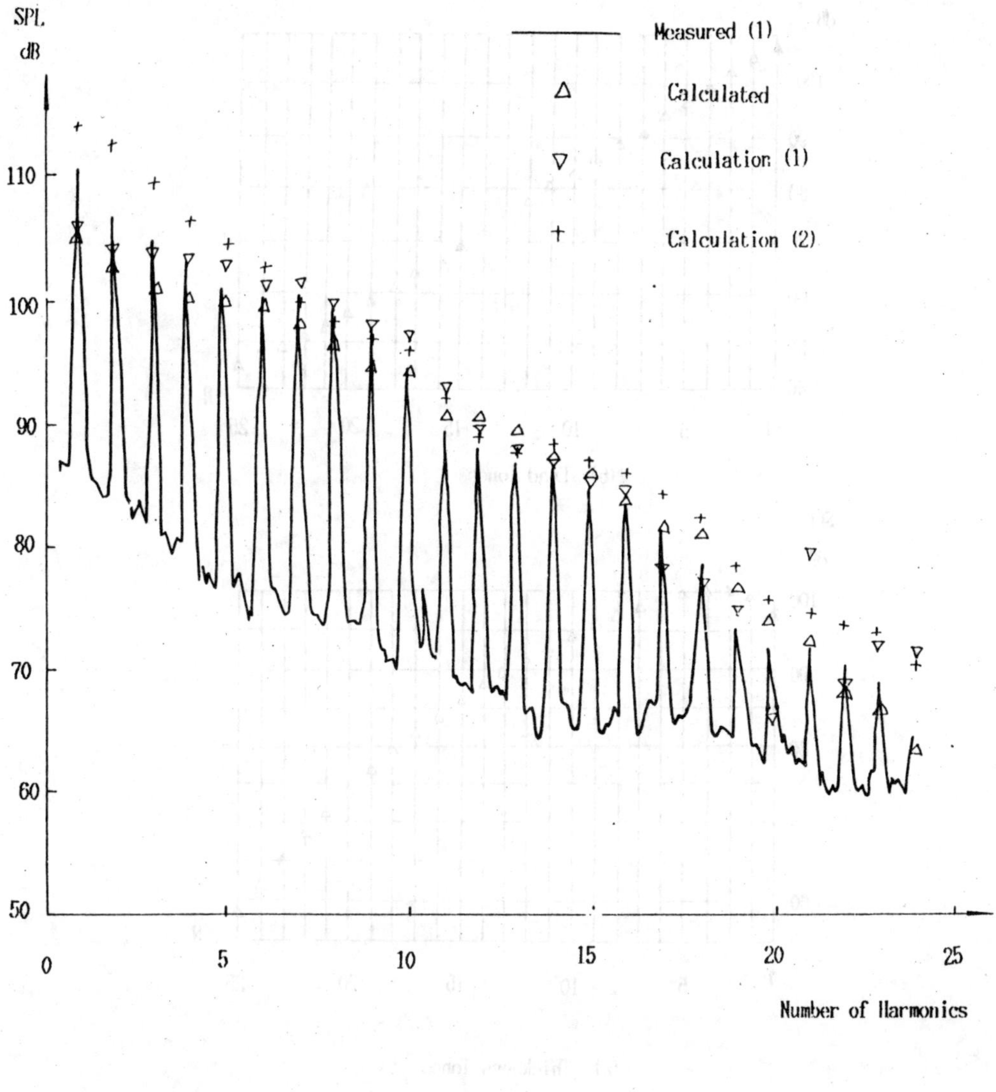

Fig.2 The Comparison of Numerical Results with Measured Data

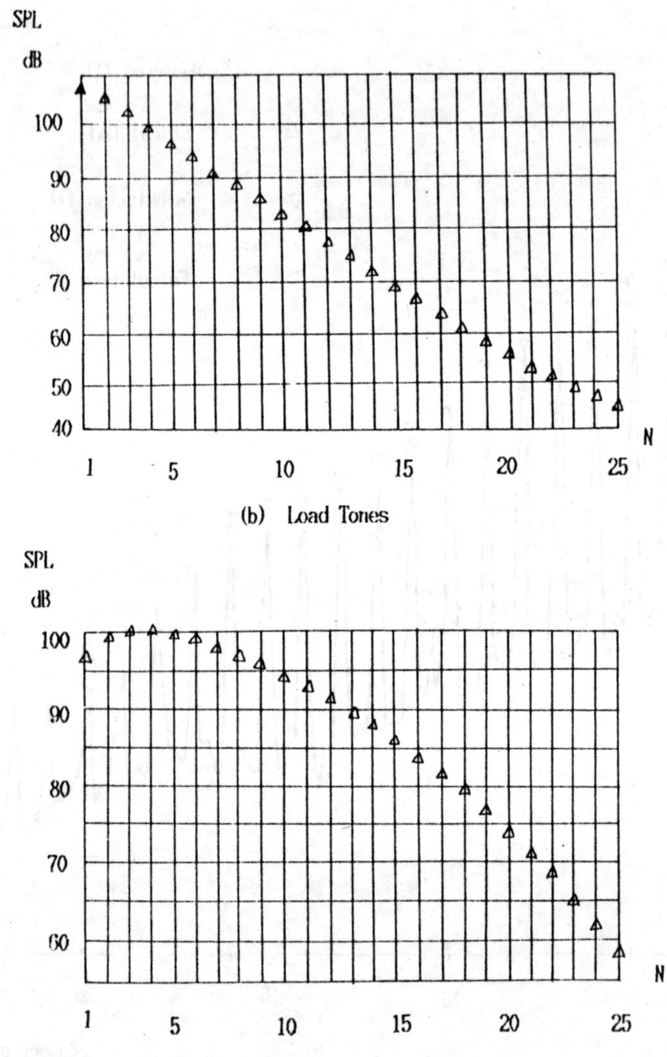

Fig. 3 Calculated SPL at each harmonics

Noise Generation by Swept Cascade

Sun Xiaofeng, Hu Zongan and Zhou Sheng
(Beijing University of Aeronautics and Astronautics)

Abstract

A theoretical model for analysing the aeroacoustic nature of swept cascade is described. Some closed form expressions for the acoustic pressure and acoustic power in far field are obtained. These expressions are used in parametric calculations of rotor viscous wake–stator vane interaction in order to study the effectiveness of sweep as a noise reduction mechanism. It is concluded that the sweep angle plays an important role in the generation of noise of cascade, in some cases, however, the sweep angle does not result in the reduction of noise generated by the interaction of rotor viscous wakes with a stator vane.

Introduction

Fan and compressor noise reduction is a current problem of interest . In general, noise reduction can be achieved in two ways: (1) reduction through the attenuation of propagating sound fields; and (2) reduction of the strength of the noise sources themselves. The design of swept cascade is an approach for the reduction of noise at the sources, which plays an important role in the aeroacoustic design of fan and compressor. Some experimental results[1,2] show that swept cascade can reduce not only the shock noise but also the rotor–stator interaction noise in turbomachines.

There have been many attempts[3–8] to understand the noise generated by a airfoil, some of which put emphasis on analysing the acoustic and flow field caused by the interaction of an airfoil with an oblique gust. For example, Filoat[3] investigated the flow and acoustic field due to the interaction of an infinite span wing with an obligue sinusoidal gust for an incompressible fluid. In 1974, Adamczyk[4] extended this model to consider the effects of compressibility, three–dimensionality and sweep, but he still made use of the assumption that the wing span is infinite. In 1983, Kerchen[7] overcame some deficiencies existing in previous works and investigated the interaction of a finite span wing with income disturbances by using Wiener–Hopf technique[8], and discussed the effects of sweep angle on the acoustic field. The wing, however, is assumed to have a semi–infinite chord in his work. In 1984, Kerchen[9] extended this model to study the noise generated by the interaction of a rotor–wake with a swept stator, but he only considered the interaction of rotor–wake with a single stator blade, i.e., the effects of cascade are not included in his analysis. In spite of this, Kerchen may be the first to give theoretical model about the aeroacoustic nature of swept cascade. In recent analysis, Namba and Fan[10] investigated the aeroelastic and aeroacoustic natures of supersonic swept cascade, and given some valuable computational results.

In this paper a new model about the acoustic field of swept cascade is presented. Some closed expressions are given by solving three dimensional linear Euler equation. Finally the effects of sweep angle on the sound generated by the cascade are discussed.

Formulation

Figure 1 shows the geometry of the problem. A swept blade of chord 2b is held rigidly on a compressible fluid which is flowing over it with uniform velocities $U_0\cos\alpha_1$ in the x_1 direction, and $U_0\sin\alpha_1$ in the x_3 direction. Superposed in U_0 in an unsteady three-dimensional disturbance u_i, the pressure and density are p_0+p, and $\rho_0+\rho$, respectively, where the subscripted part denotes the mean value, while the remaining part is associated with the disturbance.

The problem is described by the customary fluid dynamic equations for a compressible flow. The equations of conservation of mass and momentum are linearized by assuming that the perturbations are small compared to the mean values thereby allowing the squares and products of the perturbations (and their derivatives) to be neglected, while first-order terms are retained[11]. Hence,

$$\left.\begin{aligned}
&\frac{\partial\rho}{\partial\tau}+U_0\cos\alpha_1\frac{\partial\rho}{\partial x_1}+U_0\sin\alpha_1\frac{\partial\rho}{\partial x_3}+\rho_0\nabla\cdot\vec{u}=0\\
&\rho_0\left(\frac{\partial u_1}{\partial\tau}+U_0\cos\alpha_1\frac{\partial u_1}{\partial x_1}+U_0\sin\alpha_1\frac{\partial u_1}{\partial x_3}\right)+\frac{\partial p}{\partial x_1}=0\\
&\rho_0\left(\frac{\partial u_2}{\partial\tau}+U_0\cos\alpha_1\frac{\partial u_2}{\partial x_1}+U_0\sin\alpha_1\frac{\partial u_2}{\partial x_3}\right)+\frac{\partial p}{\partial x_2}=0\\
&\rho_0\left(\frac{\partial u_3}{\partial\tau}+U_0\cos\alpha_1\frac{\partial u_3}{\partial x_1}+U_0\sin\alpha_1\frac{\partial u_3}{\partial x_3}\right)+\frac{\partial p}{\partial x_3}=0
\end{aligned}\right\}\qquad(1)$$

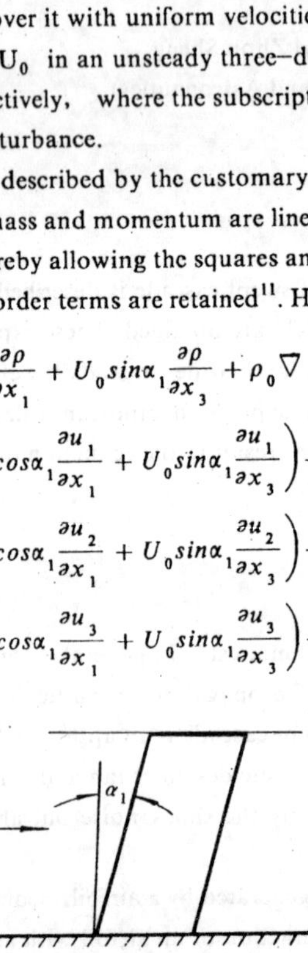

Fig.1

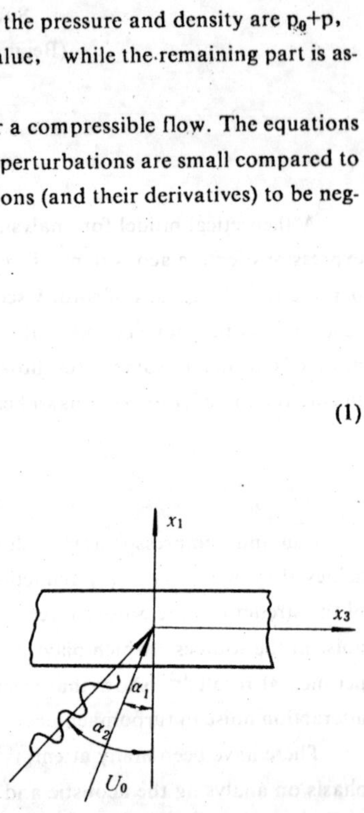

Fig.2

For small fluctuations, the change in pressure is proportional to the change in density

$$p=c_0^2\rho$$

where c_0 is the speed of sound.

By combining the equation(1) with the state equation given above, one obtains the governing differential equation for the problem

$$\frac{1}{c_0^2}\frac{D^2p}{D\tau^2}-\nabla^2p=0\qquad(2)$$

where

$$\frac{D}{D\tau}=\frac{\partial}{\partial\tau}+U_0 cos\alpha_1\frac{\partial}{\partial x_1}+U_0 sin\alpha_1\frac{\partial}{\partial x_3}$$

Equation (2) can alse be written as

$$c_0^2\left(\nabla^2 p - M_0^2 cos^2\alpha_1\frac{\partial^2 p}{\partial x_1^2} - M_0^2 sin^2\alpha_1\frac{\partial^2 p}{\partial x_3^2}\right)$$

$$-2c_0^2\left(\frac{M_0 cos\alpha_1}{c_0}\frac{\partial^2 p}{\partial x_1\partial\tau}+\frac{M_0 sin\alpha_1}{c_0}\frac{\partial^2 p}{\partial x_3\partial\tau}+M^2 cos\alpha_1 sin\alpha_1\frac{\partial^2 p}{\partial x_1\partial x_3}\right)-\frac{\partial^2 p}{\partial\tau^2}=0 \qquad (3)$$

Equation (3) is a complicated three–dimensional linear partia differential equation, and it is very difficult to solve it directly. In order to simplify the equation, suppose that the span of blade is infinite and steady attack angle of inlet flow is zero (i.e. the effects of blade mean loading on unsteady forces is not considered). Therefore a tranformation similar to Prandtl–Glauert can be intorduced in the form[11]

$$p=\bar{p}(\zeta,\ \eta)exp\left\{i[-\beta_0 M^2 cos^2\alpha_1 x_1-\bar{K}U_0 cos(\alpha_1-\alpha_2)\tau+\frac{\bar{K}bsin\alpha_2}{\beta_r}x_3]\right\}$$

where

$$\beta_0=\frac{\bar{K}bcos\alpha_2}{\beta_r^2}$$

$$\zeta=\frac{x_1}{b}$$

$$\eta=\frac{x_2\beta_r}{b}$$

$$\xi=\frac{x_3\beta_r}{b}$$

$$\beta_r=(1-M_0^2 cos^2\alpha_1)^{\frac{1}{2}}=(1-M_r^2)^{\frac{1}{2}}$$

When this is substituted into the equation(3), one obtains

$$\frac{\partial^2\bar{p}}{\partial\zeta^2}+\frac{\partial^2\bar{p}}{\partial\eta^2}+K_\theta^2\bar{p}=0 \qquad (4)$$

Where

$$K_\theta=\frac{\bar{K}b}{\beta_r^2}(M_0^2 cos^2\alpha_1-sin^2\alpha_2)^{\frac{1}{2}}$$

Taking Fourial transformation to Eq(4) yields

$$\bar{p}(\alpha,\ \eta)=\frac{\underset{\sim}{f}(\alpha)}{2}exp(-\gamma|\eta|)sgn(\eta)$$

where

$$\bar{\bar{p}}=\int_{-\infty}^{+\infty}\bar{p}(\zeta,\ \eta)e^{-i\zeta\alpha}d\zeta$$

$$\bar{p}=\frac{1}{2\pi}\int_{-\infty}^{+\infty}\bar{p}(\alpha,\ \eta)e^{i\zeta\alpha}d\alpha$$

$$sgnx=\begin{cases}1 & for\quad x>0\\ -1 & for\quad x>0\end{cases}$$

and f(α) is the Fourier transformation of the pressure difference between uper and lower surface of blade.

It is convenient to introduce for each integer m = 0, ± 1, ± 2, ± 3, ···(see Fig 3), a coordinate system

$$\eta_m=\eta-\frac{ms}{b}\beta_r$$

$$\zeta_m=\zeta-\frac{\overset{+}{ms}}{b}$$

that has its origin on the mth blade. Hence, the solution of Eq(4) becomes

$$p_m(\zeta,\ \eta)=\frac{sgn(\eta_m)}{2\pi}\int_{-\infty}^{+\infty}\frac{f_m(\alpha)}{2}exp(i\alpha\zeta_m-|\eta_n|\gamma)d\alpha$$

$$p(\zeta,\ \eta)=\sum_{m=-\infty}^{+\infty}p_m(\zeta,\ \eta) \tag{5}$$

It is noticed that in solving of Eq(5), $\bar{K}$ has been assumed to have a small positive imaginary part that will be set to zero at the end of the analysis. The net effect of this is to replace the usual outgoing–wave boundary condition at infinity by a boundedness requirement. Pysically It amounts to intorducing a small amount of damping into the system[12, 13].

Since the solution for velocity field u_i must have the same frequency and the same spanwise wave number as the pressure field, perturbation velocity component can be written as

$$u_2=\bar{u}_2(\zeta,\ \eta)exp\left[-i\bar{K}U_0cos(\alpha_1-\alpha_2)t+i\frac{\bar{K}bsin\alpha_2}{\beta_r}\xi\right] \tag{6}$$

Inserting this result into Eq(1) yields

$$\bar{u}_2=-\frac{1}{2\pi}\int_{-\infty}^{+\infty}\frac{\beta_r\gamma exp(-iK_rM_r\zeta)}{2i\left(\frac{K_r}{M_r}-\alpha\right)\rho_0U_r}\sum_{m=-\infty}^{+\infty}f_m(\alpha)exp(i\alpha\zeta_m-|\eta_m|\gamma)d\alpha \tag{7}$$

Suppose the interblade phase angle is constant, i.e.

$$f_m(\alpha)=e^{im}\Gamma f_0(\alpha) \tag{8}$$

then

$$\sum_{-\infty}^{+\infty}exp\left[im\left(\Gamma-\frac{\overset{+}{ms}}{b}\right)-|\eta_m|\gamma\right]=\frac{1}{2}\left[\frac{exp(-\eta\gamma+\frac{1}{2}\Delta_+)}{sinh\frac{1}{2}\Delta_+}-\frac{exp(\eta\gamma+\frac{1}{2}\Delta_-)}{sinh\frac{1}{2}\Delta_-}\right] \tag{9}$$

Substituting Eq(8) and Eq(9) into Eq(7) shows that the upwash velocity is now determined by

$$\bar{u}_2(\zeta, \eta) = \frac{\beta_r}{2\pi} \int_{-\infty}^{+\infty} \frac{\gamma f_0(\alpha) exp[i(\alpha - K_r M_r)\zeta]}{2i\rho_0 U_r \left(\alpha - \frac{K_r}{M_r}\right)} \times \frac{1}{2} \left[\frac{exp(-\eta\gamma + \frac{1}{2}\Delta_+)}{sinh\frac{1}{2}\Delta_+} - \frac{exp(\eta\gamma + \frac{1}{2}\Delta_-)}{sinh\frac{1}{2}\Delta_-} \right] d\alpha \tag{10}$$

Where

$$K_r = \frac{\bar{K} b U_0 cos\alpha_1 cos\alpha_2}{c_0 \beta_r^2} \tag{11}$$

$$\Delta = i\left(\Gamma - \frac{\alpha s^+}{b}\right) \pm \frac{s\beta_r \gamma}{b} \tag{12}$$

$$\gamma = \sqrt{\alpha^2 - K_\theta^2} \tag{13}$$

Eq(10) exists poles

$$1).\quad \alpha = +\frac{K_r}{M_r} \tag{14}$$

$$2).\quad \Delta_\pm = i2n\pi \qquad n = 0,\ \pm 1,\ \pm 2\cdots\cdots$$

Substituting $\Delta_\pm = i2n\pi$ to Eq(12) implies

$$\alpha_n^\pm = \Gamma_n \frac{bs^+}{(d^+)^2} \pm s\beta_r / d^+ \sqrt{K_\theta^2 - \left(\frac{b\Gamma_n}{d^+}\right)^2} \tag{15}$$

Where

$$d^+ = \sqrt{s^{+^2} + \beta_r^2 s^2}$$

$$\Gamma_n = \Gamma - 2n\pi$$

In the following section, it will be seen that the solutions of flow and acoustic field can be obtained from Eq(10).

1. Solution of the acoustic field

By analyzing the properties of Eq(10), it can be shown that the acoustic wave that can propagate to far field without decaying must meet the following condition

$$K_\theta^2 - \left(\frac{b\Gamma_n}{d^+}\right)^2 > 0 \tag{16}$$

and the remaining acoustic wave will decay exponentially rapidly at large values of $|x_1|$, in other words, they will be out off. Therefore Eq(15) that meets the condition (16) can be transformed into

$$\alpha_n^\pm = \Gamma_n \frac{bs^+}{(d^+)^2} \pm \frac{s\beta_r}{d^+} \sqrt{K_\theta^2 - \left(\frac{b\Gamma_n}{d^+}\right)^2} = K_\theta sin\theta_n^\pm$$

where

$$\gamma(\alpha_n^{\pm}) = \sqrt{\alpha^2 - K_\theta^2} = \mp K_\theta cos\theta_n^{\pm}$$

$$sin\theta_n^{\pm} = sin(x^+ \pm \delta_n)$$

$$\delta_n = cos^{-1}\frac{b\Gamma_n}{K_\theta d^+}$$

$$x^+ = tg^{-1}\frac{s^+}{s\beta_r}$$

Notice that Eq(10) can be written as

$$\bar{u}_2(\zeta, \eta) = \frac{\beta_r}{4\pi i\rho_0 U_r}\frac{\partial^2}{\partial\eta^2}\left\{\int_{-\infty}^{+\infty}\frac{f_0(\alpha)exp[i(\alpha - K_r M_r)\zeta]}{\gamma(\alpha - \frac{K_r}{M_r})}\cdot\frac{1}{2}\left[\frac{exp(-\eta\gamma + \frac{1}{2}\Delta_+)}{sinh\frac{1}{2}\Delta_+} - \frac{exp(\eta\gamma + \frac{1}{2}\Delta_-)}{sinh\frac{1}{2}\Delta_-}\right]d\alpha\right\} \quad (17)$$

It is convenient to obtain the solution of Eq(10) by means of complex variables theory. As described above, there are the poles $a = +\frac{K_r}{M_r}$ and $a_n^{\pm}$. The fromer term $a = +\frac{K_r}{M_r}$ represents a convected disturbance (vortical wave), which propagates downstream and does not contribute to acoustic pressure. Therefore the solution of the far acoustic field is

$$\bar{u}_2(\zeta, \eta) = \lim_{\zeta\to\pm\infty}\frac{\beta_r}{4\pi i\rho_0 U_r}\frac{\partial^2}{\partial\eta^2}[2\pi i Res(a_n^{\pm})] = \frac{\beta_r}{2i\rho_0 U_r}\frac{bK_\theta cos^2\theta_n^{\pm}exp[(iK_\theta sin\theta_n^{\pm} - iK_r M_r)\zeta + iK_\theta cos\theta_n^{\pm}\eta]}{d^+[M_r K_\theta sin\theta_n^{\pm} - K_r]sin\delta_n}$$

$$\times f_0(K_\theta sin\theta_n^{\pm})$$

where

$$f_0(K_\theta sin\theta_n^{\pm}) = \int_{-1}^{1}\Delta\bar{p}(\zeta)exp[-i(K_\theta sin\theta_n^{\pm} - K_r M_r)\zeta]d\zeta$$

Furthermore, the solution of acoustic pressure in far field, by applying the Eq(1), can be expressed as

$$p_n^{\pm} = \frac{bcos\theta_n^{\pm} f_0(K_\theta sin\theta_n^{\pm})}{2d^+ sin\delta_n}$$

$$\times exp[-i\bar{K}U_0 cos(\alpha_2 - \alpha_1)\tau + i(K_\theta sin\theta_n^{\pm} - K_r M_r)\zeta + iK_\theta cos\theta_n^{\pm}\eta + i\frac{\bar{K}bsin\alpha_2}{\beta}\xi] \quad (18)$$

$$p^+ = \sum_{n=m_1}^{m_2} p_n^+ \qquad \zeta\to+\infty \quad (19)$$

$$p^- = \sum_{n=m_1}^{m_2} p_n^- \qquad \zeta\to-\infty \quad (20)$$

where the "plus" sign represents the acoustic wave which propagates upstream, while "minus" sign represents that which propagates downstream, and m_1 and m_2 the minimum and maximum values of n for which the inequality (16) holds.

By considering a coordinate transformation (See Fig4)

$$y_1 = b\zeta_1 cos\theta - b\eta sin\theta$$
$$y_2 = b\zeta sin\theta - b\eta cos\theta$$

Eq(18) can be simplified as

$$p_n^+ = \frac{b}{2k_n d}\left(\frac{\sigma - 2n\pi}{d}cos\theta - \gamma_n sin\theta\right) f_0(K_\theta sin\theta_n^{\pm})$$
$$\times exp\left\{ i\left[-\bar{K}U_0 cos(\alpha_2 - \alpha_1)\tau + \frac{\sigma - 2n\zeta}{d} y_2 + \gamma_n y_1 + \bar{K} sin\alpha_2 y_3 \right]\right\} \tag{21}$$

where

$$k_n = \sqrt{k_0^2 - \beta^2\left[\left(\frac{\sigma - 2n\pi}{d}\right)^2 + \left(\frac{\bar{K}U_0 cos(\alpha_2 - \alpha_1) sin\alpha_2}{c_0 cos(\alpha_2 - \alpha_1)}\right)^2\right]}$$

$$k_0 = \frac{M_r sin\theta(\sigma - 2n\pi)}{d} - \frac{\bar{K}U_0 cos(\alpha_2 - \alpha_1) cos\alpha_2 cos\alpha_1}{c_0 cos(\alpha_2 - \alpha_1)}$$

$$\gamma_n = \frac{1}{\beta^2}[Mk_0 \pm k_n]$$

$$\beta^2 = \sqrt{1 - M_r^2 cos^2\theta}$$

2. Sound power of swept cascade

As described in ref.12 and ref.14, the intensity of sound, in moving media is defined as

$$\vec{I} = \left(\frac{p'}{\rho_0} + \vec{u}\cdot\vec{U}\right)(\rho_0\vec{u} + \rho_0\vec{U})$$

Then the sound intensity along the minus or plus direction of y can be written as

$$I_n^{\pm} = (1 + M^2) < pu_1 > + \frac{M}{\rho_0 c_0} < p^2 > + \rho_0 c_0 M < u_1^2 > \tag{22}$$

where $< pu_1 >$, $< p^2 >$ and $< u_1^2 >$ represent the time averaging values to the real part Re(p) and $Re(u_i)$ respectively.

Since

$$\rho_0\left(\frac{\partial u_1}{\partial \tau} + U\frac{\partial u_1}{\partial y_1}\right) = -\frac{\partial p}{\partial y_1}$$

substituting Eq(21) into the equation given above yields

$$u_1 = \frac{\gamma_n^{\pm}\cdot p_n}{\rho_0(\bar{K}U_0 cos(\alpha_2 - \alpha_1) - U\gamma_n)} \tag{23}$$

and

$$I_n^{\pm} = \frac{1}{\rho_0 U}\frac{M^2\beta^4\left(\frac{\bar{K}U_0 cos(\alpha_2 - \alpha_1)}{U}\right)K_y}{\left[\frac{\bar{K}U_0 cos(\alpha_2 - \alpha_1)}{c_0} - MK_y\right]^2} < \bar{p}_n^2 >$$

$$I^{\pm} = \sum_{n=m_1}^{m_2} I_n^{\pm} = \frac{1}{\rho_0 U} \sum_{n=m_1}^{m_2} \frac{M^2\beta^4\left(\frac{\bar{K}U_0 cos(\alpha_2 - \alpha_1)}{U}\right)K_y}{\left[\frac{\bar{K}U_0 cos(\alpha_2 - \alpha_1)}{c_0} - MK_y\right]^2} \bar{p}_n^2 \tag{24}$$

where

$$K_y = M\left[\frac{M_r sin\theta(\sigma - 2n\pi)}{d} + \frac{\bar{K}U_0 cos(\alpha_2 - \alpha_1) sin\alpha_2 sin\alpha_1}{c_0 cos(\alpha_2 - \alpha_1)}\right] + k_n$$

$$\bar{p}_n = \frac{b}{2k_n d}\left(\frac{\sigma - 2n\pi}{d} cos\theta - \gamma_n sin\theta\right) f_0(K_\theta cos\theta_n)$$

Apparently, the sound power can be obtained by multipling the sound intensity by the inlet or exit area of cascade. As will be seen below, the formulas given above can be used to calculate the sound power generated by swept cascade.

3. Solution of flow field

In Eq(10), let $\eta \to 0$, then

$$\bar{u}_2(\zeta, 0) = \frac{\beta_r}{2\pi}\int_{-\infty}^{+\infty} \frac{f_0(\alpha) exp[i(\alpha - K_r M_r)\zeta]}{2i\rho_0 U_r\left(\alpha - \frac{K_r}{M_r}\right)} \cdot \frac{\gamma sinh\frac{s\beta_r\gamma}{b}}{cosh\frac{s\beta_r\gamma}{b} - cos\left(\Gamma - \frac{\alpha s^+}{b}\right)} d\alpha$$

or

$$\bar{u}_2(\zeta, 0) = \int_{-1}^{1} \Delta p(\zeta) K(\zeta - x) d\zeta \tag{25}$$

where

$$K(\zeta) = \frac{\beta_r}{4\pi i}\int_{-\infty}^{+\infty} \frac{exp[i(\alpha - K_r M_r)\zeta]}{\rho_0 U_r\left(\alpha - \frac{K_r}{M_r}\right)} \cdot \frac{\gamma sinh\frac{s\beta_r\gamma}{b}}{cosh\frac{s\beta_r\gamma}{b} - cos\left(\Gamma - \frac{\alpha s^+}{b}\right)} d\alpha$$

As long as the upwash velocity is given on the surface, the pressure distrubution Δp of blade can be obtained by applying "collocation method"[16, 17]. Therefore the solutions of flow and acoustic field can also be determined. Finally, It can be demonstrated that whether the upwash velocity distribution is caused be convected disturbances or blade vibration, the flow and acoustic field can all be obtained through Eq(25), Eq(21) and Eq(24) respectively.

4. Numerical results and discussion.

The rotor wake[15] can be expanded to Fourier Series, i.e,

$$u_2 = \sum_{n=-\infty}^{+\infty} A exp\left\{i\left[KZ + n\frac{2\pi m B}{V} - mB\Omega\tau\right]\right\} \tag{26}$$

It is converient to show that $\sigma = \frac{2\pi m B}{V}$ for such disturbance waves as described above. Hence, the formulas, which are used to calculate the rotor–stator interaction noise, can be written as

$$p_n^{\pm} = \frac{b}{2k_{mn}d}\left(\frac{mB-nV}{r_m}\cos\theta - \gamma_{mn}\sin\theta\right)f_0(K_0\sin\theta_n^{\pm})$$

$$\times\exp\left\{i\left[-mB\Omega t + \frac{mB-nV}{r_m}y_2 + \gamma_{mn}y_1 + \bar{K}\sin(\alpha_2)y_3\right]\right\} \tag{27}$$

$$I_n^{\pm} = \frac{1}{\rho_0 U}\frac{M^2\beta^4\left(\frac{mB\Omega}{U}\right)K_y}{\left[\frac{mB\Omega}{c_0} - MK_y\right]^2}(\bar{p}_n^{\pm})^2 \tag{28}$$

where

$$k_{mn} = \sqrt{k_0^2 - \beta^2\left[\left(\frac{mB-nV}{r_m}\right)^2 + \left(\frac{mB\Omega\sin\alpha_2}{c_0\cos(\alpha_2-\alpha_1)}\right)^2\right]}$$

$$k_0 = \frac{M_r\cos\theta(mB-nV)}{r_m} - \frac{mB\Omega\cos\alpha_2\cos\alpha_1}{c_0\cos(\alpha_2-\alpha_1)}$$

$$\gamma_{mn} = \frac{1}{\beta^2}[Mk_0 \pm k_{mn}]$$

As an application of the expressions derived above, some computational results have been given. The main parameters which relate with the results are blade number $B=44$, vane number $V=48$, inlet speed $U=163.53\ m/s$, rotation apeed $\Omega=917.56$ rad/s, and seting angle $\theta_r=55.2°$, $\theta=14.13°$ respectively. By using these parameters, the sound power radiated for each harmonic can be calculated when the sweep angle of blade is zero, and then the reletive values of sound power can also be obtained with the change of the sweep angle. Fig 5–8 are the computational results. As shown from these figures, the main factor that affects the tendency of total sound power is the one radiated at the first harmonic, and the tendency of each harmonic is different from each other with the change of the sweep angle, only when the sweep angle approaches the cut–off angle will they show monotone decreacing tendencies. However, Aport from the noise, in the design stage the aerodynamic performance is also an important factor that affects the design. This means that designer can only choose a limited sweep angle if the aerodynamic performance is considered.

Therefore, If the design of swept cascade is used to reduce rotor–stator imteraction noise instead of the shock noise, it may be thought that designer should choose the sweep angle carefully so that the cascade has both good performance and low noise level. On the other hand, the computational results demonstrate the remarkable differiences between the single blade model as described in ref.9 and the cascade model given in present analysis, one of which is the former can not predict the "cut–off" effects, while this is an very important factor that affects the values of the sound power.

Conclusions

Several expressions for the determination of the acoustic field of swept cascade are presented herein. Although all these expressions are of theoretical value on the basis of simplized condilions, in engineering practices, it may be valuable to apply them to analyze the effects of the geometry and performance parameters of cascade on the noise generated by the interaction of a convected disturbace with swept cascade or by the vibration of blade itself. On the other hand, the computational results shows that, in some range of sweep angle, the desgin of swept blade does not certainly result in the reduction of the sound power. This means that in design stage the sweep angle should be chosen carefully so that the blade is both good aerodynamic performance and low noise level.

Nomenclature

b	semi–chord of blade
B	blade number
V	vane number
c	sound speed
i	complex $i = \sqrt{-1}$
d	blade pitch
M	Mach number $M = U_r cos\theta / c_0$
M_o	Mach number $M_0 = U_0 / c_0$
M_r	Mach number $M_r = \frac{U_r}{c_0} = \frac{U_0 cos\alpha_1}{c_0}$
r_m	characteristics radius
$\overline{K}$	wave number $\overline{K} = \frac{p\Omega}{U_0 cos(\alpha_2 - \alpha_1)}$
U_0	inlet speed of blade
U_r	$U_r = U_0 cos\alpha_1$
u_i	perturbation velocity, i = 1, 2, 3
p	sound pressure
θ	vane setting angle
θ_r	blade setting angle
ρ_0	steady average density
α_1	sweep angle
α_2	(See Fig2)
SI	sound intnsity
SP	sound Power

References

[1] Donald B et al. Analysis and design of high–speed low noise aircraft fan incoporating leading edge rotor and stator blades NASA cR B 5092, 1977

[2] Donald B et al. Design considerations for a novel low sources noise transonic fan stage AIAA 76577. 1976

[3] Filotas I. T. Response of an infinite wing to an oblique sinusoidal gust: a generation of sears problem. NASA Sp–207, 1969

[4] Adamczyk J. T. Passage of a swept aorfoil through an obligue gust. journal of Aircraft. 1974; 11: 281~287
[5] Sing Chu et al. Prediction of unsteady airloads for oblique blade–gust interaction in compressible flow., AIAA. J. 1974; 12(9): 1228-1235
[6] Rudolph Martinez. Unified aerodynamic–acoustic theory for a thin rectangular wing encountering a gust AIAA. J. 1980; 18(6): 636-645
[7] Kerchch E. J. Noise generation by a finite span swept airfoil. AIAA 83–0768, 1983
[8] Noble, Behjamin. Methods Based on Wiener–Hopf technique for the Solution of Partial Differential Equations, Pergamon Press, 1958
[9] Kerchen E. J. Noise produced by the interaction of a rotor wake with a swept stator blade. AIAA 84–2326, 1984
[10] Namba and Fan F–D. Unsteady aerodynamic characteristics of A supersonic cascade of swept blades. ISABE 85–7080, 1985: 679-686
[11] Adamczyk J.T. Analytical investigation of compressibility and three–dimeasionality of the unsteady response of an airfoil in a fluctuating flow field. AIAA 73–683, 1973
[12] Goldstem M. E. Aeroacoustics. McGRAW–HILL, 1976
[13] Lane, Frank and Manfred Friedman. Theoretical investigation of subsonic oscillatory blade–row aerodynamics. NACA TN 4136, 1958
[14] Morfey C. L. Acoustic energy in non–uniform flows. J. Sound Vib. 159-170
[15] Shaw L. M. Rotor wake characteristics relevant to rotor–stator interaction noise generation. AIAA 81–2031, 1981
[16] Sun, X.F., Hu, Z.A. and Zhou, S. Noise prediction of rotor–stator interaction for fan and compressor, ACTA Aeronautica Et Astronautica Sinica, Ser.A, Vol10, No1, 1989.
[17] Ventres, G.S. et al Turbofan noise. NASA CR–167952, 1983.

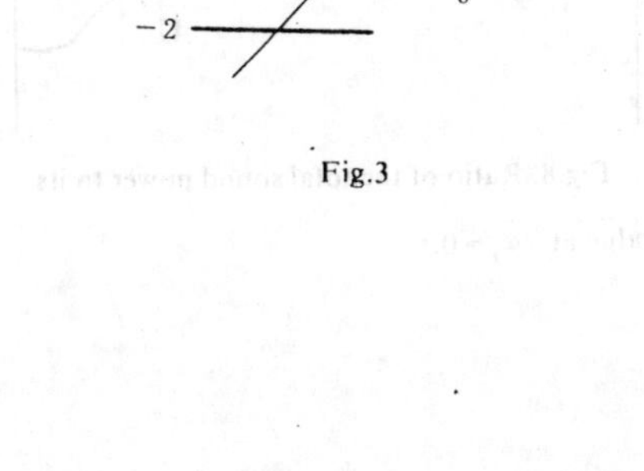

Fig.3

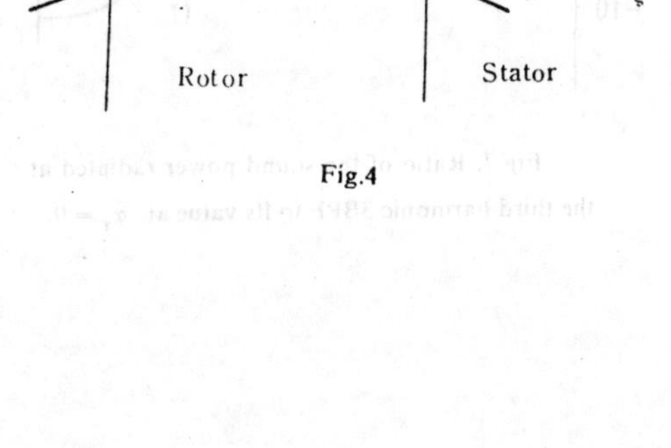

Fig.4

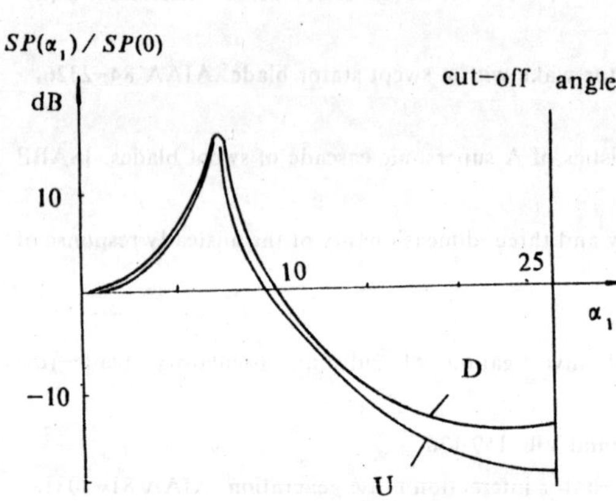

Fig.5. Ratio of the sound power radiated at the first harmonic BPF to Its value at $\alpha_1 = 0$.

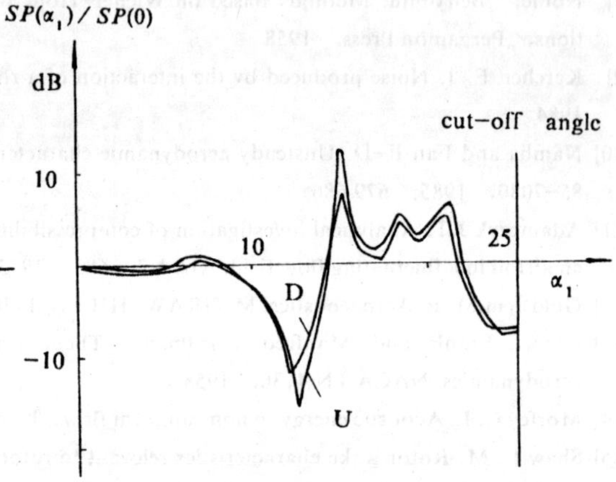

Fig.6. Ratio of the sound power radiated at the second harmonic 2BPF to Its value at $\alpha_1 = 0$

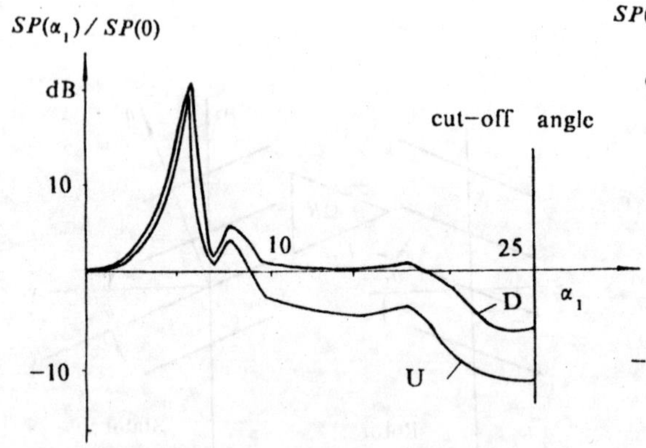

Fig.7. Ratio of the sound power r..diated at the third harmonic 3BPF to Its value at $\alpha_1 = 0$.

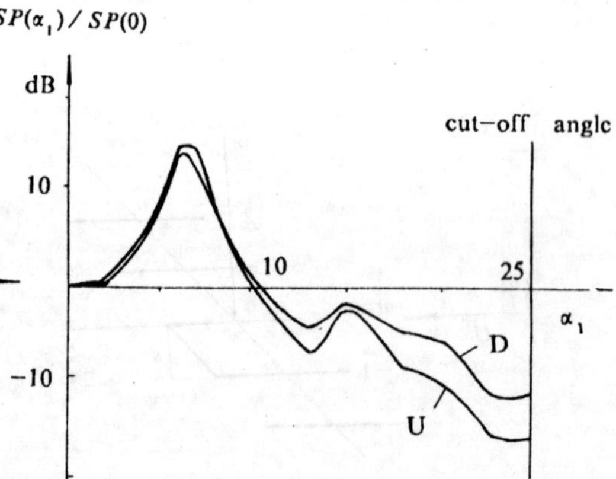

Fig.8. Ratio of the total sound power to its value at $\alpha_1 = 0$.

Part Five
Steam Turbine

AN ADVANCED ROTOR DESIGN FOR A PARTIAL ADMITTED PELTON STEAM TURBINE UNDER ASPECT OF WASTE HEAT RECOVERY

Z. Shao, D. Burhorn, M. Malobabic, M. Rautenberg
(Institut for Turbomachinery, Hannover, Fed. Rep. Germany)

ABSTRACT

The increasing shortage of raw materials dictate an improvement of their conversion efficiency and a more efficient utilization of the primary energy which they contain. Particularly in the case of heat power stations or combustion engines, which use fossil fuels as their energy source, to date only mediocre efficiencies have been achieved. In the former case - where large waste heat quantities are available - the Clausius-Rankine Process has been successfully used to improve the efficiency by recovering the waste energy (combined heat and power generation). By using a steam turbocharger (STC), the Clausius-Rankine Process can also be used to improve the efficiency of road vehicle engines. However, since they generate a relatively low exhaust gas mass flow at a very high temperature, the size and operating conditions of STCs used in vehicle applications is very different to those upon which previous experience has been based. The present paper describes the development of a small Pelton steam turbine with a power output of up to 10 kW, which is suitable for road vehicle applications, and discusses the latest developments in rotor technology. A brief summary of further improvements to the steam circulation and related components is also included.

Fig. 1 Investigated turbine rotors

NOMENCLATURE

A	Area	m^2
b	width	m
c	absolute velocity	m/s
c_p	specific heat at constant pressure	J/(kg K)
d	diameter	m
e	admission ratio	
h	enthalpy	J/kg
I	impulse	kg m/s^2
l	length	m
$\dot{m}$	mass flow rate	kg/s
Ma	Mach number	
n	rotational speed	rpm
p	pressure	Pa
P	power	W
Pi	pressure ratio	
$\dot{Q}$	heat flow	W/s
T	temperature	K
u	circumferential speed	m/s
w	relative speed	m/s
z	number	
η	efficiency	
$\varkappa$	isentropic exponent	
ν	velocity ratio	
ρ_t	density	kg/m^3
ω	angular velocity	1/s

Subscripts

ex	exit
in	inlet
no	nozzle
out	outlet
s	isentropic
sg	steam generator
st	steam
t	turbine
0	tube exit
1	rotor inlet
2	rotor outlet

INTRODUCTION

An increasing deficit of raw materials, in particular fossil fuels, has led to an increasing interest in improving their utilization. Since the usage of fossil raw materials as fuels, or their conversion to higher valued products in petrochemical plants, often leads to the dissipation of a large part of the energy they contain in the form of waste heat, ways of using the waste heat have attracted more and more attention in recent times. Due to the fact that, in some cases, the waste heat is associated with a very low mass flow but high temperature, the development of suitable methods of converting this thermal energy into mechanical or electrical energy is of particular interest.

For a number of years, the Clausius–Rankine Process (also called Rankine Process in some countries) has been successfully applied to equipment dissipating large quantities of waste heat (e.g. power stations) for the purpose of generating additional electrical energy. One particularly plant arrangement is the turbo-compound plant. The application of the Clausius-Rankine Process (CRP) to such large-scale plants has already been investigated in detail and optimized as far as possible[1].

Attention should be turned to investigating the CRP for its application to small waste heat mass flows, too. A large potential to recover waste energy is offered by exothermic chemical reactions, which occur in some manufacturing processes. However, the application of small-scale CRPs is particularly interesting in the case of combustion engines[2], which due to the high temperature of the exhaust gas, only achieve a mediocre overall efficiency. With the help of CRPs, fuel savings of up to 10–15%, for example in locomotive engines, ship's engines or vehicle engines, could well be achieved[3].

Over the last few years, the Institute for Turbomachinery of the University of Hannover (IfT) has been carrying out research on small CRPs for road vehicle applications[3,4]. New, suitably scaled, designs have been proposed for all the major components of the Clausius-Rankine Process, such as feed pump, steam generator, steam turbine and condenser[3].

Since the dynamic behaviour of such systems, and in particular the overall efficiency, is largely determined by the performance of the steam turbine, a major part of this research concerned the development of a small steam turbine for steam mass flows of 0.01 to 0.04 kg/s and waste heat mass flows of 0.1 to 0.3 kg/s and temperatures of 400-1000 °C .

The investigations at the IfT were carried out with a view to the application of CRPs in Otto-cycle (e.g. petrol) and diesel engines for both stationary and mobile applications.

Such considerations led to the development of a new type of steam turbine based on the Pelton principle (radially-ending blades), which was further developed to a Pelton-type rotor with buckets, and finally to a rotor provided with flow channels which divert the steam flow both tangentially and radially (Fig. 1)[5].

The following paper presents the initial results of experiments carried out with this third type of steam turbine rotor, and compares them to the results of previous developments at the IfT[4,6].

DESCRIPTION OF THE TEST RIG

The experimental conditions under which the steam turbine was tested, and thus the design of the test rig, were chosen to match the typical mass flow and temperature conditions of the Otto-cyle (e.g. petrol) and diesel engines used in automobiles and trucks.

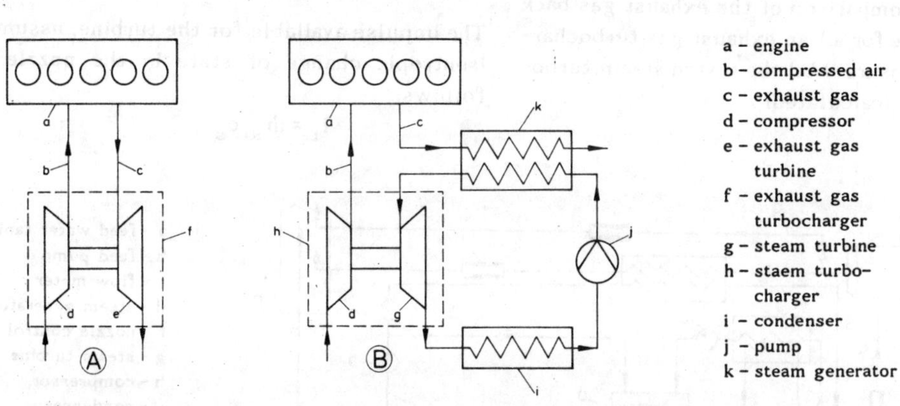

Fig. 2 Comparison of a conventional exhaust gas turbocharger system (A) and the steam turbocharger system (B)

As can be seen in Fig. 2, which compares conventional exhaust gas turbocharger system (A) with a steam turbocharger system based on the Clausius-Rankine Process (B), in the latter case the exhaust gas of the engine passes through a heat exchanger rather than through a conventional gas turbine. In contrast to exhaust gas turbocharging, where the turbine offers considerable resistance to the passage of exhaust gas, leading to a high backpressure, the heat exchanger has a very low resistance to flow and the exhaust gases are able to pass unrestricted into the atmosphere. This ensures minimum impairment to the scavenge gradient of the engine. By way of example, Fig. 3 shows typical curves for exhaust gas backpressure for an exhaust gas turbocharger with waste gate control fitted to a 2.2 l petrol engine (Audi 5-cylinder) and the predicted backpressure in the case of an STC. Between 3,000 and 6,000 rpm, the STC results in only 5% of the backpressure of a comparable exhaust gas turbocharger (ETC).

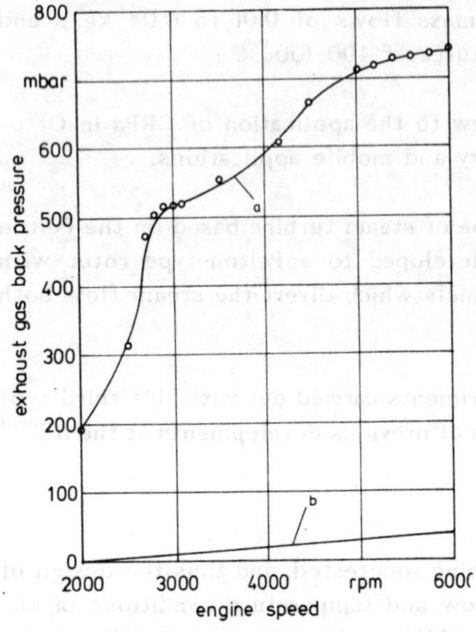

Fig. 3 Comparison of the exhaust gas back pressure for a.) an exhaust gas turbocharger (measured) b.) the tested steam turbocharger (calculated)

Fig. 4 is a schematic illustration of the test rig, showing the arrangement of the feed pump, steam generator, supersonic nozzles, turbine and condenser. On the input side, hot gas is generated by a controllable combustion chamber rather than by an engine. This has the advantage that the exhaust gas temperature and the exhaust gas mass flow are independently controllable over a wide range, and can be easily maintained at constant levels for investigating static processes.

As can be seen, the working fluid is pumped from the feed water container to the steam generator. In this arrangement, the overall process pressure is determined by the feed pump pressure. Steam leaves the steam generator in a superheated condition. After passing to the four separately controllable Laval nozzles, the steam expands in accordance with the following equation:

$$c_{s,st} = \sqrt{2\,\Delta h_{s,st}} \quad (1)$$

The impulse available for the turbine, assuming an isentropic change of state in the nozzle, is as follows:

$$I_t = \dot{m}_{st} c_s \quad (2)$$

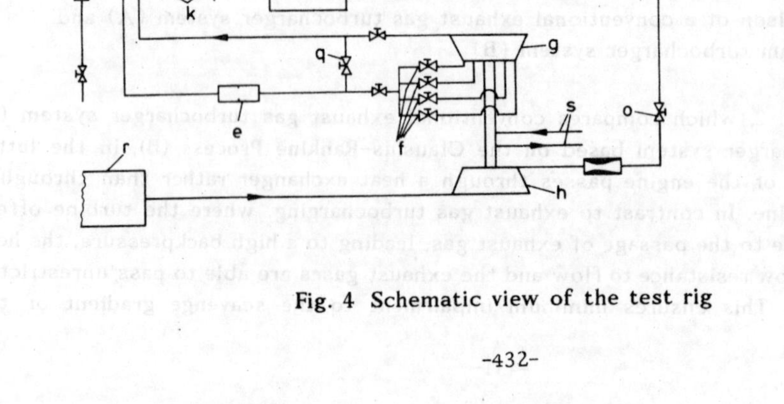

Fig. 4 Schematic view of the test rig

Steam leaving the turbine is fully condensed in the condenser and fed back to the feed water container. In the system investigated, the total pressure in the condenser and feed water container was maintained at approximately atmospheric pressure. This has the advantage that the superheated steam from the steam generator can be completely expanded in the Laval nozzles, leading to a high enthalpy difference and thus a high nozzle outlet velocity. Furthermore, operation of these components at atmospheric pressure leads to size and weight advantages which are significant for road vehicle applications. The steam expansion in the nozzles could be improved further if the turbine casing was maintained at below-atmospheric pressure (e.g. a vacuum). However, at the present time, various unsolved constructional factors (seal leakage etc.) have prevented investigations of the resulting further gain in nozzle efficiency.

An important input parameter when calculating the performance of the closed CRP circulation is the quantity of thermal energy available in the exhaust gas, which is then transferred to the working medium (water) by means of the steam generator. The power available in the steam generator can be calculated as follows:

$$P_{sg} = \dot{m}_{ex} \, \Delta h_{ex} \tag{3}$$

where

$$\Delta h_{ex} = c_p \, (T_{ex,in} - T_{ex,out}) \tag{4}$$

Assuming an average exhaust gas temperature of 500°C and an exhaust gas mass flow of 0.3 kg/s (typical values for combustion engines), and assuming that exhaust gases leaving the steam generator have a maximum temperature of 200°C , the power available to the steam generator is 100 kW.

Since, as discussed later, the tested steam turbine operates according to the Pelton principle, and thus the energy transfer between the working fluid (steam) and the turbine rotor is based on an impulse exchange (Eq. 2), a high process pressure must always be available in order to provide the necessary enthalpy difference and thus steam velocity in the Laval nozzles. Previous investigations showed that a process pressure of approx. 20 bar at a steam mass flow of 0.01-0.04 kg/s allows a particularly favorable overall performance to be achieved:

$$\dot{m}_s = \frac{\dot{Q}_{sg}}{\Delta h_{sg}} \tag{5}$$

Under these conditions the steam leaves the steam generator at 400°C .

DISCUSSION OF PREVIOUS STEAM TURBINE TYPES DEVELOPED AT THE IFT

The aim of the investigations at the IfT with respect to using steam turbochargers for waste heat recovery was the development of new systems to use much smaller waste heat quantities than was formerly the case in CRP applications. The waste heat can be utilized by converting it to mechanical or electrical energy or, in the case of combustion engines, to drive a supercharger. Advantages to be gained over conventional supercharging, based on exhaust gas superchargers, have already been discussed [3].

One of the major areas of previous investigations of small CRPs at the IfT was the development of a small high-performance steam turbine. Detailed comparisons of different turbine types [7], which are suitable for use with mass flows in the range of 0.01-0.04 kg/s at a process pressure of 5 -30 bar and a power output of 5-15 kW, clearly favoured the application of either Hero type turbines or Pelton turbines[8]. However, it was decided to base further work on the development of the Pelton turbine.

The Pelton turbine is an impulse turbine where the working medium is admitted to the rotor at a velocity c_{st}, transferring its energy without pressure change. For this reason, the working medium, which leaves the steam generator at high pressure and high temperature, is first expanded in nozzles converting the available enthalpy as far as possible into velocity (Eq. 1, 2). Accordingly, when considering the turbine in the following, it is also necessary to discuss changes that take place in the nozzles themselves, since these components, too, significantly affect the overall performance.

To date, the IfT has investigated three different steam turbine rotor types, each being a further development and improvement on its predecessor. Fig. 5a shows rotor type 1 and Fig. 5b shows the location of the two nozzles used to test this rotor. As can be seen, rotor type 1 is a classic Pelton-type turbine rotor with radial blades in the shape of a double hemisphere. Fig. 5c shows the velocity vector diagram for this rotor type.

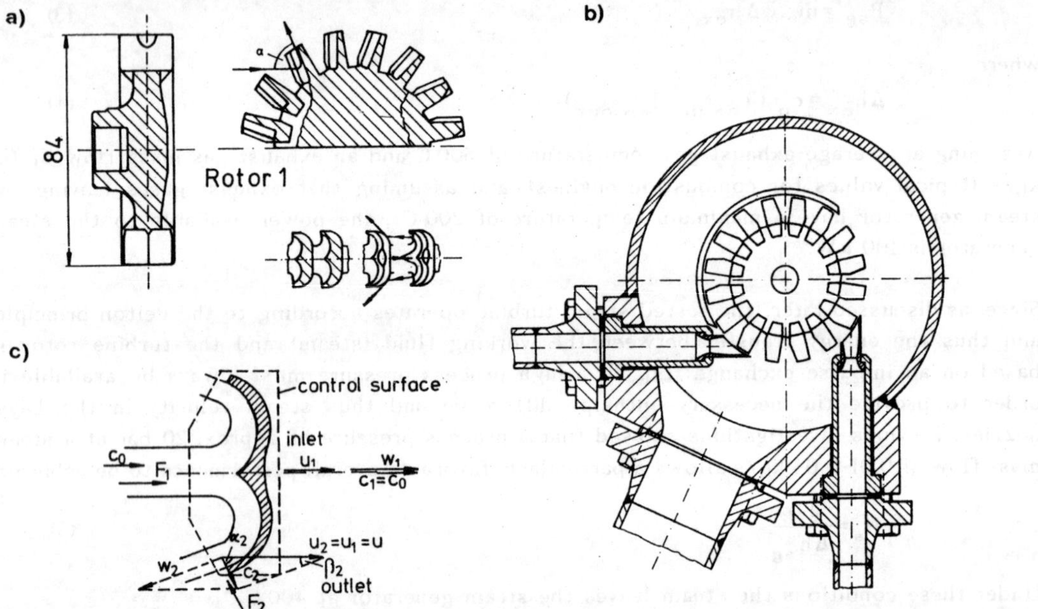

Fig. 5 Construction design of rotor type 1
a.) rotor
b.) turbine casing including the rotor
c.) Velocity vector diagram for the Pelton steam turbine

Assuming that:

$$c_1 = c_0 \tag{6}$$

or:

$$c_1 = \eta_{tu}\, c_{s,st}$$

the power of the turbine is:

$$P_t = \dot{m}\, u_1\, (c_1 - u_1)\, (1 + \cos\beta_2) \tag{7}$$

Accordingly, the turbine efficiency can be defined as follows:

$$\eta_t = \frac{P_t}{c_1^2} = 2(1+\cos\beta_2)\left[\frac{u_t}{c_1} - \left(\frac{u_t}{c_1}\right)^2\right] \tag{8}$$

if Eq. 8 is differentiated in accordance with:

$$\frac{d\,\eta_t}{d\left(\frac{u_t}{c_1}\right)} = 0 \tag{9}$$

it can be seen that the maximum power and thus the maximum efficiency of the turbine occurs at the following velocity ratio[6]:

$$u_t / c_1 = 0.5 \tag{9a}$$

The velocity c_1 is determined by the Laval nozzles, which were designed for supersonic flow. The design point of the nozzles used for rotor type 1 was :

$p_{st} = 5$ bar
$T_{st} = 300$ °C
$\dot{m}_{st} = 0.03$ kg/s

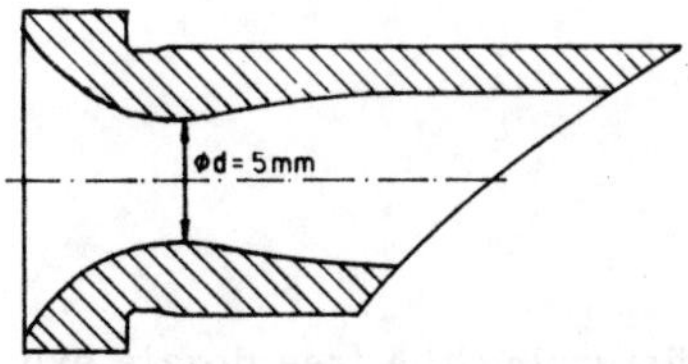

Fig. 6 Design of the supersonic nozzles

As mentioned earlier, in the CRP arrangement under consideration, since the pressure within the turbine casing is practically atmospheric, steam expansion in the nozzles is maximised. Accordingly, all of the enthalpy available before the nozzles is converted to velocity in accordance with Eq. 1.

In order to ensure supersonic velocity at the nozzle outlet, a Laval-type design is required (Fig. 6). Assuming isentropic conditions, the speed of sound in the smallest cross-section d of the nozzles is:

$$c^* = \left(\frac{2\,\varkappa_{st}}{\varkappa_{st}+1} R_{st}\, T_{0,st}\right)^{1/2} \tag{10}$$

The density of the working medium in the smallest cross-section of the nozzles is:

$$\rho^* = \frac{p_{0,st}}{R_{st}\, T_{0,st}} \left(\frac{p^*_{st}}{p_{0,st}}\right)^{1/\varkappa_{st}} \tag{11}$$

where:

$$p^*_{st} = p_{0,st} \left(\frac{2}{\varkappa_{st}+1}\right)^{\frac{\varkappa_{st}}{\varkappa_{st}-1}} \tag{12}$$

This enables the dimensions of the smallest cross-section to be calculated by using the equation of continuity as follows:

$$A^* = \dot{m}_{st} / (\rho^* c^*) \tag{13}$$

In the case of rotor 1, and using the nozzle design point stated earlier, the optimum diameter of the smallest cross-section of the nozzles was calculated to be 5 mm. The diameter of the nozzle outlet, which must be slightly larger than the smallest cross-section to ensure supersonic velocity at the outlet, can be calculated from Eq. 1 together with the continuity equation.

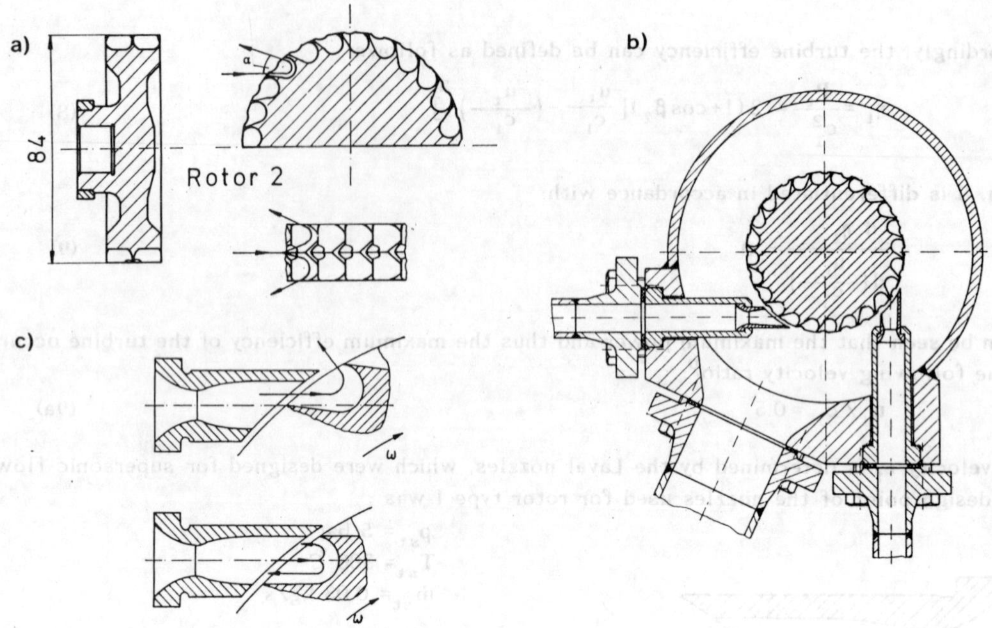

Fig. 7 Construction design of rotor type 2
a.) rotor
b.) turbine casing including the rotor
c.) Single bucket passing through the steam jet disturbing the free nozzle exit

Experimental measurements carried out on rotor 1 using nozzles as described above resulted in the only mediocre efficiency of η_t = 20–30%. This poor efficiency is mainly due to high ventilation losses and rotor losses. The losses consist of lateral rotor friction losses, blade losses and nozzle losses[5]. Furthermore, due to the shape of the blades, part of the steam is pumped away from the rotor in a radial direction, consuming energy (Fig . 5). At the same time, on leaving the blades some of the steam is reflected on adjacent blades (Fig. 5), leading to further losses.

An improvement on this unfavorable situation led to the development of rotor type 2. With this rotor, the radial blades have been replaced by buckets, considerably reducing some of the disadvantages of rotor 1. For example, there is an increased deflection of the steam stream within the buckets, both in a radial and a tangential direction (Fig. 7a). This avoids reflection on adjacent blades. The efficiency of this rotor type was much better than with rotor 1, reaching η_t = 40–60%.

However, new losses arose, because, after the deflection within the buckets, part of the steam was reflected in such a way that it impaired the nozzle outlet flow (Fig. 7c)[5].

Fig. 8 shows the further-developed rotor type 3, with associated turbine casing. In contrast to previous turbine types, four nozzles were used. As can be seen in Eq. 8 and 9, the turbine efficiency and turbine power reaches a maximum when the deflection of the steam stream approaches 180°, since in this case the angle of the relative velocity (Fig. 5c) approaches zero and thus the expression $(1+\cos\beta_2)$ is a maximum. At the same time, Eq. 9 and 9a show that maximum efficiency is reached when u_t/c_1 = 0.5. When developing rotor 2, the angle β_2 was considerable improved in comparison to rotor 1, with the disadvantage, however, of the impairment of the nozzle outlet flow due to steam reflection. Rotor type 3, with an entirely

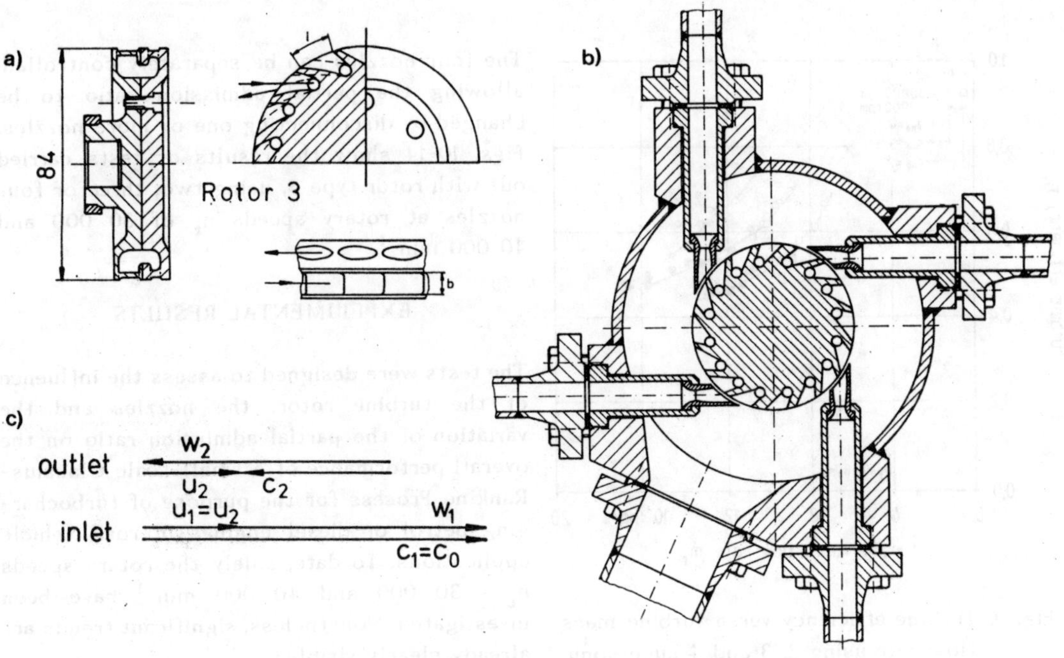

Fig. 8 Construction design of rotor type 3

a.) rotor

b.) turbine casing including the rotor

c.) Velocity vector diagram for the rotor type 3

different construction and a newly developed nozzle system, was designed to minimize these disadvantages. Rather than using blades or buckets, the rotor is constructed with flow channels, allowing the steam to leave the rotor on a plane which is displaced to one side, thus avoiding impairment to the nozzle flow (Fig. 8a). At the same time, this arrangement allows the steam stream to be deflected by 180° both in radial and tangential direction (Fig. 8c). As discussed above, this results in a maximum utilization of the available impulse (Eq. 8).

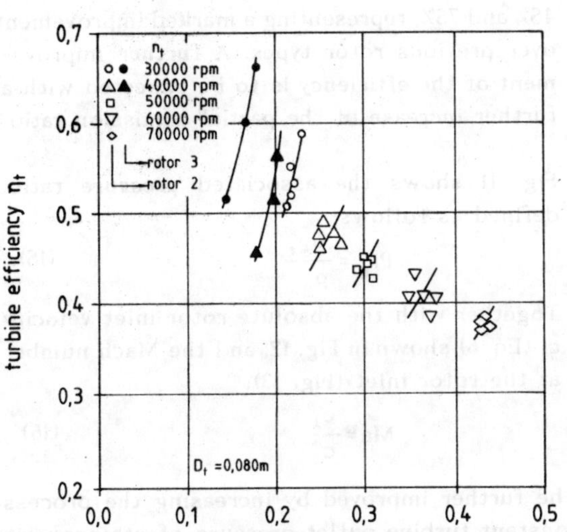

Fig. 9 Comparison of turbine efficiency between rotor type 2 and rotor type 3 run by two nozzles

With two nozzles, the new rotor type results in an increase of the efficiency to 68% (Fig. 9).

An increase in the number of nozzles increases the partial admission ratio of the rotor as follows:

$$e = \frac{A_{no}\, z_{no}}{A_c\, z_c} = \frac{A_{no}\, z_{no}}{b_c\, l_c\, z_c} \qquad (14)$$

Since the turbine was designed for the same process conditions as test rotors 1 and 2 ($\dot{m}_{st}$ =0.01–0.03 kg/s, p_{st} = 5 bar), increasing the number of nozzles to four requires a reduction in the nozzle dimensions. In accordance with the law of similarity, the required diameter of the smallest nozzle cross-section was calculated to be d = 3 mm (Fig. 6).

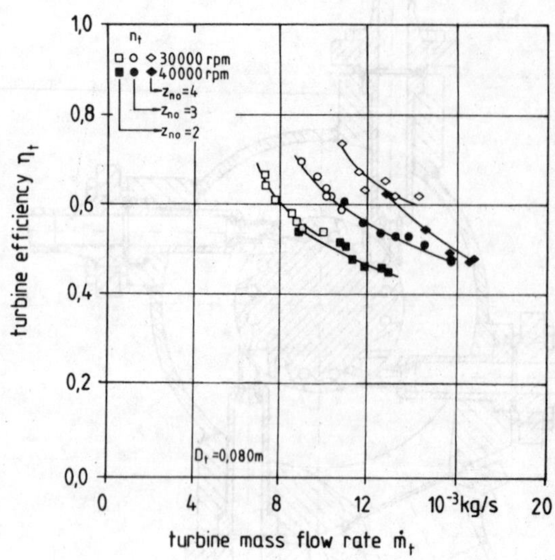

Fig. 10 Turbine efficiency versa turbine mass flow rate using 2, 3 and 4 supersonic nozzles

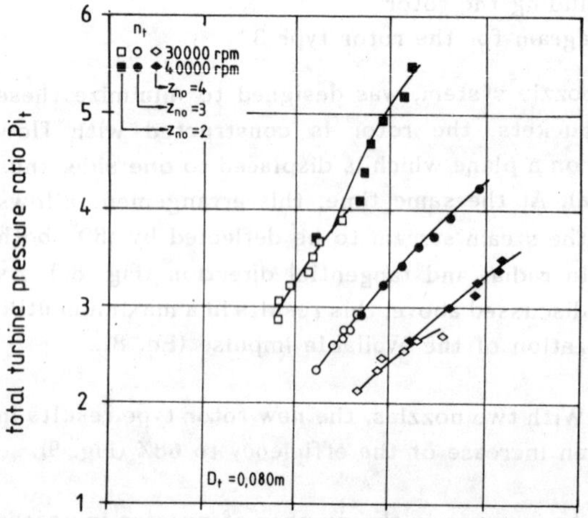

Fig. 11 Total turbine pressure ratio vs. turbine mass flow rate with different nozzle numbers

The four nozzles can be separately controlled, allowing the partial admission ratio to be changed by disconnecting one or more nozzles. Figs. 10–14 show the results of tests carried out with rotor type 3, using two, three or four nozzles at rotary speeds n_t of 30 000 and 40 000 min^{-1}.

EXPERIMENTAL RESULTS

The tests were designed to assess the influence of the turbine rotor, the nozzles and the variation of the partial admission ratio on the overall performance of a small-scale Clausius–Rankine Process for the purpose of turbocharging petrol or diesel engines in road vehicle applications. To date, solely the rotary speeds n_t = 30 000 and 40 000 min^{-1} have been investigated. Nonetheless, significant trends are already clearly visible.

Referring to Fig. 10, which shows the turbine efficiency, it can be seen that an increased partial admission ratio results in an increase in both the efficiency and the mass flow[9]. With two nozzles, the mass flow varied from 0.007–0.013 kg/s, whereas with four nozzles, the mass flow increased to 0.009–0.016 kg/s. In this mass flow range, the turbine efficiency varies between 45% and 75%, representing a marked improvement over previous rotor types. A further improvement of the efficiency is to be expected with a further increase of the partial admission ratio.

Fig. 11 shows the associated pressure ratio, defined as follows:

$$Pi_t = \frac{p_{st}}{p_2} \tag{15}$$

Together with the absolute rotor inlet velocity c_1 (Eq. 6) shown in Fig. 12, and the Mach number at the rotor inlet (Fig. 13):

$$Ma = \frac{c_1}{c^*} \tag{16}$$

it is apparent that the turbine efficiency can be further improved by increasing the process pressure. If this is possible, and assuming a constant turbine outlet pressure of atmospheric pressure, Eq. 10 to 12 and 15, 16 show that the pressure ratio of the nozzles and thus the velocity c_1 also increases. In accordance with Eq. 2, this is equivalent to an increased impulse, which under consideration of Eq. 7 and Eq. 8, results in an improvement of the turbine power and thus efficiency.

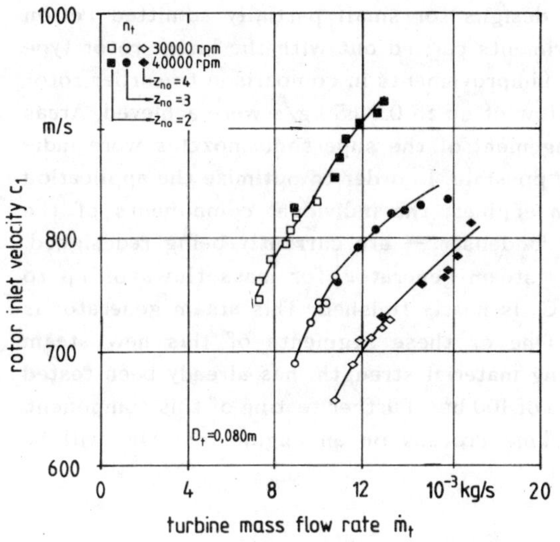

Fig. 12 Turbine inlet velocity vs. turbine mass flow rate with different nozzle numbers

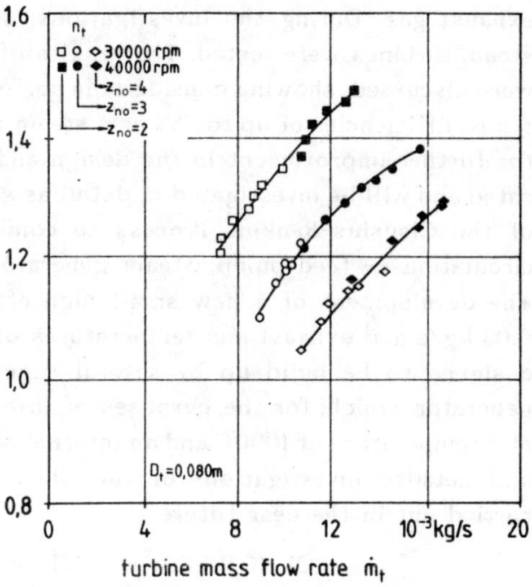

Fig. 13 Mach number c_1/c^* vs. turbine mass flow rate with different nozzle numbers

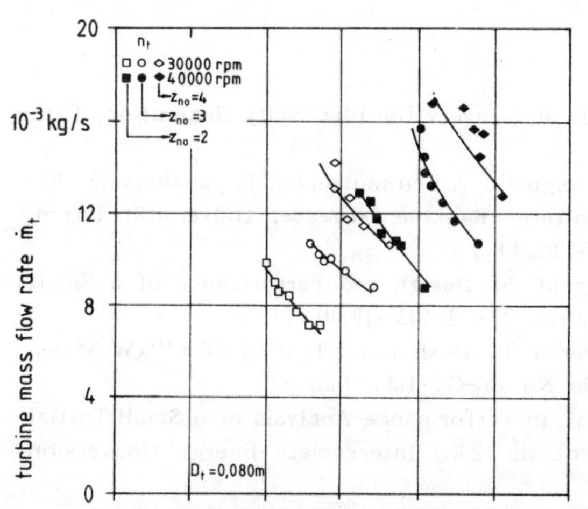

Fig. 14 Turbine mass flow rate vs. velocity ratio u/c_s with different nozzle numbers

As can be seen in Fig. 13, the Mach numbers achieved in this test series vary from Ma=1–1.55, and are substantially lower than the expected Mach numbers of Ma = 2.5–3. An increase of the Mach number and thus of velocity c_1 appears necessary. This can only be achieved by an improved design of the Laval nozzles and a simultaneous increase of the pressure ratio.

Fig. 14 shows the relationship between the turbine mass flow and the velocity ratio ν, in accordance with the following equation:

$$\nu=\frac{u_t}{c_{1,s}} \qquad (17)$$

Up to now, tests have been made at a velocity ratio range of ν = 0.15–0.225. In order to get a further efficiency improvement, it is necessary to increase the velocity ratio under consideration of Eq. 9a. However, an increase of the peripheral velocity u_t and thus the velocity ratio ν can only be achieved by increasing the pressure ratio and simultaneously improving the nozzles and the partial admission ratio. Mach numbers of about 2.5 – 3 should be realised.

CONCLUSION AND PREVIEW

The theoretical and experimental investigations described in this paper were carried out with the aim of utilizing small waste heat quantities by applying the Clausius–Rankine Process to the steam turbocharging of combustion engines, the heat energy being obtained from the exhaust gas. During the investigations, several designs for small partially admitted Pelton steam turbines were tested. The results of experiments carried out with the latest rotor type were discussed, showing considerable performance improvements in comparison to earlier rotor types. Efficiencies of up to 75% at a steam mass flow of up to 0.0165 kg/s were achieved. Areas for further improvement in the design and arrangement of the supersonic nozzles were indicated and will be investigated in detail as soon as possible. In order to optimize the application of the Clausius–Rankine Process to combustion engines, the individual components of the circulation -- feed pump, steam generator and condenser -- are currently being redesigned. The development of a new small high efficiency steam generator, for mass flows of up to 0.03 kg/s and exhaust gas temperatures of 1000℃, is nearly finished. This steam generator is designed to be build up by several segments. One of these segments of this new steam generator, which, for the purposes of investigating material strength, has already been tested at a temperature of 1000℃ and an internal pressure of 100 bar. Further testing of this component, and detailed investigations of the Clausius–Rankine Process on an engine test rig, will be carried out in the near future.

ACKNOWLEDGEMENT

The research work described in this paper was funded by the German Research Association (DFG). The authors would like to thank them for their support. The authors also wish to express their appreciation for the efforts of Mr. Foss and Mr. Quietmeyer, who prepared and skilfully operated the test rig. Furthermore, the authors are thankful to Mr. D. Long and Mr. T. Harder for their assistance in preparing and laying out the manuscript.

REFERENCES

1. Papperitz C, Pichler W, Roemheld , In: Waste Heat Recovery for Electricity Generation, Turbo machinery International, Nov./Dec. 1981
2. Steimle F, In: Experimentelle Untersuchung geeigneter Arbeitsmittel und Expansionsmaschinen für den Verbrennungsmotor nachgeschalteter Rankine–Prozesse, Universität Essen, Zwischenbericht über das Vorhaben AIF–Nr. 5420, 1983
3. Shao Z, Malobabic M, Burhorn D, Rautenberg M, In: Design and Performance of a Small Power Clausius–Rankine Process, ASME Paper No. 89–GT–215 (1989)
4. Rautenberg M, Malobabic M, Mobarak A, Kader A, In: Design and Testing of a 10kW Steam Turbine for Steam Turbochargers, ASME Paper No. 85–GT–113 (1985)
5. Malobabic M, Shao Z, Burhorn D, Rautenberg M, In: Performance Analysis of a Small Partial Admitted High Speed Turbine, Proceedings of the 23rd Intersociety Energy Conversion Conference (IECEC) 1988
6. Rautenberg M, Malobabic M, Mobarak A, Kader A, In: A Small Pelton Turbine for Steam Turbocharger, Proceedings of the 19th Intersociety Energy Conversion Conference (IECEC) 1984
7. Rautenberg M, Mobarak, Kader A, Meyer C, In: Comparison of Different Small Steam Turbines for Steam Turbocharger, ASME Paper No. 84–GT–239 (1984)
8. Dusnatan AT, In: The Design, Testing and Application of a Small High Speed Inward–Flow Radial Turbine, Thesis submitted for PhD, Oxford University (1981)
9. Korematsu K, Hirayama N, In: Performance Estimation of Partial Admission Turbines, ASME Paper No. 79–GT–123 (1979)

THE NON-SYNCHRONOUS VIBRATIONS OF THE SHROUDED BLADE ROW OPERATING UNDER STEAM CONDENSATION

Shun Chen
(Power Generation Operations Division, Westinghouse Electric Corp., Orlando, FL. USA)

ABSTRACT

Non-synchronous blade vibration is one of the most troublesome problems in turbine design and operations. The problem is further complicated in low pressure steam turbines in which flow instabilities may be exacerbated by steam condensation. Most of the next to last stage (L-1 stage) blades in a reheat fossil low pressure steam turbines operate near steam saturation line in which steam may undergo super-saturation and sudden condensation. The steam condensation has been known to degrade turbine performance and also been shown in laboratory to cause flow instabilities. A telemetry turbine test was conducted in the field for which the L-1 rotating blades had experienced distress previously. A non-synchronous vibration of these blades was observed at a frequency for which the blades were well tuned. A test was run to map out the instability zones and to identify the possible causes of the blade vibrations. Further tests in the scaled model turbine also showed instability at about the same frequency though with blade stresses much lower than that of the field turbines. An experimental investigation was then conducted in a stationary linear cascade to assess the interactions of shocks, boundary layers and spontaneous condensation. It is shown that the increased stresses due to these interactions are small based on this stationary cascade test indicating that the non-synchronous blade vibration observed in the field turbine is aeroelastic in nature that a dynamic cascade, with the blade section vibrating, is needed for investigations of such a phenomenon.

1. INTRODUCTION

Blade distress of low pressure (LP) steam turbine is one of the most frequently reported causes of outages and downtime in a large power-plant. Among the reported LP turbine blade distresses, 40% of them occurred in the L-1 (next to last) stages according to the 1981 EPRI survey (Ref. 1). Most of the L-1 stage blades in a reheat fossil LP steam turbines operate near steam saturation lines in which steam may undergo super-saturation and sudden condensation. the transition of steam from superheated state to wet steam determines the influence of stress corrosion and of unsteady spontaneous condensation.

The present paper is concerned with the L-1 stage shrouded rotating blades of the fossil LP turbines which had reported incidents of airfoil and tenon distresses. The majority of these incidents were corrosion induced. Two units that experienced non-corrossive type of distress on L-1R blades were chosen for a field test to find the causes of the blade distress. The field tests uncovered a nonsynchronous blade vibration on these tuned blades. The non-synchronous blade vibration is one of the most troublesome problems in turbine design and operations. The problems are further complicated in the L-1 stages in which flow instability may be enhanced by steam condensation. Steam condensation has been known to degrade turbine performance and also been shown in laboratory to cause flow instabilities. A non-synchronous vibration of these blades was observed at a frequency for which the blades were well tuned. A comprehensive test was then run to map out the instability zones and to identify the possible causes of the blade vibrations. This second mode blade vibration, mainly in axial direction, was found to occur only in a narrow zone characterized by steam wetness and in transonic exit flows.

The phenomenon strongly suggested the participation of condensation instability and its interaction with the blade vibration. Further tests in the scaled model turbine also showed blade vibrations at about the same frequency though with blade stresses much lower than that of the field turbines. An experimental investigation was then conducted in a stationary linear cascade to assess the interactions of shock, boundary layer and spontaneous condensation. It is shown that the increased stresses due to these interactions are small based on this stationary cascade test indicating that the non-synchronous blade vibration observed in the field is aeroelastic in nature that a dynamic cascade, with the blade section vibrating, is needed for investigation of such a phenomenon.

In what follows, the phenomenon associated with flow instability due to steam condensation is reviewed briefly. Detailed field telemetry test results are described next, followed by the scaled model turbine test and a stationary cascade test results.

2. FLOW INSTABILITY DUE TO CONDENSATION

The pressure and Mach number variations in a one-dimensional flow with heat addition can be shown to be:

$$\frac{dM}{M} = \frac{1 + \frac{\gamma-1}{2} M^2}{M^2 - 1}\left(\frac{dA}{A} - \frac{1 + \gamma M^2}{2}\frac{dq}{q}\right) \qquad (1)$$

and

$$\frac{dp}{p} = -\frac{\gamma M^2}{M^2 - 1}\left(\frac{dA}{A} - \frac{\gamma-1}{a^2}\,dq - \frac{M^2-1}{\gamma M^2}\frac{dm}{m}\right) \qquad (2)$$

where M = flow Mach number; γ = specific heat ratio; A = cross sectional area; a = speed of sound; q = heat input or condensation heat release; p = static pressure; m = mass flow rate.

The heat addition (condensation heat release) to the flow is seen to have a similar effect as the area variation (-dA/A). In a convergent-divergent nozzle, when condensation occurs in the divergent portion with supersonic fluid flow, the heat addition will reduce Mach number and increase pressure of the flow as can be seen from Eqs. (1) and (2). The flow will be thermally choked at M=1 and $q=q_{crit}$. For super-critical heat addition $q > q_{crit}$, a shock will form in the condensation zone. Its location depends on the nozzle geometry and heat addition process. When the shock affects the heat addition process the flow becomes unstable. This interaction between heat addition and flow will result in cyclic formation of condensation shock in the supersonic flow region with the shock wave moving toward and past the throat and decay in the subsonic flow region. The unsteady flow caused by the cyclic formation of condensation shock waves have been observed in Laval nozzle (Refs. 2,3 & 4) and cascade tests (Ref. 5). Pressure variations up to 30 ~ 40% in a frequency range of 300 ~ 2000 Hz have been reported. The intensity of this condensation shock is found to depend on the flow Mach number, Reynolds number, blade profile, and most importantly, the inlet flow condition relative to equilibrium saturation state. The unsteady flow induced by the condensation instability also will be influenced by the interaction with boundary layer, adiabatic shock, disturbances caused by the wakes and their oscillations. The frequency of condensation shock formation and oscillation depends on blade geometry and boundary layer characteristics. A wide spectrum of frequencies has been measured for this self-oscillations of the condensation shock waves. Non-synchronous vibration will occur on the tuned blades if the blade natural frequencies coincide with the flow oscillation frequency or its multiples.

The interaction of blade vibration and the condensation shock may create a flutter-like blade excitation. When the condensation shock position is affected by the blade vibratory motion the blade vibration amplitude may be magnified due to energy input into the blade as a result of the condensation shock oscillation and its interaction with blade vibration.

3. FIELD TURBINE TEST

TEST SET-UP

The L-1R blade rows of two low pressure steam turbines were tested via radio telemetry. These double flow LP turbines were arranged in tandem compound as shown in Fig. (1) with the LP2 closest to the generator end and the LP1 located in the governor end. Two rows of shrouded blades, three in a group, and one row of free standing blades were installed and instrumented for the test. The following table shows the types and numbers of strain gauges and pressure transducers used:

Type of Gauge	LP1 Governor End (Free Standing)	LP1 Generator End (Shrouded)	LP2 Governor End (Shrouded)
Dynamic Strain	17	13	15
Static Strain		2	
Total Pressure	2	2	2
Static Pressure	1	1	1

Table 1. Type and number of gauges used for the test

The shrouded blades were installed on the generator end of LP1 and also on the governor end of LP2. In the governor end of LP1 the free standing blades were installed for the test.

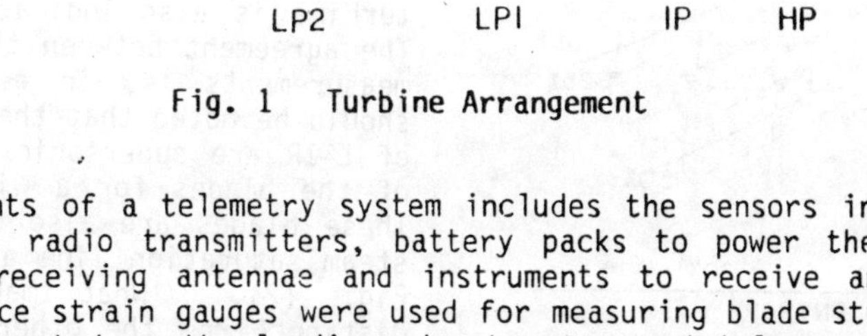

Fig. 1 Turbine Arrangement

The basic components of a telemetry system includes the sensors installed in the turbine, miniature radio transmitters, battery packs to power the transmitters, transmitting and receiving antennae and instruments to receive and process the signals. Resistance strain gauges were used for measuring blade strain. Pressure transducers were mounted on the leading edge to measure total pressure and on the suction surface near the throat to measure static pressure. The adhesives used to secure the sensors must withstand a loading of 20,000 g's at a maximum temperature of 260°C (500°F) for the turbine operating at 3600 RPM. The temperature of the transmitter and battery power supply was kept under 150°C (300 F) and centrifugal loads of 8600 g's. The transmitters operated in the FM band between 88 and

108 MHz. A time code generator was used to supply a time signal which is recorded on tape. With time as reference, data from various channels can be analyzed and compared during a given time interval in the case of a transient operation.

In addition to the normal dynamic strain gages installed to measure the blade response due to dynamic excitation, two static strain gages were also installed to measure potentially large quasi-static stresses introduced when the unit was synchronized. Both static and total pressure transducers were installed to measure flow excitation as a potential source of blade excitation. They were used to monitor non-uniform flow amplitudes such as the disturbance from wakes. Only dynamic pressures can be measured from these transducers. The on-line data was displayed continuously by oscilloscopes during the tests. The strain gage and pressure transducer signals were analyzed with fast Fourier transform-based spectrum analyzers for the frequency response measurements. Correlations of the flow disturbances with the strain gage signals were made by comparing the spectrum of these signals.

STEADY THERMODYNAMIC DATA OF THE TEST TURBINES

The turbine generator unit is rated 665 MW at design condition. The low pressure turbine average expansion lines for various load conditions are shown in Fig. (2). The 100% line corresponds to 665 MW load. During the test, the LP turbine inlet condition (crossover pressure and temperature) and the condenser pressure were measured and the values displayed continuously by digital counters. The approximate flow conditions at the L-1R exit can be estimated from Fig. (2) once the knowledge of the inlet and exit conditions are known from the measurement. Detailed flow distribution of flow properties has been calculated for this LP turbine by solving the axisymmetric flow equations using a matrix method. Steam table was used to represent the equation of state in the calculation. Fig. (3) is the distribution of stream lines for the last three stages of this LP turbine. Fig. (4) shows the relative exit Mach number distribution of the L-1 blade row at full and one half loads. The test data obtained through traversing on the generator end of LP2 turbine is also indicated in the figure. The agreement between the predictions and measurements is, in general, good. It should be noted that the flows at the exit of L-1R are supersonic in the upper half of the blades for a wide range of load. These blades are also operating near the steam saturation line as can be seen from Fig. (2). What makes these blades distinct from the other is that they may be operating in both dry and wet steam during the load change with the flow passages in transonic conditions.

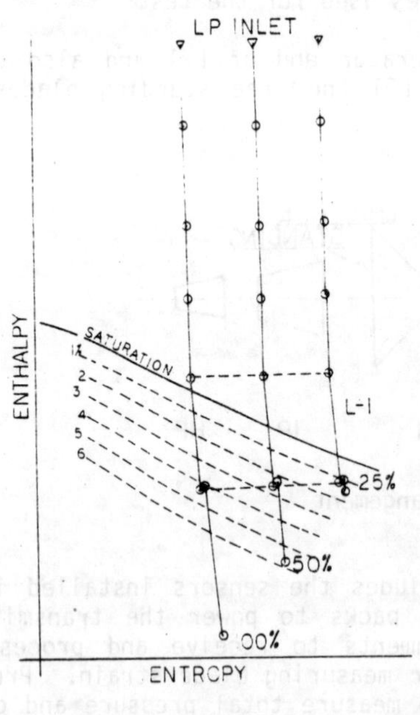

Fig. 2 Condition Lines of the Field LP Turbines

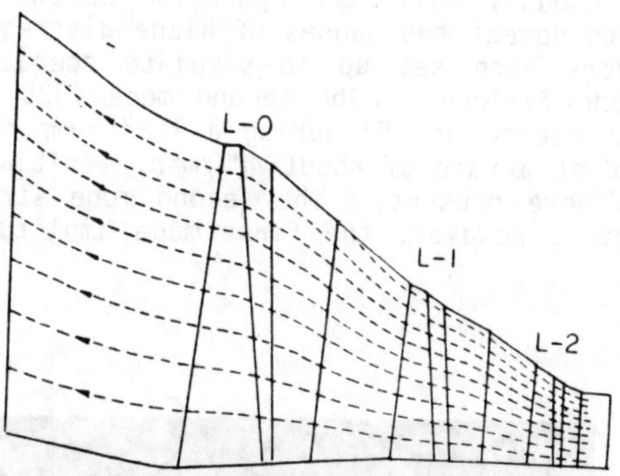

Fig. 3 Stream Line Distribution by Flow Field Calculation

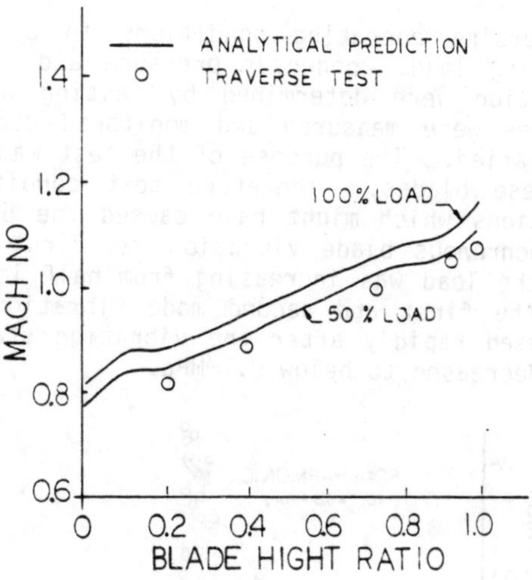

Fig. 4 Relative Mach Number Distribution at L-1R Exit

TURBINE OPERATIONS AND THE BLADE VIBRATION

The first three modes of the shrouded L-1R blades were well tuned. Under normal operating conditions the measured gage stresses corresponding to the first three modes were below 0.7 MPa (approximately 100 psi). The largest forced excitation forces were due to the nozzle wakes, which induced an 80th harmonic dynamic pressure (80 vanes). Fig. (5) shows the total pressure spectra measured from three instrumented blades. The vane passing frequency amplitudes were about the same for all three LP ends. These amplitudes increased with load as can be seen from Fig. (6).

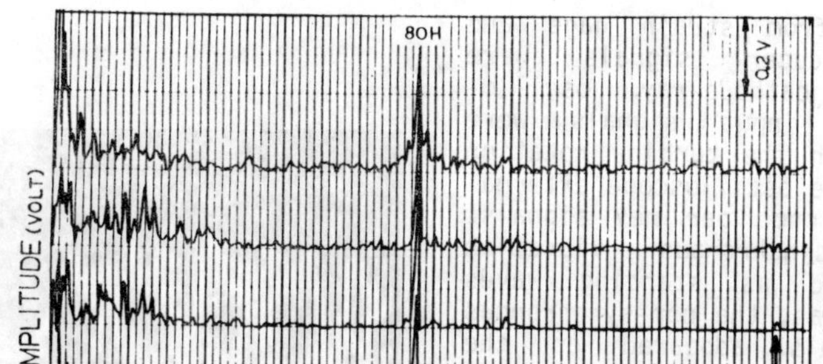

Fig. 5 Total Pressure Spectra for three LP Ends

The analytically predicted value for the wake excitation agree well with the test data (Ref. 6 and Appendix A). The dynamic pressure signals were monitored continuously to detect any unusual flow disturbances which may be present during load and other parameter changes.

The turbine operating conditions in a telemetry test were normally set up by adjusting load, condenser pressure and reheat temperatures. The effects of flow extraction were determined by testing with or without heater in service. Blade stresses were measured and monitored continuously while the operation parameters were varied. The purpose of the test was to detect the causes of blade distresses on these blades. Therefore test conditions were set up to simulate operating conditions which might have caused the blade failures. The second mode (325 Hz) nonsynchronous blade vibration was first uncovered on LP2 during a load ramp test when the load was increasing from half load at a rate of about 4MW/min. Initially both the first and second mode vibrations were present. The second mode stress increased rapidly after the vibration started, however, the first mode amplitudes soon decreased to below 0.7 MPa.

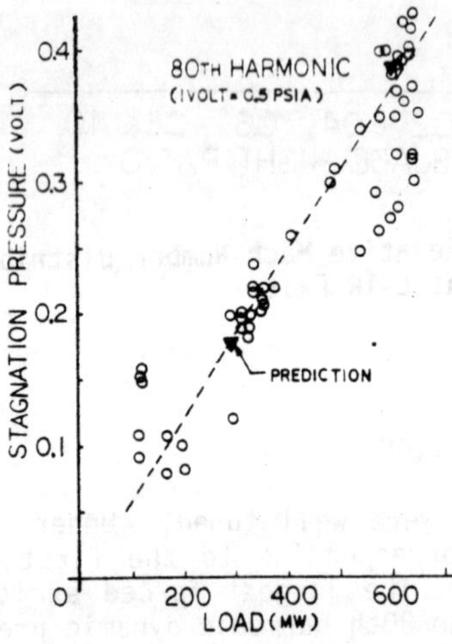

Fig. 6 Stagnation Pressure Amplitudes

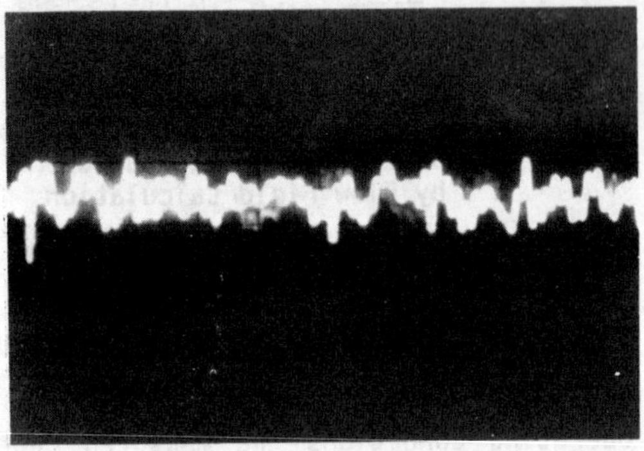

Fig. 7 Steady Synchronous Vibration Signal - Strain Gage.

The raw data from strain gage and pressure transducer signals were monitored by cathode-ray oscilloscopes. The input signals were synchronized with the turbine rotation by triggering the oscilloscope external sweep with the once per rev signals. The steady synchronous vibration signals appeared as standing waves (Fig. 7). Whereas, the non-synchronous vibration signals appeared on the screen in multiple wave forms with phase lag between two successive waves. Fig. (8) shows the pressure signals when the blades were vibrating at the second mode. Fig. (9A) and Fig. (9B) show the corresponding strain gage signals at two instances. The unsteadiness of the vibration can be seen by comparing the stress amplitudes between these two figures.

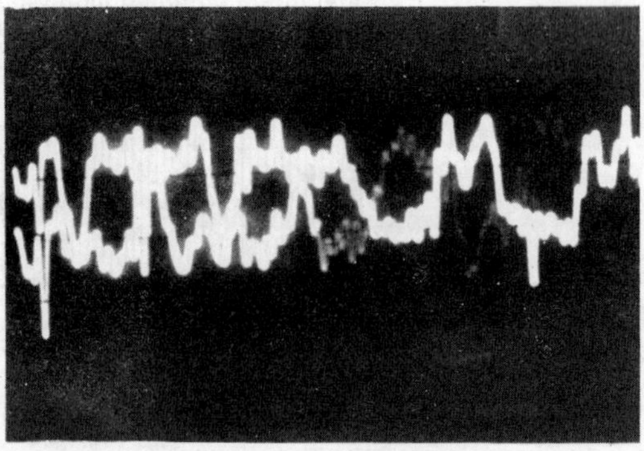

Fig. 8 Non-Synchronous Vibration Signal - Pressure Transducer

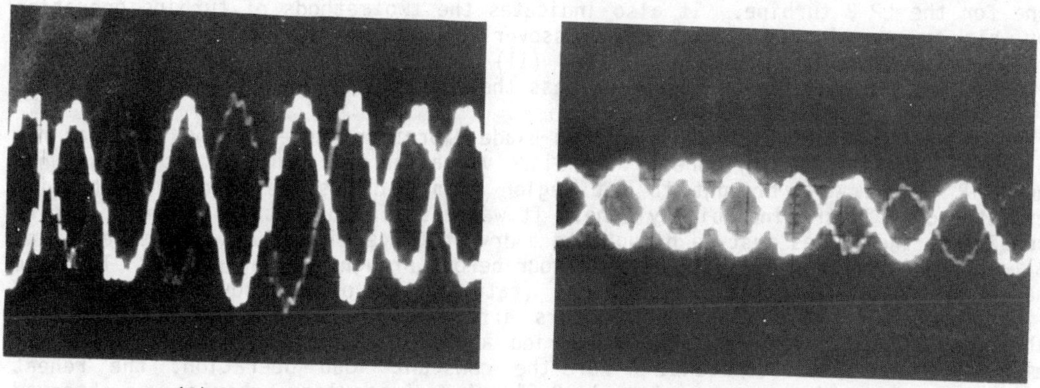

(A) (B)

Fig. 9 Strain Gauge Signals Under Non-Synchronous Non-Steady Vibration at Two Separate Instants

After a thorough investigation of the phenomenon, it was found that the non-synchronous vibration occurred only under certain operating conditions. There existed a narrow zone, in an area surrounded by constant moisture lines, for which blade vibration occurred and the stress reached a maximum. The stress decreased drastically outside this instability zone.

Two methods were adopted to bring the blades to cross the instability zone. First, the turbines were operated at constant load. For such operations the inlet temperature to the LP turbines (crossover temperature) was varied while the inlet pressure (crossover pressure) was kept roughly constant. In the second method the crossover temperature was kept constant while the crossover pressure was varied. The second operation is equivalent to keeping reheat temperature constant while varying load. In both operations the blades were excited and the second mode amplitude increased when the blades were passing through the instability zone (at

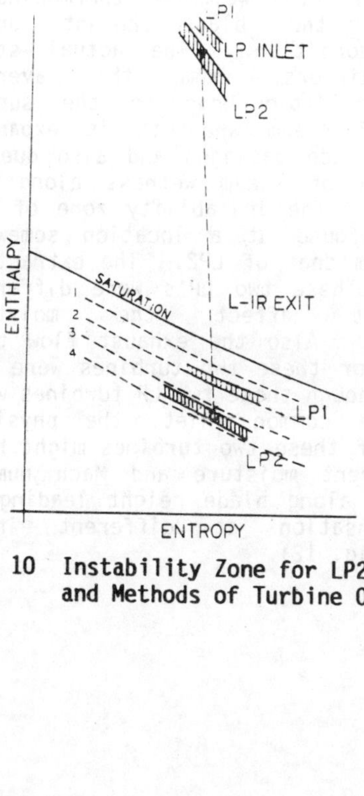

Fig. 10 Instability Zone for LP2 and Methods of Turbine Operation

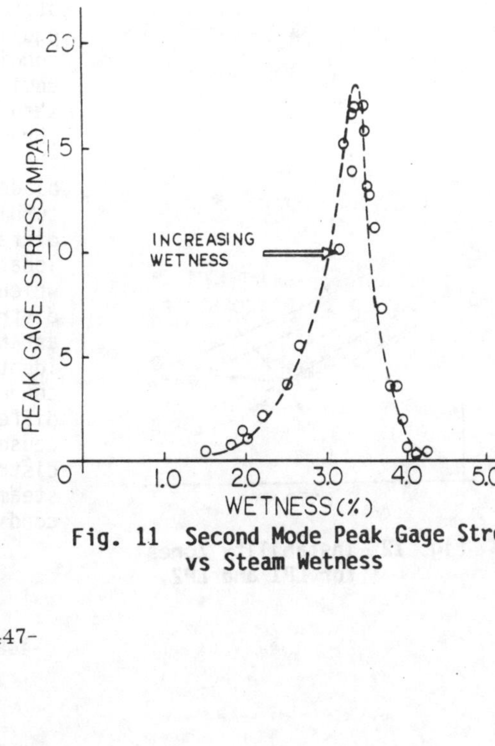

Fig. 11 Second Mode Peak Gage Stress vs Steam Wetness

average moisture of about 3 to 4% for the LP-2). Fig. (10) shows the instability zone for the LP-2 turbine. It also indicates the two methods of turbine operation for which wetness increases when the crossover temperature was decreased or when the crossover pressure was increased. Fig. (11) shows a typical plot of stresses vs wetness when the wetness was increased pass the instability zone.

After the discovery of the nonsynchronous blade vibration, an extensive test was run to map out the instability zones. The blade vibration was found to occur only when the L-1R blades were brought to wet region from a dry state, and the vibration persisted only in the instability zone. It was also found that, in order to excite the blades, the blades must be brought to a dry state (equilibrium wetness less than 2%) and remain dry for at least half an hour before brought to wet region. The rate change of load and reheat temperature (rate change of crossover pressure and temperature) were also important factors affecting the initiation of the blade vibration. Typically, the load was varied at 4MW/min. or lower when the reheat temperature was kept constant. For the constant load operation, the reheat temperature was varied at approximately 0.5°C/min. A peculiar behavior was observed in that the non-synchronous vibration was not present when the blades were brought from wet to dry states even though they were passing through the instability zone.

The fact that the non-synchronous blade vibrations occurred only in an instability zone characterized by steam wetness and that the blades were operating at transonic flow region suggest that the phenomena were the results of steam condensation, as described in Section 2, and its interaction with the blade vibratory motion. The phenomena may also be the results of shock, condensation and boundary layer interactions. The dependence of the phenomena on the rate change of pressure and temperatures, and the differences in phenomena between wet-to-dry and dry-to-wet operations seem to indicate the non-linear nature of the vibrations (Ref. 7).

INSTABILITY ZONES

The instability zone of L-1R for the LP2 turbine is shown in Fig. (10). Corresponding to the instability zone for the L-1R exit condition, there is an instability zone for the LP inlet condition defined by the crossover temperature and pressure. It should be noted that the condition line, which was obtained by the simple one dimensional calculation, represents the flow weighted thermodynamic state along the blade height under equilibrium condition. The actual steam condition differs from the average equilibrium condition due to the super-saturation of steam when it is expanded through the blade passages and also due to the variation of steam wetness along the blade height. The instability zone of LP1 turbine was found at a location somewhat different from that of LP2. The extraction locations of these two LP's were different which might affect the moisture distributions. Also the exhaust flow pipe arrangement for these two turbines were not identical. Though these two LP turbines were connected to a common inlet, the physical differences of these two turbines might have caused different moisture and Mach number distributions along blade height leading to steam condensation at different inlet conditions (Fig. 12).

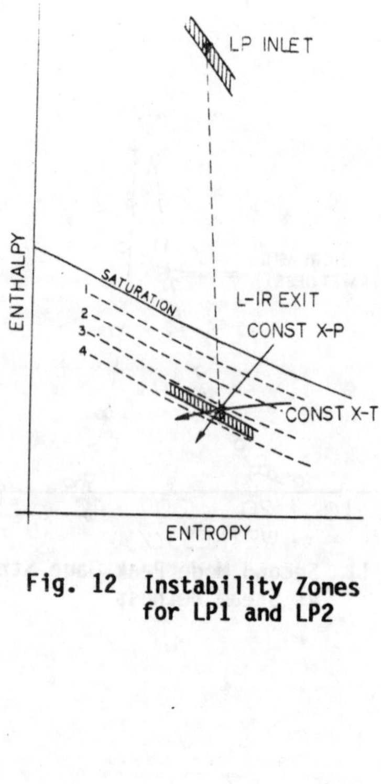

Fig. 12 Instability Zones for LP1 and LP2

The instability zones in the Mollier diagram (Fig. 12) can be mapped to a plane relating reheat temperature and load. Fig. (13) shows the instability zones of LP1 and LP2 in this plane. The blade excitation occurs only when the load is increased in a certain direction as indicated by the arrows. It is possible to operate the turbines in the instability zone without initiating blade excitation if the load is decreased at a constant reheat temperature or the reheat temperature is increased at a constant load. Since the occurrence of blade excitation depends on time history of rate change of the operation parameters, the blade could be operated in the instability zones without building up high stresses if the blade vibration was not initiated by a proper procedure. Combining the two instability zones of Fig. (13), a more general operation map can be constructed as shown in Fig. (14). Within the instability region, both LP turbines can be operated without initiating the blade vibrations regardless of the operation procedure.

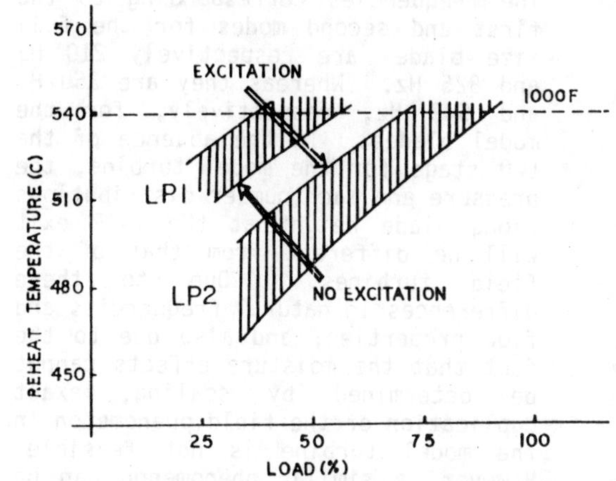

Fig. 13 Instability Zones of LP1 and LP2

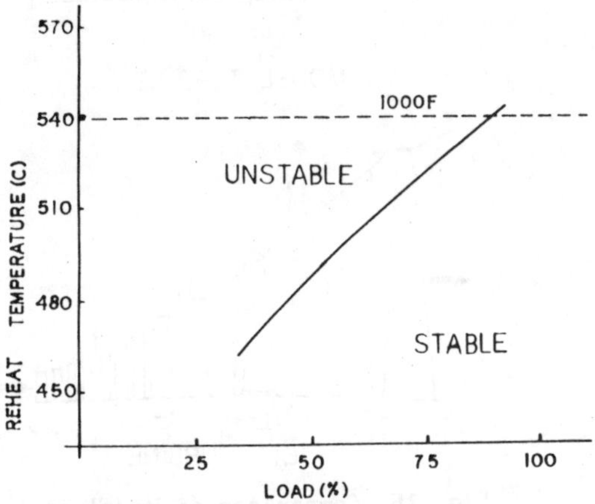

Fig. 14 Instability Zone for Safe Turbine Operation

Returning to Fig. (13), the instability zone for LP2 is seen to be in a region at lower reheat temperature than that of LP1 for a given load. This turbine generator unit has been operating at either 50% or 100% load most of the time at a reheat temperature near 538°C (1000°F). For such an operation LP1 is more susceptible to the non-synchronous vibration than LP2, since LP1 will be in the instability zone at 50% load, whereas LP2 will be in the stable zone at both 50% and 100% loads. This explains why the blade distresses had been found on LP1 rather than on LP2.

Although the shrouded blades exhibited the second mode non-synchronous vibration at certain operation conditions, the free standing blades, did not show similar blade vibration in this test. Studies on similar free standing blades by Whitehead's method on transonic flow unstall flutter (Ref. 8) indicated that these blades may encounter unstalled flutter under certain operating conditions. Recent field tests by the blade vibration monitor (BVM) did show non-synchronous vibration at higher load, however, moisture seems to have damping effects on the blade vibration.

4. LABORATORY MODEL TURBINE TESTS

A 0.6 scale turbine consisting of the L-1 and L-2 stages was built for the laboratory test. The L-1 stage served as a test stage and L-2 stage as the flow conditioning stage. The maximum end loading attainable for the facility is about 66,390 $Kg/hr\text{-}m^2$ (13,600 $lb/hr\text{-}ft^2$) based on the L-OR (last row blade) annulus area which corresponds to about 90% of the design full load. This test turbine differs some what from the field turbine in that a constant radius cylinder flow guide was adopted for the test turbine whereas it was a cone surface with flare as shown in Fig. (15).

Strain gages were installed in the L-1R blades, and telemetry tests were conducted on this model turbine. Wherever required, the flow measurements were performed by traversing behind the rotating blade row. Like the field tests, the turbine inlet and exit conditions were monitored continuously. Similar instrumentation as in the field tests was used for gathering and analyzing the strain gage signals.

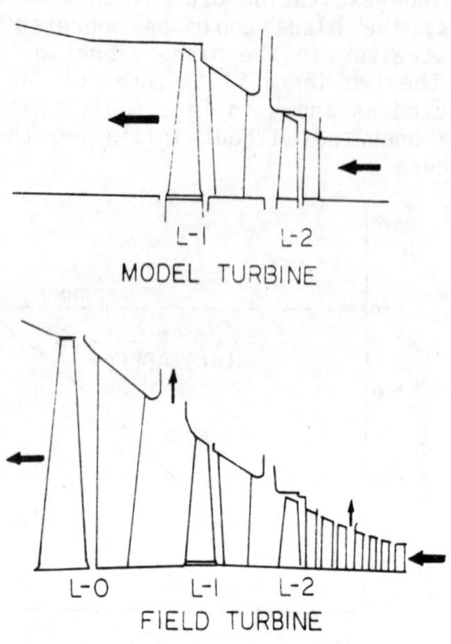

Fig. 15 Comparison of Model Test and Field Turbines

The frequencies corresponding to the first and second modes for the full size blades are respectively 210 Hz and 325 Hz. Whereas they are 350 Hz and 542 Hz, respectively, for the model blades. In the absence of the L-0 stage for the model turbine, the pressure and Mach number distributions along blade height at the L-1R exit will be different from that of the field turbines. Due to these differences in natural frequencies and flow properties, and also due to the fact that the moisture effects cannot be determined by scaling, exact duplication of the field phenomenon in the model turbine is not feasible. However, a similar phenomenon can be simulated by carefully manipulating the test conditions.

TEST PROCEDURES AND FLOW PROPERTIES AT THE EXIT OF L-1R BLADE ROW

As mentioned previously, the model test turbine used a zero tapered contour at the O.D. of diffuser (flow guide). This zero tapered straight contour tends to raise steam pressure near the blade tip and thus lower the flow exit Mach number in that region. Analytical predictions of the pressure and relative Mach number distributions at the exit of the L-1R blade row for the model turbine are shown in

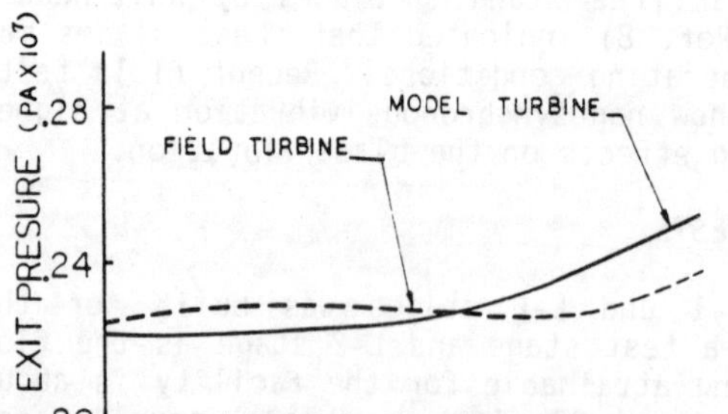

Fig. 16 Exit Pressure Distribution of Model Test Turbine

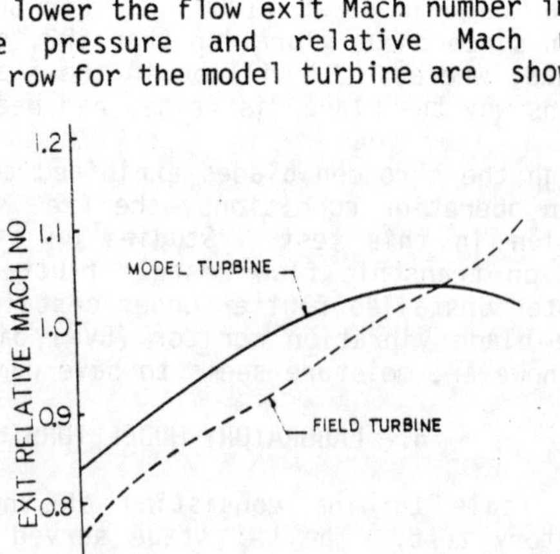

Fig. 17 Exit Relative Mach Number Distribution of Model Test Turbine

Figs. (16) and (17). These flow properties are compared to that of the field turbines in the same figures. For the field turbine, the calculation was performed for the last three stages with the L-0 stage in place. Both calculations were based on the same inlet condition and the same average pressure at the L-1R exit. It can be seen that, although the average pressure may be the same for both turbines, the model turbine gives lower pressure than the field turbines in the lower 60% of the blades; in the upper 40% of the blade the pressure is higher for the model turbine. Due to these deviations of pressure distribution the Mach number at L-1R exit of the model turbine is different from that of the field turbines. It is seen in Fig. (17) that Mach numbers near the blade tip are lower than that of the field turbines. This means that the exit average pressure must be kept lower than that of the field turbine if the exit Mach numbers in a region near the blade tip are to be simulated as in the field turbines.

Basically, the same turbine operations as in the field turbine tests were adopted for investigation of the non-synchronous blade vibration. Fig. (18) shows the three tests conducted to bring the blades to cross the constant moisture lines. Test 2 and 3 were, respectively, constant temperature and pressure operations. These two tests were both started at a wetness below 2%, which is in a dry state due to supersaturation. Test 1 is constant pressure operation but with the initial condition superheated at exit of the turbine.

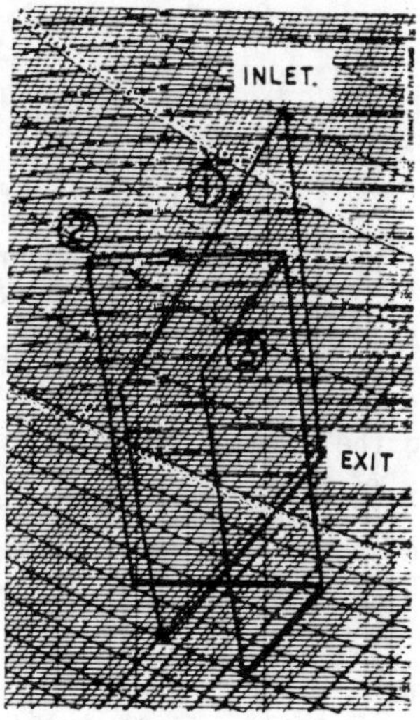

Fig. 18 Model Turbine Test Procedures

When the tests were run with the pressure ratio (inlet pressure/exit pressure) of the model turbine set equal to that of field turbine, all three tests (Test 1, 2, and 3) did not show any stress build-up. The peak stress levels were about 0.7 MPa (100 psi) for the first mode (350 Hz) and 0.5 MPa (70 psi) for the second mode (542 Hz). An average pressure ratio of 5.2 was used for these tests, for which the exit Mach number near the tip of L-1R exit was substantially lower than that of the field turbines. Apparently, a larger pressure ratio is required for the model turbine test to produce Mach number near the tip region comparable to the field turbine.

TEST IN DRY STEAM

The next test was run with the L-1R blades in a dry environment (corresponding to initial condition of Test 1). Fig. (19) shows the result of this test. The L-1R exit Mach number increased with reduction of condenser pressure. The first mode peak stress started to increase as the back pressure was decreased. The first mode stress reached a maximum of 1.3 MPa (190 psi) at pressure ratio of about 6.9. Further reduction of back pressure again reduced the stresses. At the maximum stress condition a traverse was made at L-1R exit to measure the Mach number distribution, which is shown in Fig. (20). It is shown that the maximum stress occurred when the exit Mach number is about 1.1 at the blade tip. Since the blades were operating in a dry state and the only change made was exit Mach number the instability encountered is due to Mach number effects. Also shown in Fig. (20) are the exit Mach number measured for the pressure ratio corresponding to the design condition ($\bar{p}$ = 5.2). The exit Mach number distribution predicted analytically for the field turbine is also shown for comparison. It is seen that the transonic instability occurred when the exit Mach number at the blade tip for the model was comparable to that of the field turbines. The instability observed is probably associated with the shock oscillation and its interactions with boundary layer (Refs. 9 and 10). It is to be noted that the second mode stress was low and no significant changes have been observed.

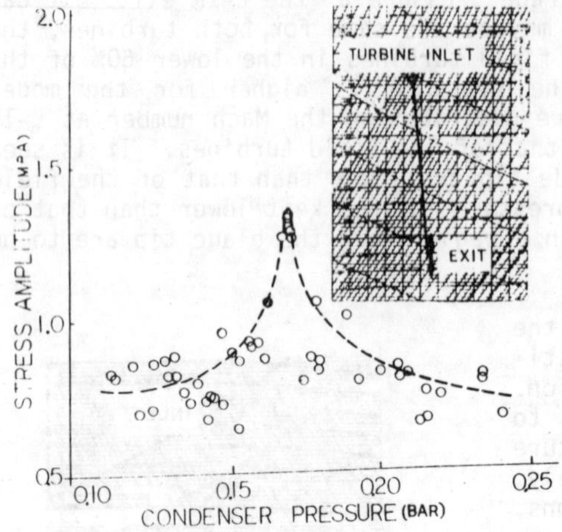

Fig. 19 Stress vs Condenser Pressure with Model Turbine Operating in Dry Steam

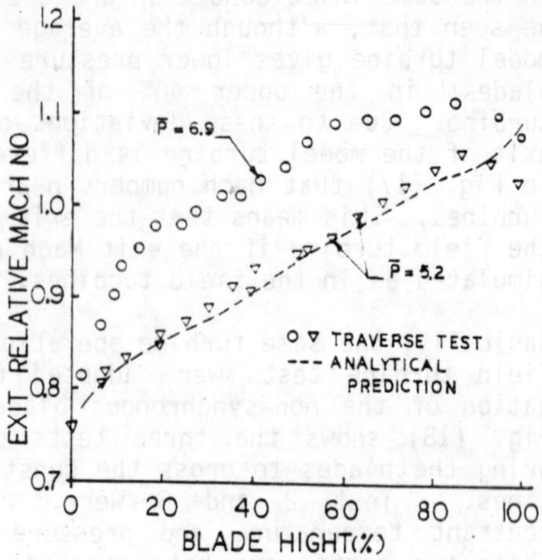

Fig. 20 Comparison of Pressure Distribution

TEST IN WET REGIONS

Using the same inlet pressure 1.17 bar (17.0 psia) and a condenser pressure of about 0.17 bar (2.46 psia) the inlet temperature was decreased while the pressure ratio was kept unchanged (kept exit Mach number unchanged). In this process the wetness increased as the inlet temperature was decreased. At the inlet temperature of about 124°C (255°F) a sudden increase of peak stress, corresponding to the first mode (350 Hz), was encountered (Fig. 21). At this operating condition, the wetness at the L-1R exit is about 4%. The stress decreased when the inlet temperature was reduced further. The sudden build-up of stress is a phenomenon quite similar to that observed in the field turbines. It occurred at a frequency (350 Hz) close to that for the field test (325 Hz) except in the model turbine it is the first mode rather than the second mode for the field turbine. The instability zone is also very narrow as was observed in field test. In the field test the nonsynchronous vibration occurred only when the blades were brought from dry to wet. In the model tests, the stress build-up of the first mode occurred in both directions. The maximum stress was about 2.0 MPa (280 psi) when the blades were brought from dry to wet. When going from wet to dry the peak stress was only about 1.2 MPa (170 psi). The phenomenon appeared to be very sensitive to the flow condition. It occurred only under a carefully controlled flow condition and when the blades were crossing the instability zone very slowly as was experienced in the field test. No appreciable change in the second mode stress was noticed during this test.

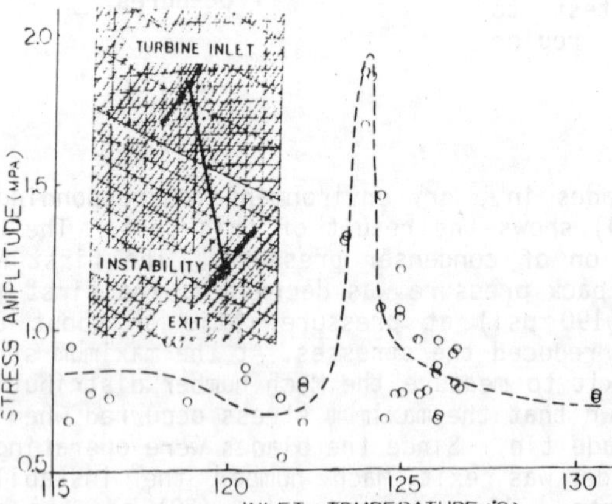

Fig. 21 First Mode Stress vs Turbine Inlet Temperature for the Model Turbine Test (P_{in} = 1.17 bar)

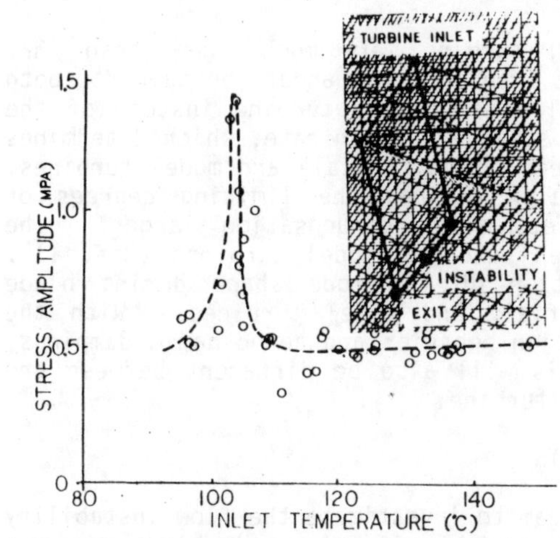

Fig. 22 First Mode Stress vs Turbine Inlet Temperature for the Model Turbine Test ($P_{in.}$ = 0.76 bar)

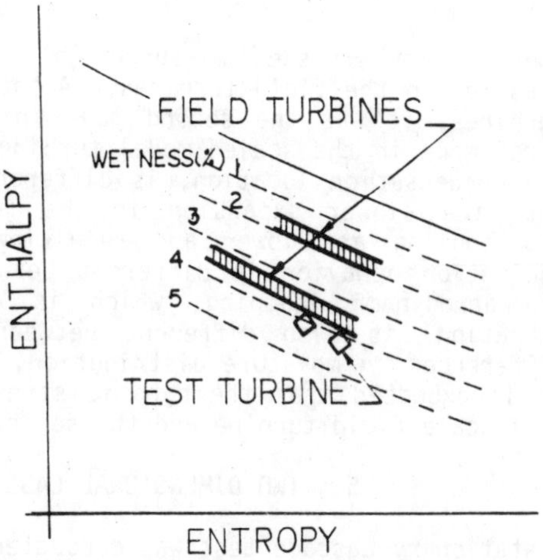

Fig. 23 Instability Zones of Field Turbines and the Model Turbine

The last test was conducted at an inlet pressure of 0.76 bar (11.0 psia). The condenser pressure was determined by optimizing the first mode vibratory stress. A condenser pressure of about 0.13 bar (1.82 psia) had been identified to give transonic instability in the dry state. Keeping the above pressures fixed, tests were run by changing the inlet temperature. Again, at about the 4% moisture line (L-1R exit) a sudden increase of stress was encountered as shown in Fig. (22). The maximum stress at 350 Hz was about 1.4 MPa (200 psi) changing from about 0.48 MPa when the blades were brought from dry to wet. The maximum stress was only about 0.8 MPa (120 psi) when going from wet to dry.

The instability zone found for the two stage model turbine agrees well with that obtained in the field test of LP2 turbine. Shown in Fig. (23) is a comparison of these results. In the model test the instability zone for the L-1R exit lies close to a 4% moisture line, whereas the field test showed a region between 3.0 to 4.0% moisture lines for LP2 and 1.5 to 2.0% for LP1. In the model tests the back pressure must be lowered to produce a higher exit Mach number for the blade tip. If the model turbine was built with the L-0 stage in place, the back pressure for both the model and field turbines would have been the same and the instability zones for both turbines would have coincided with each other.

Although the non-synchronous vibration phenomenon observed in the field tests has been reproduced at approximately the same frequency in the model turbine, it is the first mode for the model and second mode for the field turbines. The fact that the frequencies of vibration for both the model and field turbine are about the same suggests that the flow instability itself has a frequency band in the neighborhood of 325 Hz. The measured stresses for the model turbine tests were much lower than the field test results, possibly due to the difference in aerodynamic damping between the model and field turbines as a result of the condensation phenomena which are not scaled and of the vibration modes which are different (first vs second mode). The phenomenon has been found occurring at an exit Mach number in the neighborhood of 1.1 near the blade tip and at 4% moisture based on the average exit condition. The phenomenon appeared to be very sensitive to the flow conditions and the rate change of operation parameters at which the blades were brought to cross the instability zone as was observed in the field test. The instability zone of model turbine was more close to LP2 than LP1 indicating that moisture distribution of the model turbine was more similar to that of LP2.

The maximum stresses measured in the model turbine were much lower than that measured in the field turbines. Although the frequency was about the same for both turbines, it was the second mode in the full scale field turbine instead of the first mode in the scaled model turbine. The steam expansion rate, which determines the condensation location, is different between the full scale and model turbines. With the slower expansion in the full scale turbine, the limiting degrees of supercooling are lower and resulting droplets after condensation larger. The nucleation behavior is different between the field and model turbines (Ref. 11). The aerodynamic damping, which is a function of the mode shape during blade vibration, is also different between the field and model turbines. With the differences in moisture distribution, nucleation behavior and aerodynamic dampings, it is expected that the dynamic stress levels will also be different between the full scale field turbine and the scaled model turbine.

5. TWO DIMENSIONAL CASCADE TESTS

A stationary cascade test was conducted in steam to investigate the flow instability due to the interaction of condensation, shocks and boundary layer in blade passages (Ref. 12). The cascade was made up of six full scale L-1R tip sections and two end pieces, corresponding to suction and pressure surfaces (Fig. 24). The blades were enclosed between two 12.7 mm (0.5 inches) thick optical grade Lexan windows. Miniature pressure transducers, eight on pressure surface and fifteen on suction surface, were used on the blade surfaces of passage #3 to measure the unsteady static pressure. Wall taps were also placed upstream and downstream of the blades for measuring the inlet and exit static pressures.

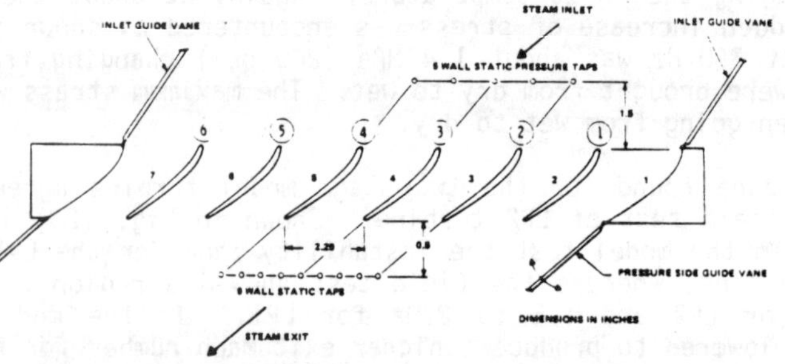

Fig. 24 Cascade Test Section Details

The test was performed to assess the unsteady forces arising from the interactions of shock, condensation and boundary layer in a stationary cascade. In this test, the interaction of blade movement with the unsteady shock/condensation (aeroelastic) has been neglected. The only unsteady forces measured from the test are those due to instability caused by flow which will result in forced vibration of blade. In the first test a steady inlet pressure was employed. The tests covered a wide range of flow exit Mach numbers. The unsteady pressures measured by the transducers embedded on the blade surface show that the interaction of condensation and boundary layer interaction was not significant. The condensation, however, seemed to enhance shock/boundary layer interactions.

In the subsequent tests, transient inlet conditions were employed. In this test the steam temperature in the plenum was established in the superheated region at P_t = 35KP$_a$ (5 psia) and T_t = 127°C (260°F) with the exit pressure provided desired exit Mach number. The inlet total pressure was slowly lowered until the inlet conditions

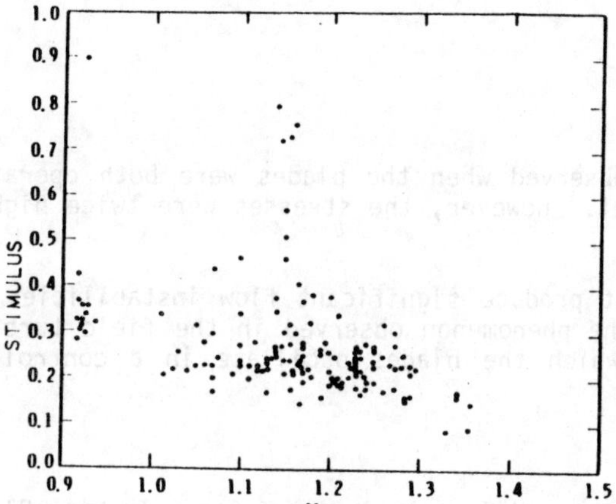

Fig. 25 Stimulus vs Isentropic Exit Mach Number

were near saturation and then raised until the original conditions were reached. This operation properly simulated the constant load operation of the field turbine. The unsteady signals from the minature transducer were integrated spatially to find unsteady forces acting on the blade surface. Highly variable activities were observed at exit Mach number around 1.15 although the stimulus are generally low. Shown in Fig. (25) is the stimulus (unsteady force amplitude/steady force) from all sets of transient data. This high activity at transonic exit condition is consistent with the model turbine test in the dry state.

Although the cascade tests, both steady and transient inlet pressure, show low unsteady forces due to the shock, condensation and boundary layer interaction, it should be emphasized that the stationary cascade tests do not simulate the effects of three dimensional flow in field turbine nor take into account the interaction between the blade vibratory motion and the unsteady flow (aeroelastic flutter phenomenon).

6. CONCLUDING REMARKS

A non-synchronous blade vibration (325 Hz, second mode) has been detected in the field turbines on two rows of well tuned shrouded L-1R blades. The tests conducted for the field turbines, a scaled model turbine and a stationary cascade lead to the following conclusions:

(1) The high stress due to the non-synchronous blade vibration is a major contributor to the blade distresses.

(2) The non-synchronous blade vibration occurred only in a narrow region characterized by constant moisture lines. The phenomenon is aeroelastic in nature and the vibrations are caused by the interaction of blade vibratory motion with the condensation, and possibly with the shock and boundary layer.

(3) The onset of the non-synchronous blade vibration depends on the mode of turbine operation. The rate change of load or turbine inlet temperature must be kept slow in order to initiate the vibration and to amplify the blade excitation.

(4) The field test showed a peculiar behavior in that the blade excitation could be initiated only when the blades were operated from dry to wet state. Blade excitations have not been detected when the blades were operated going from wet to dry state.

(5) The fact that the occurrence of blade vibration depends on the rate change of operation parameters and on the direction of operation strongly suggest that the phenomenon is non-linear in nature.

(6) The scaled model turbine showed similar phenomena. However, the blade vibration occurred at the first mode (350 Hz) rather than the second mode (325 Hz) for the field turbines. The measured stresses for the model turbine were much lower than that in the field turbines due to effect of scaling.

The stress built-ups have been observed when the blades were both operated from wet to dry and from dry to wet. However, the stresses were twice higher when going from dry to wet.

(7) The stationary cascade test did not produce significant flow instabilities to excite the blades. To simulate the phenomenon observed in the field turbine test a dynamic cascade test, in which the blades oscillate in a controlled mode, must be conducted.

REFERENCES

1. Davey, R.P., McCloskey, T.H. and Rieger, N.F., "Analysis of Steam Turbine Blade Failure in the Utility Industry," ASME paper 83-JPGC-Pwr-20, 1983.

2. Deich, M.E., Kurshakov, A.V., Leonov, V.M. and Tishchenko, "Pulsation Conditions and Resonance Phenomena with Spontaneous Condensation in Laval Nozzles," Thermal Engineering, 31 (10), 1984.

3. Yousif, F.H. Campbell, B.A. and Bakhtar, F., "Instability in Condensing Flow of Steam," Proc. Inst. Mech. Engrs., Vol. 186, 37/72, 1972.

4. Moore, M.J. and Sieverding, C.H., "Two-Phase Steam Flow in Turbines and Separators," McGraw-Hill Book Company, 1976.

5. Deich, M.E., Kurshakov, A.V., Tishchenko, A.A., Leonov, V.M. and Emets, O.Z., "Condensation Instability in Supersonic Turbine Cascades," Teploenergetika, 34, (11) 48-52, 1987.

6. Chen, S., "Application of the Rotating Water Table to Nozzle Wake Excitation in Low Pressure Turbines," ASME paper No. 85-IGT-45, Int. Gas Turbine Symposium and Exposition, Beijing, PRC, Sept. 1-7, 1985.

7. Evan-Iwanowski, R.M., "Resonance Oscillations in Mechanical Systems" Elsevier Scientific Publishing Company, 1976.

8. Whitehead, D.S., "The Calculation of Steady and Unsteady Transonic Flow in Cascades," Cambridge University Engineering Dept., Rept. CUED/A-Turbo/TR118, 1982.

9. Kobayashi, H., "Effect of Shock Wave Movement on Aerodynamic Instability of Annular Cascade Oscillating in Transonic Flow," ASME paper No. 88-GT-187, Gas Turbine and Aeroengine Congress, Amsterdam, the Netherland, June 6-9, 1988.

10. Surampudi, S.P. and Adamczyk, J.J., "Unsteady Transonic Flow Over Cascade Blades," AIAA J., Vol. 24, No. 2, Feb., pp 293-302, 1986.

11. Bakhtar, F. and Heaton, A.V., "On the Dissimilarities in Wet Steam Behavior in Model and Full Scale Turbines," Technology of Turbine Plant Operating with Wet Steam, BNES, London, 1988.

12. Whirlow, D.K., McCloskey, T.H., Davids, J., Chen, S., Kadamki, J.R. and Farn, C.L.S. "Flow Instability in Low Pressure Turbine Blade Passage," Paper 84-JPGC-GT-15, Joint Power Generation Conference, Toronto, Canada, Oct. 1984.

APPENDIX A

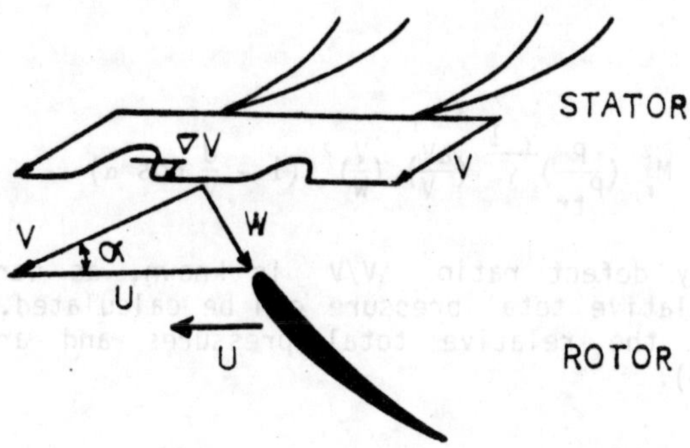

Fig. A-1 Flow due to wakes

The velocity defects due to the wakes of the stator vanes can be expressed in terms of the axial distance from the vane trailing edges as follows (Ref. 12).

$$\frac{\Delta V}{V} = \frac{K}{(\frac{x}{c} + 0.025)^{1/2}} \qquad \text{(A-1)}$$

where ΔV is the velocity defect amplitude; x is the distance from the stator trailing edge and c is the chord length of the stator vane. K is the factor determined from the test.

The rotating blades see a relative velocity change ΔW when they are passing the wakes. From the velocity triangle, we have

$$W^2 = V^2 + U^2 - 2UV \cos \alpha \qquad \text{(A-2)}$$

From the above equation, the velocity change ΔW can be derived. It is given by

$$\frac{\Delta W}{W} = \frac{\Delta V}{V} \left(\frac{V}{W}\right)^2 \left(1 - \frac{U}{V} \cos \alpha\right) \qquad \text{(A-3)}$$

The stagnation pressure transducers rotating with the blades measure the total pressure relative to the rotating blades. The relative total pressure is defined by

$$\frac{P_{tr}}{p} = \left(1 + \frac{\gamma-1}{2} M_r^2\right)^{\gamma/\gamma-1} \qquad \text{(A-4)}$$

where M_r is the relative flow inlet Mach number, an γ is the specific heat of steam.

Assuming that the pressure and the temperature has been kept constant, then the total pressure change by the wake can be determined from Equations (A-2), (A-3) and (A-4). The result is

$$\frac{\Delta P_{tr}}{P_{tr}} = \gamma M_r^2 \left(\frac{P}{P_{tr}}\right)^{\frac{\gamma-1}{\gamma}} \left(\frac{\Delta V}{V}\right) \left(\frac{V}{W}\right)^2 \left(1 - \frac{U}{V} \cos \alpha\right) \qquad \text{(A-5)}$$

When the wake velocity defect ratio $\Delta V/V$ is known, as can be estimated from Equation (A-1), the relative total pressure can be calculated. This equation has been used to predict the relative total pressures and are compared to the measurements in Fig. (6).

MATHEMATICAL SIMULATION OF UNSTEADY COMPRESSIBLE 3-DIMENSIONAL FLOW THROUGH TWO-STAGE TURBINE SECTION WITH THE INTERMEDIATE EXTRACTION OF A WORKING MEDIUM

V. I. Gnesin
(Institute of Problems in Machinery, Ukr. SSR Academy of Sciences, 310046, Kharkov, USSR)
V. G. Solodov
(Automobile and Highway Institute, 310078, Kharkov, USSR)

ABSTRACT

A general method of a direct mathematical simulation of unsteady 3- dimensional flow at the turbine stages, aerodynamically interacting with circuits of non-regulated heating steam extraction is proposed. Different mathematical models of a turbine stage functioning in the section with heating extraction are discussed. The generation of method for the modeling of aerodynamical interaction of stage and different non-axisymmetric flow path elements: inlet and outlet pipes, combustion chambers are considered. The mathematical model of steady 3-dimensional flow through the power and heat extraction circuit is proposed. Some attention is paid to a practical realization of algorithms for flow simulation with allowance for capabilities of modern computers. The results of the flow simulation through two-stage nearextraction section of the powerful heating turbine are presented. The data on flow structure and performances of nonstationary processes in channels of blading rows and axial spacing are given. The effect of geometric performances of the extraction circuit on the value of low-frequency pulsations of aerodynamic forces at blades of nearextraction blading rows is discussed.

NOMENCLATURE

ω	angular velocity	u	velocity of rotor blade points
Z_S, Z_R	number of stator and rotor blades	C_o	maximal velocity
r, φ, z	coordinates of cylindrical system	P_o	stagnation pressure
t	time	l	blade length
P	static pressure	D	stage average diameter
ρ	density	δ_z	inter-row (stage) axial spacing width
a	speed of sound	b	blade width
γ	angle of flow path expansion	A	pulsation amplitude
P_i^k	pressure instant value in point K		**subscripts and superscripts**
s	arc coordinate	—	average by time value
W	flow relative velocity	$\wedge$	dimensionless value
k	harmonic number	e	outlet boundary

F_t	instant value of full aerodynamic force	*	critical value
α	whirl angle between the radial direction and velocity vector		

INTRODUCTION

The development of the blading of modern powerful steam and gas turbines brings about the problem of aerodynamic impact of bladeless non-axisymmetric flow path elements such as combustion chamber, inlet and outlet pipes, connections and circuits of non-regulated working medium extraction on the operation of turbine stages.

Here of great importance are the problems arising from the impact of the irregularity of pressure patterns, whirls, periodical non-stationarity upon the development of flow separation in the blading, the values of non-stationary aerodynamic forces at the rotor blades of adjacent stages. In particular, the available experimental data[1~5] for full-scale turbines and models testify to the possibility of occurrence during extractions for power-and-heat generation, of considerable circular and radial pressure gradients within the blading caused largely by the diagonal mode of gas flow through the axial spacing, by the extraction pipes arrangement, configuration of the interyoke slot and the extraction chamber itself. The structure of the flow becomes complicated by possible redistribution of flow rate through the connection as a result of the flow whirl in the axial spacing between the stages. The enumerated above peculiarities of the flow in the axial spacing bring about considerable non-uniformiity in the parameters distribution and lead to low-frequency aerodynamic forces pulsation at the rotor blades.

The analysis of the literature on the subject testifies to the absence of a systematic approach to the study of aerodynamic influence of bladeless flow path elements upon the operation of a turbine stages nowadays. The available experimental data[2-5] for the full-scale turbines are incomplete and generally relate to the regenerative bleed-off. The study of models are not numerous either[1,4,6] and as a rule they don't effects in the stages near the extraction zone and in the extraction circuits have not been made.

The problem of interaction of the stage with other flow path elements, and outlet pipes, in particular, has been studied to the same extent. New studies relating to the pipe design with consideration of the complicated flow structure in the inlet part and analysis of the "last stage-exhaust hood" complex have been made recently[7,8]. But it should be mentioned that all the calculations were made on the assumption of the flow axial symmetry. At the same time the experience, e. g. ref. 9, shows considerable variation in static pressure (up to 20 per cent) in the circular direction behind the blade row resulting from the exhaust hood assymmetry and leading to periodical variation in flow in the last stage.

Up to now analytical study of such flows was impossible since it required principally new approach to tne formulation and numerical realization of 3-dimensional unsteady problems in trubine gas dynamics. The formulation and numerical solution algorithm of a direct problem on the 3-dimensional transonic gas flow through relatively moving spatial cascades having been developed in ref. 10 the computations of unsteady flow in a turbine stage under arbitrary conditions at the inlet and outlet became possible.

Here below basing on ref.. 10 we propose an approach towards a numerical computation of a 3-dimensional flow in a stage aerodynamically interacting with the non-axisymmetric flow path elements.

METHOD

The formulation of a problem for the stage interacting with the ono-axisymmetric flow path element is based on a model of 3-dimensional periodically unsteady flow of inviscid non-heat conductive gas through two relatively moving spatial blade rows with consideration of their aerodynamical interaction and rather optional assumptions concerning their geometry.

The ideal gas flow can be described by a complete system of unsteady gas dynamics equations presented in a conservation law from, integrated by the Godunov's difference scheme of the 2-nd order accuracy.

The integration of gas dynamics set of equations, and also the region grid generation are based on the property of spacing of gas flow through the blade-to-blade channels of the interacting blade rows as a whole at the fixed moment of time and its periodicity for a fixed channel during a rotation of rotor.

In the first case the computational region includes a group of rotor and stator channels, the latter making up the stage spatial period. In the second case the computational region consists of an channel of the rotor and the one of the stator, and the algorithm is built up on the fact that the flow structure in each channel is periodical with the time interval of $\tau_S = 2\pi/(Z_R\omega)$ for the stator and $\tau_R = 2\pi/(Z_S\omega)$ for the rotor. The stage common time interval is the least common multiple of the periods of both stator and rotor $T = \tau_S K_S = \tau_R K_R$. In both cases the computational region is complemented with the periodicity regions in the vicinity of the stage inlet and outlet as well as with the ones of the axial spacing fragments between the blade rows.

In the present paper the computational region including a channel of the stator and the one of the rotor has been chosen as it is preferbale from the point of view of economizing of computer resources, when the stage period has a big value.

The number of boundary conditions in the stage inlet and outlet results from analysis of perturbation diffusion in unsteady flow. According to the quasi-one-dimensional theory of characteristics applied in the vicinity of the boundary for correct formulation of problem when the axial velocity vector component is subsonic* it is necessary to assign four gas-dynamics parameters or their relationships at the inlet and the outlet. These conditions are complemented by relationships at the characteristics.

The boundary conditions to be assigned are as follows: the distribution of total enthalpy and entropic function, the direction of the velocity vector at the inlet boundary; the static pressure distribution—at the outlet one. At the rigid boundaries of the computational region, the condition of non-leakage is realized. For the axial spacings are assigned the generalized conditions of spacing and periodicity (GSP conditions) which take into account the flow fluctuation phase shift in the adjacent channels of rows under discussion caused by the difference between the cascades pitches. The gas dynamical parameters as a boundary conditions may be different[11,12].

The boundary conditions type in the inlet and outlet boundaries depend upon the turbine stage model to be computed. For a stage model operating separately from stages and any other flow path elements there may be assumed uniform by the distance to the stage, distribution of gas-dynamical parameters in the circular direction. For a stage operating as a part of a section the boundary conditions are determined both by the characteristics of the connected elements e. i. inlet valve, combustion chamber, extraction circuit, outlet pipe etc. and by the characteristics of other stages. Therefore, the flow simulation through the turbine stage, the latter interacting with the flow path elements, must be based either on the experimental data or the results of the aerodynamic calculations of the above mentioned elements.

The influence of the flow path elements on the flow in a stage may be assigned by functional dependance desribing the variation in the distribution of the flow static and kinematic parameters in the circular direction on the whole circumference.

At the same time the arbitrary law of parameters variation in time at the computational region boundaries violates the GSP conditions which occur for the separated case (when the steadiness in circular direction of the boundary condition functions is given). To acquire a periodic solution in the channel of the rotor and stator with a period equal to the one of the stage, the functions approximating the boundary conditions at the inlet, outlet and axial spacing boundaries of the stage should meet the requirements of the generalized spacing and periodicity[10]. This stringent limitation in the law of variation simulating the influence of the flow path elements upon the stage, generally speaking, is not realized in practice. The violation of the GSP conditions brings about the expansion of the flow path in the general case up to 2π which leads to the necessity of flow computations either in all the stage channels or all periodicity pictures during a complete revolution of the rotor.

However, the experience shows that the flow path elements such as combustion chamber, inlet and outlet pipes, extraction chamber circuits create circular large-scale variation in gas-dynamic parametes in the axial spacing with a characteristic scale several times the period of a typical turbine stage.

The circular largescale parameters variation created by the flow path elements is approximated by a piece-wise-constant function, the sections of constancy being not less than 2-3 stage periods. At every boundary conditions functions section of constancy there is made the computation of the flow through the stage, the GSP conditions being used.

The general solution of unsteady flow through the stage during a complete revolution of rotor under the non-uniform boundary conditions at the inlet and outlet is the linear interpolation between the solutions at the sections. This approach is substantiated by the fact that the physical time of the stabilization of the flow through the stage from a certain initial approximated solution is comparable with the time of the turn of the working blade to an angle equal to 2-3 angular periods of the stage. The spproximation of the approach lies in the negligence of flow fluctuation of non-periodic character in channels occurring at the transition intervals between the constancy sections of the boundary conditions.

When unsteady boundary conditions both at the inlet and outlet are taken into account together, e. g. for a one-stage turbine, the mathematically correct solution of the problem in periodically unsteady flow through the stage channels is possible with the application of the mentioned approach if the intersections intervals of the constancy sections of the inlet and outlet boundary conditions piecewise constant approximation satisfy the conditions of the physical stabilization of the flow formulated above.

The general solution of the problem on flow in a stage makes it possible to get the space fields of gas-dynamical parameters in the stage channels and also the circular dependence of non-stationary aerodynamical forces influencing the rotor blade as a result of both the pitch non-uniformity of parameters in axial spacing and largescale variation at the inlet and the outlet.

Within the limitations on formulation the boundary conditions for the stage caused by the wish to make the flow simulation only at the stage time period we shall consider below different variants of formulation of the problems on the interaction of a stage with the parameter field variation from the flow path elements.

1. The Stage Interacting with the Consequent Extraction Chamber or the Exhaust Hood

A. At the inlet we consider the flow with uniform distribution of parameters, at the outlet—the counterpressure distribution along the radius and the circular coordinate is assigned.

B. For a particular case of the pre-extract stage it can be joined by the volume of the inter-stage axial spacing with the geometrically assigned the extraction chamber slot at the periphery. The static counterpressure is assigned in the interstge axial spacing plane perpendicular to the turbine axis downstream the flow extraction; in the extraction chamber slot there assigned the distribution either of the static pressure or flow rate.

2. The Stage Interacting with the Pre-connected Inlet Pipe or the Extraction Chamber

A. At the inlet we consider the flow with the non-uniform distribution of parameters along the circumference due to the either non-axisymmetric character of the working medium feeding from the inlet pipe, or as a result of the extraction impact. The non-uniform are assigned also the distribution along the radial coordinate of total enthalpy and velocity vector angles, they being matched within phase and period. With a small level of non-uniformity, e. g. for the regenerative bleed-offs, the flow angle in the meridional plane is determined by the flow path contour. Such formulation of problem is grounded upon the results of tests with model stage[1,2,4] and related to the governing impact of the total enthalpy distributions and velocity angles upon the flow structure in the postextraction stage. In the outlet section at a distance from the stage the statement of the constant counterpressure of the distribution in accordance with the radial equilibrium law is physically justifiable.

B. For a particular case of the postextraction stage its simulation can be made with joining of the inter-stage axial spacing volume with the extraction chamber slot geometrically assigned at the periphery. In this case the boundary conditions in the inlet to the computational region are determined by the blade row of the pre-extraction stage. At the conical surface of the computational region corresponding to the inlet of the extraction chamber slot, there assigned the static pressure distribution, subordinate to the GSP conditions mentioned above.

3. Problem Formulation on a Flow in the Nearextraction Section

The most comprehensive here is the formulation of a problem on a gas flow in the computational region comprising the rotor and stator channels sets making up the complete two-stage section period[15]. If the stages being a part of section have equal ratios Z_S/Z_R then the section period coincides with the stage one.

The problem can also be solved when the GSP conditions in the interstage axial spacing are not observed. In this case the flow parameters in the inter-stage spacing in the circular direction are averaged; the GSP conditions are rigorously observed in the inter-rows axial spacings.

Of some interest from the point of view of methods and practice is a simplified formulation of the problem on a flow through the nearextraction section. In this case the computational region embraces the pre-extraction stage rotor channel, inter-stage spacing fragment and postextraction stator channel.

Depending on the aim of the research the analysis through the combination "rotor-stator" can be carried out both with the help of an algorithm based on the GSP conditions in the inter-stage axial spacing with consideration of the piecewise-constant static pressure distribution or the flow rate in the extraction interyoke slot, and with the application of the pitch averaging of the parameters in the circular direction. The latter case may happen to be expedient when there take place big heating steam extractions from the wide inter-stage spacings.

In the rotor computational region inlet there may be assigned the distributions of total enthalpy, entropy function and flow velocity angles in the rotating reference system related to the rotor, and in the outlet from the postextraction stage stator channel — the counterpressure distribution. The above mentioned distributions are determined by the pre- extraction stage rotor, post-extraction rotor and also by the amount of extraction.

Physically, more natural is the model of one-and-a-half type section with extraction, comprising the pre-extraction stage channel and the post-extraction stage stator channel.

To complete the problem formulation on a flow in the nearextraction section there have been made a mathematical model of 3-dimensional steady flow in the extraction chamber of a steam trubine[14]. The computational region includes 3-dimensional annular header volume with an annular inlet slot for the working medium and several optionally oriented pipes which can form the receiver. In the computational region the inviscid compressible gas flow is considered, which is described by a set of Euler's unsteady 3-dimensional equations in a divergent form[10]relatively to the unknown values of the static pressure, density, and three components of velocity.

The flow in the annular volumes of the header and the interyoke slot is formulated in the cylindrical coordinate system; to describe the flow in pipes (the latter having rectangular cross-section) the Cartesian system of coordinated is used. To integrate the set of Euler's equations the Godunov's difference scheme is completed by corrections[10]increasing the accuracy to the 2nd order. The steady flow in the computational region we have as a result of the time-marching technique. At every time step the flow in the slot , header and in pipes is computed separately. At the boundaries where the regions join the linear interpolation algorithm is applied.

As the stationary boundary conditions the following is stipulated: at the cylindrical surface of the inlet to the interyoke-slot this is the velocity vector direction, the distribution of the enteropy function and the flow total enthalpy; in the outlet from the pipes there is stipulated the counterpressure or flow rate distribution; at the rigid boundaries the conditions of non-leakage are assigned.

There can be simplified 2-dimensional formulatons of the flow computation problem and the algorithms in the extraction circuit: the problem on the flow in the inter-stage axial spacing plane perpendicular to the turbine axis; the problem on the flow in the spatial extraction circuit with the averaged parameters along the axial coordinate with due regard for the chamber variable dimension in the axial direction. These models are suitable for flow description in the circuits with a wide interyoke slot.

The ideal gas model, used in the problem, approximately describes the flow core structure but it does not contain solutions representing separation effects at the velocity vector abrupt turns and the flow expansion. For a closer relation of the used model to the real flow in channels with the sudden expansion of the flow path there is effectively used the artificial viscosity of scheme available during numerical simulation which develops as the error of Euler's equations approximation. In a number of researches the artificial viscosity is used for an adequate simulation of viscosity effect such as the creation of circulation zones over backward-facing steps, entropy wakes behind rows, etc.[10,13].

NUMERICAL RESULTS AND DISCUSSIONS

The gas dynamics of extraction circuits with different pipes configuration and of the stages adjacent to the extraction circuits and exhaust hood have been studied, the developed models and numerical method being applied.

1. Flow Simulation in the Extraction Chamber

The study of a circuit comprising an interyoke slot, annular header and two abstraction pipes (Fig. 1). has been made with the application of a 3-dimensional mathematical model of the flow.

The geometric characteristics of the basic configuration are:
$r/R=09$, $Z/R=0.3$, $z/R=0.03$, $\varphi_1=30°$, $\varphi_2=60°$
The study has been made in the regime $P_e/P_o=0.95$ with the stagnation parameters at the inlet to the slot being constant along the circular coordinate.

Some results of the flow computation in the different configurations of circuit are given in Fig. 2, 3. The flow in the header is characterized by considerable circular non-uniformity; the availability of intensive extraction arcs in the vicinity of the pipes and a wide area of a low-speed flow. At the peripheral limiting wall of the header there are areas of flow branching, i. e. between the pipes and in the diametrically opposite zone. When coming about sharp edges in the place of connection of pipes with the header, the flow separates, which leads to narrowing of the effective flow section and reducing of flow rate through the pipes.

The flow structure in the circuit can be conventionally divided into a largely radial flow in the interyoke slot of jet character and a developed tangential flow in the header outside the slot plane. In the meridional planes there occur weak secondary flows whirled from the flow plane towards the header periphery. At the axis of symmetry in the place of the flow separation there have been found out vortexes of low intensity occurring as a result of interaction between opposite tangential flows of different intensity in the header areas outside the slot plane. The shift of the slot in the axial direction from the symmetric position increases the vortexes intensity due to the flow rate redistribution in tangential slots outside the slot plane. When there is no whirl at the inlet the flow structure is symmetric relatively to the construction plane symmetry. The appearance of the whirl $\alpha(\varphi)=\text{const}$ at the inlet breaks the flow symmetry in the circuit and brings about the effect of flow rate redistribution through the pipes caused by different conditions of flow inlet into different pipes. Alongside, in the further pipe in the flow there appears a wide flow separation zone with reduced stagnation pressure in the core which jams the passage flow section. In the nearest, by the flow, pipe there is also a flow separation zone of the same kind. Its intensity considerably decreases with the increase of the whirl. The calculation shows that the dimensions of the flow separation zone do not exceed 1.5~2.0 diameters of the pipes; the results presented here are for pipes with $L/d=2:5$. The whirl angle variation in the range of $\alpha\leqslant 45°$ for the circuit configuration 1, 2 (Fig. 3) enabled us to find out the flow rate redistribution through the extraction pipes and acquire satisfactory coincidence with the results of the experiment[1]. The static pressure distribution analysis at the inlet of the header and in the slot plane shows that the whirl increases the pressure non-uniformity value $P_{max}-P_{min}$. The value of the pressure local minimum decreases sharply at the extraction arc to the first pipe and maintains at the same level at the extraction arc to the second pipe. From the point of view of nearextraction stages operation reliability an important characteristic of any extraction circuit is a spectral composition of the static pressure distribution at the header inlet, the expansion into series according to the rotor revolution frequency harmonics provided. At Fig. 3 one can see amplitude-frequency characteristics (AFC) of these distributions for configurations 1, 2, 3, 4 the flow whirl at the inlet being equal $\alpha=45°$. A specific feature of the given AFC is low values of harmonic amplitudes bearing numbers over 6. The mentioned phenomenon is characteristic for circuits having no local jams in the header. The flow whirl being increased, all the harmonics amplitudes bearing numbers under 6 increase uniformly too. The amplitudes rise of the Ist and 2nd harmonics is especially sharp.

The comparative analysis of the header slot pressure fields for configurations with the pipes oriented radialy shows, that the pipe axis turns, their radial orientation being maintained and the angle being increased φ_2 from 90° to 180°, leads to the dissappearance of the circuit flow rate redistribution. At the same time there decreases the inlet static pressure non-uniformity and the even harmonics amplitudes, up to the 5th inclusive increase notably. The nonuniformity spectrum on the even harmonics for configuration 6, φ_2 being equal 180°, is similar to the non-uniformity spectrum at the inlet for configuration 7 (with one pipe).

2. Flow Simulation in the Nearextraction Section

The aerodynamic computation method for turbine stages with flow path elements can be illustrated by an example of numerical study of unsteady flow in a typical nearextraction section stages of a 1000 MW steam turbine designed for power-and-heat generation. The section is made with the flow path expansion of $\gamma=20°$ and $\gamma=30°$ for stages A and B correspondingly; the blade number ratio is equal for both stages $Z_S/Z_R=140/160$. The section average diameter is 3200mm, the D/l is 10 for the stage A and 6 for the stage B. The inter-row axial spacing δ_z/b is 0. 5 for both stages and the inter-stage one is 1. 5.

The influence of the static pressure non-uniformity in the extraction chamber and inter-stage axial spacing upon the non-stationary characteristics of stages A and B during nominal regime of operation of the section with extraction, equalling 35 per cent of steam flow rate through the stage A, is considered. The extraction chamber has 3 non-symmetrically arranged extraction pipes depictured at Fig. 1. The stages A and B are in the subsonic regime of operation and according to the heat calculations the post-stage static pressure to the total pressure before the stage ratios are equal to: $P_{eA}/P_{oA}=0.36$, $P_{eB}/P_{oB}=0.51$, respectively.

The unsteady flow computations in the nearextraction section follow the steady flow computations in the extraction circuit. The inital data for the circuit computation are the heat calculations for the intermediate pressure cyllinder as a whole. At the same time the pre-extraction stage is assumed to operate in the nominal regime with the ratio u/C_o being close to the optimal one. In this case the condition of the whirl absence at the interyoke slot inlet $\alpha=0$ is assumable. Besides, in the computations the extraction chamber local jams were neglected and there was used the condition of the total energy constancy along the circular coordinate at the extraction slot inlet. In reality, as the experiment[4] showed, there is negligible total enthalpy circular non-uniformity at the circuit inlet. However it does not bring about any considerable change in the flow structure in the given circuit model. The non-uniformity curve of the static pressure δP in the extraction slot (Fig. 4a), being a result of the circuit computation, is presented as a difference of static pressure without extraction and with it reaching the value of 35 percent of flow rate through stage A.

The static pressure distribution in the inter-stage axial spacing along the radius from the periphery to the root is supposed to be with a constant positive gradient, the value of the latter being based upon the experimental data[2] on the flow structure study in the spacing with extraction. The dependance of pressure along the radius in the characteristic points of the non-uniformity diagram Ⅰ, Ⅱ, Ⅲ, is presented at Fig. 4b.

The unsteady flow computation in the pre-extraction stage is based on the application of boundary conditions formulation according to the model I. A, the flow through the post-extraction stage is computed in accordance with the model 2. B with stipulation of the total pressure distribution at the inlet to the computations region which is similar to the static pressure distribution (Fig. 4). Some computation results relating to the flow structure in the working channels, and also to the flow dynamic influence upon the rotor blades are given at Fig. 5, 6. It follows from the computations that the stage A blade row operates in the transonic regime, periodically with the decrease of pressure ratio on the upper sections. In the vicinity of points Ⅱ, Ⅲ the flow velocity in the channel reaches the value of the local sound velecity, a supersonic zone appears and the non-uniformity expansion $P(r, \varphi)$ is limited largely by the region of the channel bevel cut.

In the selected working channel of the stage B during the rotor revolution, there were noted considerable low-frequency (LF) pulsations of the static pressure and velocity vector on the whole surface of the blade. It is demonstrated by the dimensionless static pressure distributions along the profiles contours in the root and peripheral sections of the stages A and B rotor blades when they pass in the vicinity of points Ⅰ, Ⅲ of the diagram δP. On the curves there is also shown the high frequency (HF) pressure pulsation peak-to-peak value from the flow pitch non-uniformity in the inter-row spacings prevailing in the vicinity of the rotor blade leading edges.

The stage B is loaded less and characterized by a subsonic regime with the diagonal flow character in the working channels especialy at the places of maximal pressure ratio (points Ⅲ). In the flow about the diaphragm channels situated in the vicinity of these points, at the periphery of the flow path the computation displayed the regions of contraflow from the diaphragm channels counter to the main flow; their length being varied in the time in the axial direction and reaches 25 per cent of the inter-stage spacing length.

The comparison of curves at Fig. 5 testifies to considerable flow in the stage B rotor channels during a revolution. Due to the pressure radial gradients at the stage inlet the groups of diaphragm channels bear the impact of developed secondary flows, which influence upon the blade force and aerodynamic characteristics, and lead to overrating of the angles of attack. The last fact seems to serve an explanation to the increase of HF pressure pulsation in the vicinity of the leading edge of the rotor blade of the stage B during a turn from point Ⅰ to point Ⅲ.

By means of the unsteady fields integration of static pressure on the blading surface during a rotor revolution there was achieved a full aerodynamic force (FAF) change on blades of every stage. The results of the FAF harmonic analysis are presented as their amplitude-frequency characteristics (AFC) in Fig. 6. The FAF amplitude is made dimensionless by a reference to $\bar{\rho}_* a_*^2 l^2$ product and given as a function of discrete harmonic series multiple to the speed of rotation.

The acquired data point to the considerable exceeding of LF amplitude of FAF pulsation (Ist — 3rd harmonics) over HF pulsation from the pitch non-uniformity (140th, 280th harmonics) which can be explained by big relative axial spacings in stages. At the same time LF amplitude of FAF pulsation as referred to the average force value, influencing upon the corresponding rotor blade profile in several times less than FAF relative peak-to-peak value at a rotor revolution since it characterizes only the contribution of a concrete harmonic into the full force peak-to-peak value. The silmilarity of the force AFC harmonic composition in the stage A rotor blade periphery profile and of static pressure distribution in the inter-stage axial spacing, and also the absence of this similiarity for FAF at stage B rotor blades are worth attention too. Besides, the force AFC for after-extraction stage rotor blade profile have a similar harmonic composition determined by the flow total restructuring in the radial direction due to the influence of stage B diaphragm. The comparison of AFC's points to the phase coincidence of FAF change by profiles for postextraction stage rotor blade unlike the character of FAF change along the stage A blade.

So, the transonic character of flow in the stage A blade row leads to the concentration of intensive LF pulsations of FAF on the rotor blade suction side within the channels bevel cut, there occurs the separation of LF and HF pulsations influence, by the leading and trailling edge regions, accordingly, which increases the rotor blade vibro-loading. The stage B rotor blade load is more aerodynamically uniform the flow about it is of essentially diagonal character, the static pressure LF upon the pre- and post-extraction stages blading is different both by the amplitude change along the blade length and by the phase characteristics. The FAF peak-to-peak value in the section rotor blade peripheral profile reaches 10~15 per cent of the average force influencing the one. This corresponds with results of the experiment[3] on model stages with the same paramethers.

3. Flow Simulation through the Last Stage

This method is illustrated by computations of a 3-dimensional flow in the last stage of a powerful steam turbine interacting with the exhaust hood in the nominal regime of operation. As follows from the experiments[9], the relative static pressure lower values correspond to the lower part of the exhaust, and their circular distribution character is close to the harmonic law. Therefore, during the simulation of the stage operation together with the exhaust hood for this example, the static pressure behind the rotor was maintained constant along the radius and variable in the circular direction according to the law: $P_z/P_o = 0.18 + 0.02 \cdot \cos 2\pi\omega t$. The stage flow path is performed with the expansion angle of about 35°, $Z_S/Z_R = 56/112$; $D/l = 2.86$; the inter-row axial spacing δ_Z/b variates within the range 0.15~0.8 from the root to the periphery.

The flow about the blade row is characterized by LF pulsation of its parameters caused by non-axisymmetric pressure distribution behind the stage and HF pulsations from the pitch non-uniformity, determined by the stator-rotor blades number ratio. By the calculated unsteady pressure fields with the help of the harmonic analysis, there were received the values of amplitudes and phases of main harmonics of pressure on the convex and concave blade surfaces. In Fig. 7 the character of pressure coefficient pulsation amplitude distribution along the rotor blade chord $\bar{A} = (P_j^k - \hat{P}^k) / \rho_* a_*^2$ is presented.

The principal contribution in the static pressure LF pulsation is made by the Ist and 2nd harmonics, and in the HF one the 56th and 112th harmonics. Unsteady disturbance, caused by the pitch non-uniformity and inter-shift of cascades, are displayed to the most extent at the inlet part of the rotor and diminish towards the outlet. The amplitude of the 56th harmonic is maximal in vicinity of the root section leading edge at the suction side and makes 10~15 per cent of the dynamic mean head. The radius being increased, the 56th harmonic decreases monotonically. It can be explained by the increase of the relative axial spacing to the periphery and the decrease of the profile curvature and blade thickness. At the rotor inlet the amplitude of the 112th harmonic is 10~40 per cent of the amplitude of the 56th harmonic and decreases towards the chord.

The analysis shows that in the transonic stage the HF perturbation induced by the cascade interaction display largely in the downstream row and decrease towards the flow. The character of LF perturbation upstream distribution is determined by the relative velocity at the cutlet from the rotor row. In the considered stage the velocity is supersonic, and the axial component is subsonic. Therefore the perturbation spreads upstream within the bevel cut. In the root section the LF Mach wavefront hits the blade convex surface at the distance equal to 60~70 per cent of the chord length from the leading edge. This region expands with the radius increase and embraces the whole convex surface of the profile and a part of the concave one in the peripheral section.

By means of the unsteady pressure distribution integration by the blade surface were determined unsteady aerodynamic forces influencing the blade.

In Fig. 8 there are displayed the plots of the radial $\overline{F}_x$, circular $\overline{F}_y$ and axial $\overline{F}_y$ unsteady force pulsation components on the root (a) and peripheral (b) sections.

The ordinates correspond to the pulsation components, and the absissas — to a revolution time. The dimensionless values of unsteady

$$\overline{F}_x = \frac{F_x^i - \hat{F}_x}{\rho_* a_*^2 l^2} \quad , \quad \overline{F}_y = \frac{F_y^i - \hat{F}_y}{\rho_* a_*^2 l^2} \quad , \quad \overline{F}_z = \frac{F_z^i - \hat{F}_z}{\rho_* a_*^2 l^2}$$

The components of the full non-stationary force influencing the blade section are presented in the right Cartesian system of coordinates, th OX—axis being oriented int he radial direction, OZ—towards the turbine axis, along the flow, OY—in the circular direction counter to the rotor blading shift.

As appears from the analysis, the lift pulsation is determined largely the 56th harmonic (with the blade shift frequency of ωZ_S), the pulsation component amplitude being diminished from the root to the blade periphery, making 60 per cent of the force average value in the root section and 20 per cent in the peripheral one. The axial force pulsation contains both the higher 56th and the Ist harmonics with the rotor revolution frequency. The amplitude of the 56th harmonic reaches 30 per cent of the mean force in the root section and 15 per cent in the peripheral. The amplitude of the Ist harmonic increases towards the periphery and for the peripheral section reaches 5～10 per cent of the axial mean force. The influence of higher harmonic on the vibration of the last stage rotor blades in axial turbomachines is negligible. At the same time the frequency of the aerodynamic distrubing forces equal to the rotation frequency can be close to the natural frequency and bring about the dangerous level of vibrations and the destruction of blading.

The authors of paper are scientists.

REFERENCES

1. Gogolev I. G. et al., Teploenergetika, 3 (1979) 48.
2. Lapin N. V. and Marchenko Y. A., TsKTI Papers, 184 (1981), 106.
3. Silber T. M. et al., Teploenergetika, 8 (1987), 14.
4. Gogolev I. G. et al., Teploenergetika, 6 (1976), 53.
5. Prokopenko A. G. et al., Teploenergetika, 5 (2969), 29.
6. Kosyak Y. F. et al., Teploenergetika, 9 (1988), 29.
7. Troyanovsky B. M. and Radionova L. K., Teploenergetika, 8 (1979), 62.
8. Stastny M., Machinenbautechnik, 4 (1982), 154.
9. Lagun V. P. et al., Teploenergetika, 8 (1971), 35.
10. Sokolovsky G. A. and Gnesin V. I., Unsteady Transonic and Viscous Flows in Turbomachines, Naukova Dumka, 1986, 256.
11. Fung K. V. AIAA Journal, 2 (1981), 182.
12. Bayliss A., J. Comput. Phys., 2 (1982), 182.
13. Belotserkovsky O. M., Numerical Simulation in Continuous Medium Mechanics, Nauka, 1984, 520.
14. Belovol A. V., Gesin V I and Solodov V. G., Proc. of Conf. on Math. Simulation of Processes

and Structures of Energy and Transport Turbomachinery, Gotvald, 1 (1988), 173.
15. Gnesin V. I. and Solodov V. G., Proc. of 11th Conf. on Turbomachine Aeroelasticity, Kiev, 11 (1987), 6.

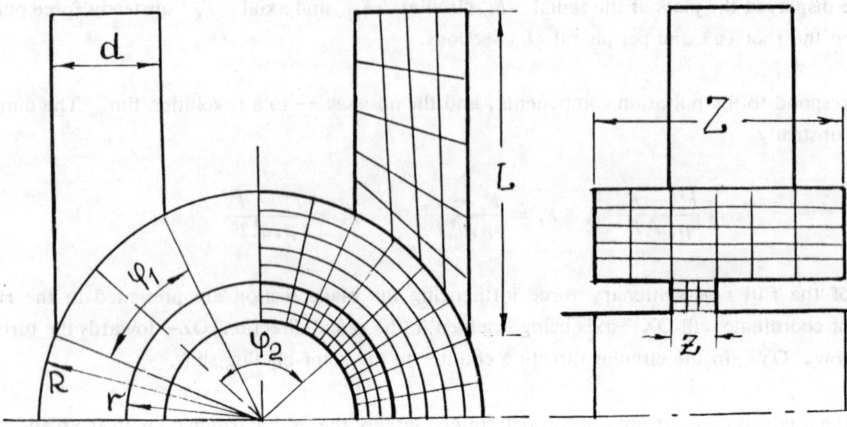

Fig. 1 Computational region of extraction circuit.

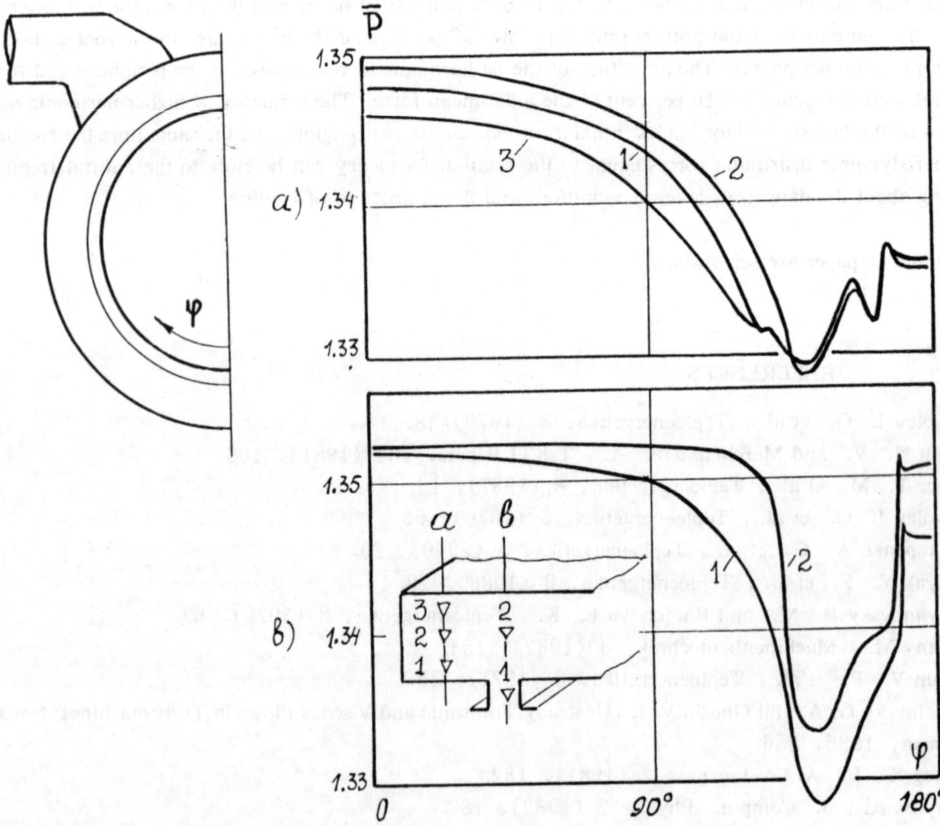

Fig. 2 Static pressure in header space.

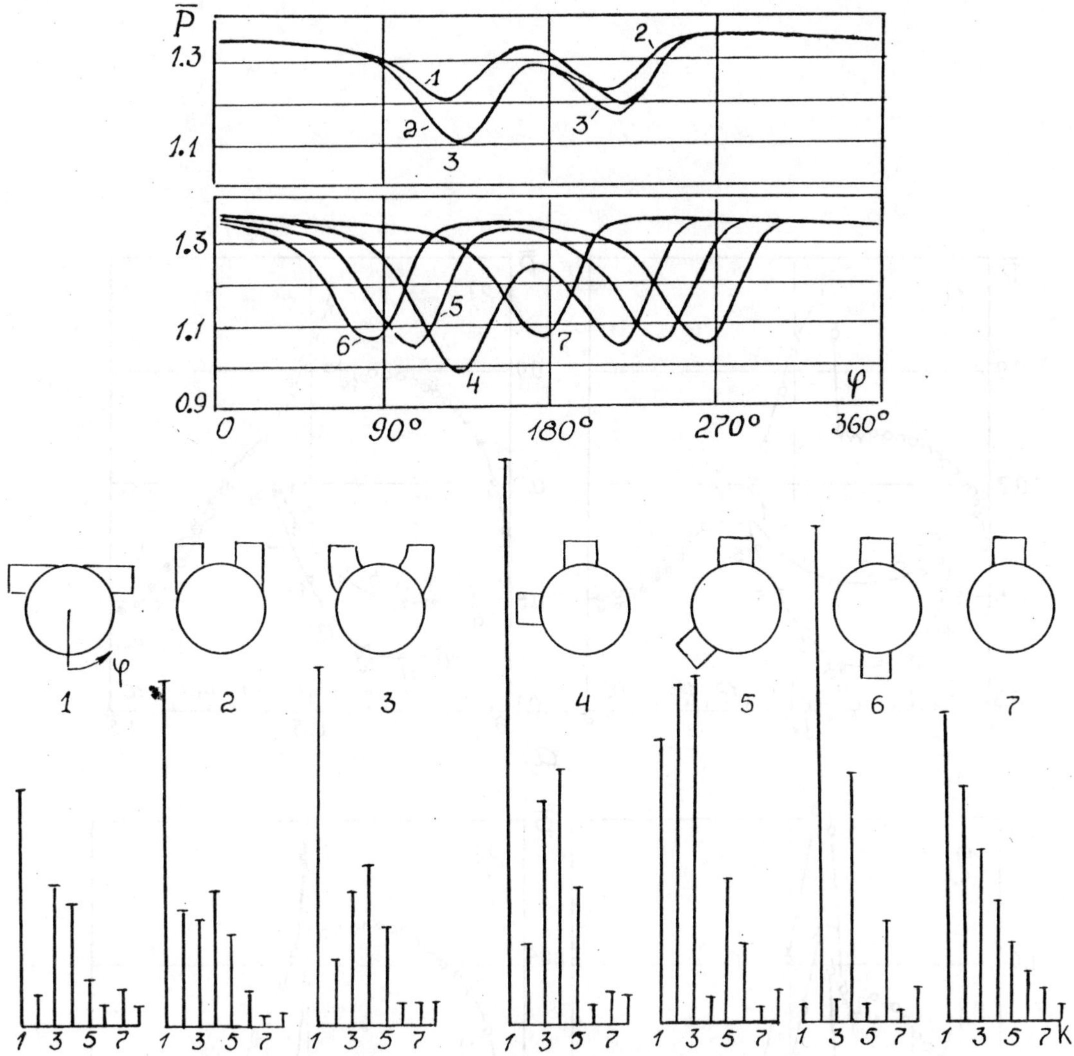

Fig. 3 Static pressure in head inlet boundary for different circuit configurations; $\alpha = \varphi_1 = 45°$

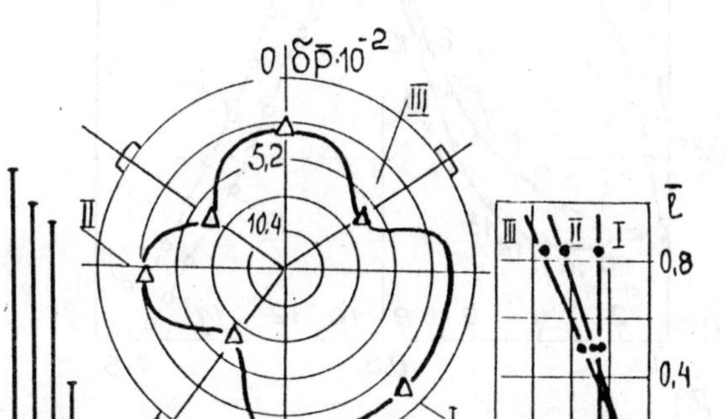

Fig. 4 Static pressure non-uniformity in plane of inter-stage axial spacing.

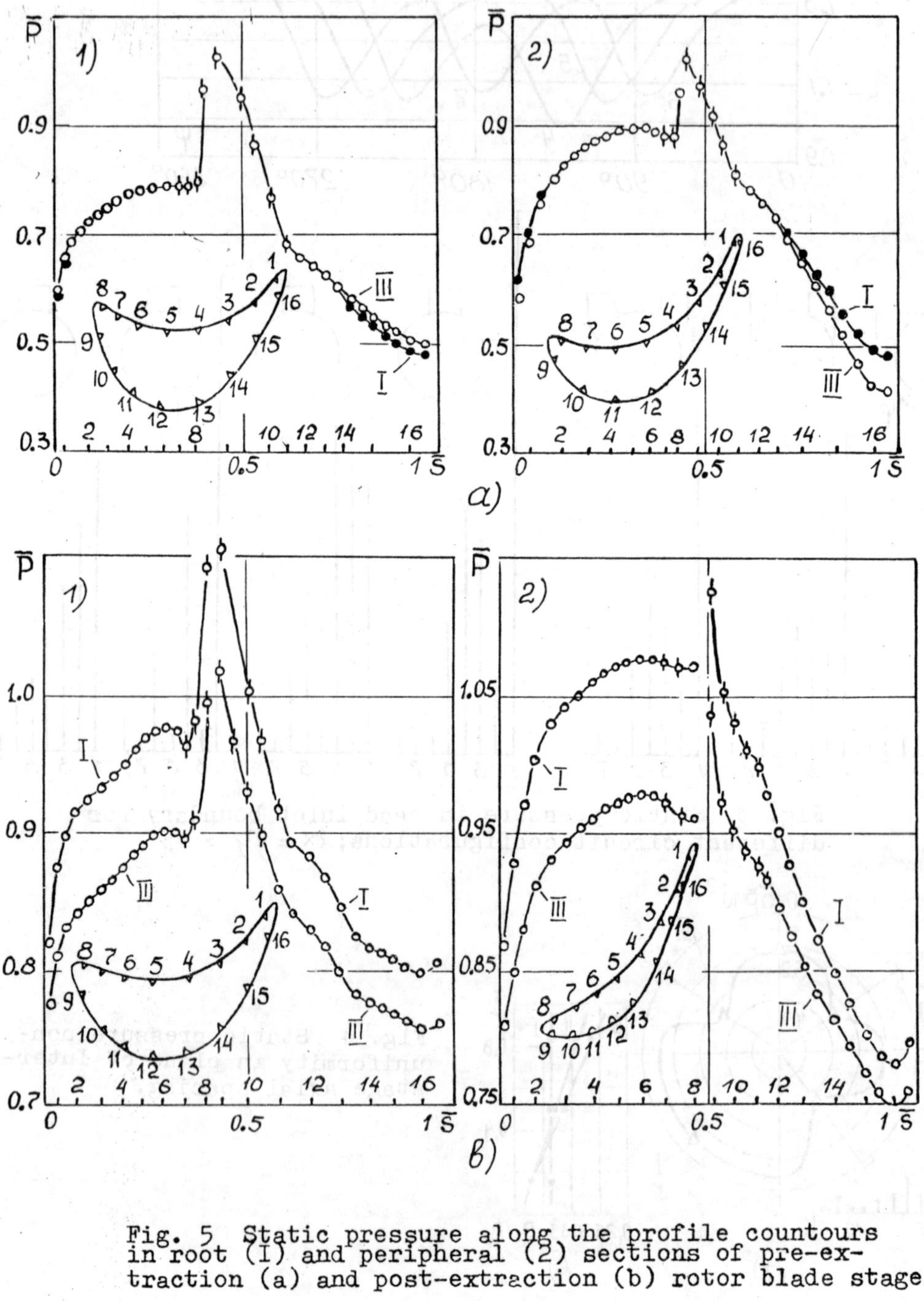

Fig. 5 Static pressure along the profile countours in root (I) and peripheral (2) sections of pre-extraction (a) and post-extraction (b) rotor blade stages.

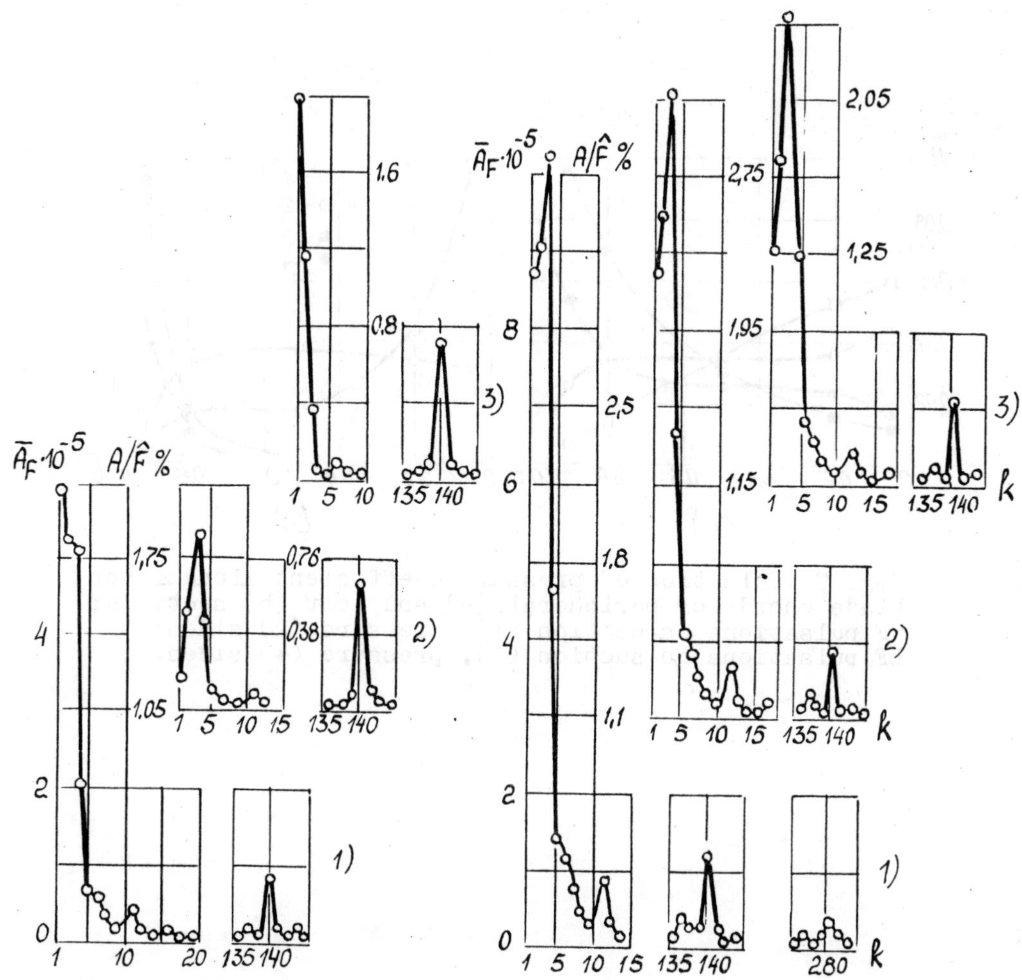

Fig. 6 Aerodynamic force AFC rotor blade peripheral(1), intermediate (2), root (3) sections of pre-extraction (a) and post-extraction (b) stages.

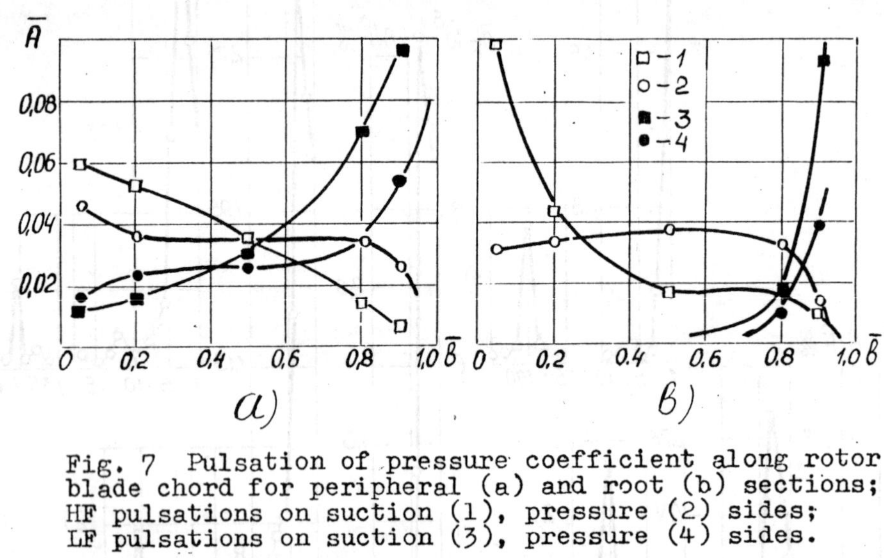

Fig. 7 Pulsation of pressure coefficient along rotor blade chord for peripheral (a) and root (b) sections; HF pulsations on suction (1), pressure (2) sides; LF pulsations on suction (3), pressure (4) sides.

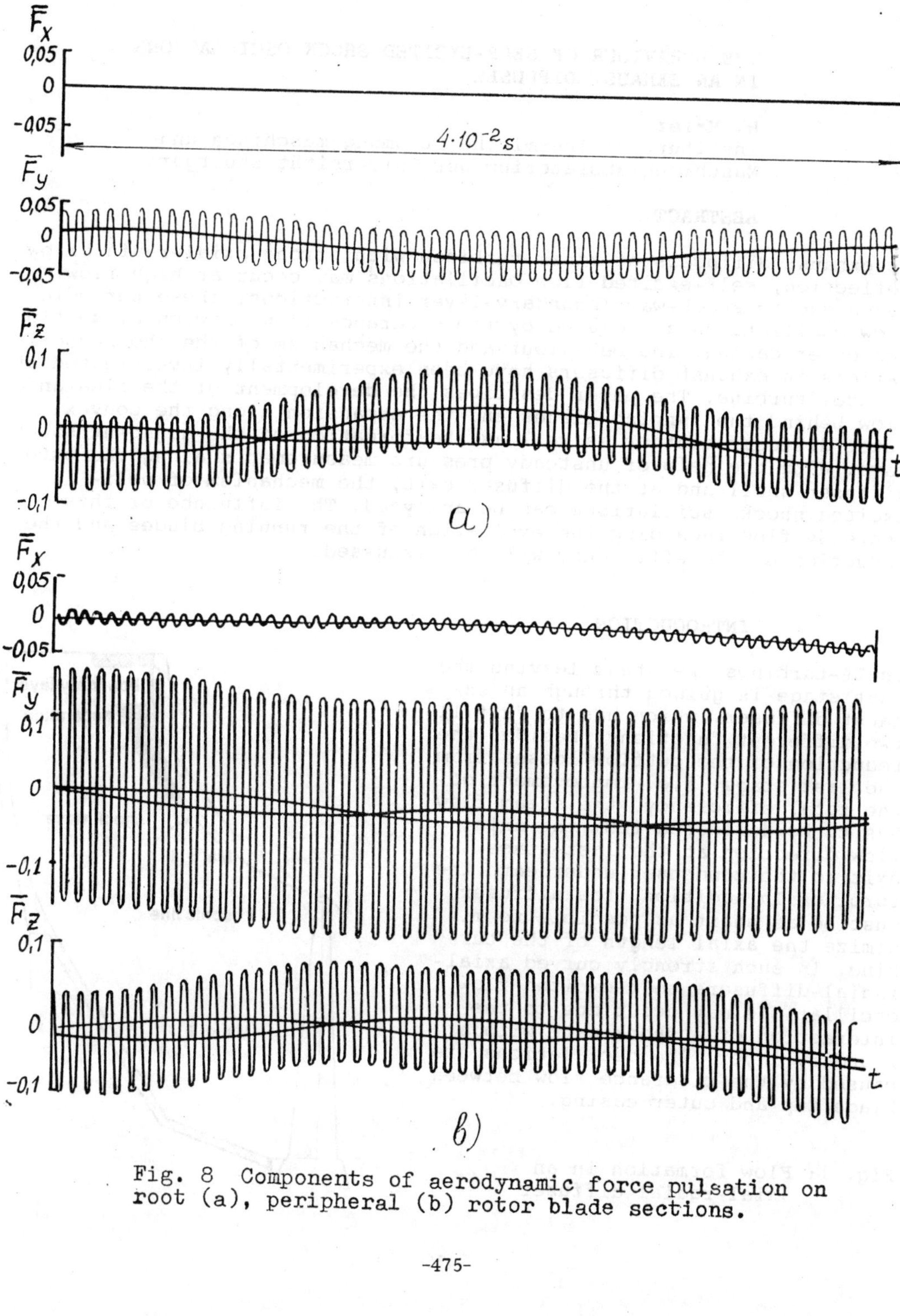

Fig. 8 Components of aerodynamic force pulsation on root (a), peripheral (b) rotor blade sections.

THE BEHAVIOUR OF SELF-EXCITED SHOCK OSCILLATIONS IN AN EXHAUST DIFFUSER

R. Maier
Institut für Thermische Strömungsmaschinen und Maschinenlaboratorium der Universität Stuttgart

ABSTRACT

In exhaust diffusers of steam turbines with strong axial-radial flow deflection, self-excited flow oscillations may occur at high flow rates due to shock-wave/boundary-layer interactions. These periodic flow oscillations are caused by the clearance flow between blade tip and outer casing. The behaviour and the mechanism of the shock oscillations in exhaust diffusers have been experimentally investigated at a model turbine. The paper describes the development of the clearance flow behind the running blades and the expansion along the convex outer diffuser contour as well as the formation of the adjacent shock wave. With the help of unsteady pressure measurements along the outer diffuser shell and at the diffuser exit, the mechanism of self-excited shock oscillations can be analyzed. The influence of this unsteady flow regarding the excitation of the running blades and the reduction of the efficiency will be discussed.

INTRODUCTION

In LP-turbines the steam leaving the last stage is guided through an exhaust diffuser, which has to fulfill two different functions. Besides the reduction of the outflow losses from the last stage, i.e. improvement of the total efficiency, the steam flow has to be deflected from the axial flow direction at the last-stage exit to the condenser underneath the turbine. In addition, the diffuser must be as short as possible to minimize the axial length of the turbine. In such strongly curved axial-radial-diffusers self-excited flow oscillations may occur due to the interaction of a shock wave with a boundary layer. The oscillations are caused by the clearance flow between blade tip and outer casing.

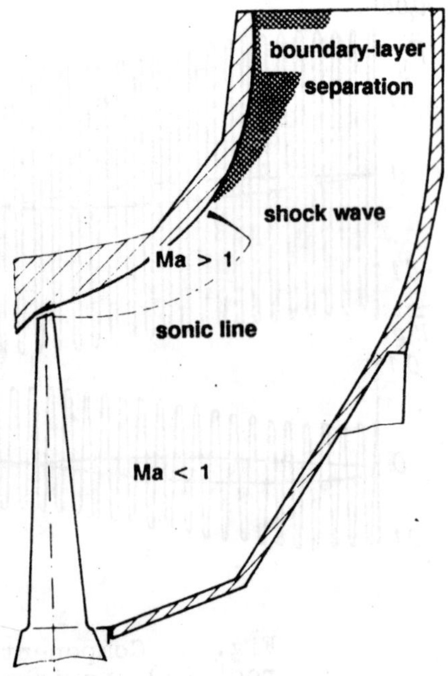

Fig. 1: Flow formation in an axial-radial diffuser

The strong pressure gradient over the running blades accelerates the clearance flow at the rotor exit by an overexpansion and leads to a high energetic supersonic flow region at the outer convex-curved diffuser shell (Fig. 1). With rising flow rate the supersonic clearance flow is additionally accelerated by a Prandtl-Meyer expansion. Downstream the clearance flow is limited by a normal shock wave. If the shock wave reaches a certain strength, the boundary layer separates from the wall. Subsequently, the back-pressure behind the shock wave rises and leads to an unstable local shock position. The shock wave is moving upstream dragging its own separation region behind. If the shock strength decreases during the upstream movement or if shock foot and separation point are not identical any more, the shock movement stops and the original condition is formed again. This mechanism becomes noticeable by a buzzing noise, which therefore is called diffuser buzz.

The self-excited shock-boundary-layer oscillations were investigated at different diffusers at a last-stage test stand /1,2/. The measurements revealed shock oscillations starting just behind the design point at a volumetric flow rate of 0.6 and frequencies of about 120 Hz. With rising flow rate, which may occur in winter time due to low cooling water temperatures, the frequency of the self-excited shock oscillations is steadily increasing. At extremely high flow rates, which can not occur in regular operating conditions of the turbine, the pulsations stop at frequencies of about 176 Hz (Fig. 2).

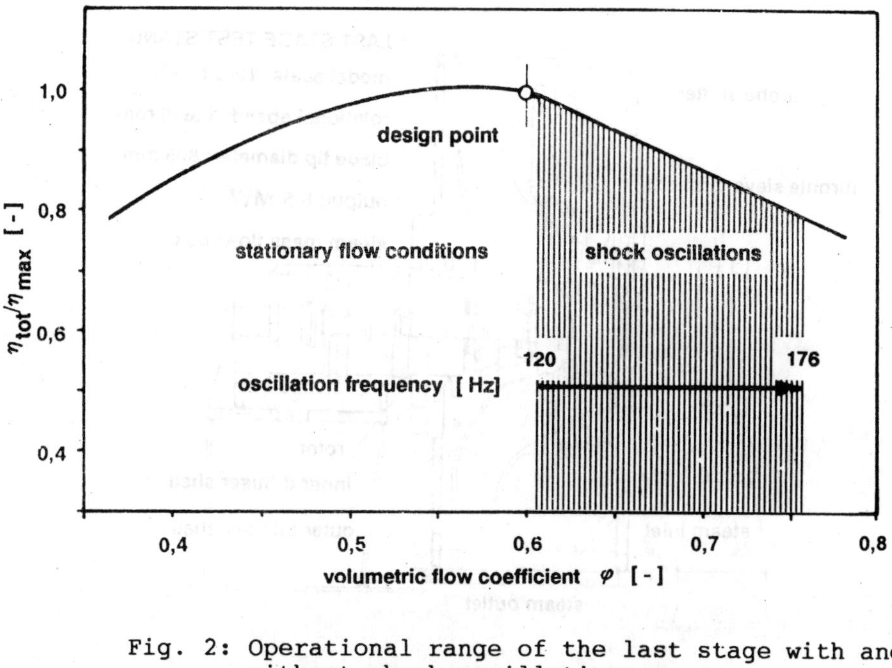

Fig. 2: Operational range of the last stage with and without shock oscillations

Similar phenomena are known from buffeting of airfoils /3,4,5/, shock oscillations in the blading of turbomachinery /6,7/, buzz of supersonic intakes of aero engines /8,9/, shock oscillations in Laval nozzles /10,11/, and in ducts with an abrupt increase in cross section /12, 13/. An extensive survey about these self-excited shock-wave/boundary-layer oscillations is given in /14,10,2/.

In this paper some measurement results are presented describing both the formation of the tip-clearance flow as well as the behaviour and mechanism of the shock-boundary-layer oscillations in an exhaust diffuser. Furthermore, the effects on the running blades and the efficiency of the turbine are discussed followed by methods to prevent this flow instability.

EXPERIMENTAL SET-UP

The measurements of the following investigations were accomplished at the last-stage test stand of the Institut für Thermische Strömungsmaschinen at the University of Stuttgart (Fig. 3). The rig was designed for the purpose of analyzing the flow and vibration characteristics of the last stages of large low-pressure steam turbines under practical conditions. The investigations were carried out with a single stage version using a 1:4.21 scale model of a 10 m^2 last stage. The pre-stages were simulated by throttle sieves at the inlet.

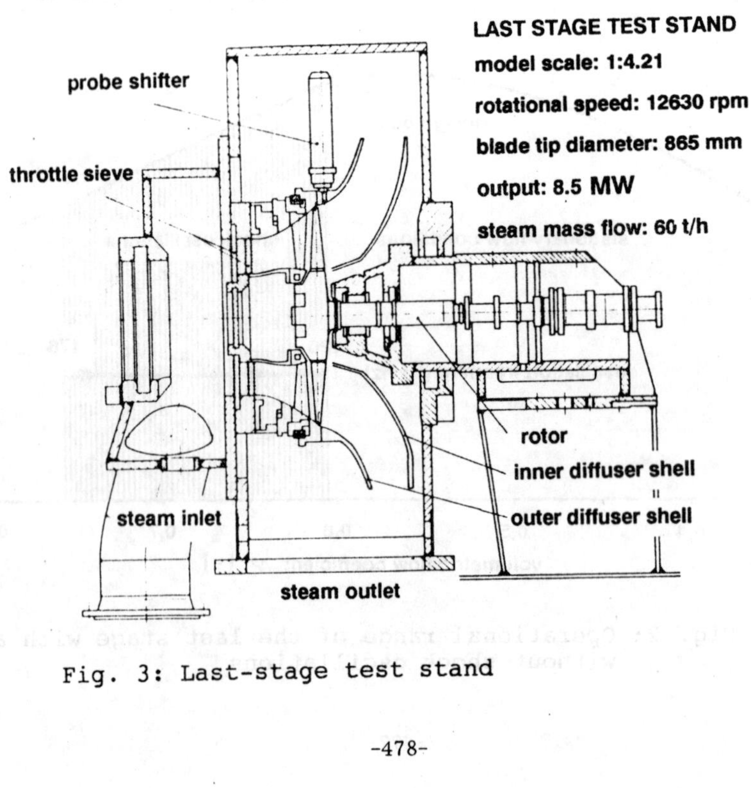

Fig. 3: Last-stage test stand

For the presented measurements a $15^0/15^0$- diffuser was used referring to both the inlet angle at the outer wall of the diffuser and the blade-tip angle. Figure 4 shows the investigated diffuser and its cross section along the diffuser length.

To get an insight into acceleration and expansion of the clearance flow, 10 pressure taps were placed at the outer wall between the trailing-edge of the guide vanes and the diffuser inlet, while 30 pressure taps were applied along the outer diffuser contour (Fig. 5).

The profile of the clearance flow in radial direction was determined by a flat pitot probe at three different measurement positions in the diffuser (Fig 6).

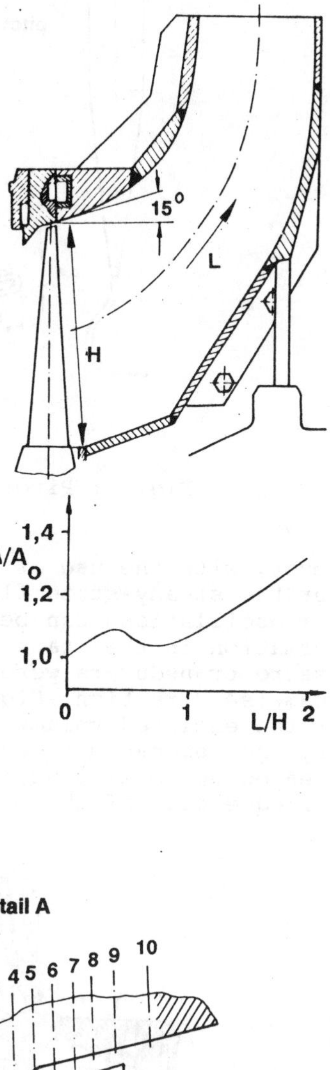

Fig. 4: Investigated $15^o/15^o$-diffuser with cross-section distribution

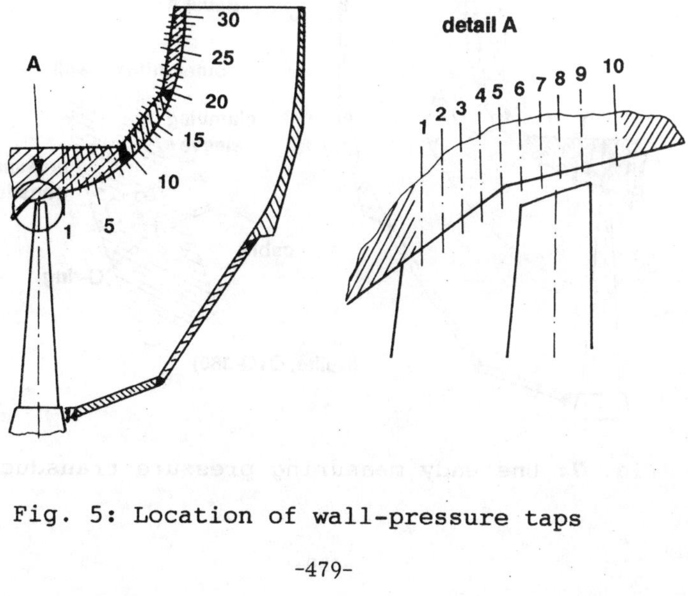

Fig. 5: Location of wall-pressure taps

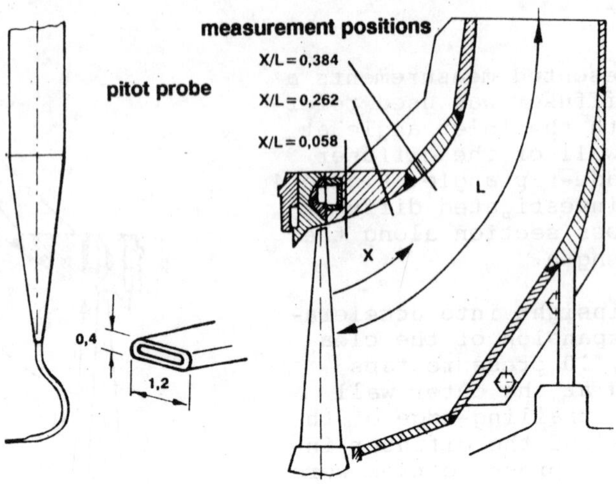

Fig. 6: Pitot probe and measurement positions

However, with the use of this measurement technique, only information about the steady-state flow conditions up to the beginning of the shock oscillations can be recorded. In order to obtain additional information in the state of shock oscillations, 11 high-response pressure transducers were mounted into the outer diffuser wall in a streamwise direction (Fig. 7). The transducers were installed such that the enclosed volume in front of the diaphragm was minimized. Thus, the resonance frequency of the pressure transducers could be increased up to 2200 Hz, which is an order of magnitude higher than the frequencies of the measured signals.

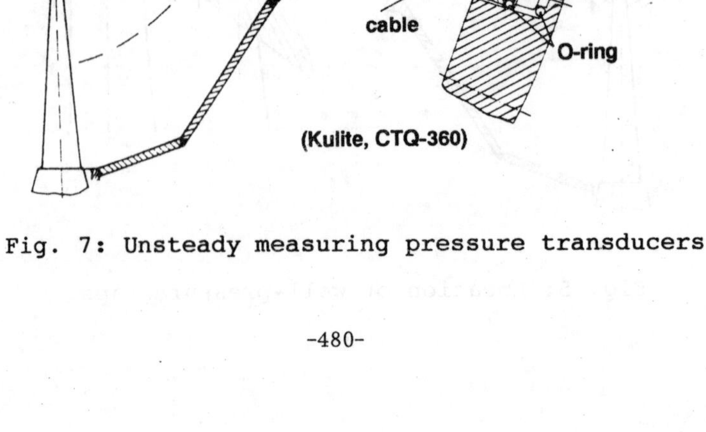

Fig. 7: Unsteady measuring pressure transducers

The shock oscillations are coupled with an oscillation of a separation region at the convex diffuser leading to a pulsation of the whole diffuser flow. In order to study the characteristics of the oscillating separation region at the diffuser exit and the pressure fluctuations at the rotor exit, a high response pitot probe was used (Kulite, XCS-062-5D).

EXPERIMENTAL RESULTS

1. Development of the Tip-Clearance Flow

To ensure safe operation of turbomachines a gap has to be designed between blade tip and outer casing. In the last stages of LP-turbines the tip clearance is relatively large. Besides the elongation of the blades caused by rotation and temperature, mainly the axial displacement of the rotor with respect to the divergent casing has to be taken into account. At the same time, the pressure difference over the running blades is relatively high. Thus, a strong clearance flow from the pressure side to the suction side of the blade may develop. This flow generates clearance vortices inside the passage altering the outlet angle of the flow and consequently producing losses. In the case of the heavily twisted blades of the last stages, the clearance flow additionally causes a jet flow by overexpansion which is accelerated along the curved outer diffuser shell due to a Prandtl-Meyer-expansion (Fig. 8).

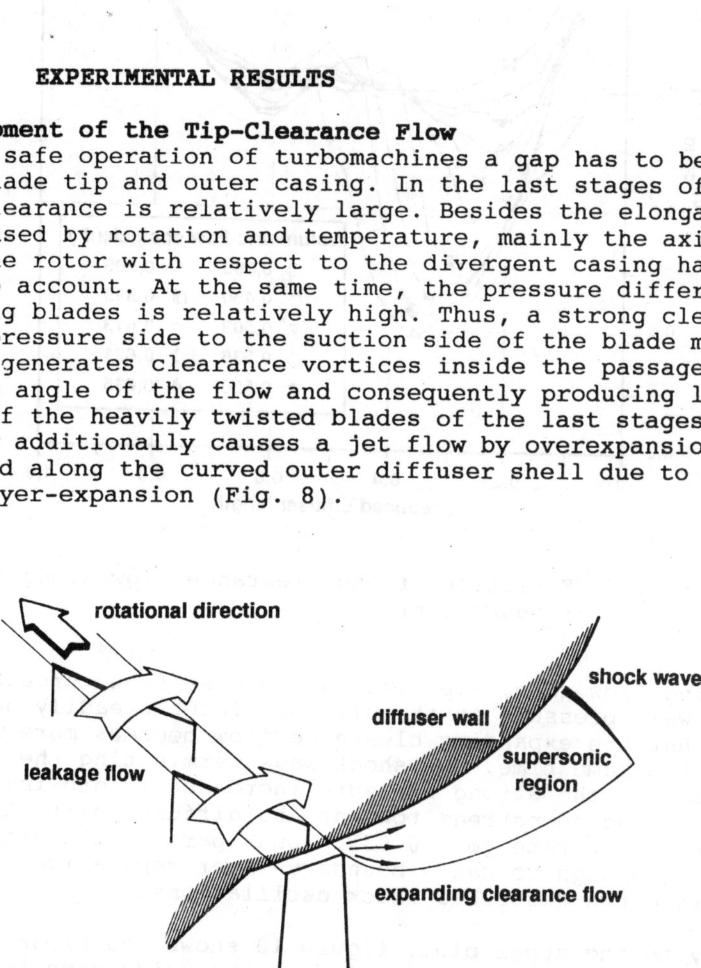

Fig. 8: Tip-clearance flow in a last stage

The formation of the clearance flow behind the running blades along the diffuser shell is plotted for different volumetric flow rates in figure 9.

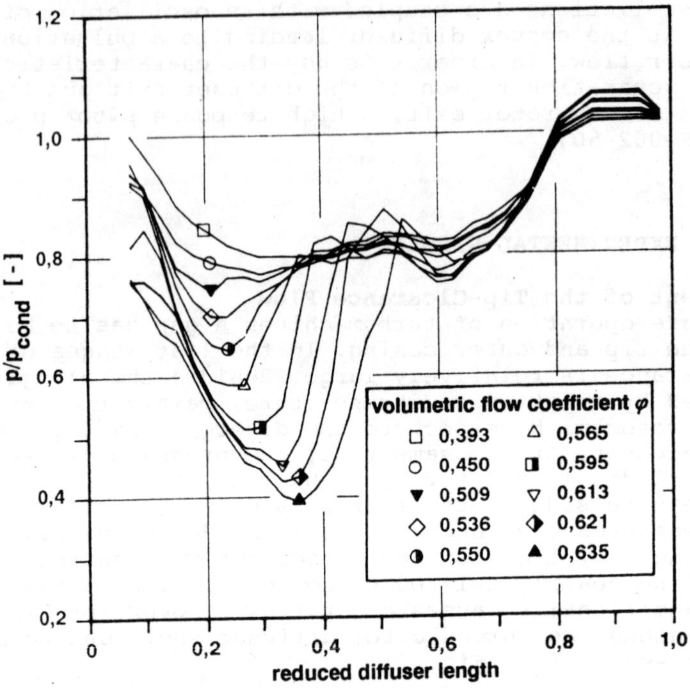

Fig. 9: Formation of the clearance flow along the diffuser contour

With increasing flow rate, i.e. increasing mass flow through the turbine, the wall pressure at the diffuser inlet steadily decreases, which means that the expanding clearance flow becomes more and more dominant. At the same time, the shock wave terminating the supersonic region (visible at the strong pressure increase) is growing in strength and moving downstream towards the diffuser exit. At a certain volumetric flow rate (φ = 0.635, upper part of the diffuser) the shock is strong enough to cause boundary-layer separation and therefore to initiate the described shock oscillations.

Supplementary to the upper plot, figure 10 shows the pressure distribution along the outer wall, beginning at the guide vane trailing-edge over the blade tip of the running blades to the diffuser exit. Additionally, the critical pressure p^*, for which the flow velocity is equal to the speed of sound, is plotted into the diagram. It is cearly visible, that most of the flow acceleration takes place in the clearance above the blade tip and short behind the rotor exit. The critical pressure exactly appears at the clearance exit which means that the clearance flow is accelerated inside the tip clearance up to the speed of sound while further acceleration due to overexpansion occurs directly behind the clearance exit. Hence, the tip clearance can approximately be considered as a convergent nozzle.

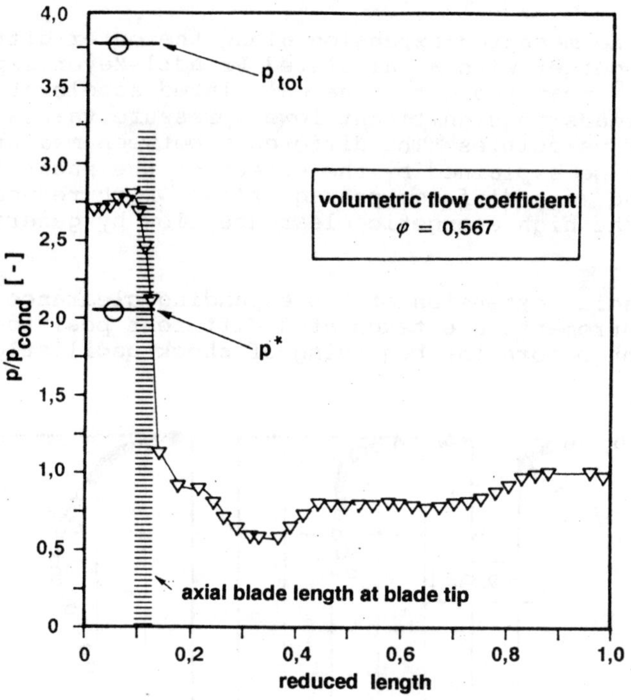

Fig. 10: Wall-pressure distribution at the outer diffuser wall

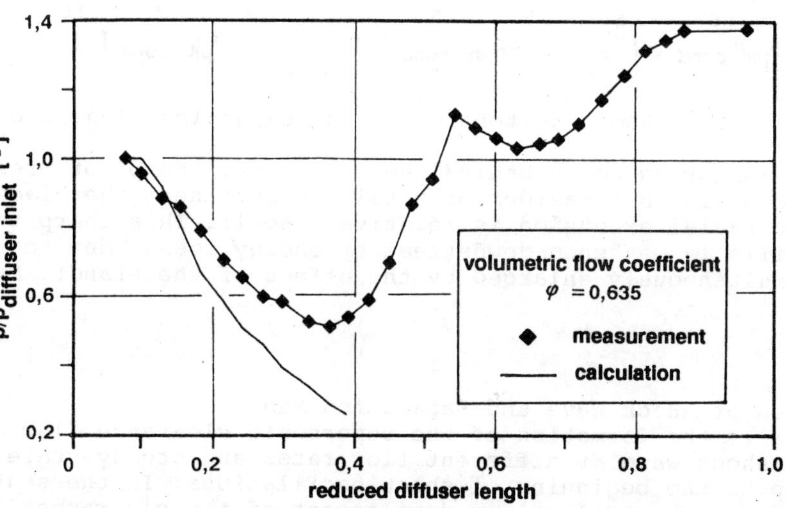

Fig. 11: Measured and calculated expansion of the tip-clearance flow along the convex diffuser wall

Figure 11 depicts the measured expansion along the outer diffuser shell (φ = 0.63) together with a calculated Prandtl-Meyer expansion. In comparison to the measured data, the calculated acceleration along the diffuser wall leads to significant lower pressure values and consequently to higher velocities. The difference between measurement and calculation can be explained by the effect of the shear layer between jet flow and main diffuser-passage flow, which reduces the total pressure of the high energetic clearance flow by generating vortices.

In figure 12 the radial extension of the expanding clearance flow is displayed. The measurements are taken at 3 different positions in streamwise direction before the beginning of shock oscillations.

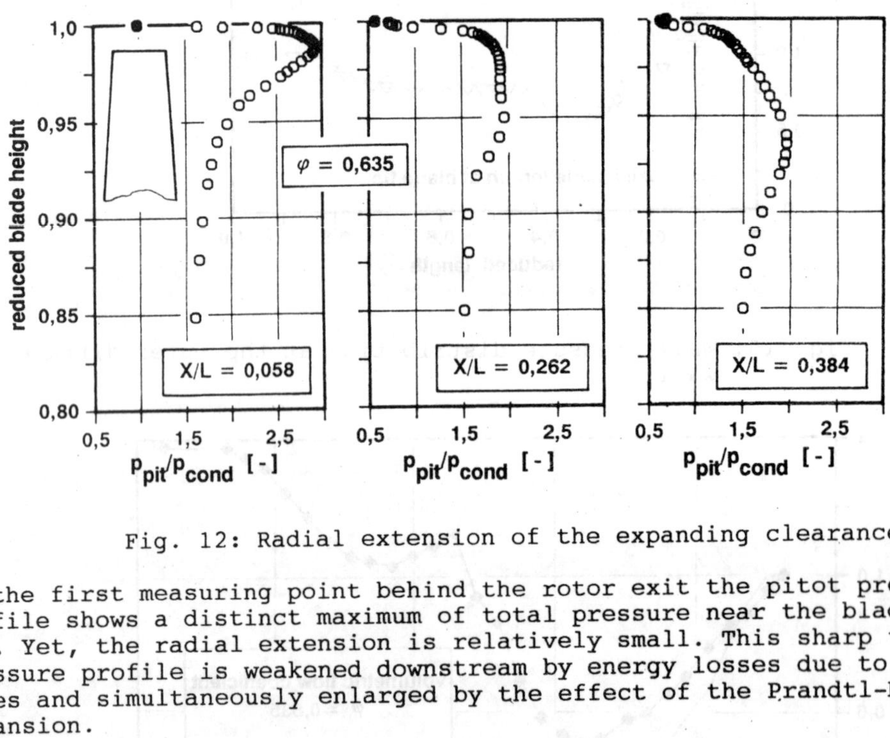

Fig. 12: Radial extension of the expanding clearance flow

At the first measuring point behind the rotor exit the pitot pressure profile shows a distinct maximum of total pressure near the blade tip. Yet, the radial extension is relatively small. This sharp total pressure profile is weakened downstream by energy losses due to vortices and simultaneously enlarged by the effect of the Prandtl-Meyer expansion.

2. Oscillation of Shock Wave and Separation Region

Figure 9 depicts the formation of the supersonic clearance flow and the adjacent shock wave at different flow rates and steady-state flow conditions up to the beginning of shock oscillations. In the state of shock-boundary-layer oscillations development of the clearance flow, formation of the shock wave, boundary-layer separation as well as shock movement were recorded by 11 high response pressure transducers mounted in the outer diffuser shell.

The time history of the 11 pressure transducers are given for an oscillation frequency of 160 Hz in figure 13. The plots illustrate extremely regular shock-boundary-layer oscillations. The shapes of the pressure curves recorded from pressure transducer 3 to 6 show a distinct decrease of the wall pressure, describing the development of the supersonic region (1), and an adjoining sudden sharp pressure increase due to the shock movement over the single pressure transducers (2). The connection of the points of beginning pressure increase of PT 2 to 6 (a) gives an idea about the upstream traveling of the shock wave. The velocity of the movement is between 26 and 46 m/s. In contrast, the time history at PT 7 to 11 are not directly caused by the shock movement but by the pressure increase during the growing of the separation region downstream of the moving shock wave.

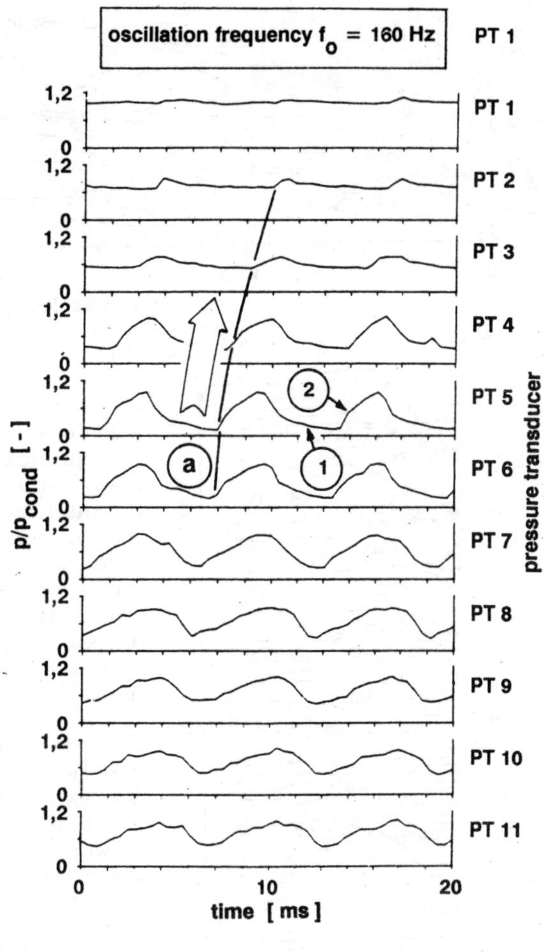

Fig. 13: Time histories of shock oscillations

In order to gain a deeper understanding about the formation of the supersonic region, the adjacent shock wave, and its upstream movement, the pressure signals of the 11 unsteady measuring pressure transducers, recorded on tape, were digitized with a fast A-D-converter. The connected pressure data of all pressure transducers at the same scan time give a momentary pressure distribution along the diffuser wall which can be plotted as a function of time.

Such a plot is given in figure 14 for three different oscillation frequencies. The development of the supersonic region by a steady pressure decrease (1), the formation of a shock wave (2), and the upstream shock movement (3) can be observed. The behaviour of self-excited shock oscillations, however, is not exactly the same for

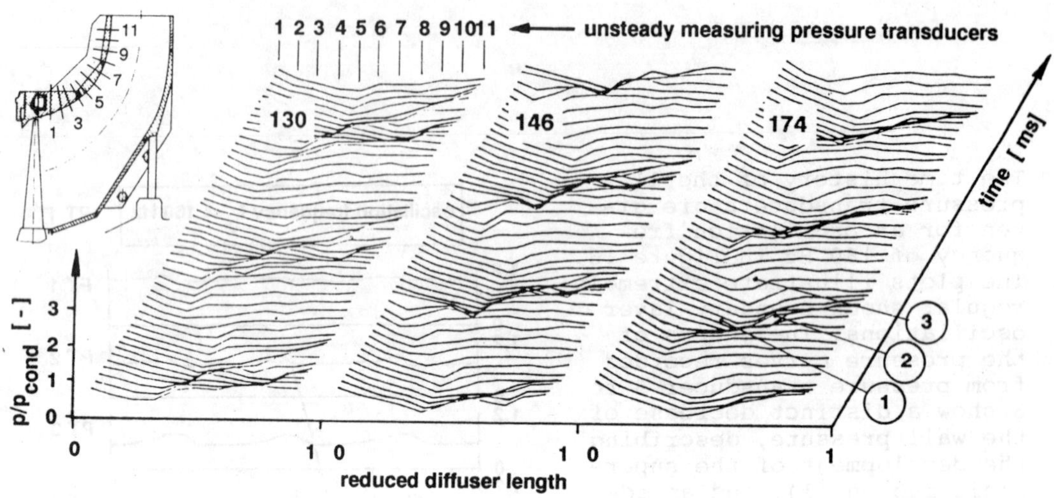

Fig. 14: Shock wave oscillations at three different frequencies

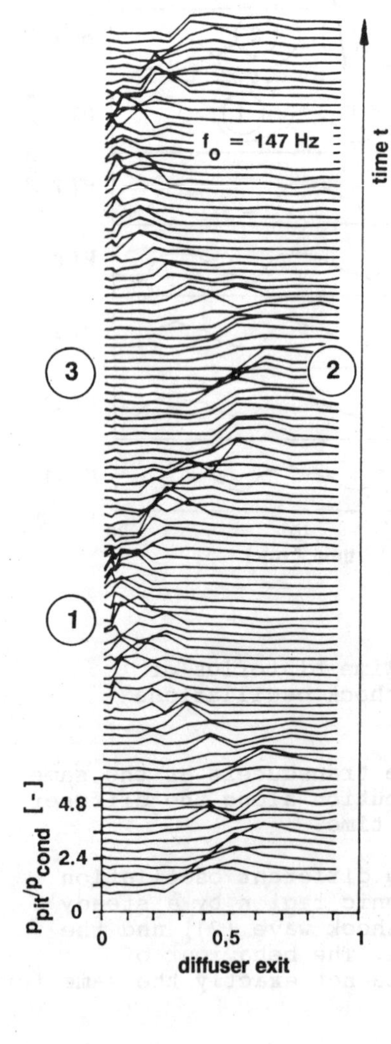

different oscillation frequencies. At frequencies of about 146 Hz the shock wave moves up almost to pressure transducer 1, located just behind the rotor exit, while at higher and lower frequencies the moving shock is not able to reach this measuring position. In addition, the development of the supersonic region and the formation of the adjacent shock wave are quite different.

Besides the development of a considerable supersonic region and an adjacent shock wave, both formation and behaviour of the boundary-layer separation downstream of the shock wave are other decisive phenomena. More information about the extension of the separation region with its strong influence on the entire diffuser flow are obtained by taking measurements at the diffuser exit with a fast-response pitot probe. The total-pressure signals at different measuring positions were treated similar to the above mentioned wall-pressure signals and displayed as momentary pressure distributions at the diffuser exit versus time. For an oscillation frequency of 147 Hz such a plot is shown in figure 15.

Fig. 15: Oscillation of separation region at diffuser exit (f_o=147 Hz)

When no boundary-layer separation exists, which is equivalent to the development of supersonic region and shock wave, the maximum total pressure at the diffuser exit can be found at the convex-curved outer diffuser shell (1). If the boundary layer separates, the total pressure maximum shifts towards the middle of the diffuser (2). The region of flow separation is characterized by a constant pressure of $p_{pit}/p_{cond} = 1$ (3). As illustrated in the figure, the separation region occupies up to 40 % of the entire diffuser cross-section. Therefore, the self-excited shock oscillations at the curved outer diffuser shell, induced by a very small tip-clearance flow, have an enormous effect on the general diffuser flow. They finally lead to blade vibrations and efficiency decrease which will be discussed later.

3. The Frequency Behaviour of Shock Oscillations

The described shock oscillations start developing just behind the design point with an oscillation frequency of about 120 Hz, which goes up when increasing the flow rate. It is of interest to know why the frequency of the shock oscillations is dependent on the volumetric flow rate and why those oscillations occur in such a regular way. It seems reasonable to assume that these shock pulsations are triggered, with respect to time, by a certain mechanism for which the mean flow velocity is mainly responsible.

According to McCroskey /15/, large-scale unsteady effects can become important when the time scale of the physical motion (shock-boundary-layer oscillation) is comparable in size to the basic fluid-dynamic time scale (diffuser-passage flow), i.e. when $\omega \cdot L/c$ is in the order of 1. This ratio is called the reduced frequency k and can be expressed in the case of the diffuser flow as follows

$$k = \frac{2 \cdot \pi \cdot f_o \cdot L}{c} \quad ,$$

where f_o is the oscillating frequency, L the length of the outer diffuser shell (as a characteristic length), and c the mean flow velocity at the diffuser inlet calculated from the volumetric flow coefficient $\varphi = c/u$. Thus, the shock oscillations are triggered by the proper dynamics of the diffuser itself, if the cycle frequency of the self-excited shock oscillations $2 \cdot \pi \cdot f_o$ is equal to the eigenfrequency c/L of the corresponding diffuser.

Figure 16 displays the reduced frequency versus the oscillation frequency f_o. In the whole range of shock-boundary-layer oscillations with oscillation frequencies between 120 and 176 Hz the reduced frequency k is nearly 1. Hence, the time behaviour of these oscillations is basically defined by the dynamics of the diffuser. The frequency of possible shock-boundary-layer oscillations can therefore be predicted from the length of the diffuser and the mean flow velocity.

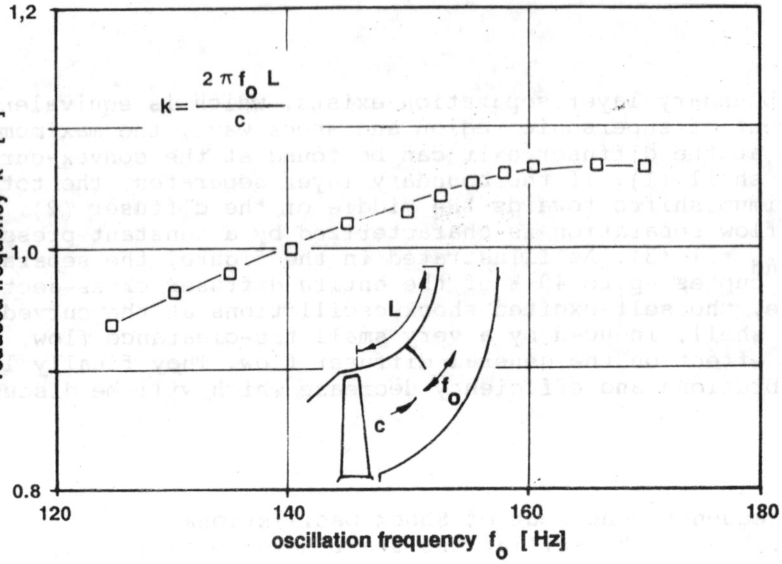

Fig. 16: Reduced frequency versus oscillation frequency

4. Effects of Self-Excited Shock-Boundary-Layer Oscillations

The measurement results discussed so far illustrated the phenomenon of shock oscillations coupled with the fluctuation of extensive separation regions taking up to 40 % of the diffuser cross-section. As a result, strong pressure oscillations occur in every point of the diffuser flow, leading to an excitation of the running blades as well as to a decrease of the efficiency.

For a frequency of 153 Hz the time history of the static pressure at the rotor exit is depicted at different radial positions in Fig. 17. It can be concluded that the excitation of the running blades due to variation of the back-pressure is impulse-shaped and in phase to the blade height. The peak to peak pressure amplitudes are about 20 % of the condenser pressure; the pressure oscillations themselves are not strong enough to cause noticeable forced vibrations of the running blades. However, the existance of an oscillating flow field, acting on a rotating bladed disc, can create a great number of possible excitation frequencies according to the following equation

$$f_e = \left| i \cdot f_o \pm k \cdot f_r \right|, \qquad i, k = 1, 2, 3 \ldots$$

where f_o represents the oscillation frequency, f_r the rotational frequency, and f_e the excitation frequency. This relationship defines an entire spectrum of excitation frequencies created by a single oscillation frequency. Considering the enormous bandwidth of the diffuser buzz between 120 and 176 Hz, the blades are exposed to a multitude of excitation frequencies. Consequently, resonances are unavoidable when the turbine is operated within the state of shock oscillations.

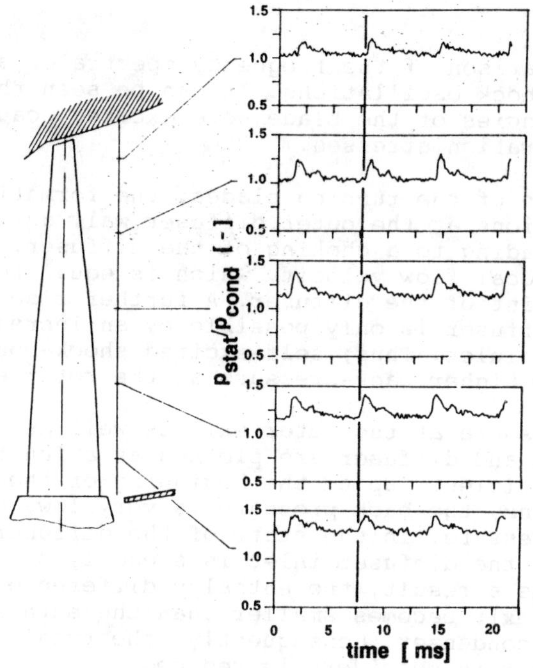

Fig. 17: Blade excitation forces due to shock oscillations

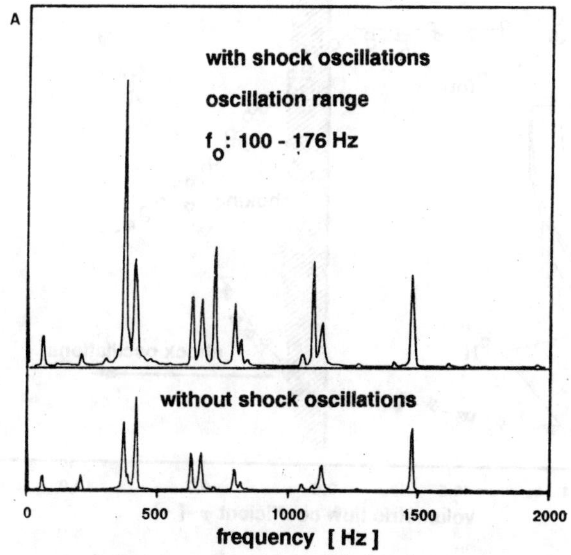

Fig. 18: Frequency spectra of a single blade with and without shock oscillations

Figure 18 shows the comparison of the frequency spectra of a single blade with and without shock oscillations. It can be seen that several resonance frequencies of the blade were excited, causing distinctly increased vibration stresses.

Apart from the excitation of the running blades, the formation of extensive separation regions at the outer diffuser wall during the shock oscillations is leading to a choking of the diffuser. This condition is defined by a local flow velocity which is equal to the speed of sound at any point of the diffuser. A further rise of the mass flow through the diffuser is only possible by an increase of the pressure at the diffuser inlet. Thus, self-excited shock-boundary-layer oscillations cause higher back-pressure at the rotor exit.

In figure 19 the hub pressure at the rotor exit as well as the total efficiency of last-stage and diffuser are plotted as a function of the volumetric flow coefficient. Up to the beginning of the self-excited shock oscillations, the back-pressure is very low, corresponding to the condenser pressure. In the state of the diffuser buzz, however, the pressure at the diffuser inlet is steadily moving up with rising flow rate. As a result, the enthalpy difference between turbine inlet and rotor exit becomes smaller than the enthalpy difference between inlet and condenser. Consequently, the total efficiency of last-stage and diffuser is considerably reduced.

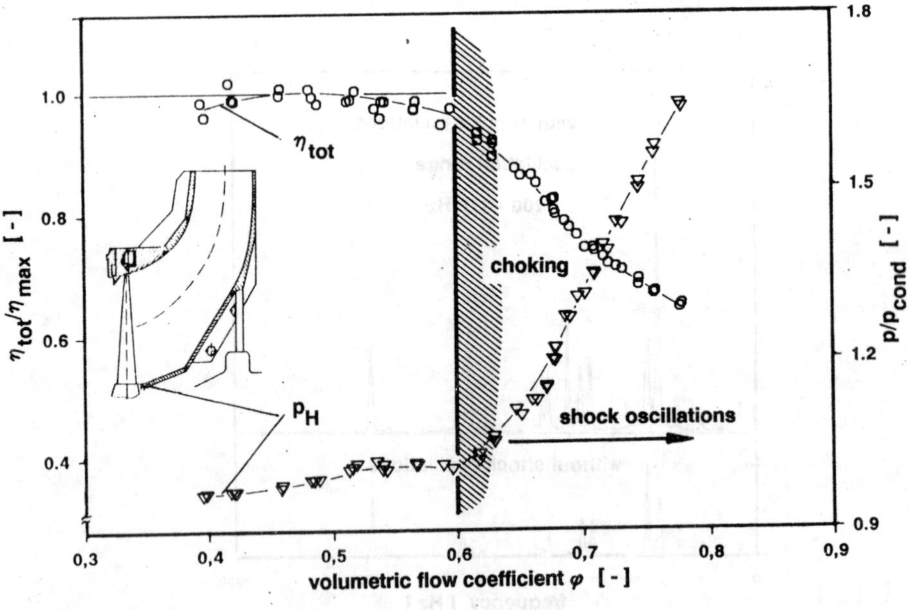

Fig. 19: Diffuser inlet pressure and total efficiency versus volumetric flow coefficient

5. Prevention of Shock Oscillations

The negative effects of self-excited shock-boundary-layer oscillations in exhaust diffusers, regarding blade excitation and efficiency decrease, require structural modifications in the diffuser design to avoid this flow instability. Different methods were investigated at the last-stage test stand /2/.

The use of turbulence wires, disturbing the supersonic flow region in front of the shock wave, proved to be a less suitable measure. At low flow rates, i.e. for partial load of the turbine, the wire caused strong steady-state flow separation at the convex outer diffuser contour, leading to a significant decrease of the efficiency. At higher flow rates, the described oscillations were unavoidable. Obviously, the mechanism of shock-boundary-layer oscillations is extremely dominant and consequently, the used wire had no marked influence on the development of this self-excited flow instability.

Single reduction of the clearance gap up to 80 % of the original height is not sufficient to eliminate the shock oscillations either. This measure would only be successful for a reduction of nearly 100 %, which may be achieved with tip shrouds.

Another method exists in the generation of several oblique shock waves leading to a deceleration of the supersonic flow region. This can, for example, be realized by small grooves in the diffuser surface. Yet, an even stronger effect can be achieved by implementing a chamber in the convex diffuser wall under the shock foot, which is covered with a perforated sheet /16/. Due to the pressure difference at the normal shock wave, a self-excited reverse flow can develop inside the chamber. Downstream, behind the shock wave, higher pressure forces the fluid to enter the chamber. It then moves upstream and escapes in front of the shock wave into the supersonic flow region on a low pressure level. As a result, the outflowing medium forms a ramp for the arriving supersonic flow and generates a weak oblique shock wave at every single bore. This mechanism leads to both a deceleration of the supersonic flow upstream of the shock wave and a sucking-away of low-energy boundary-layer material. Hence, the normal shock wave is weakened and fixed to the perforated sheet.

The influence of various perforated sheets with different hole diameters and open areas were investigated and compared to a smooth surface contour at the last-stage test stand. Figure 20 illustrates the construction of the double-walled, perforated outer diffuser shell.

Best results were obtained with holes of 1 mm in diameter and an open area of 22.5 %. Except for the lower part of the diffuser, where in a very restricted operational range real low amplitude shock oscillations occur, no marked self-excited shock oscillations were found. The total efficiencies of the diffuser with perforated wall and the diffuser with smooth wall were identical in this case /2/. Therefore, the double-walled, perforated diffuser shell seems to be an acceptable measure to prevent the development of shock-boundary-layer oscillations without decreasing the efficiency.

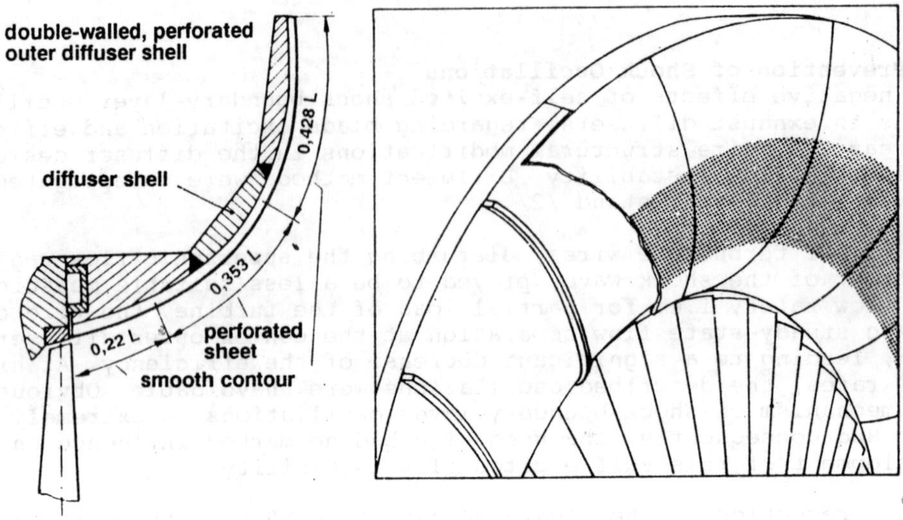

Fig. 20: Double-walled, perforated outer diffuser shell

SUMMARY

The mechanism and behaviour of self-excited shock-boundary-layer oscillations in an exhaust diffuser have been investigated at a last stage test stand. This flow instability is caused by the high-energetic clearance flow between blade tip and outer casing, leading to a supersonic flow region at the convex-curved diffuser shell. The interaction of the terminating shock wave with the boundary layer creates self-excited flow oscillations. The shock pulsations develop just behind the design point of the turbine at high flow rates and frequencies of about 120 Hz. With rising flow rate the oscillation frequency increases steadily. According to the measurements, the frequency is triggered by the proper dynamics of the exhaust diffuser. Therefore, the oscillation frequency is predictable from diffuser length and mean flow velocity.

The shock oscillations are coupled with periodic boundary layer separations acting on the entire diffuser flow. This way, the shock oscillations cause an excitation of the running blades and a reduction of the total efficiency. The use of a double-walled, perforated outer diffuser shell seems to be a successful measure to prevent the development of the self-excited shock oscillations without decreasing the efficiency. By a skillful optimization of hole diameter and open area of the perforated sheet, even an increase of the efficiency may be possible in the range of high flow rates.

REFERENCES

1. MAIER, R., WACHTER, J.: Shock-Induced Flow Oscillations in Steam Turbine Diffusers, Transactions of the ASME, Journal of Turbomachinery, Vol. 110, No. 2, April 1988.
2. MAIER, R.: Experimentelle Untersuchung selbsterregter Stoß-Grenzschicht-Pulsationen in Abdampfdiffusoren, Dissertation Universität Stuttgart, 1989.
3. LEVY, L. L., BAILEY, H. E.: Computation of Airfoil Buffet Boundaries, AIAA Journal, Vol. 19, No. 11, 1981.
4. REDEKER, G.: Calculation of Buffet Onset for Supercritical Airfoils, IUTAM Symposium Transonicum, ed. by K. Oswatitsch and D. Rues, Springer-Verlag, Berlin 1976.
5. THOMAS, F.: Die Ermittlung der Schüttelgrenzen von Tragflügeln im transsonischen Geschwindigkeitsbereich, WGLR Jahrbuch, 1966.
6. HENNE, J. M., BROICHHAUSEN, K.-D., KAUKE, G., GALLUS, H. E.: Instationäre Reibungs- und Stoßvorgänge bei trans- und supersonischen Strömungen, Teil I: Instationäre Stoß-Grenzschicht-Wechselwirkung an Einzelprofilen und Gittern, VDI-Bericht Nr. 572, 1985.
7. ARAKI, T., OKAMOTO, Y., OHTOMO, F.: Self-Excited Flow Oscillation in the Low Pressure Steam Turbine Cascade, Proc. of the 2nd Int. Symp. on Aeroelasticity in Turbomachinery, ed. P. Suter, Juris Verlag Zürich, 1981.
8. SEDDON, J., GOLDSMITH, E. L.: Intake Aerodynamics, Collins Professional and Technical Books, London 1985.
9. BOGAR, T. J.: Structure of Self-Excited Oscillations in Transonic Diffuser Flows, AIAA Journal, Vol. 24, No. 1, 1986.
10. MEIER, G. E. A.: Ein instationäres Verhalten transsonischer Strömungen, Mitteilung aus dem MPI für Strömungsforschung und der AVA, Göttingen, 1974.
11. EVANS, N. A.: Shock Wave/Boundary Layer Interaction in Dry and Wet Transonic Steam, Transactions of the ASME, Journal of Engineering for Power, Vol. 105, No. 10, 1983.
12. MEIER, G. E. A., GRABITZ, G., JUNGOWSKI, W. M., WITCZAK, K. J., ANDERSON, J. S.: Oscillations of the Supersonic Flow Downstream of an Abrupt Increase in Duct Cross-Section, Mitteilung aus dem MPI und der AVA, Nr. 65, Göttingen, 1978.
13. MEIER, G. E. A., GRABITZ, G., JUNGOWSKI, W. M., WITCZAK, K. J., ANDERSON, J. S.: Oscillations of the Supersonic Flow Downstream of an Abrupt Increase in Duct Cross-Section, AIAA Journal, Vol. 18, No. 4, 1979.
14. FINKE, K.: Stoßschwingungen in schallnahen Strömungen, VDI-Forschungsheft 580, 1977.
15. McCROSKEY, W. J.: The Challenge of Unsteady Separating Flows, Proceedings of the American Society of Civil Engineers, ASCE, Vol. 107, No. EM3, June 1981.
16. KROGMANN, P.: Untersuchungen an einem transsonischen Profil mit 30 mm breiter Perforation/Kammer-Anordnung, Forschungsbericht der Deutschen Forschungs- und Versuchsanstalt für Luft- und Raumfahrt, DFVLR-FB 86-59, 1986.

Part Six
Engineering Design

AERODYNAMIC LOSSES IN CONVENTIONAL FAN BLADES OF HIGH BY PASS TURBO ENGINE

Sohail Mahmood
(PIA Training Centre Karachi, Pakistan)

ABSTRACT

Existing conventional Fan Blades of High By Pass Turbo Fan Engines have different Aerodynamic losses namely, Secondary loss, shock loss, profile loss, claper loss, and losses due to Blade boundry layer seperation. These losses can be minimized by adopting optimised tip geometry, supercritical aerofil, reduced blade diffusion rate, snubberless design, wide chord and improved diffusion through blade root. By minimizing the above describe losses Fan efficiency can be improved which results in increase of thrush, hence reduction in specific fuel consumption.

INTRODUCTION

In this article, the details of the aerodynamic losses in conventional Fan blade of High By Pass Turbo fan Engines and their solutions are given. High By Pass Tarbo fan engines are very successfull in the Civil Aviation Industry due to its High efficiency, and the specific fuel consumption (SFC) of High By Pass engines used in sub sonic Civil Aircrafts is very low. A tipical SFC Value on ground 0.03 Kg/N Hr. From the SFC formula given below it is evident that as the By Pass ratio increases the SFC decreases

SFC = FAR/(Vj + m x Vf)

where FAR = Fuel Air Ratio
Vj = Jet Velocity
m = By Pass Ratio
Vf = Fan Exit Velocity

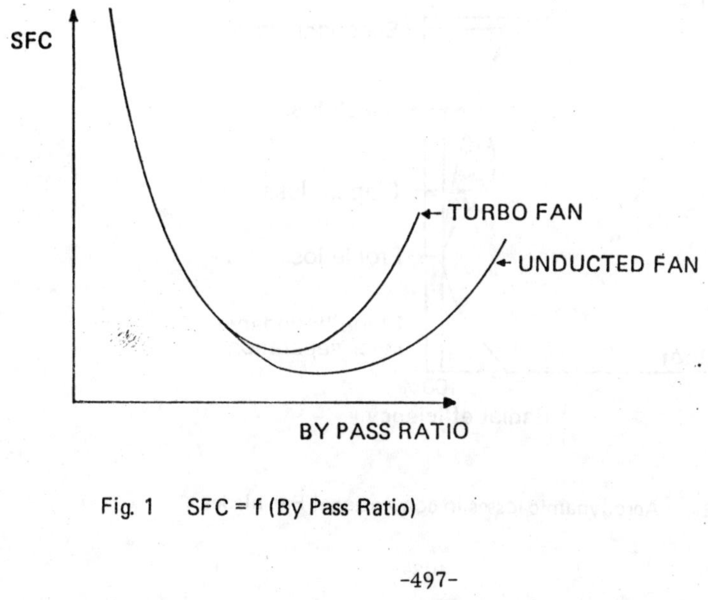

Fig. 1 SFC = f (By Pass Ratio)

However the By Pass Ratio can only be increased upto a certain extent Primarily due to limitation imposed by high angular velocity and fixed pitch, it means that fan efficiency decreases above the critical By Pass Ratio hence the SFC increases. Fan efficiency is a very importent factor in High by Pass Turbo fan Engines, because in a High By Pass Turbo fan Engines (about 5:1) about 80% of total thrust can be provided by the secondary (By pass airflow) through the secondary duct therefore if Fan efficiency decreases it effects on the 80% thrust hence thrust decreases, SFC increases. It means by improving the fan efficiency SFC can be decreased. There are two ways to decrease the SFC and increase the fan efficiency.

1. By increasing the By Pass Ratio
2. Without increasing the By Pass Ratio.

1. By increasing the By Pass Ratio :

There are limitations of increase in By Pass Ratio due to the high angular velocity and fixed Pitch, these limitations can be inimize by decreasing the RPM of the Fan and allowing variable pitch mechanism, it may be called as prop fan or unducted fan.

2. Without increases in By Pass Ratio :

Fan efficiency can be increased without increase in By Pass Ratio, by minimizing the fundamental aerodynamics losses in conventional fan blades of High By Pass Turbo fan engines.

AERODYNAMIC ANALYSIS

In the conventional fan blade of High By Pass Turbo fan Engines due to the High By Pass Ratio Diameter of the fan blade is large. Therefore to restore the vibration of the Fan blades it requires snubbers (clappers) at the mid span. There is no variable pitch mechanisum & the RPM of the fan is also very high and there is no reduction gear, therefore angular velocity is also very high at the tip of the fan blade and high centrifugal force. Therefore from root to tip there are different aerodynamic losses, by minimizing there losses fan efficiency can be improved. Fundamental aerodynamic losses are shown in fig. 2.

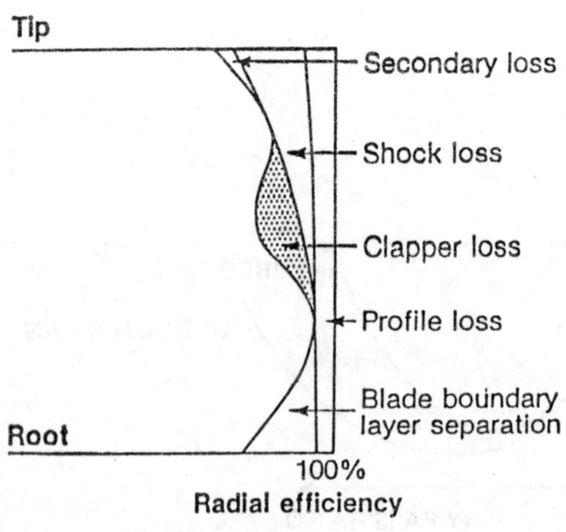

Fig. 2 Aerodynamic losses in conventional Fan blades.

Description of each fundamental aerodynamic loss is as follows :

SECONDARY LOSS

Due to the tip geometry there is annulus wall boundary layer saperation. Laminer flow changes into the turbulent flow, it is maximum at the tips.

SHOCK LOSS

Due to the large diameter of fan blades of high PRM angular velocity at the tip is very high and when the angular velocity at the tip increases above the Mach critical, shock losses are there. Critical Mach number is the minimum value of the free - streem Mach number for which the local flow becomes supersonic. As this value increases the shock wave strenghs sufficiently to cause the drag to rise suddenly.

PROFILE LOSS

Profile loss depends upon the blade geometry, main factors in the blade geometry are thickness at the leading edge and the diffusion rate.

CLAPPER LOSS

In the Conventional Nerrow Chord Fan Blades at the Mid span there are Shrouds (Snubbers Clappers) to control the vibration of large fan blades, but these Snubbers casue blockage and loss of efficiency it means Drg increases Laminer Flow changes into the Turbulent flow, viration increases, effective Area decreases, pressure ratio decreases, thrust decreases, specific fuel consumption increases, sound level increases, less flexibility therefore less resistence to foriegn object damage. Losses due to midspan shroud is called clapper losses, and these losses are maximum at the shroad area.

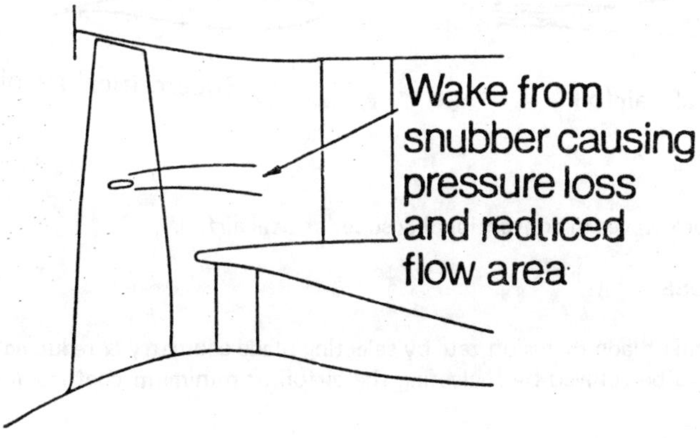

Fig. 3. Clapper loss.

BLADE BOUNDARY LAYER SEPERATiON

Laminer flow from root of the fan blade changes into the turbulant flow towards the blades, therefore boundry layes from the root saperates towards the blades section, hence working area decreases. Turbulant flow due to the boundry layes saperation can cause stalling of the compressor.

SOLUTIONS

Aerodynamic losses in the Fab Blade can be minimize by the methods as follows.

SHOCK LOSS

This could be minimize by reducing blade surface Mach number by using super critical airfoil in the fan blade, due to this advance airfoil critical Mach number can be increased & Fan can be operated with weak shock wave at higher Mach number & high angular velocity, therefore fan efficiency can be increased. In the super critical airfoil flow accelerates to locally supersonic conditions at a free streem Mach number. However it is shaped, so that around its design lift coefficient, the flow decelerates to subsonic conditions through a distribution of weak compression waves instead of one strong one.

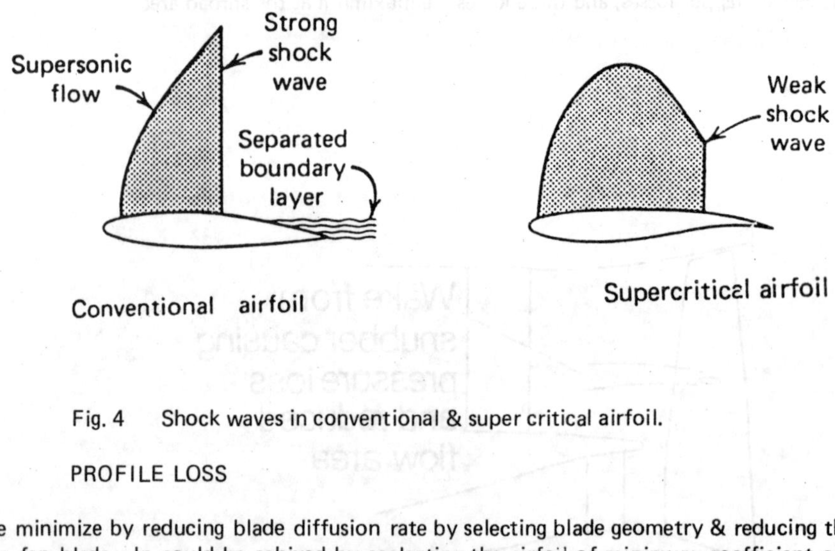

Fig. 4 Shock waves in conventional & super critical airfoil.

PROFILE LOSS

It could be minimize by reducing blade diffusion rate by selecting blade geometry & reducing the thickness of the fan blade. In could be achived by seclecting the airfoil of minimum coefficiant of drag & higher lift to drag ratio.

CLAPPER LOSS

Clapper losses can be eliminated by delating the Mid Span Shrouds and to restroe the vibration chord of the Fan Blade can be increased by decreasing the Aspect ratio. This removes the fundamental aerodynamic loss, togetheir with the Shock induced interactive losses on the surrounding airfoil and gives in over all improvement by some 5% relation to the Conventional high aspect ratio Snubbered Fan. Due to the wide Chord Snubberless Fan Blade, has less blockage of the air, less turbulence, more laminar flow, increased pressure ratio, can pass more air for a given inlet area and greater Thrust Potential, Fan efficiency can be increased.

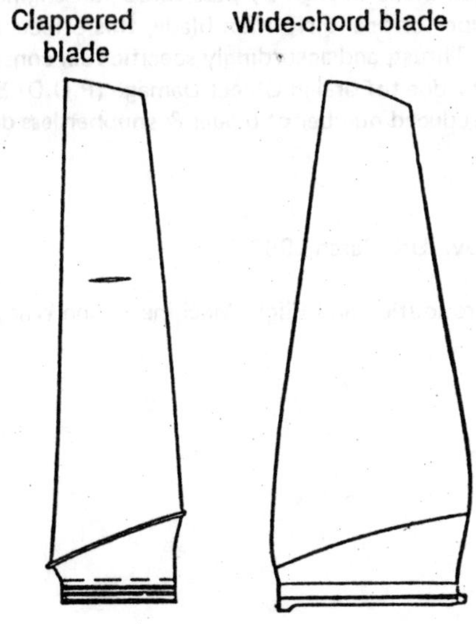

Fig. 5 Comparision of fan blades.

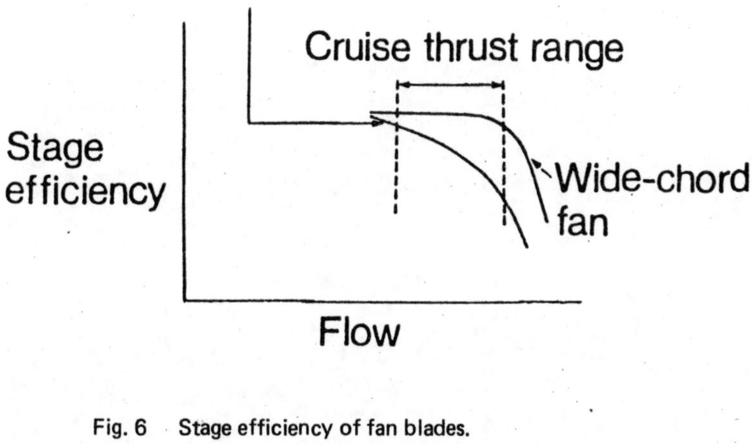

Fig. 6 Stage efficiency of fan blades.

BLADE BOUNDARY LAYER SEPARATION

It can be minimize by designing new root sections with improved diffusion through passage, therefore there will be more laminar air & more effective area & less turbulance.

CONSLUSION

The fundarnental aerodynamic losses in the fan blade of high by pass turbo fan engine can be minimize by adopting wide chord snubburless with super critical airfoil fan blade, this effects at increase in Fan efficiency by 2% at low thrust and 6% at high Thrust, and accordingly specific fuel consumption decreases by 40% at maximum cruise, excellent integrity due toForeign Object Damage (F.O.D) & flexibility which absorbs impact energy and low weight due to reduced number of blades & snubber less design.

REFERENCES

1. Rolls - Rayce Plc Ts 20442 Issue 2. Derby. UK March 1988.

2. Barnes W. McCormick Aerodynamics, Aeronoutics, and Flight Machines. John Wiley 1979.

AEROSTRUCTURAL CONSIDERATIONS FOR THE POWER PLANT OF OVERLAPPING WING CONFIGURATION

M.A.Y.KAMAL, OPERATIONS TECHNICAL INSTRUCTOR
DR.S.A.HASAN, CHIEF INSTRUCTOR OPERATIONS
(PAKISTAN INTERNATIONAL AIRLINES, TRAINING CENTRE
KARACHI AIRPORT, PAKISTAN)

ABSTRACT

Overlapping Wing Configuration is a special case of staggered bi-planer arrangement of an aircraft wings, with forward sweep given to the near wing and backward sweep given to the front wing, both joined at the tips.

Conventional propulsion systems can be used for this configuration, however,considering the peculiar arrangement of the wing, the stresses, and essentially the bending moments, it is determined that the structural design of the wing may be made lighter and efficient using tapering techniques & thus minimizing the structural thickness along the normal axis.

The Thrust line and power plant location on the aircraft can also be determined keeping the peculiar wing set-up in view. In relation to the C.G. location and landing gear arrangement it becomes essential to carefully locate the power plant. The down-wash effect on the rear wing and the location of the propulsion system on the wing changes the performance characteristics of the Aircraft.

Future development of the concept of Overlapping Wing Aircraft is discussed considering various power Plant/Structural integrations.

INDRODUCTION

The Study presented here is a part of a detailed research study of the Overlapping Wing Air Craft concept[1]. The NASA/Lockheed Trisonic Bi-plane design,modified on lines of Dornier Corporation Overlapping Wing design with addition of Wing Tip Winglets was used to evolve a new Configuration. Study has been conducted to evaluate various characteristics of the design. This study primarily discusses the approach necessary to select the propulsion system position on the design keeping its constraints in view.

O W CONCEPT

The Overlapping Wing Aircraft Concept evolved out of the research conducted to study the feasibility of a trisonic bi-plane. Initial studies include the workdone by Lockheed under NASA contract in 1974,evaluating certain configurations consisting of a forward mounted swept-back low wing and an aft mounted swept-forward upper wing, connected by a vertical fin at the tip (fig-1). This design proved to reduce induced drag and allowed a constant section fuselage shape maintaining a close agreement with the ideal area distribution curve for the transonic regime Dornier Corporation in West Germany later further developed the concept by designing an aircraft with unconventional double airfoil rehombid wing. It was observed that the transonic and supersonic induced drag reduces with a significant increases in the maximum coefficient of Lift, manoeuerability and turn capability. The Wing planform was essentially a delta-wing shape with a high shoulder mounted forward wing joined to a belly mounted aft wing at the tip. Similar experiments were conducted in England developing the Warren-Young configuration for low induced drag. (fig-2)

MODIFIED CONCEPT

The modified concept of Overlapping Wing Configuration, developed for this study program consisted of two primary lifting surfaces. The swept-back wing attached to the forward fuselage at the top and swept-forward wing attached to the lower portion of the aft fuselage. The two wings were connected at the tips and NASA Winglets were incorporated at the tips (fig-3) with this arrangement following improvements can be predicted over a similar mono-plane configuration of the same Aspect ratio:[2]

- Light Weight
- High Stiffness
- Low induced drag
- High C_L max
- Direct lift control
- Direct side force control
- Good stability and control
- Reduced wetted area

SPECIAL CHARACTERISTICS AND CONSIDERATIONS

Overlapping Wing aircraft configuration is a special case of staggered Bi-plane. Due to the peculiar nature of the bi-planer arrangement around the fuselage, it becomes very important to locate the propulsion system of the aircraft at a suitable location in order to obtain the best possible results. Moreover, the parameters of wing design and configuration determines the engine pylon design, length & sweep, nacelle area, fuel capacity and engine location. The Aerodynamic flow charecteristics over the wing and around the fuselage, in overlapping wing configuration, differs from a conventional aircraft. At the same time, they also differ from the conventional Bi-plane configuration due to a high degree of stagger, sweep & wing tip bracing.[3] The non-conventional flow conditions necessitates special considerations for the powerplant location for the arrangement. Conventionally, powerplants can be located in-side the aft fuselage, inline with the longitudinal axis, out-side the aft fuselage, under the main wings or at the wing tips. Each position has it implications on aircraft stability due to the moment armed formed between the thrust line any C.G. of the aircraft (fig-4). Also, ground clearness and down-wash effects are essential to evaluate in each case. The structural strength of the supporting member, bending and torsional moments also dictates the final decision along with its relevence to weight distribution. Due to the high forward wing attachment to the front fuselage and low attached rear wing with sweep-back & sweep-forward, respectively the rearwing falls under the down-wash effects of the front wing. The adverse effects, however, can be controlled by the aerodynamic twist incorporation in the wings/The mid-span or wing-root location of an engine in this case will disturb the flow over the rear-wing as the jet down-wash will produce a turbulent flow.(fig-5,. The stability of the O.W. aircraft is shown in the fig-6. Thrust line effects for each position under consideration can be calculated from there. Another possibility is to locate the engines at the tip of the joining wings. Tip mass concentration e.g due to tip tanks winglets etc. provide valuable inertia relief for maneuvering airloads on both conventional and O.W. configurations On the other hand, tip masses provide a difficult stressing condition for landing and taxi loads on a conventional cantiliver wing. The situation is different for the joined wings arrangements because of its high resistance to loads applied at the apex of the truss. This results in a favourable design criteria for tip masses in an O.W design. The tip joing members tends to be heavily loaded and therefore requires a large structure.[4] It is benificial to employ a winglet for this purpose. The joining structure will therefore, give aerodynamic benefits by reducing induced drag. Excellent resistance to air-loads & moments exists in the O.W. arrangement as the winglets are located at the a pex of the truss. Integration of engine necelle with the winglet(Fin) can save wet area by elimination of pylon. Necelle integrating is best suited for O.W. as winglet is not a cantiliver structure (Fig-7).

The most suitable location for the powerplant in O.W. design is inside the rear fuselage with intake duct at the tip or above the aft wing, attached to the fuse lage(fig-8). The two possible locations impose no drage, penalities or flow disturbances in the arrangement and the thrust line can also be maintained inline with the longitudinal axis. The location of engines on the rear wing is not practicable due to low ground clearence and small landing gear struts located in the aft fuselage.

CONCLUSIONS

It is evident from the discussion that the structural as well as aerodynamic characteristics of the overlapping wing arrangement requires aft location of the propulsion system. A more thorough and indepth study is required to evaluate comprehensive results for a given design. It is recommended that such study may be carried out inorder to design military, commercial or R.P.V. version of the O.W. concept.

REFERENCES

1. Cahill, J.F., & Stead, D.H., "Preliminary Investigation at Subsonic & Transonic speeds and the Aerodynamic Charecteristics of a Biplane Composed of a swept back and a swept forward wing joined at the tips" NACA RM L53 L24 b 1954.

2. Ried, E.G., "Applied Wing Theory", Mcgraw Hill, 1932.

3. Lange, R.H. et al, " Feasibility Study of the Transonic Biplane concept for Transport Aircraft Application", NA CA CR-132462, June 1974.

4. Aviation week & Space Technology, April 24,1978.

5. Addoms , R.B., "Aerodynamic & Structural Design Considerations for High lift Biplane wing system", Ph.D. Thesis, Nov.1971, Dept. of Mechanics & Structures, University of California, Los Angeles, Ca.

CONFIGURATION DIAGRAM
BOXPLANE

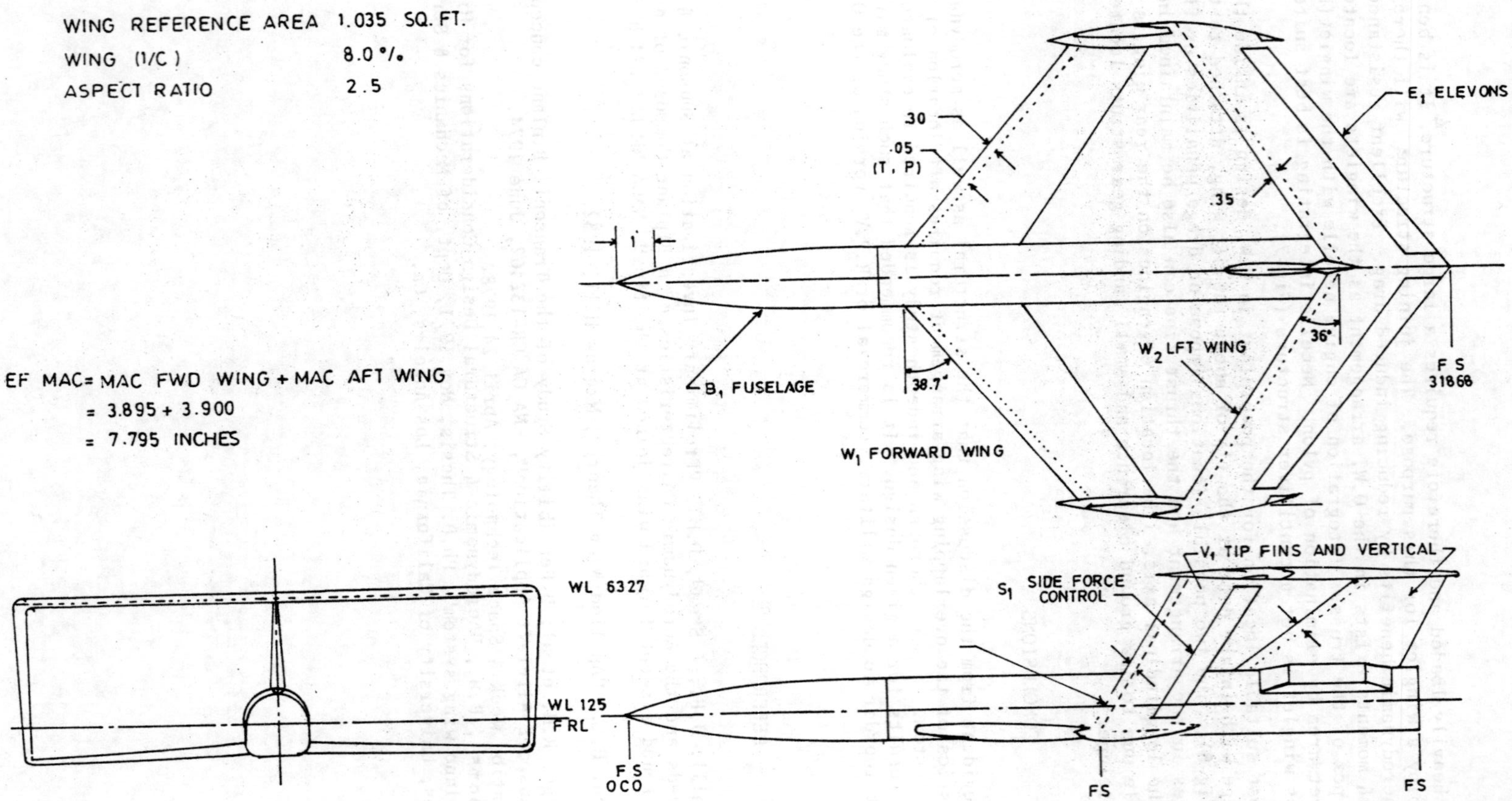

FIG. 1 LOCKHEED BOXPLANE WIND TUNNEL MODEL

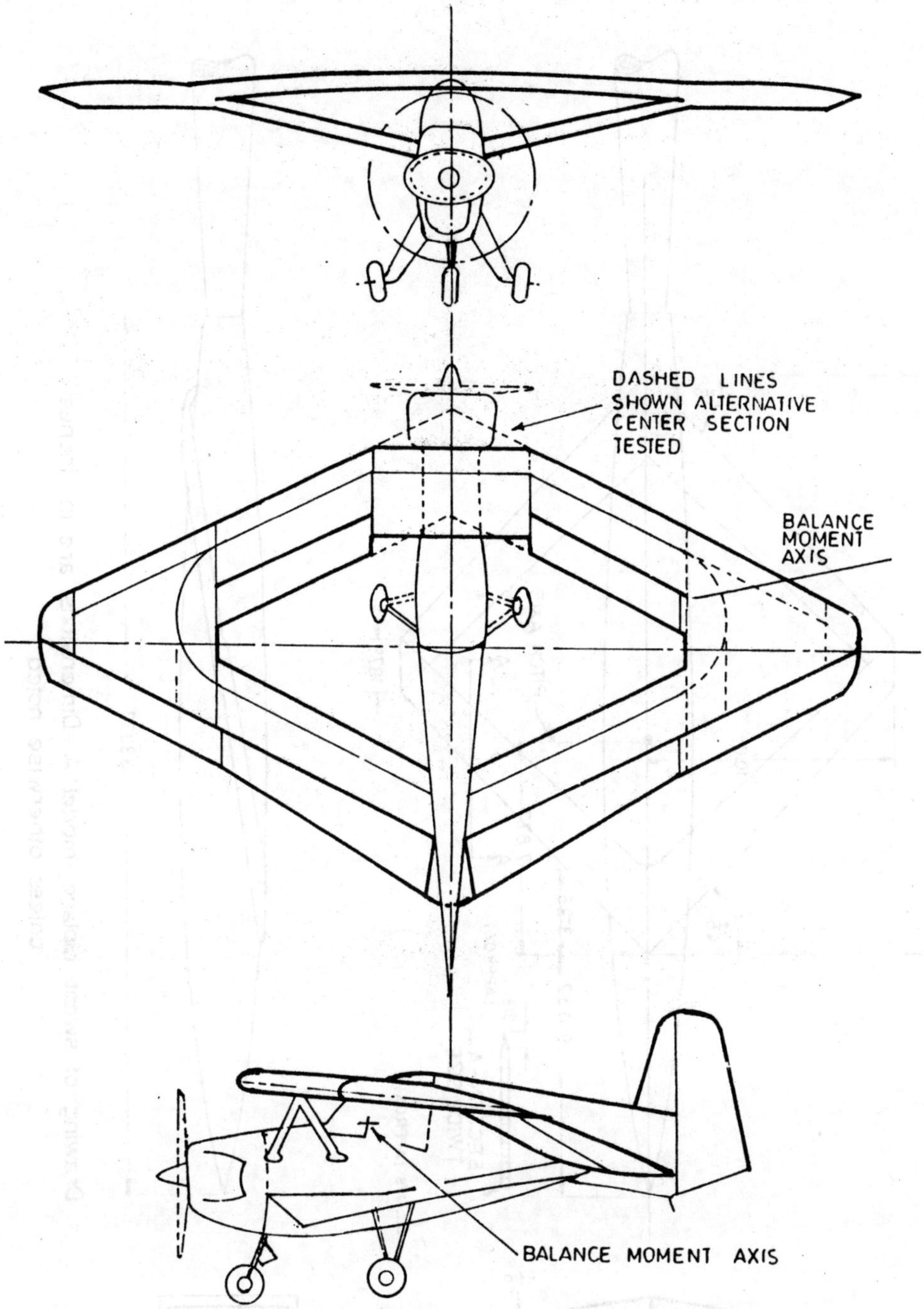

FIG. 2 WARREN-YOUNG WING TUNNEL MODEL

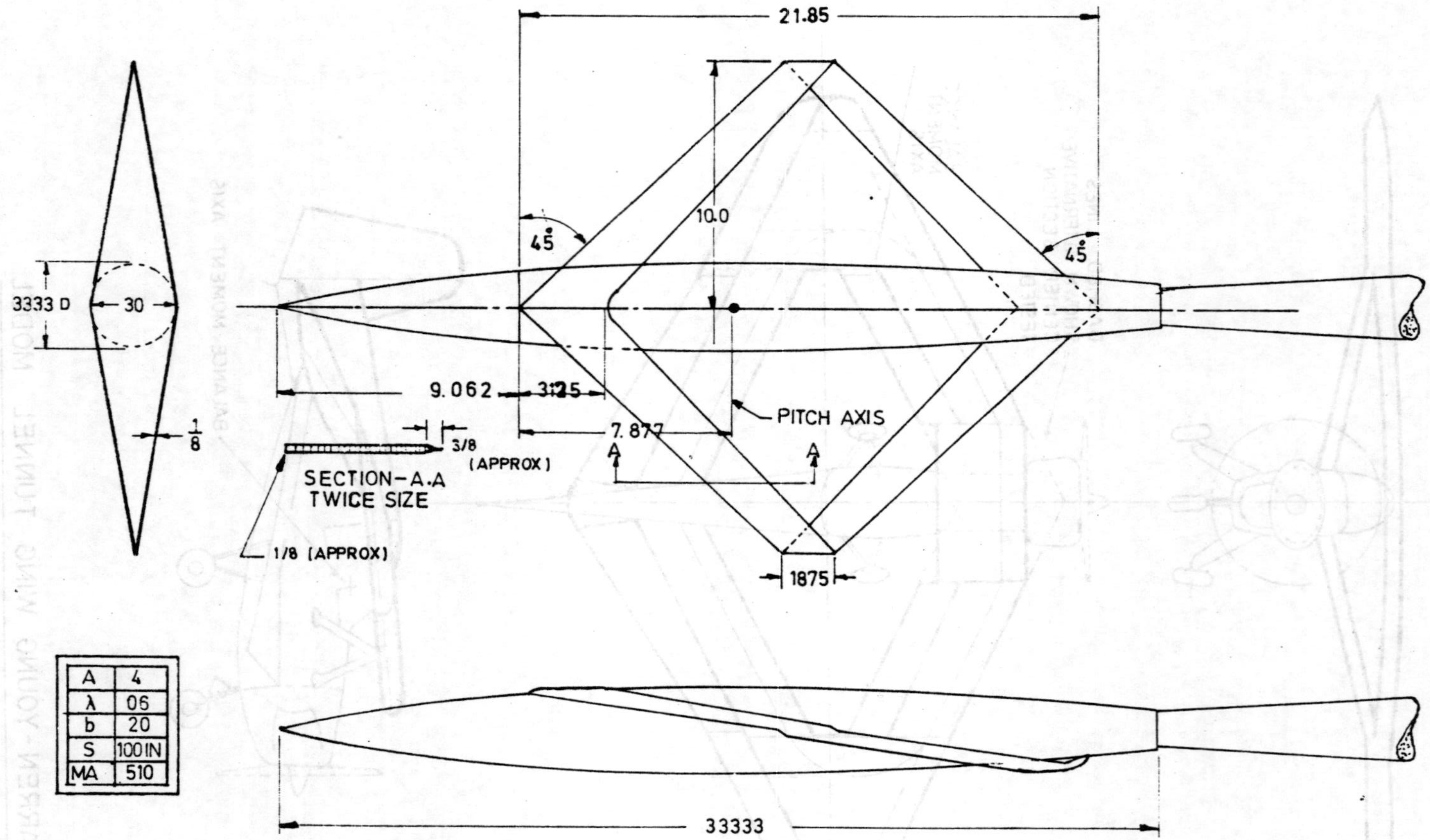

Drawing of swept biplane model A. Dimensions are in inches unless otherwise noted.

FIG. 3- WIND TUNNEL MODEL OF O.W. AIRCRAFT

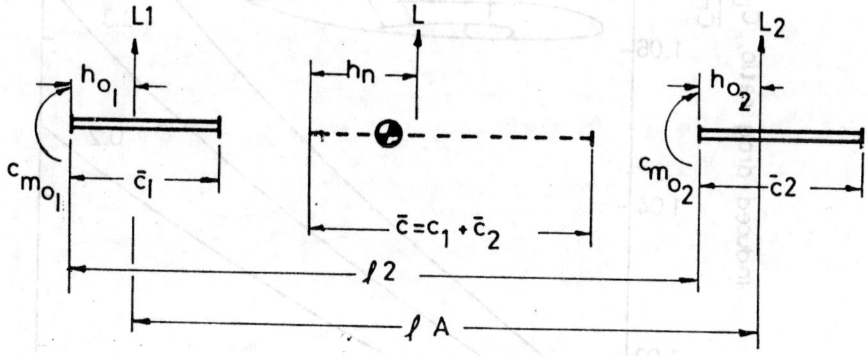

FIG. 4- FORCE AND MOMENT REPRESENTATION.

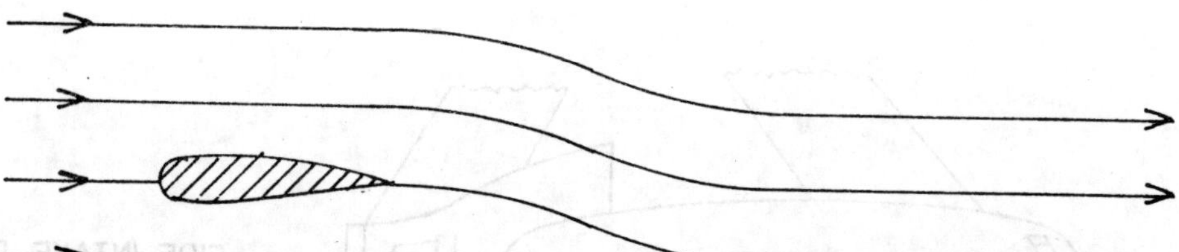

FIG. 5- REAR WING IN THE FLOW OF FRONT WING.

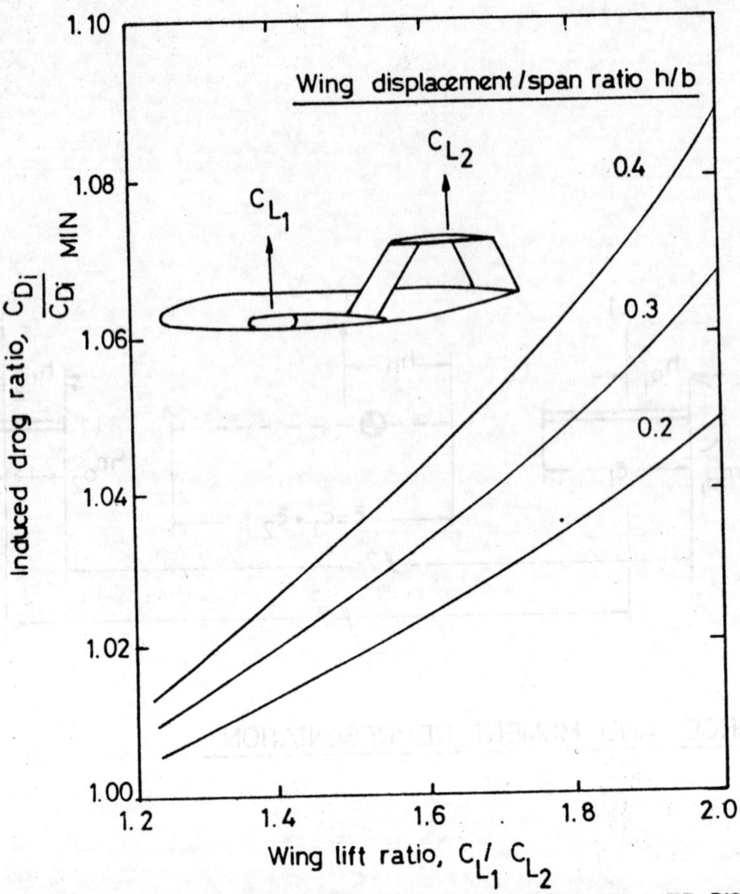

FIG. 6 - SENSITIVITY OF INDUCED DRAG TO NON-OPTIMUM LIFT DISTRIBUTIONS.

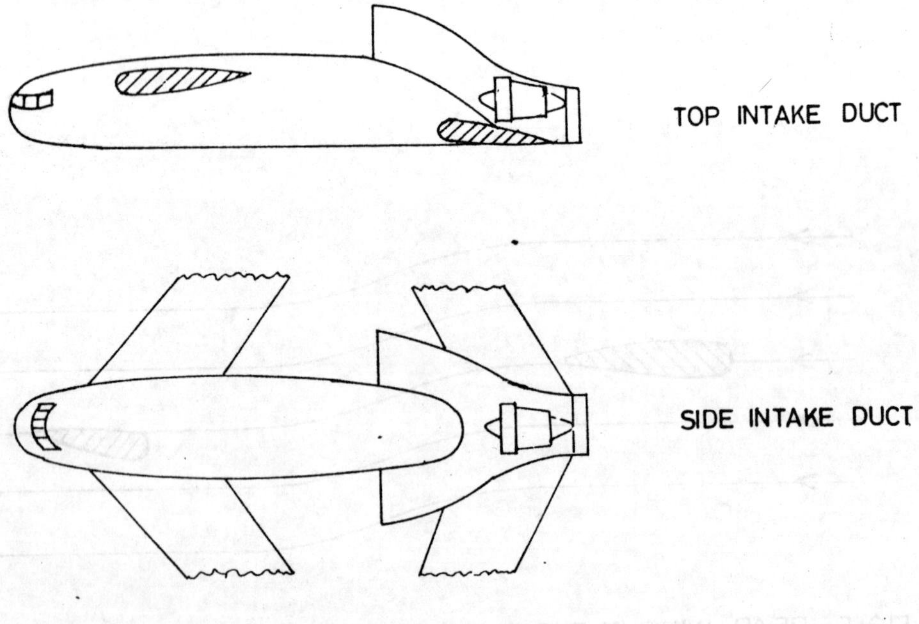

FIG. 7 - POSITIONS OF POWERPLANT

Comment

On the Fifth Symposium of Unsteady Aerodynamics & Aeroelasticity in Turbomachinery & Propellers (UAATP) from the Angle of Engineering Background

Zhou Sheng, Cao ChuanJun
(Beijng University of Aeronautics & Astronautics)

In this symposium 43 papers are accepted that equal to the number of papers presented in last symposium and three more than those in 3rd symposium. As a series of symposiums in the field of technological science, they have important engineering background. In the table shown as following, the papers in recent three symposiums are classified from the angle of engineering background.

Tab. 1 Classification of the papers according to engineering background (% of total)

Symposium	Aeroengine	Steamturbine	Propeller	Others	total
3rd	30 (75%)	5 (12.5%)	1 (2.5%)	4 (10%)	40
4th	27 (62.8%)	13 (30.2%)		3 (7%)	43
5th	27 (62.8%)	9 (20.9%)	4 (9.3%)	3 (7%)	43

(Remark: the papers about turbine are grouped in "Steamturbine", those about centrifugal compressor-in "Other".)

From the broad point of view, what are presented in these symposiums are researches on the mechanism in the flow induced vibrations of the open/closed cycle turbomachinery. It is the margin of disciplines-fluid mechanics and solid mechanics. As in the case of turbomachinery, it forms the overlapping of unsteady aerodynamics and structure dynamics in turbomachinery. From 3rd symposium which was held in 1984, it was entitled as recently UAATP, instead of the title "Aeroelasticity in trubomachinery" used in 1st and 2nd symposiums. The change of title reflects the fact that this branch of science becomes matured. Because, in a broad sense, "Aeroelasticity in Turbomachinery" deals with problems concerning not only "Aeroelastic stability" but also "Aeroelastic Dynamic Response". And "Unsteady Aerodynamics in Turbomachinery" is closely related to the structure integrity and the performance improvement of turbomachines, because the "unsteadiness" reflects the essence of turbomachinery. Another unsteady aerodynamic phenomenon is the noise emitted by the turbomachines, which is intensely linked to vibration. In the following table, the papers are classified according to this point of view.

Tab.2 Classification of the papers according to the nature of problems

Symposium	Aeroelastic Stability	Dynamic Response	Aeroacoustics	Others	total
3rd	30 (75%)	8 (20%)	1 (2.5%)	1 (2.5%)	40
4th	26 (60.5%)	13 (30.2%)	3(7%)	1 (2.3%)	43
5th	20 (46.5%)	15 (34.9%)	5 (11.6%)	3 (7%)	43

The technological science, as a link between the production domain and the fundamental science contains three elements:

1 those related to engineering background, facts or phenomena;
2 those related to concepts, models or theories.
3 corresponding researching means, techniques and tools.

That accounts for the development of a branch of technological science is surmounting the contradictories occuring unceasingly among those elements mentioned above, as shown in the following diagram.

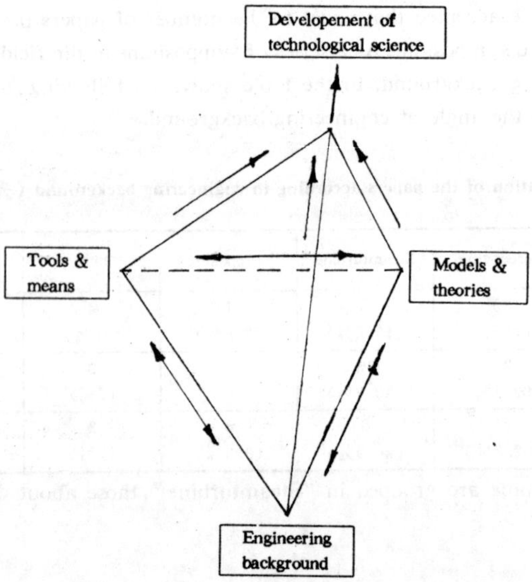

The engineening background of the discipline—Aeroelasticity—is related to many important industries about energy resources, transportation and national defence. While increasing of the capacity of single turbomachine, lightening of its structure and improving of its operating regime, more attention will be paid to Aeroelasticity. Although not perfectly developed, the scientific concepts and theories of Aeroelasticity are related to abtruse theories of mathematics and mechanics, it belongs to a rigorous branch of technological science. To do the theoretical or experimental researches on Aeroelasticity very high techniques, means and tools are required and sometimes great expense is needed.

In this symposium, 43 papers covering three above-mentioned categories advance some new ideas in every aspect. For example, the problems of correlation between different unstable factors in aeroengines are mentioned in several papers, they are certainly worthy of attention. The phenomenon existing are always so vivid and complex, that in the course of researching not only analysing but also synthesizing are required. In the papers about steamturbine, the problems on the correlation of condensation and aeroelastic stability, on the blade vibration in the thermal electric power plant, and on the self-excited oscillation of shock waves in exhaust diffuser are very interesting, warranting a further study. And in a paper of summary on partial admitted Pelton steam turbine in the aspect of waste heat recovery, there are quite a lot of new ideas. As more and more composite materials are used today in building both conventional and advanced propellers, there will be many new problems to be considered in the Aeroelasticity.

In papers presented in this symposium, some new models, means and concepts in the theory of aeroelasticity of turbomachinery are advanced. For example, in the works on the correlation between different unstable factors and its effect on blade vibration as stated above; on the aeroelastic stability of 3-D rotating cascade; on closely interrelating the unstabilities in aerodynamics and structure dynamics while solving the problems of aeroelastic stability, etc.

Some papers about experiments and measuring methods are presented, such as those on some researching methods about the unsteady aerodynamic factors affecting 2-D linear cascade wind tunnel; on the combination of noncontact measurement of blade vibration with the application of microcomputer in order to greatly increase the function of measurement, and on monitoring of the stability of the operating turbomachine according to the acoustic signals generated.

Some papers introduce a series of experimental works, which are apparently very important due to complication of the phenomena and multiplicity of influencing factors in aeroelasticity of turbomachinery. Such as those on a series of fundamental experiments performed in the NASA Lewis transonic oscilating cascade facility with a plenty of valuable data obtained; on 2-E forcing function effects on turbomachinery gust unsteady aerodynamics performed in the multistaged axial flow compressor test facility in Purdue Universty; and on the experiments on flutter of BF-1 rotor performed in Beijing University of Aeronautics & Astronautics.

In this symposium, the various and interesting papers presented reflect the tremendous development of this discipline. As a branch of technological science, it has its own stages of emerging, developing, transforming and even declining. How to estimate the nowaday tendency of development of this discipline will be stated below from the point of view stressing on the engineering background.

The thrust-weight ratio of aeroengines will reach 15. 0 to 20. 0 around the year 2000 instead of 8. 0 today, that will definitely influence on the development of Aeroelasticity in turbomachinery. There is a great need for steam turbine powerplant in China. In the region as Shanxi or Inner Mongolia, where there is rich coal resources but short of water supply, the requirement of steam turbine with aircooling instead of water cooling and the neccesity of increasing capacity of single machine provide the motive force for the development of this discipline. Realizing the use of propfan as power of passenger/transport airplane of new generation in the near future, the development of both conventional and advanced propellers, and the extensive application of various wind turbines and other turbomachines offer a lot of new problems to this discipline.

In short, there is quite a broad engineering background of the branch of science on which we are working and there will be new areas to be explored. I do believe our series of symposiums will continue, and our international family of UAATP will be prosperous and flourishing. Now I'd like to anounce, the 5th symposium of UAATP is successfully concluded.

Author Index